Studies in Mechanobiology, Tissue Engineering and Biomaterials

Volume 26

Series Editor

Amit Gefen, Department of Biomedical Engineering, Tel Aviv University, Ramat Aviv, Israel

The series **Studies in Mechanobiology, Tissue Engineering and Biomaterials** is intended for a general international professional audience with interest in advancing the fields of **mechanobiology, tissue engineering and biomaterials**, including for example scientists and practitioners in the fields of bioengineering, biomedical engineering, mechanical engineering, materials engineering, biophysics, biology, and medicine.

Indexed by SCOPUS.

Tuhin Subhra Santra ·
Ashwini Uma Surendra Shinde
Editors

Advanced Drug Delivery

Methods and Applications

Springer

Editors
Tuhin Subhra Santra
Department of Engineering Design
Indian Institute of Technology Madras
Chennai, Tamil Nadu, India

Ashwini Uma Surendra Shinde
Department of Engineering Design
Indian Institute of Technology Madras
Chennai, Tamil Nadu, India

ISSN 1868-2006 ISSN 1868-2014 (electronic)
Studies in Mechanobiology, Tissue Engineering and Biomaterials
ISBN 978-981-99-6563-2 ISBN 978-981-99-6564-9 (eBook)
https://doi.org/10.1007/978-981-99-6564-9

This Springer imprint is published by the registered company Springer Nature Singapore Pte Ltd.
The registered company address is: 152 Beach Road, #21-01/04 Gateway East, Singapore 189721, Singapore

Preface

In the realm of pharmaceutical research and development, the intention for more effective and precise drug delivery systems has led to groundbreaking advancements in both in vitro and in vivo applications. Advanced drug delivery approaches are now at the forefront of modern medicine, offering tailored solutions for targeted delivery and treatment, personalized therapies, and improved patient outcomes.

In vitro drug delivery systems involve designing and characterizing drug carriers and formulations outside the living organism, typically in laboratory settings. These technologies play a pivotal role in early-stage drug development, allowing researchers to screen and optimize candidate drugs, assess their interactions with various carrier systems, and evaluate their efficacy and safety profiles. In vitro models also enable the investigation of drug release kinetics, stability, and compatibility with different biological environments, providing crucial insights into the behavior of drug candidates before proceeding to animal or clinical studies.

On the other hand, in vivo drug delivery refers to the deployment of advanced drug carriers and formulations within living organisms, where the therapeutic agents can directly interact with their intended targets. These systems are designed to navigate biological barriers, evade clearance mechanisms, and selectively accumulate at specific disease sites. By precisely delivering therapeutic agents to the desired locations, in vivo drug delivery minimizes off-target effects, reduces systemic toxicity, and enhances the therapeutic index of drugs, resulting in enhanced treatment efficacy and improved patient compliance.

This comprehensive exploration of advanced drug delivery for in vitro and in vivo applications aims to illuminate cutting-edge technologies, novel approaches, and emerging trends transforming modern medicine, spanning from nanoparticle-based carriers, liposomes, gene editing tools, to cell-based therapies, expanding the possibilities for enhanced drug delivery. Furthermore, the challenges and opportunities of translating these advancements from the lab to clinical practice, including regulatory considerations, safety assessments, scalability, and cost-effectiveness, empower researchers, clinicians, and pharmaceutical companies to revolutionize patient care and elevate medical standards. The intricate world of advanced drug delivery is recognized for its transformative impact on health care, ushering in personalized medicine and a new era of precision therapeutics. Through innovation, collaboration, and dedication, the full potential of these systems is being unlocked, forging a path toward a healthier and more resilient future.

Advanced Drug Delivery: Methods and Applications is a comprehensive exploration of cutting-edge drug delivery systems, offering insights from leading experts worldwide in this field. The book covers controlled drug delivery systems, biological vectors, liposomes, mechanoporation, and nanotechnology-based platforms, along with intelligent drug delivery and stimuli-sensitive approaches. Chapters also delve into cytokine responses, polymeric systems, microfluidics, and magnetic nanoparticles.

Chivte et al. explore drug delivery systems (DDS) using biological vectors or biomimetic nanoparticles (BNPs). These vectors mimic normal cell features, offering high biocompatibility and specificity with minimal immune responses. They can carry significant cargo and undergo surface alterations. Challenges include stability, large-scale production, and efficacy. Understanding the biology behind these vectors is crucial for successful clinical applications. The chapter highlights promising biological vectors, drugs loaded, clinical trials, and limitations of this drug delivery system.

Shetye et al. explore liposomes as targeted drug delivery systems, highlighting their advantages, synthesis methods, and mechanisms of drug release. The chapter discusses their applications in treating microbial diseases and cancer, emphasizing specific ligand binding for efficient targeting. Commercial examples like DOXIL and MYOCET are presented, showcasing the potential of liposomes in enhancing therapeutic efficacy and minimizing side effects.

Kumar et al. emphasize liposomes as superior delivery systems for drugs, nutrients, vaccines, and bioactive substances. These spherical vesicles with phospholipid bilayers offer efficient drug encapsulation using various techniques. Liposomal nanoformulations enable active and passive targeting. Stimuli-sensitive liposomes provide controlled drug release in response to internal and external cues. They are being explored for cancer treatment due to their non-toxic, biodegradable, and biocompatible properties. Liposomal formulations are undergoing clinical trials, and innovative applications are on the horizon.

Pabi et al. discuss tailored treatment, needing efficient drug delivery into single live cells. Electroporation, a popular technique using electrical pulses to permeabilize cell membranes, offers advantages like easy operation, high efficiency, and controllable delivery. The chapter covers bulk electroporation (BEP), single-cell electroporation (SCEP), and localized single-cell electroporation (LSCEP). SCEP assesses cell-to-cell variance, while LSCEP utilizes organelles and biochemical effects. The chapter concludes with future aspects of electroporation.

Haider et al. discuss mechanoporation-based techniques for safe and rapid intracellular cargo delivery. Mechanoporation deforms cells temporarily, enabling efficient diffusion of macromolecules. The chapter covers historical background, benefits, and drawbacks of these processes. Advancements in microfluidic and nanotechnological techniques offer better control over membrane disruption. Future research may focus on creating multifunctional platforms for drug delivery.

Rohit and Raj discuss drug delivery into cells through thermoporation and photothermal techniques. Thermoporation uses heat to generate tiny openings in the cell membrane, facilitating drug diffusion. The chapter elaborates on thermoporation mechanisms, applications, and limitations. Photothermal methods involve excited photosensitizers, near-infrared radiation, and nanoparticles for cell transfection. Membrane abnormalities caused by rapid temperature fluctuations are discussed as a limitation. The chapter benefits readers interested in cell transfection and cell lysis using photothermal methods.

H. Manoj and M. Shanmugam discuss microinjection as a powerful technique for understanding single-cell processes and delivering therapeutic drugs. Microneedles achieve 100% transfection efficiency by directly penetrating cells, and their development allows efficient drug delivery with minimal discomfort. The chapter covers microinjection basics, microneedle devices, applications, and advantages and disadvantages comprehensively.

James F. Leary discusses magnetic nanoparticles as an attractive option for advanced drug delivery. Magnetic fields can concentrate nanoparticles at specific sites in the body, enabling controlled drug release and single-cell hyperthermia. Superparamagnetic ferric oxide nanoparticles offer low nanotoxicity, easy biodegradation, and serve as MRI and X-ray contrast agents for non-invasive tumor imaging. Magnetic fields can also be used for magnetoporation, drug release, gene transfection, and single-cell lysis, allowing multi-step targeting of drugs. These nanoparticles' magnetic properties simplify manufacturing and purification processes, making them highly practical for nanomedical systems.

Solovchuk and Hsu provide an overview of microfluidic technology in drug delivery and diagnostic testing for bacterial diseases. Microfluidics enables precise control of small volumes of fluids, benefiting drug preparation, formulation, testing, and controlled delivery. The chapter explores advanced diagnostics for viable bacteria testing, high-throughput testing, point-of-care testing (POCT), and rapid antibiotic susceptibility testing. Single bacterium microfluidics and wearable/implantable devices for drug delivery are also discussed. Microfluidics has the potential to revolutionize bacterial disease diagnosis and treatment by offering rapid and precise diagnostic tools and targeted drug delivery to infected tissues.

Sundaram et al. discuss non-viral gene delivery systems' promise for cancer and disease treatment. Factors affecting their success include cationic charge, nuclease degradation protection, intracellular delivery, and endosomal escape. The ability to remain "stealth" without causing adverse reactions is crucial. The chapter summarizes recent literature on cytokine response to nanoparticulate delivery systems, emphasizing structure-function studies and implications for specific applications. The goal is to familiarize readers with cytokine-related issues and intervention strategies for cytokine-mediated disorders.

Pillai et al. explore the fundamental aspects of controlled drug delivery systems, advancements, and future possibilities. It emphasizes their importance in optimizing therapy and patient compliance. Various systems, such as oral, injectable, transdermal, and implantable, are discussed, along with their mechanisms and advantages. Challenges include formulation complexities and regulatory considerations. Recent advancements include intelligent drug delivery, nanotechnology-based targeting, and bio-responsive systems. These advancements hold promising applications in personalized medicine and targeted drug delivery, revolutionizing drug therapy and improving patient outcomes.

Naik et al. explore drug delivery advancements and the use of biomaterials to enhance various delivery methods. Factors like cellular efficiency, pharmacokinetics, and toxicity are crucial for drug delivery. Polymeric systems, both naturally and synthetically driven, are commonly used. The chapter provides a comprehensive analysis of polymeric systems in drug delivery, including types, formulations, advancements, and drawbacks.

Kumar et al. discuss transdermal drug delivery systems (TDDS) as effective alternatives to oral and parenteral routes. TDDS offers several benefits like self-administration, minimal side effects, and improved patient compliance. Advancements include nanovesicles, polymeric nanocarriers, iontophoresis, and microneedle technology for delivering macromolecular biomolecules. The chapter covers mechanisms, factors, formulations, and permeation promoters of TDDS, as well as existing patches and latest technologies. Expectations for innovative tools in transdermal drug delivery are highlighted.

This book's specific objectives include providing a comprehensive exploration of cutting-edge drug delivery systems, covering various approaches and technologies, highlighting recent advancements and potential future possibilities, showcasing the significance of controlled drug delivery systems in optimizing therapy and enhancing patient compliance, emphasizing improved drug efficacy with minimized side effects, inspiring researchers, practitioners, and students to explore innovative solutions, and encouraging contributions to advance drug delivery for improved patient outcomes and better health care.

This book aims to inspire researchers, practitioners, and students to contribute to drug delivery science, improving patient outcomes and health care. The book's diverse content promises to be a valuable resource for all those interested in advancing drug delivery techniques and applications.

Chennai, India

Tuhin Subhra Santra
Ashwini Uma Surendra Shinde

Contents

Biological Methods for Drug Delivery

Prajkta Chivte, Vinal Pardhi, and Akhilraj Pillai

Abstract Drug delivery systems (DDS) are widely employed to enhance the bioavailability of a drug while alleviating its potential toxicity. Biomedical engineers across the globe have been attempting to design efficacious drug delivery vectors that allow patient-tailored therapies for various diseases. One such broad class of drug delivery systems uses biological vectors or biomimetic nanoparticles (BNPs), an evolving field in nanotechnology, wherein the surface is integrated or doctored with biomaterials that mimic the natural features and functions of normal cells. Such vectors exhibit high biocompatibility and specificity, extended retention times and minimal unsought immune responses. Biological vectors include a range of organism-derived particles such as endogenous cell membranes, extracellular vesicles and exogenous substances such as viruses, virosomes and virus-like particles (VLP). These vectors allow the integration of substantial amounts of cargo while simultaneously providing multiple options for surface alterations with different functional agents. However, a few major challenges for successfully translating this drug delivery system from laboratory to clinical use include stability, large-scale production and high efficacy. Continual probing of the biology behind the usage of such vectors is imperative for attaining clinical success. This chapter provides a list of promising biological vectors, the type of drugs loaded, information on vectors that underwent clinical trials and the limitations of this drug delivery system.

P. Chivte
Department of Chemistry and Biochemistry, Northern Illinois University, Dekalb, IL, USA

V. Pardhi (✉)
Scientity Services, Thane, Maharashtra, India
e-mail: vinal@scientityservices.com

A. Pillai
Department of Biological Sciences, St. John's University, Queens, NY, USA

T. S. Santra and A. U. S. Shinde (eds.), *Advanced Drug Delivery*,
Studies in Mechanobiology, Tissue Engineering and Biomaterials 26,
https://doi.org/10.1007/978-981-99-6564-9_1

1 Introduction

The discovery and development of new drug entities is known to be a laborious and expensive process. From the initial identification of the target to securing approval for its marketing, the timeline could extend beyond a decade. Hence, improvising the efficacy of the previously established drugs by using approaches such as dose titration, therapeutic drug monitoring, controlled release rates and targeted drug delivery have been attempted by the researchers. The drug delivery system (DDS) is one such approach that has been successful in enhancing therapeutic efficacy, increasing patient compliance and reducing the overall toxicity of the drug for the treatment of numerous diseases (Tiwari et al. 2012; Kiriiri et al. 2020; Mohs and Greig 2017).

The coupling of nanotechnology with DDSs has further shown great potential to improve the solubility and stability of encapsulated drugs, cellular uptake, circulation durations and target specificity. DDSs have also steered the evolution of new genres of therapeutics such as peptides, proteins, antibodies and even cells as these systems can be easily modified to meet specific delivery needs. As of 2023, we have witnessed three generations of nanoparticles (Fig. 1). The early generation of nanoparticles had non-fouling coatings to minimize their interaction with the surrounding cells and hence were biologically inert. The second generation of nanoparticles is known to actively target disease sites and interact with the local milieu. The third and most recent generation of nanoparticles is inspired by biology and is known to mimic living cells/viruses (Sushnitha et al. 2020; Gao et al. 2023; Waheed et al. 2022). At this point, nanoparticle DDSs are being globally investigated for the treatment of various diseases in the forms like liposomes, polymeric nanoparticles, metallic nanoparticles, hybrid nanoparticles and more recently introduced forms called biomimetic nanoparticles (BNPs) or biological vectors (Tiwari et al. 2012; Zhang et al. 2020).

BNPs are a promising class of nanoparticles whose surface is integrated or doctored with different types of biomaterials that mimic the biological characteristics and functions of the human body. As a result, the biomimetic nanoparticles have tremendously bettered the biocompatibility, target specificity, retention times and undesired immune responses of the encapsulated drugs (Szliszka et al. 2009; Chen et al. 2021). These biological methods for DDSs can be further classified based on the type of biological transporting vector used, such as

(i) cell membranes (endogenous)
(ii) various types of viral vectors, virus-like particles (VLPs) and virosomes (exogenous)
(iii) extracellular vesicles.

The functionalization of nanoparticles primarily consists of three steps: extraction of biological vectors, deciding the functional drug, and finally camouflaging the functional drugs within the biological vector. Overall, BNPs have been extensively developed to deliver various types of cargo and are expected to serve as a competent candidate for therapeutics and imaging, particularly for achieving high target specificity (Chen et al. 2021).

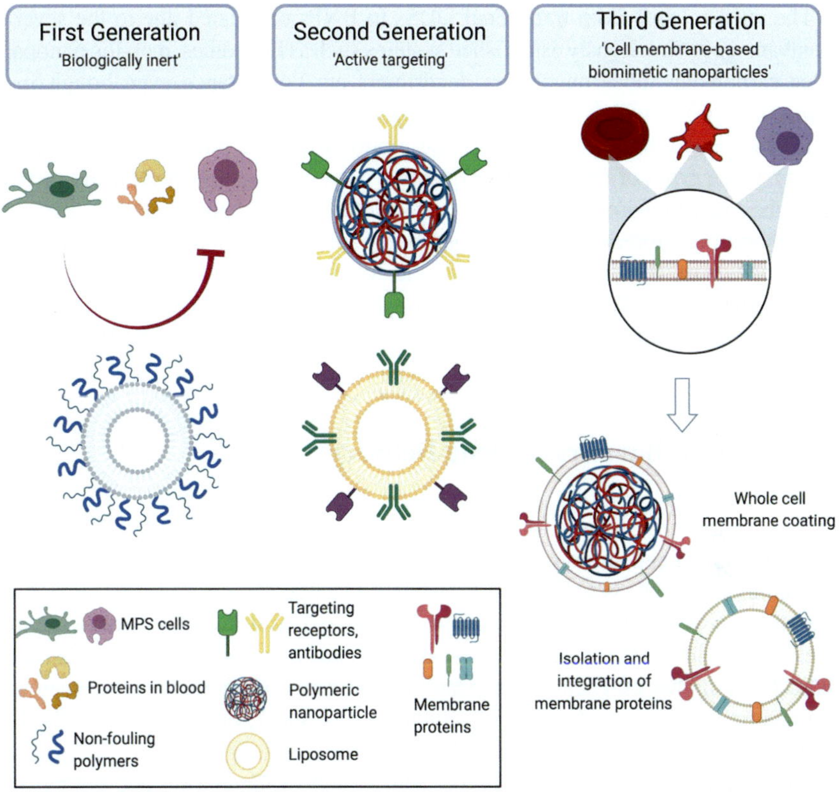

Fig. 1 A diagrammatic representation of the evolution of the generation of nanoparticles. The first generation of nanoparticles was biologically inert whereas the second generation of nanoparticles is known for involving active targeting molecules for targeted delivery. The third generation is inspired by nature and includes the different types of BNPs. Copyright © 2020 Sushnitha et al. (2020)

Additionally, based on the mechanisms through which the target is reached, these BNPs can also be classified into passive and active targeting. The size and shape of BNPs determine the biodistribution and accumulation at the tissue levels. This forms the basis of passive targeting wherein the half-lives of BNPs are prolonged. Hence, passive targeting is mostly used in the case of malignant tumors and is known to cause enhanced permeability and retention effect (EPR). Passive targeting has the disadvantage of low target specificity as there is no biochemical identification of the target. On the other hand, active targeting has high target specificity and comprises engineered BNPs surfaces (ligands, peptides, proteins, antibodies, aptamers) that bind to cell surface receptors. The targets could also be intracellular components such as nuclei, mitochondria and lysosomes (Chen et al. 2021; Soprano et al. 2022).

The need to shift from traditional DDSs to BNPs originated due to the several disadvantages of previously established systems such as liposomes, metallic nanoparticles, carbon nanotubes, nanofibers, dendrimers, etc. For instance, even though liposomes, one of the widely explored DDS, have self-assembly capabilities, are less toxic, and have a high payload, they are reported to have poorly controlled drug release. Other types of DDSs such as quantum dots, metal nanoparticles, and carbon nanotubes have high immunogenicity, high cytotoxicity, and poor aqueous solubility and absorption. Moreover, some nanoparticles are known to cross the blood–brain barrier and cause neurotoxicity when the intended target is not the brain (Adepu and Ramakrishna 2021). This chapter focuses on the types of biological methods of drug delivery, types of drugs loaded, clinical trials, and the limitations of BNPs.

2 Biological Barriers to Drug Delivery

BNPs or rather any type of drug have to face several biological barriers before reaching their intended target. Biological barriers are considered one of the major limitations of the DDSs to effectively deliver the cargo. The complexity of each of these barriers is affected by various other factors such as drug administration route, disease type and disease stage (Blanco et al. 2015). In order to overcome these barriers and achieve higher efficacies, understanding the types of interactions between the biomimetic materials with the extracellular matrix and various types of cells and tissues is indispensable (Gao et al. 2023).

The immune system, which has the primary job of identifying and getting rid of foreign objects, prevents BNPs from being localized to the target site. BNPs are swiftly eliminated from circulation after injection via two main pathways: the mononuclear phagocyte system (MPS) and natural clearance by filtering organs. The former involves a direct connection between BNPs and immune cells, whereas the latter is controlled by particle size. MPS, which involves a system of phagocytic cells, primarily the macrophages in the spleen, lymph nodes and liver, is known to sequester nanoparticles immediately after injection. The sequestration process involves the opsonization of circulating BNPs which is nothing but the adsorption of plasma proteins such as serum albumin, apolipoproteins, and immunoglobulins onto its surface. Moreover, the formation of such a protein corona around the BNPs is influenced by its size, hydrophobicity, surface charge as well as its chemistry (Sushnitha et al. 2020; Blanco et al. 2015).

Researchers are constantly striving to better the BNPs to overcome the immune system barrier. For instance, some researchers are engaged in developing BNPs that would exhibit surface proteins which would ultimately prevent them from being identified as foreign. Such surface proteins will avoid rapid recognition of the reticuloendothelial system (RES) and slow the BNPs clearance from the system by circumventing the body's immune response. Moreover, there are numerous natural interactions such as receptor-ligand binding and adhesion that can be engineered as surface proteins. Another important factor to consider here is the charge of the surface proteins which could impact the biodistribution and pharmacokinetics of BNPs. For example, it is reported that highly cationic nanoparticles tend to be cleared rapidly from circulation as compared to highly anionic nanoparticles. Whereas neutral nanoparticles or nanoparticles with slightly negative charge show significantly prolonged circulating half-lives. Hence, such BNPs with surface modifications will possess higher capacities to enter the disease target sites and thus increasing their therapeutic efficacy in vivo (Sushnitha et al. 2020; Blanco et al. 2015).

3 Biological Methods for Drug Delivery

In the presence of high-specificity interactions between biological molecules and their target site, complex reactions can occur, and selecting an appropriate sequence is the first step in the development of a biomimetic system (Sabu et al. 2018). To achieve maximum therapeutic efficacy, with minimal adverse effects, drug delivery involves directing therapeutic agents to target tissues, cells, and utilizing natural elements that have evolved to carry out this function (Yoo et al. 2011). There has been a significant advancement in DDS to meet the specific needs of the therapeutic landscape, which has shifted from small-molecule pharmaceuticals to a new generation of treatments, including proteins, peptides, monoclonal antibodies, nucleic acids, and even live cells (Gao et al. 2023). Since biological vectors inherit the structural and functional complexity of their original donors, they limit undesirable immune responses and prevent immediate removal (Chen et al. 2021). Understanding the molecular architecture and functional dynamics of these delivery vesicles, as well as the interactions between drug and carrier, could enable these nanoparticle vehicles to be further improved with amended bioavailability, distribution, longer retention, drug release, stronger target specificity, and extended formulation stability. In addition, the highly customizable structure and dynamic nature make protein aggregation likely, which is a major risk that impedes or hinders the development of biological products (Phyo et al. 2021). Bioactive agents, such as therapeutics, imaging agents and reporter molecules, are frequently internalized into cells because they are able to bind to intracellular targets to perform their functions. Large hydrophilic molecules can enter cells but are prevented by the plasma membrane's impermeable structure (Su et al. 2013). Considering the limitation of traditional DDSs, this sophisticated class of BNPs emerged that has enabled the development of customizable vehicles with explicit control over their shape, size and surface properties. Here, we have briefly described biological methods for drug delivery i.e., BNPs derived from endogenous cell membrane vectors, extracellular vesicles, and exogenous substances: viruses, virosomes and VLPs. A schematic representation of various types of BNPs and their development methods is shown in Fig. 2.

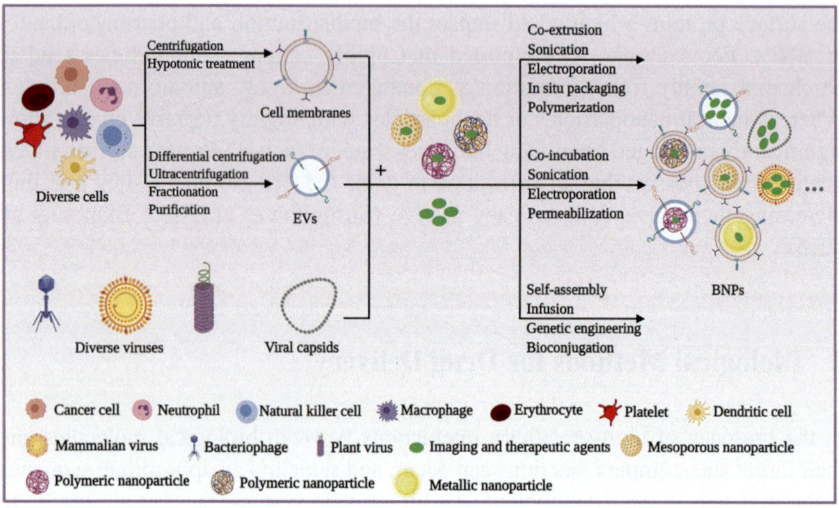

Fig. 2 A schematic representation of various procedures to develop BNPs. Biomimetic nanotechnology utilized these biological vectors to camouflage the loaded drugs for direct or indirect targeted delivery. Copyright © 2022, Chen et al. (2021)

3.1 *Endogenous Cell Membranes Vectors*

A cell membrane is the outermost layer of a cell that is known to provide a cell its identity and is composed of various lipids, proteins and carbohydrates. They are usually 5–10 nm in thickness and perform certain crucial functions such as structural rigidity, fluidity, adhesion, cellular recognition and signaling (Marrink et al. 2019; Chugh et al. 2021). One of the main types of BNPs is derived from mammalian cell membranes where the drug is camouflaged with cell-derived membranes. Various types of membranes have now been used for diverse applications. They include red blood cell membranes, cancer cell membranes, mesenchymal cells, immune cell membranes (neutrophils, NK cells, macrophages, platelets, lymphocytes) and hybrid cell membranes (Chen et al. 2021; Soprano et al. 2022). The presence of surface proteins on these cell membranes is responsible for recognition and interaction which makes it suitable for diverse applications such as the treatment of various diseases as well as delivering vaccines (Chugh et al. 2021).

The latest progress made in the field of BNPs is expected to improve the efficacy of drugs used for cancer immunotherapy. Moreover, incorporating cancer cell membranes in BNPs provides the required neoantigens for targeting highly mutagenic tumors. Yu et al. and Chug et al. summarize several studies where cell membrane bases BNPs were used for cancer treatment. It is established that BNPs can enhance our immune response and actively target the antigen presenting cells which would ultimately lead to suppressing tumor growth (Chugh et al. 2021; Yu et al. 2022). A study by Zhang et al. reported the use of neutrophil-based BNPs that inherited the antigenic exterior and associated membrane functions and made an ideal decoy to neutralized proinflammatory cytokines such as IL-1β and TNF-α for the treatment of rheumatoid arthritis. This study was conducted in a mouse model wherein the neutrophil-based BNPs showed reduced concentrations of cytokines, suppressed synovial inflammation and chondroprotection against joint damage when compared to the control groups, signifying successful reduction of arthritis at the systemic level (Zhang et al. 2018).

With the rise in antibiotic resistance and various infectious diseases, researchers have been testing BNPs to treat these infectious diseases. It is reported that BNPs fight the pathogens in three steps: they target the source of the infection, neutralize the pathogen's mechanism to avoid the immune cells and finally regulate the immune response for anti-pathogen activity. There are reports of using cell membrane-based BNPs for the delivery of antibiotics as well as using them as toxin neutralizing platforms which would prepare the immune cells to neutralize the pathogens (Fang et al. 2015; Angsantikul et al. 2018) A recent study also demonstrated the use of cell membrane-based BNPs derived from human lung epithelial type II cells and human macrophages (also referred to as cellular nanosponges) to target COVID-19 infection in a mouse model. It was concluded these cellular nanosponges were safe and were able to neutralize SARS-CoV-2 effectively (Zhang et al. 2020).

However, cell membrane-based BNPs have several limitations as well. These types of BNPs involve complex preparation protocols and have relatively low production yields. Also, there is a possibility of the formation of uneven or incomplete coverage of cell membrane coating over the nanoparticles which could lead to undesired side effects in blood circulation. The biological characterization methods of these BNPs are also limited. Moreover, various cell membranes present a plethora of proteins on the cell membrane surface. Thus, optimizing protocols to selectively retain the proteins of interest and remove the nonessential proteins from the cell membrane surface is another huge challenge (Chugh et al. 2021; Yu et al. 2022).

3.2 *Exogenous Substances: Viruses, VLPs and Virosomes*

In the last two decades, several new viruses have emerged that have caused pandemics and epidemics, and particularly SARS-CoV-2 has ravaged the world and demanded advanced drugs and vaccines to curb the infection. Due to the simplistic assembly of any given virion and the presence of certain surface proteins on the viral membrane surface, viruses can evade the immune system. Considering the characteristics such as site-specific targeting, bypassing the host immune system and the ability to enter the cells made viruses a promising candidate for developing BNPs for delivering varied types of cargo (Mougenot et al. 2022).

This class of BNPs can be majorly classified into three groups: viral vectors, VLP and virosomes. As viruses have naturally evolved to transfer genes into the host for their propagation, they can be easily engineered as vehicles to deliver specific genes. VLPs are self-assembled structures formed by capsid or envelop proteins that mimic the original virus whereas the virosomes are phospholipid bilayer spherical vesicles that are integrated with glycoproteins derived viruses. While both of these are spherical nanostructures with hollow cores, they can deliver various chemical drugs as well as biomacromolecules. Moreover, as they lack genetic material, they are non-contagious and range in size from 20 to 200 nm. Such types of BNPs can originate from either mammalian viruses, bacteriophages, or plant viruses (Mougenot et al. 2022). Among the mammalian viruses, adenovirus, adeno-associated viruses and retroviruses are the most commonly used viral vector systems (Yoo et al. 2011).

BNPs based out of mammalian viruses are gaining attention due to reduced potential side-effects while providing an opportunity to modify them using peptides, aptamers and antibodies. For instance, Sun et al. reported BNPs made from Siman vacuolating virus 40 to deliver Hirulog (a synthetic peptide) for the treatment of atherosclerotic plaques in an in vivo model. These BNPs were exceptionally specific and were found to be localized in the areas of plaque formations in the diseased animal models. Such BNPs models can be considered great "theranostic" models that could be used for both therapeutic drugs as well as for diagnosis (as imaging probes) (Sun et al. 2016). Performing gene delivery using virus derived BNPs is an emerging field for the treatment of cardiovascular diseases, infectious diseases, autoimmune diseases, etc. Bulcha et al. provided an extensive summary of different types of viral platforms that are used in gene therapy (Bulcha et al. 2021). Martinez-Navio et al. established treatment for HIV infection by delivering anti-HIV monoclonal antibodies using the adeno-associated virus in rhesus monkeys (Martinez-Navio et al. 2019). Another example is Ad26.COV2.S, which is a recombinant adenovirus serotype 26-based vaccine encoding a full-length and stabilized SARS-CoV-2 spike glycoprotein developed by Johnson & Johnson, USA against COVID-19. This study confirmed the safety and efficacy of this vaccine in 593 individuals from wide age groups (Sadoff et al. 2021). The Oxford University-AstraZeneca group from the UK was also involved in developing the ChAdOx1 nCoV-19 vaccine (AZD1222) that consisted of SARS-CoV-2 spike protein in a chimpanzee adenoviral vector (Voysey et al. 2021).

Several studies have used various types of bacteriophages such as Qβ, P22, MS2, filamentous bacteriophage, etc. for targeted drug delivery. Schwarz et al. substantiated the delivery of Cas9 protein to targeted cells using P22 bacteriophage, derived from *Salmonella typhimurium* (Schwarz et al. 2015). Plant-based viruses such as tobacco mosaic virus are also being enormously used for molecular imaging as well as for drug delivery. Czapar et al. reported the successful delivery of Phenanthriplatin, a DNA-binding anticancer drug using the tobacco mosaic virus to various anti-cancer cell lines (Czapar et al. 2016). Another plant virus called cowpea mosaic virus is also regularly used as it is merely 30 nm in size and has a large surface area to volume ratio. A study by Wen et al. described the incorporation of zinc ethynylphenyl porphyrin, a photosensitizer into the cowpea mosaic virus which was modified by alkyne-functionalized carboxyl dendrons to perform dual delivery cancer cell as well as macrophages (Wen et al. 2016).

Unlike viral vectors, VPLs are known to be highly resistant to any degradation or denaturation and can withstand harsh purification processes. To mimic the live attenuated viruses, empty VPLs were developed as they showed similar antigenicity (Yoo et al. 2011). The HPV vaccine developed by Merck and GlaxoSmithKline which was approved by US Food and Drug Administration is one early example of using VLPs for vaccine purposes (Villa et al. 2005). Since then, several VLP vaccine candidates have been under investigation for emerging infectious diseases. More than 110 viral proteins from 35 different viral families have demonstrated the capabilities of self-assembly to form VPLs. Vaccines based on VLPs derived from hepatitis B virus, human papillomavirus, hepatitis E virus, human parvovirus, norovirus, arenavirus, bunyavirus, filovirus, paramyxovirus, coronaviruses, etc. are a few of the examples (Nooraei et al. 2021). GlaxoSmithKline along with Medicago also developed CoVLP (Covifenz) in Canada, which is a COVID-19 vaccine. This is a plant-based VLP of SARS-CoV-2 derived from *Nicotiana benthamiana*, a close relative of tobacco plants. It had achieved an efficacy of 71.0% for the SARS-CoV-2 delta variant in the Canadian population (Dyer 2022).

Virosomes are recently gaining clinical approvals as nanocarriers/nanovaccines against various viral infections such as respiratory syncytial virus, HIV-1, hepatitis A, influenza and SARS-CoV-2. The most common example of virosomes for vaccine delivery is Inflexal®V for influenza which was first developed by the Swiss Serum and Vaccine Institute, Berne, Switzerland in 1997 and later adopted by many countries. This is a trivalent influenza virus virosome vaccine that includes neuraminidase and haemagglutinin glycoproteins as its membrane components. This vaccine revolutionized the field of vaccine development and to date, there have been more than 18 clinical trials to confirm the safety and efficacy of Inflexal®V (Asadi and Gholami 2021; Mischler and Metcalfe 2002). A patent has also been approved for a virosome containing fusion peptides for the treatment/prevention of breast cancer which shows overexpression of Her2/neu protein. The virosome consisted of 10 different peptides whose combination resulted in a synergistic enhancement of the overall antibody response in cancer patients (Kammer et al. 2011).

3.3 Extracellular Vesicles

Extracellular vesicles are membrane-enclosed BNPs that participate in intercellular communications such as those concerning physiological or pathological status. They are produced by mainly all cell types in vivo or in vitro and can be broadly classified into exosomes (40–100 nm), microvesicles (50–1000 nm), and apoptotic bodies (1–5 μm) as per their biogenesis. These vesicles are made up of membrane proteins, phospholipid bilayers, nucleic acids and metabolites that reflect their biological origins and hence can bypass the immune system. Both exosomes and microvesicles are gathering increased attention as DDS for the delivery of imaging agents as well as therapeutic molecules such as synthetic drugs, proteins, peptides, miRNA and siRNA due to their desirable physiochemical characteristics (Chen et al. 2021; Mougenot et al. 2022).

Remarkably, extracellular vesicles have proven to have lower toxicity for carrying out brain-targeted delivery as they can cross the blood–brain barrier which is uncommon in membrane vectors (Chen et al. 2021). Exosomes being nanosized vesicles are known to act as a channel for long-distance intercellular communication and biomaterial transfers without any cell-to-cell connection. Numerous studies have demonstrated the application of exosomes as they can effectively enter cells with minimal immune response and have lowers side effects even with repeated doses (Zhang et al. 2020). A study by Wei et al. compared the activity of free doxorubicin with exosomes loaded with doxorubicin for the treatment of osteosarcoma. Here, the loaded exosomes showed good loading capacity, targeted cytotoxicity against the cancer cells and low cytotoxicity against the healthy cells (Wei et al. 2019). A separate study by Alvarez-Erviti et al. showed the delivery of short interfering (si)RNA by exosomes produced by dendritic cells for the treatment of Alzheimer's disease. The therapeutic efficacy of these exosomes was supported by strong mRNA and protein knockdown of *BACE1* which is a target for Alzheimer's disease, when tested in mouse models (Alvarez-Erviti et al. 2011).

Microvesicles or ectosomes which are produced by outward budding of the plasma membrane can be found in varying sizes ranging from nanoscale to microscale. These microvesicles have sparked interest among bioengineers as they have the potential to deliver not just chemical drugs but also huge biomacromolecules (Chen et al. 2021). Ran et al. utilized these capabilities to deliver oncolytic adenoviruses using microvesicles-based BNPs to selectively target human lung cancer cells. These microvesicles were also derived from tumor cells and were responsible for the mediated entry of the adenovirus into pulmonary carcinoma cells without eliciting an immune response (Ran et al. 2016).

Similar to cell membrane-based BNPs, extracellular vesicle production also faces challenges concerning low production yield and optimization of isolation protocols. Extracellular vesicles are produced by culturing specific types of cells and collecting spent media. Techniques such as ultracentrifugation, ultrafiltration, or tangential flow filtration are utilized to isolate these vesicles. Scientists are working to enhance the extracellular vesicles-based BNPs research on dual levels: first by scaling the cell culturing step and second by increasing the vesicle production yield per cell (Mougenot et al. 2022).

4 BNPs Production Methods

The production methods of nanoparticles can be classified under two methods: Top-down Method and Bottom-up method. In the case of the top-down method, the synthesis starts at a macro or micro scale and various techniques are involved to reduce the particle size to the nanoscale. BNPs produced by the top-down method are generally derived from extracellular vesicles, hybrid nanovesicles, and mammalian and bacterial cell membranes. This approach is quite common and easily scalable for industrial production. However, there is very little control over the final size and the size distribution of BNPs produced by this method. The techniques involved in this methodology include extrusion, high-shear mixing and sonication (Mougenot et al. 2022).

By contrast, the bottom-up approach of producing nanoparticles takes advantage of supramolecular interactions which perform self-assembly of the nanostructures at specific conditions and parameters. The assembly starts from the atomic or molecular scale and goes up to the nanoscale. Thus, this method is not used for synthesizing BNPs. Another way of synthesizing BNPs is by using microfluidics-based technology. Molinaro et al. demonstrated the incorporation of membrane proteins within the bilayer of leukosomes using a microfluidic-based platform. This study provides a system for the scalability of BNPs and in turn, provides a cost-effective method while increasing the yield and consistency (Molinaro et al. 2018).

4.1 Synthesis of Cell Membrane-Based BNPs

Cell membrane derived BNPs synthesis comprises a multistep process that involves lysis of parent cells, extraction and purification of cell membrane, and formation of the final vesicles/BNPs. The parental cells are broken down by digesting in a hypotonic buffer solution. Cellular components like the cytoplasmic organelles and the nucleus are separated from the cell membrane mainly by centrifugation. Again, based on the type of cell, the centrifugation protocols may vary. For instance, the cell membranes of nucleus-free cells such as red blood cells (RBCs) are extracted by simple centrifugation starting with the whole blood, followed by separation of RBCs using hypotonic lysis treatment and finally removal of hemoglobin. On the other hand, the cell membrane of the nucleus containing cells like stem cells, leukocytes and tumor cells require discontinuous sucrose gradient centrifugation or differential centrifugation to remove the intracellular components. These nucleated cells have specialized surface proteins which should be retained throughout the extraction process to utilize their bio-interacting capabilities. After successful extraction, the cell membranes are either stored at 4 °C for short term use or undergo lyophilization for long term usage. Such lyophilized cell membrane pellets can be easily rehydrated by techniques such as extrusion and sonication (Le et al. 2021; Sevencan et al. 2020).

Once the cell membranes are isolated, they must be coupled with the drug/cargo to form the desired BNPs. As these cell membrane derived vesicles have a hollow core, they are an exemplary model for coating diverse kinds of therapeutic drugs. Several methods such as ultrasonication, extrusion, electroporation, graphene nanoplatforms-based coating and in situ encapsulation are reported for coating the cell membranes with the nanoparticles. Extrusion was the first reported method for cell membrane coating whereas sonication is now being widely used particularly to scale up production. The selection of coating strategies highly depends on the type of cells and the cargo to be coated. Nonetheless, care must be taken to preserve the original architectural orientation of the cell membrane in order to maintain the surface proteins and achieve targeted drug delivery (Le et al. 2021; Sevencan et al. 2020).

Interestingly, the functions of the cell membrane can also be tweaked for achieving enhanced drug delivery. There are two strategies to manipulate the functions of the cell membrane derived vesicles: pre-modification and post-modification methods. In the pre-modification method, changes are made before the lysis of parental cells i.e., at the genetic or metabolic level. The parent cells receive certain treatments that would alter the lipid composition in the cell membrane or the expression levels of surface proteins on the cell membrane (Le et al. 2021). A study by Bose et al. reported the modification of cell membranes extracted from adipose-derived stem cells to express a C-X-C chemokine receptor type 4 in order to penetrate the endothelial barrier to treat severe hindlimb ischemia (Bose et al. 2018). Contrary to the pre-modification methods of cell membranes, the post-modification method involves the introduction of new components in the cell membranes after the extraction and collection from the parental cells. Numerous post-modification protocols have been explored as they are convenient and diverse types of modification materials are available to alter the cell membrane. These modification materials include proteins, lipids, nucleic acids as well as some synthetic materials such as polyethylene glycol (Le et al. 2021). A research study by Peng et al. demonstrated post-modification by using a 26-mer G-quadruplex oligonucleotide on the cancer cell membrane to target nucleolin protein that is specifically expressed in tumor cells (Peng et al. 2015).

Scientists have also explored the capabilities of BNPs by fusing cell membranes from two different parent cells. Here, the derived vesicle inherits the characteristics from both parental cell membranes. This technique is called cell membrane hybridization which involves cell membranes exhibiting multiple functions synergistically to achieve superior results. Dehaini and colleagues reported "biomimetic dual membrane-functionalized nanoparticles" derived from RBCs as well as platelet cell membranes. The RBCs are known to express the CD47 receptor on its surface which can avoid clearance by RES (immune barrier) whereas the platelet cell membrane has P-selectin protein, which is a ligand for CD44 receptor, particularly expressed on tumor cells (Dehaini et al. 2017). Hence, such hybrid cell membranes can perform dual functions and overall, positively impact the drug delivery efficacy.

4.2 Synthesis of Exogenous Substances (Viruses, VLPs and Virosomes) Based BNPs

Viruses and VLPs have been successfully synthesized to perform gene therapy and drug delivery, to perform high throughput screening of gene functions, to design antiviral drugs, and for vaccine development. The production of viral vectors was originally carried out by using natural hosts of the viruses. With the advent of cell culturing techniques in the 1950s, cell cultures steadily replaced the use of live animals for the synthesis of viral vectors. The very first vaccine that was generated using the tissue culture was the poliovirus vaccine. The adenovirus vectors are produced by the HEK 293 cell lines. The retrovirus vectors are produced in cell lines such as HT1080, TE671, NIH 3T3, HEK293 and CEM cell lines. The adeno-associated viral vectors are produced by 293 or A549 cell lines by transient transfection (Roldão et al. 2019).

The production of VLPs can be broadly divided into three phases: (i) upstream processing i.e., production of VLPs using various expression platforms, (ii) downstream processing involving purification techniques to isolate and concentrate the VLPs, and (iii) formulation of BNPs (majorly vaccines). The viral structural proteins of interest that have self-assembly abilities are cloned in expression systems. Various expression systems such as eukaryotic (transgenic plants, mammalian cells and baculovirus/insect cell system) or prokaryotic (yeast and bacteria) systems are used depending on the final application. Based on the type of expression systems, the quaternary structure of the viral capsid proteins in the VLPs can vary as certain post-translational modifications may or may not be included. Moreover, these post-translational modifications such as glycosylation and phosphorylation play an important role in determining the immunogenicity of the VLP-based vaccines. Nowadays, cell-free protein synthesis systems are also being used for the production of VLPs. These systems are time-efficient and contain minimum cellular contaminants. However, they are expensive in nature and show poor scalability. The purification strategies used for VLPs depend on whether the formed VLPs are readily released in the extracellular medium or cell lysis is required to extract the VLPs. The methods used for clarification and concentration of VLPs include tangential flow filtration, depth filtration and cell sedimentation. The VLPs can also be selectively captured by performing ion-exchange chromatography, affinity chromatography, or hydrophobic-interaction chromatography. Finally, the residual contaminants are removed and the VLPs are sterilized by filtration. During the vaccine formulation process, the VLPs should be shielded from physical, chemical, enzymatic and immune barriers. Hence, excipients such as buffers, preservatives (glycerol, sucrose, and trehalose) stabilizing compounds and adjuvants (chitosan, aluminum salt-based (Alum) adjuvant, bacterial toxin, pattern recognition receptors (PRRs) agonist adjuvant) are added to the final formulation. Characterization of VLPs is a critical step to determine their stability, potency and functionality before commercializing the VLP-based vaccines. Biological (enzyme-linked immunosorbent assay, surface plasmon resonance) biochemical (mass spectrometry, gel electrophoresis, liquid chromatography) and biophysical (types of electron microscopy, size exclusion chromatography, differential scanning calorimetry, dynamic light scattering) characterization methods are used for synthesized VLPs (Nooraei et al. 2021; Roldão et al. 2019).

The preparation of virosomes is a simple and cost-effective process and the structure and composition of virosomes are quite similar to that of liposomes. The initial step involves solubilization of an enveloped virus of choice by detergents such as Octaethylene glycol monododecyl ether or Triton X-100 followed by removal of nucleocapsid structures as well as viral genetic material by ultracentrifugation. The detergent is also removed, whereas the viral proteins and lipids are extracted. Finally, the lipids, viral fusion proteins and viral antigens epitopes undergo a self-assembly phenomenon to form virosomes. Furthermore, the virulence of virosomes can be enhanced by the addition of viral components (proteins, DNA, RNA) and anti-virus drugs for specific applications. The formed virosomes are then separated from the unused viral debris by performing discontinuous sucrose density gradient ultracentrifugation (Asadi and Gholami 2021; Ali et al. 2023).

4.3 Synthesis of Extracellular Vesicles-Based BNPs

The synthesis protocols of extracellular vesicles currently have no consensus among bioengineers. The manufacturing of extracellular vesicles includes several steps: selection of parental cells and culture conditions, isolation of extracellular vesicles, drug loading, purification and removal of contaminants and finally formulation and storage. The exosomes that formed as intraluminal vesicles are derived from the endosomal systems whereas the microvesicles are derived through the outward budding and fission of the plasma membrane. The extracellular vesicles in general are separated from the parental cells by conventional procedures such as differential ultracentrifugation, size exclusion chromatography, precipitation, affinity chromatography and tangential flow filtration. The drug/cargo is loaded into the formed extracellular vesicles by diffusion through the vesicle membrane using chemical or mechanical techniques such as passive incubation, electroporation, saponin treatment and sonication. Next, additional purification steps may also be required to remove the free drug. Lyophilization of the extracellular vesicles based BNPs is the widely employed way to store for long term usage. Several studies have shown enhanced efficacies of these types of BNPs by utilizing engineering approaches that impact circulation kinetics, biodistribution, specific targeting and internalization (Herrmann et al. 2021; Abels and Breakefield 2016; Murphy et al. 2019).

5 Clinical Trials of Biomimetic Nanoparticles

The prospects of BNPs for the treatment of various diseases are highly positive and the scientific literature shows several proof-of-concept reports. Clinical trials for vaccinations, medicines, and gene delivery utilizing BNPs started in the early 2000s. The FDA granted approval to several drugs inspired by biomimetic viruses throughout this decade (Hecolin®, Gardasil®, Invivac®, Recombinax HB®, Epaxal®, Engerix-B®, Cervarix®, and Inflexal®V). These drugs certainly proved the concept of using virus-inspired nanoparticles for various purposes. Over the past two decades, scientists have shown continued interest in developing all types of BNPs (Mougenot et al. 2022). The clinical trials for BNPs for various human diseases are summarized in Table 1.

Table 1 A list of clinical trials of BNPs for various human diseases (Chen et al. 2021)

Type of BNPs	Type of cargo	Disease	Identifiers
Adeno-associated virus	Survival motor neuron	Spinal muscular atrophy 1	NCT02122952, NCT03421977, NCT03306277
Adeno-associated virus	Ocular gene	X-linked juvenile retinoschisis	NCT02317887
Adeno-associated virus	PG9 antibody gene	Immune prophylaxis	NCT01937455
Retrovirus	Cytocidal cyclin G1 construct	Osteosarcoma, sarcoma, pancreatic cancer	NCT00505271, NCT00572130, NCT00505713
γ retrovirus	COL7A1 gene	Recessive dystrophic epidermolysis bullosa	NCT02984085
Adenovirus	Spike protein gene of SARS-CoV-2	COVID-19	NCT04341389, NCT04400838
Oncolytic adenovirus	TMZ-CD40L and 4-1BBL genes	Pancreatic cancer	NCT02705196
Microvesicles derived from tumor cells	Cispatin	Malignant ascites or pleural effusion II	NCT01854866
Microvesicles derived from tumor cells	Methotrexate	Malignant pleural effusion	NCT02657460, ChiCTR-ICR-15006304
Microvesicles derived from tumor cells	Methotrexate	Advanced bile duct cancer	ChiCTR-OIB-15007589
Exosomes	Stimulator of interferon genes agonist	Advanced solid tumor	NCT04592484
Plant exosomes	Curcumin	Colon cancer	NCT01294072
Exosomes derived from dendritic cells	Tumor antigen	Non-small cell lung cancer	NCT01159288
Exosomes derived from allogeneic mesenchymal stem cells	miR-124 (microRNA)	Acute ischemic stroke	NCT03384433
Exosomes derived from mesenchymal stem cells	$Kras^{G12D}$ siRNA	Metastatic pancreatic cancer	NCT03608631

6 Challenges and Future Directions

Compared to synthetic nanoparticles, BNPs face different challenges in their transition from the laboratory to the clinic as these vectors are complex entities produced by living things. Due to a number of technical and practical constraints, the design and application of most of the BNPs for payload delivery are still in the experimental stage. The current difficulties with using these vectors in clinical settings primarily relate to several factors. First, it is important to note that BNPs are unstable moieties. Second, the complicated methods of preparation make it difficult for BNPs to be mass-produced, which limits their clinical translation. Furthermore, the biological safety and bioactive effect of BNPs in humans remain unclear (Chen et al. 2021; Soprano et al. 2022).

Consequently, BNP studies require a multidisciplinary team wherein the biologists are aware of the functional biomolecules from the biological sources, biotechnologists and nanotechnology specialized engineers are involved in the synthesis of BNPs and to develop production processes, whereas medics are required to evaluate the effects of the synthesized BNPs on the human body. Hence, a practicable strategy to address the main challenges faced by pharmaceutical industries to deliver drugs or vaccines via BNPs would be combining the advantages of DDSs such as mass production and controllability and high throughput production with the inherent characteristics of natural vesicles. Eventually, researchers should be able to develop something like a magic bullet that will overcome all the existing challenges (Mougenot et al. 2022).

Despite all the benefits that cell membrane-based BNPs offer, there are a number of obstacles that must be overcome before they can be successfully applied in clinical settings. Specifically, the selection of the cell source can have an impact on how successfully identifying and isolating cell membranes can be accomplished over a lengthy, multi-step process. For instance, while several cell sources may exhibit significant heterogeneity, it can be quite difficult to obtain stem cells from a patient. Moreover, erythrocytes and platelets can be frequently obtained through transfusion, but since they lack a nucleus, it is challenging to manipulate genes ex vivo using these cells. Therefore, the source of cells must be taken into account in order to obtain appropriate quantities of starting material for the successful implementation of these strategies in the clinic (Sushnitha et al. 2020).

The primary aim of any DDS will always remain to deliver the cargo to specific sites while maintaining its therapeutic concentrations. It is predicted that in the coming future, BNPs will be able to cross several systemic and cellular barriers while providing higher specificity toward the target site. There is also a possibility of developing BNPs that have improved pharmacokinetics and could achieve preprogrammed pulsatile release of the drug over a period of time, particularly as personalized medicines. Moreover, it is also predicted that BNPs could be coupled with soft electronics or with artificial intelligence to achieve higher efficacies. To re-emphasize, such types of enhanced BNPs will require efforts from multidisciplinary groups of researchers as well as strong funding support (Gao et al. 2023).

7 Conclusion

To summarize the chapter, BNPs can be defined as nanomaterials synthesized by using biological materials to mimic endogenous features found in the human body. The development of BNPs is an interdisciplinary field involving biology, chemistry, medicine, and bioengineering aspects to design, synthesize, characterize, and finally use the drug for its intended purpose. There certainly is an upward trend in this BNP research to deliver various types of drugs, genes as well as vaccines.

By transporting across membranes and enhancing the solubility and stability of encapsulated cargos, BNPs have the potential to persist in systemic circulation for longer durations. These delivery nanoplatforms feature intelligent designs with complex, well-organized self-assembled architectures that have a range of functional specialties and integrated stability. In order to overcome the limitations of lipid-based, polymeric, and inorganic nanoparticles, it is now possible to recreate complex architecture and cellular functionalities by using naturally derived nanocarriers like extracellular vesicles, VLPs, virosomes or cell membranes.

Although BNP-related drug delivery has been studied for several years, it has only been in the last two decades that research has expanded dramatically. It has indeed turned out to be a rewarding strategy to use these naturally derived vesicles to enhance biospecific characteristics, site-specific targeting, and loading capacity of BNPs to extend circulation time. Thus, a thorough understanding of the source properties of the BNPs as well as of the target site/disease is crucial for successful drug delivery.

Declaration of Competing Interest The authors declare that they have no known competing financial interests or personal relationships that could have appeared to influence the work reported in this paper.

Acknowledgements We would like to express our sincere gratitude to all authors for their contribution in the creation of this book chapter.

References

Abels ER, Breakefield XO (2016) Introduction to extracellular vesicles: biogenesis, RNA cargo selection, content, release, and uptake. Cell Mol Neurobiol (Springer New York LLC) 36(3):301–312. https://doi.org/10.1007/s10571-016-0366-z

Adepu S, Ramakrishna S (2021) Controlled drug delivery systems: current status and future directions. Molecules (MDPI) 26(19). https://doi.org/10.3390/molecules26195905

Ali H et al (2023) Virosome: an engineered virus for vaccine delivery. Saudi Pharm J (Elsevier B.V.) 31(5):752–764. https://doi.org/10.1016/j.jsps.2023.03.016

Alvarez-Erviti L, Seow Y, Yin H, Betts C, Lakhal S, Wood MJA (2011) Delivery of siRNA to the mouse brain by systemic injection of targeted exosomes. Nat Biotechnol 29(4):341–345. https://doi.org/10.1038/nbt.1807

Angsantikul P et al (2018) Coating nanoparticles with gastric epithelial cell membrane for targeted antibiotic delivery against Helicobacter pylori infection. Adv Ther (Weinh) 1(2). https://doi.org/10.1002/adtp.201800016

Asadi K, Gholami A (2021) Virosome-based nanovaccines; a promising bioinspiration and biomimetic approach for preventing viral diseases: a review. Int J Biol Macromol (Elsevier B.V.) 182:648–658. https://doi.org/10.1016/j.ijbiomac.2021.04.005

Blanco E, Shen H, Ferrari M (2015) Principles of nanoparticle design for overcoming biological barriers to drug delivery. Nat Biotechnol (Nature Publishing Group) 33(9):941–951. https://doi.org/10.1038/nbt.3330

Bose RJ et al (2018) Bioengineered stem cell membrane functionalized nanocarriers for therapeutic targeting of severe hindlimb ischemia. Biomaterials 185:360–370. https://doi.org/10.1016/j.biomaterials.2018.08.018

Bulcha JT, Wang Y, Ma H, Tai PWL, Gao G (2021) Viral vector platforms within the gene therapy landscape. Signal Transduct Target Therapy 6(1) (Springer Nature). https://doi.org/10.1038/s41392-021-00487-6

Chen L, Hong W, Ren W, Xu T, Qian Z, He Z (2021) Recent progress in targeted delivery vectors based on biomimetic nanoparticles. Signal Transduct Target Therapy (Springer Nature) 6(1). https://doi.org/10.1038/s41392-021-00631-2

Chugh V, Vijaya Krishna K, Pandit A (2021) Cell membrane-coated mimics: a methodological approach for fabrication, characterization for therapeutic applications, and challenges for clinical translation. ACS Nano (American Chemical Society) 15(11):17080–17123. https://doi.org/10.1021/acsnano.1c03800

Czapar AE et al (2016) Tobacco mosaic virus delivery of phenanthriplatin for cancer therapy. ACS Nano 10(4):4119–4126. https://doi.org/10.1021/acsnano.5b07360

Dehaini D et al (2017) Erythrocyte–platelet hybrid membrane coating for enhanced nanoparticle functionalization. Adv Mater 29(16). https://doi.org/10.1002/adma.201606209

Dyer O (2022) Covid-19: WHO set to reject Canadian plant based vaccine because of links with tobacco industry. BMJ (Montreal)

Fang RH, Luk BT, Hu CMJ, Zhang L (2015) Engineered nanoparticles mimicking cell membranes for toxin neutralization. Adv Drug Deliv Rev (Elsevier B.V.) 90:69–80. https://doi.org/10.1016/j.addr.2015.04.001

Gao J, Karp JM, Langer R, Joshi N (2023) The future of drug delivery. Chem Mater (American Chemical Society) 35(2):359–363. https://doi.org/10.1021/acs.chemmater.2c03003

Herrmann IK, Wood MJA, Fuhrmann G (2021) Extracellular vesicles as a next-generation drug delivery platform. Nat Nanotechnol (Nature Research) 16(7):748–759. https://doi.org/10.1038/s41565-021-00931-2

Kammer A, Amacker M, Zurbriggen R (2011) Multiepitope vaccine for Her2/neu-associated cancers

Kiriiri GK, Njogu PM, Mwangi AN (2020) Exploring different approaches to improve the success of drug discovery and development projects: a review. Futur J Pharm Sci 6(1). https://doi.org/10.1186/s43094-020-00047-9

Le QV, Lee J, Lee H, Shim G, Oh YK (2021) Cell membrane-derived vesicles for delivery of therapeutic agents. Acta Pharmaceutica Sinica B (Chinese Academy of Medical Sciences) 11(8):2096–2113. https://doi.org/10.1016/j.apsb.2021.01.020

Marrink SJ, Corradi V, Souza PCT, Ingólfsson HI, Tieleman DP, Sansom MSP (2019) Computational modeling of realistic cell membranes. Chem Rev (American Chemical Society) 119(9):6184–6226. https://doi.org/10.1021/acs.chemrev.8b00460

Martinez-Navio JM et al (2019) Adeno-associated virus delivery of anti-HIV monoclonal antibodies can drive long-term virologic suppression. Immunity 50(3):567-575.e5. https://doi.org/10.1016/j.immuni.2019.02.005

Mischler R, Metcalfe IC (2002) Inflexal ® V a trivalent virosome subunit influenza vaccine: production

Mohs RC, Greig NH (2017) Drug discovery and development: role of basic biological research. Alzheimer's Dement Trans Res Clin Interv (Elsevier Inc.) 3(4):651–657. https://doi.org/10.1016/j.trci.2017.10.005

Molinaro R et al (2018) Design and development of biomimetic nanovesicles using a microfluidic approach. Adv Mater 30(15). https://doi.org/10.1002/adma.201702749

Mougenot MF et al (2022) Biomimetic nanovesicles—sources, design, production methods, and applications. Pharmaceutics 14(10) (MDPI). https://doi.org/10.3390/pharmaceutics14102008

Murphy DE et al (2019) Extracellular vesicle-based therapeutics: natural versus engineered targeting and trafficking. Exp Mol Med (Nature Publishing Group) 51(3). https://doi.org/10.1038/s12276-019-0223-5

Nooraei S et al (2021) Virus-like particles: preparation, immunogenicity and their roles as nanovaccines and drug nanocarriers. J Nanobiotechnol (BioMed Central Ltd) 19(1). https://doi.org/10.1186/s12951-021-00806-7

Peng LH et al (2015) Cell membrane capsules for encapsulation of chemotherapeutic and cancer cell targeting in vivo. ACS Appl Mater Interfaces 7(33):18628–18637. https://doi.org/10.1021/acsami.5b05065

Phyo P, Zhao X, Templeton AC, Xu W, Cheung JK, Su Y (2021) Understanding molecular mechanisms of biologics drug delivery and stability from NMR spectroscopy. Adv Drug Del Rev (Elsevier B.V.) 174:1–29. https://doi.org/10.1016/j.addr.2021.02.007

Ran L et al (2016) Delivery of oncolytic adenovirus into the nucleus of tumorigenic cells by tumor microparticles for virotherapy. Biomaterials 89:56–66. https://doi.org/10.1016/j.biomaterials.2016.02.025

Roldão A, Silva AC, Mellado MCM, Alves PM, Carrondo MJT (2019) Viruses and virus-like particles in biotechnology: fundamentals and applications. Compr Biotechnol (Elsevier) 633–656. https://doi.org/10.1016/B978-0-12-809633-8.09046-4

Sabu C, Rejo C, Kotta S, Pramod K (2018) Bioinspired and biomimetic systems for advanced drug and gene delivery. J Control Release (Elsevier B.V.) 287:142–155. https://doi.org/10.1016/j.jconrel.2018.08.033

Sadoff J et al (2021) Interim results of a phase 1–2a trial of Ad26.COV2.S covid-19 vaccine. N Engl J Med 384(19):1824–1835. https://doi.org/10.1056/nejmoa2034201

Schwarz B et al (2015) Symmetry controlled, genetic presentation of bioactive proteins on the P22 virus-like particle using an external decoration protein. ACS Nano 9(9):9134–9147. https://doi.org/10.1021/acsnano.5b03360

Sevencan C et al (2020) Cell membrane nanotherapeutics: from synthesis to applications emerging tools for personalized cancer therapy. Adv Ther (Weinh) 3(3):1900201. https://doi.org/10.1002/adtp.201900201

Soprano E, Polo E, Pelaz B, del Pino P (2022) Biomimetic cell-derived nanocarriers in cancer research. J Nanobiotechnol (BioMed Central Ltd.) 20(1). https://doi.org/10.1186/s12951-022-01748-4

Su Y, Li S, Hong M (2013) Cationic membrane peptides: atomic-level insight of structure-activity relationships from solid-state NMR. Amino Acids 44(3):821–833. https://doi.org/10.1007/s00726-012-1421-9

Sun X et al (2016) In vivo targeting and imaging of atherosclerosis using multifunctional virus-like particles of Simian virus 40. Nano Lett 16(10):6164–6171. https://doi.org/10.1021/acs.nanolett.6b02386

Sushnitha M, Evangelopoulos M, Tasciotti E, Taraballi F (2020) Cell membrane-based biomimetic nanoparticles and the immune system: immunomodulatory interactions to therapeutic applications. Front Bioeng Biotechnol (Frontiers Media S.A.) 8. https://doi.org/10.3389/fbioe.2020.00627

Szliszka E, Czuba ZP, Domino M, Mazur B, Zydowicz G, Krol W (2009) Advances in biomimetic nanoparticles for targeted cancer therapy and diagnosis. Molecules 14(2):738–754. https://doi.org/10.3390/molecules

Tiwari G et al (2012) Drug delivery systems: an updated review. Int J Pharm Investig 2(1):2. https://doi.org/10.4103/2230-973x.96920

Villa LL et al (2005) Prophylactic quadrivalent human papillomavirus (Types 6, 11, 16, and 18) L1 virus-like particle vaccine in young women: a randomised double-blind placebo-controlled multicentre phase II efficacy trial. Lancet Oncol 6(5)

Voysey M et al (2021) Safety and efficacy of the ChAdOx1 nCoV-19 vaccine (AZD1222) against SARS-CoV-2: an interim analysis of four randomised controlled trials in Brazil, South Africa, and the UK. Lancet 397(10269):99–111. https://doi.org/10.1016/S0140-6736(20)32661-1

Waheed S, Li Z, Zhang F, Chiarini A, Armato U, Wu J (2022) Engineering nano-drug biointerface to overcome biological barriers toward precision drug delivery. J Nanobiotechnol (BioMed Central Ltd.) 20(1). https://doi.org/10.1186/s12951-022-01605-4

Wei H et al (2019) A nanodrug consisting of doxorubicin and exosome derived from mesenchymal stem cells for osteosarcoma treatment in vitro. Int J Nanomed 14:8603–8610. https://doi.org/10.2147/IJN.S218988

Wen AM et al (2016) Utilizing viral nanoparticle/dendron hybrid conjugates in photodynamic therapy for dual delivery to macrophages and cancer cells HHS public access. Bioconjug Chem 27(5):1227–1235. https://doi.org/10.1021/acs.bioconj-chem.6b00075

Yoo JW, Irvine DJ, Discher DE, Mitragotri S (2011) Bio-inspired, bioengineered and biomimetic drug delivery carriers. Nat Rev Drug Discov 10(7):521–535. https://doi.org/10.1038/nrd3499

Yu H, Wu M, Chen S, Song M, Yue Y (2022) Biomimetic nanoparticles for tumor immunotherapy. Front Bioeng Biotechnol (Frontiers Media S.A.) 10. https://doi.org/10.3389/fbioe.2022.989881

Zhang Q et al (2018) Neutrophil membrane-coated nanoparticles inhibit synovial inflammation and alleviate joint damage in inflammatory arthritis. Nat Nanotechnol 13(12):1182–1190. https://doi.org/10.1038/s41565-018-0254-4

Zhang Q et al (2020) Cellular nanosponges inhibit SARS-CoV-2 infectivity. Nano Lett 20(7):5570–5574. https://doi.org/10.1021/acs.nanolett.0c02278

Zhang M, Du Y, Wang S, Chen B (2020) A review of biomimetic nanoparticle drug delivery systems based on cell membranes. Drug Des Dev Therapy (Dove Medical Press Ltd.) 14:5495–5503. https://doi.org/10.2147/DDDT.S282368

Liposome-Based Drug Delivery—A New Therapeutic Paradigm

Laukik Shetye, Amrita Sherlekar, and Vijay Mendhulkar

Abstract Targeted drug delivery systems have evolved over the years to improve the therapeutic efficacy of drugs and reduce their side effects. In this context, a range of nanoscale delivery vehicles like liposomes have been developed. Liposomes are spherical particles made up of phospholipids, fatty acid esters, and fat alcohol ethers of phosphatides. They consist of a hydrophilic core and an amphipathic outer lipid bilayer. The advantages of using liposomes for drug delivery include biocompability, feasibility to deliver hydrophilic and hydrophobic drugs, low toxicity and wrapping lipids around inorganic and polymeric nanoparticles to form supported monolayers or bilayers for enhanced permeability and retention effect. Different methods of synthesis of liposomes have been developed over the years and they are based on a common principle of hydration of lipids dissolved in organic solvents. The various properties that play a decisive role in efficient drug delivery from liposomes include size, charge and fluidity of liposomes. The mechanism of drug delivery by liposomes involves either fusion of the liposome with cell membrane or by ingestion and degradation of liposomes inside the cell that leads to release of the loaded drug into the cell. Further, liposomes can be efficiently targeted by specific ligand binding. The current chapter discusses the rationale of liposome synthesis, advances in the synthesis process, current state-of-art of liposome drug delivery systems for ther apeutic applications in treatment of microbial diseases and cancer, and regulations related to commercialisation of liposomes with examples like DOXIL and MYOCET.

L. Shetye · V. Mendhulkar (✉)
Department of Botany, The Institute of Science, Homi Bhabha State University, Mumbai 400032, India
e-mail: profmendhulkar@gmail.com

A. Sherlekar
Department of Biotechnology, R.D. & S.H. National College, Mumbai, India

T. S. Santra and A. U. S. Shinde (eds.), *Advanced Drug Delivery*,
Studies in Mechanobiology, Tissue Engineering and Biomaterials 26,
https://doi.org/10.1007/978-981-99-6564-9_2

1 Introduction

Liposomes targeted drug delivery systems (DDS) are one of the most promising systems of drug delivery currently available, providing a host of advantages over traditional single-agent dosing of drugs. Liposomal DDS utilizes specially designed liposomes, which are self-assembled particles constructed from naturally occurring phospholipids, to encapsulate, protect and specifically deliver drug molecules to target cells and tissues. It provides efficient transport across the cell membrane and improves the bioavailability of drugs in comparison to traditional delivery methods (Goyal et al. 2005; Large et al. 2021; Lian and Ho 2001). It can also reduce systemic toxicity and improve the therapeutic outcomes of drugs by protecting them from the acidic environment of the digestive tract (Guimarães et al. 2021). This technology has revolutionized the drug delivery field and is quickly becoming one of the most preferred DDS for a wide range of therapeutic applications.

The rationale for using liposomes for drug delivery can be found in the fact that most drug molecules are unstable in the human body and are degraded before it can be safely delivered and absorbed. Liposomes encase and protect such molecules and deliver them to their target sites (Pattni et al. 2015). Furthermore, liposomes can be engineered to display various characteristics such as porosity, drug release kinetics, target cell specificity and degree of cross-membrane permeability, enabling superior control over drug delivery. Compared to traditional single-agent dosing, liposomes can extend the drug's shelf-life, provide greater sustainability and reduce the frequency of dosages, as well as increase the drug's efficacy, safety and accuracy of therapeutic action (Guimarães et al. 2021; Large et al. 2021; Pattni et al. 2015).

Liposomal drugs provide several advantages over traditional drug delivery methods. The primary advantages of liposomal DDS are that it can deliver drugs specifically to target cells, enabling better control of their actions, and that it can reduce harmful side effects to other cells. Targeted DDS can selectively deliver drugs to the desired location, reducing the patient's exposure to drugs and associated toxic effects, and thereby improving overall efficacy (Kim 2016; Pattni et al. 2015; Sapra and Allen 2003). Liposomes have also been found to increase the solubility of hydrophobic drugs, releasing them in a sustained and controlled manner that reduces the toxicity usually associated with high drug concentrations. Furthermore, liposomal DDS can be designed for timed release of drugs, allowing for sustained levels of drug concentration in the body. This controlled delivery is beneficial when drug doses need to last over an extended period of time, as well as when a drug's pharmacokinetics need to be better understood and regulated to maximize effectiveness (Daraee et al. 2016; Kim 2016). Liposomes can penetrate biofilms, which are a major cause of antibiotic-resistant infections, thus providing a means to improve their eradication (Alhariri et al. 2013; Chalmers et al. 2021). Lastly, liposomal DDS are also suitable for illnesses that require more potent drugs, such as those associated with cancer, in which case the liposomes would carry a higher drug dosage that is geared towards cancerous tissues (Kumari et al. 2016; Sapra and Allen 2003). Liposomal DDS have enabled improved bioavailability, sustained and targeted drug delivery and improved circulation time of drugs. Additionally, they are versatile and can be tailored to various drug types and therapeutic purposes (Bulbake et al. 2017; Guimarães et al. 2021; Sapra and Allen 2003).

Despite these advantages, there are several potential drawbacks to the use of liposomes as a targeted drug delivery system. Liposomes can require expensive processes to manufacture and may need to be stored at specific temperatures to remain stable. Additionally, as liposomes can merge with other cells, this feature can cause unwanted immune responses or even autoimmune diseases. Furthermore, the free drug molecules released by a liposome can be biodegraded in the body, much like a traditional drug. As such, their effects are not as long-lasting and may require frequent doses to maintain their effect (Antimisiaris et al. 2021; Daraee et al. 2016; Kim 2016).

The major applications of liposomal drug delivery systems are cancer treatments, ophthalmic drugs, topical agents and systemic infections (He et al. 2019; Lamichhane et al. 2018; Large et al. 2021; Li et al. 2019; Lian and Ho 2001; Maurer et al. 2001). In cancer chemotherapy, liposomal drugs can provide enhanced and localized drug concentrations, thus reducing systemic toxicity and potentially leading to better antineoplastic efficacy. In particular, tumour transfection is a promising strategy for targeted and localized delivery of drugs (Zununi Vahed et al. 2017). For ophthalmic drugs, liposomes can help to sustain drug release and enable improved passage across the cornea (Mishra et al. 2011). Furthermore, liposomes can be used to improve the performance of topical agents, due to the sustained release of drugs and improved penetration of the skin (Pierre and dos Santos Miranda Costa 2011). Lastly, liposomes can help to achieve higher drug concentrations in the blood plasma, which could be effective in treating systemic infections (Nisini et al. 2018; Nsairat et al. 2022). This technology is also promising for vaccines, with the potential for faster, more efficient and cost-effective vaccine production. Additionally, nano-liposomes are currently being studied for their suitability in delivering nucleic acids, such as DNA and RNA, to the body (Kulkarni et al. 2021).

In conclusion, liposomal targeted DDS are a promising technology that has the potential to revolutionize the way drugs are delivered to target cells. While such systems offer many advantages over traditional single-agent dosing of drug molecules, such as extended drug shelf-life, greater targeting accuracy and controlled drug release kinetics, there are also potential drawbacks that must be taken into account. Nevertheless, the potential applications and possibilities for personalized therapies that can be enabled by this technology, and the fact that it is becoming increasingly cost-effective, suggest that liposomal targeted DDS are one of the most promising technologies available today. As research continues to be conducted, and the technology is further developed and refined, it is likely that the liposomal DDS will become an integral part of the medical care of the future.

2 Synthesis Techniques

Liposomes have emerged as an effective drug delivery system in recent years due to their increased stability and prolonged duration for release of the drug. Liposomes are typically composed of biocompatible phospholipids, such as phosphatidylcholine (PC), phosphatidylserine (PS), phosphatidylinositol (PI), and cholesterol, which are organized into a bilayer lipid membrane wall (refer Fig. 1).

Liposomes can also incorporate a variety of drugs including peptides, proteins, and chemicals (Alavi et al. 2017; Kim 2016; Sercombe et al. 2015). There are a variety of liposomal vesicles, ranging from traditional unilamellar to multilamellar and from conventional to nonconventional liposomes (refer Fig. 2). Each type of liposomal vesicle has different characteristics and is useful for different applications. For example, traditional unilamellar vesicles can achieve sustained drug delivery, while multilamellar vesicles are ideal for a high drug loading. In contrast, conventional liposomes are used for the efficient encapsulation and protection of drugs, whereas nonconventional liposomes can target specific cell types. Other liposomal vesicles include fusion liposomes, drug-lipid conjugates, and immunoliposomes. In particular, immunoliposomes are able to interact with specific antigens on the surface of invaders and target them for drug delivery. Additionally, nanoliposomes, nanoliposome hybrids, and dendrimer-liposomes are being investigated for their potential applications in drug delivery (Antimisiaris et al. 2021; Pattni et al. 2015; Sercombe et al. 2015). By understanding the properties of the different types of liposomal vesicles, their advantages and limitations, researchers can make informed decisions about which type of vesicle to use in a drug delivery system.

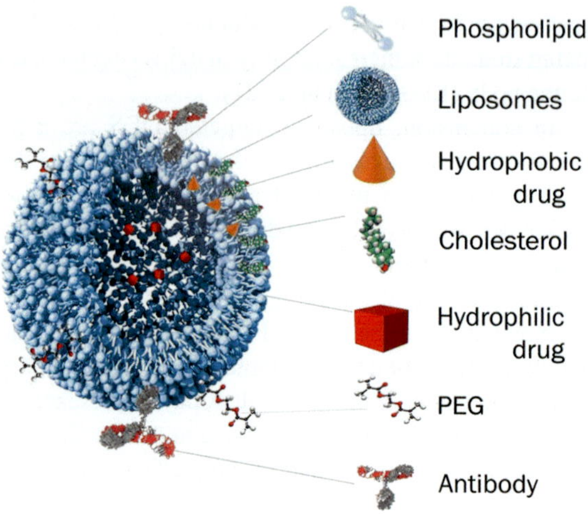

Fig. 1 Structure and composition of lipososme (Beltrán-Gracia et al. 2019)

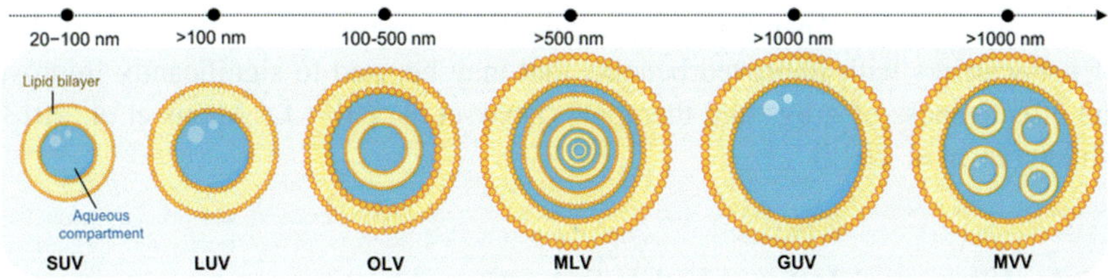

Fig. 2 Types of liposomes based on size and lamellarity. SUV (small unilamellar vesicle), LUV (large unilamellar vesicle), OLV (oligolamellar vesicle), MLV (multilamellar vesicle), GUV (giant unilamellar vesicle), MVV (multivesicular vesicle) (Luiz et al. 2023)

There are numerous synthesis techniques that can be used to prepare liposomes. Physical methods and chemical methods are two of the most common methodologies. Physical methods include polymer-lipid hybrid vesicle formation, ultrasound-induced vesicle formation, sonomechanical vesicle formation, and electroporation. Chemical methods used to form liposomes include the thin film method, the reverse phase evaporation method, the desolvation method, and the stable freeze–thaw method. Each of these methods has advantages and disadvantages that must be taken into consideration prior to synthesis.

2.1 *Physical Methods*

Physical methods involve the physical manipulation of the lipid and other components used in LDDs, such as disruption of lipid bilayers, resealation of lipid vesicles, and the use of nanoparticles and other materials for targeting and drug encapsulation. These physical methods can be used to control the size and shape of the liposomes, as well as their content. It is also possible to use these physical methods to target specific areas or to provide sustained drug release.

2.1.1 Polymer-Lipid Hybrid Vesicle Formation

Polymer-lipid hybrid vesicle (PLHV) formation is a method used for preparing vesicles with polymeric layers that confers improved properties such as enhanced drug loading capacity, improved stability of the vesicles, and improved circulation time in vivo. PLHV formation involves an oil-in-water emulsion that contains both liposomal and polymeric components. A surfactant, such as sorbitan monooleate, is typically added to the mixture to stabilize the emulsion and form the hybrid vesicles. Studies have revealed that hybrid vesicles can form stable encapsulating structures with enhanced biocompatibility, allowing drug molecules to be released in a near-pulse fashion. In addition, hybrid vesicles can incorporate different targeting agents, such as antibodies, peptides, and proteins, to achieve enhanced targeting and precise delivery of payloads. Therefore, polymer-lipid hybrid vesicles represent a novel class of nanocarriers with improved benefits that may be used to significantly improve the effectiveness of drugs and therapies (Ishak et al. 2017; Le Meins et al. 2013; Mukherjee et al. 2019).

2.1.2 Ultrasound-Induced Vesicle Formation

Ultrasound-induced vesicle formation (UVF) is a fast and economical method for forming liposomes. This method involves the use of an ultrasound probe to induce cavitation of a mixture of a lipid of interest, an aqueous solution, and a surfactant. Cavitation causes the lipid to undergo solubilization and form vesicles, which can then be isolated by one of a variety of techniques available. The size and drug loading capacity of the vesicles can be tailored by adjusting the components of the lipid suspension. These vesicles can then be triggered with ultrasound waves, causing them to release the drug material into the body. These liposomal systems have been found to have greater pharmacodynamic and pharmacokinetic benefits than other delivery methods, with the potential to increase the drug potency, reduce toxicity, and extend its shelf life (Huang et al. 2010; Richardson et al. 2007; Silva et al. 2010).

2.1.3 Sonomechanical Vesicle Formation

Sonomechanical vesicle formation (SMVF) is another ultrasound-based method of forming liposomes. This technique involves the use of ultrasound to break down the phospholipid molecules in solution in order to form liposomes. This process is advantageous due to its low cost and scalability, and it has the potential to improve the delivery of drugs which could previously not be encapsulated in traditional liposomes. Studies have demonstrated that sonomechanical vesicle formation provides particle sizes suitable for drug delivery and also increases the drug to lipid ratio, making it a potential alternative for liposome encapsulation. Additionally, it has been shown that this method is also applicable to numerous types of drugs and molecules, including hydrophobic and hydrophilic drugs, peptides, proteins, and polysaccharides, making it a viable option for drug delivery. Furthermore, sonomechanically formed liposomes are more stable compared to free drugs and show increased efficacy in targeting desired sites and releasing drugs when compared to other lipid-based drug delivery systems (Huang et al. 2010; Lombardo and Kiselev 2022; Silva et al. 2010, 2011).

2.1.4 Electroporation

Electroporation is a method used to form vesicles on the nanoscale. This technique involves the application of a pulsed electric current to a suspension of lipids in an aqueous solution. The electrical field causes the formation of nanovesicles with increased drug loading capacities compared to conventional liposomes. This method is widely used for the delivery of drugs and for gene therapy (Deshmukh et al. 2021; Lamichhane et al. 2015).

2.2 *Chemical Methods*

Chemical methods involve chemical modification of the lipid and other components of the LDDs for drug encapsulation or binding, stabilization, and controlled release of drugs. These methods also include the use of additional chemical species for targeted and improved drug delivery. These chemicals can be used to form covalent linkages and to conjugate proteins to lipids, as well as to form linkers and complexes to latch onto liposome components. Chemical modification can be used to alter the characteristics of the liposomes, such as their size, shape, release rate, and targeting ability. This allows for improved accuracy and precision in drug delivery. Chemical techniques are also used to facilitate controlled release of drugs from the liposomes, as well as to protect them from being quickly cleared by the body.

2.2.1 Thin Film Hydration Method

The thin film hydration method is the most widely used method for the synthesis of LUVs (refer Fig. 3). This method is useful in maintaining drug integrity and preventing its degradation. It consists of dispersing the liposome vesicle forming components in a solvent, forming a thin film, and then dispersing them in an aqueous phase to form liposomes. Various lipid compositions and solvents have been explored to optimize the formulation and properties of the liposomes, including the use of phospholipids, cholesterol, polyethylene glycol, and others. Recent studies have focused on the use of novel lipids, improving the endocytosis of liposomes, and formulations that withstand harsh environments. Additionally, pre-formulation studies such as stability and drug encapsulation efficiencies are important in developing effective liposomal drug delivery systems.

The combination of thin film method and other technologies, such as ethanol injection technique, has enabled the preparation of liposomes with improved structural, dispersional, and functional properties (Guimarães et al. 2021; Large et al. 2021; Liu et al. 2022).

2.2.2 Reverse Phase Evaporation Method

The reverse phase evaporation method is another widely used technique for the synthesis of liposomes. This method works by dispersing the active ingredients into an aqueous solution before adding lipid powders. The lipid-based solution is then either inserted into vials or if necessary formed into various shapes by spray-drying. In this method, the lipids are homogenized at temperatures above the melting points of

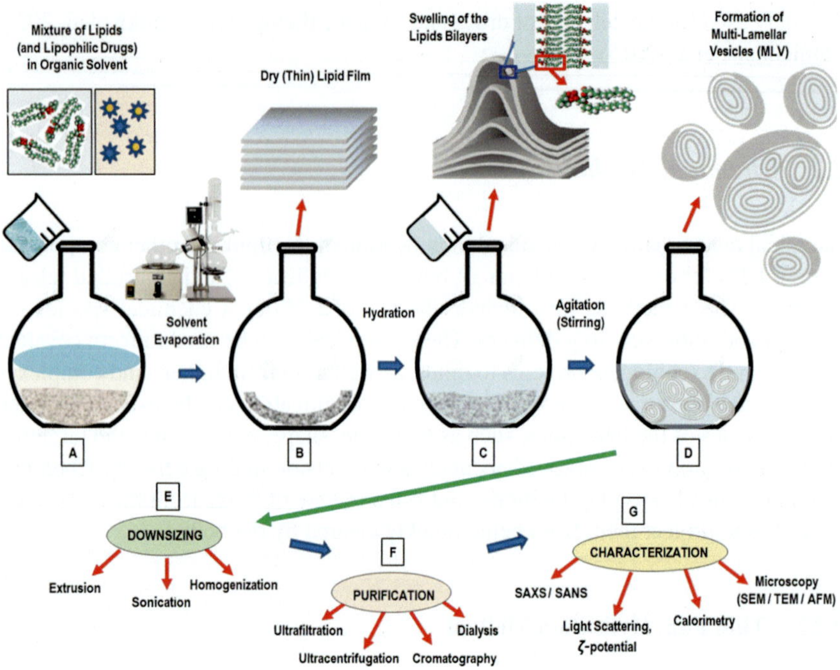

Fig. 3 Schematic representation of thin film hydration method (Lombardo and Kiselev 2022)

the components. This homogenization procedure is used to bring about the fusion of the lipid components, which result in the formation of liposomes. The effectiveness of this method in loading large amounts of drug substances into the liposomes has been extensively studied. In addition, the reverse phase evaporation method yields liposomes with increased payloads of drug substances, decreased lipid entrapment, and higher drug loading efficiency, which provides considerable advantages compared to other current drug delivery systems (Guimarães et al. 2021; Kim 2016; Large et al. 2021) (Fig. 4).

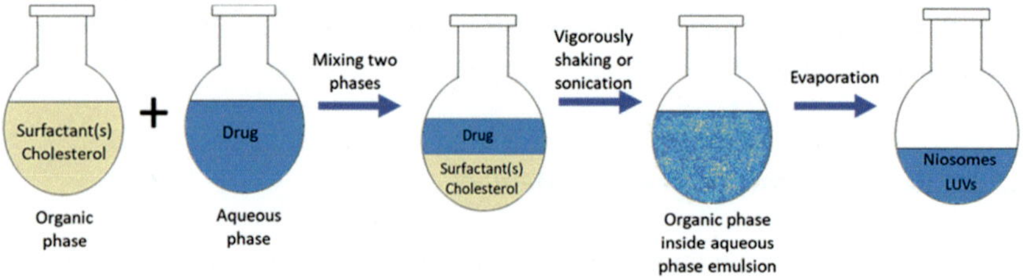

Fig. 4 Schematic representation of reverse phase evaporation method (Durak et al. 2020)

2.2.3 Freeze–Thaw Method

The freeze–thaw method is commonly used in the formulation of liposomes for drug delivery as it produces liposomes with more uniform size distribution. This method involves freezing a liposome suspension and then thawing it, which causes the liposomes to rupture and reform, thus resulting in smaller liposomes. During this process, additives such as lipids, surfactants, and other chemicals may be added in order to further influence the drug delivery. Studies have shown that this method results in improved physical stability and enhanced drug encapsulation efficiency. Additionally, improvements in drug loading capacity and sustained release of drugs have been achieved by using this method. Furthermore, different kinds of liposomes can be produced through the freeze–thaw method, such as multilamellar vesicles, small unilamellar vesicles, and multivesicular liposomes, allowing for a wide array of drug delivery possibilities (Dissanayake et al. 2022; Large et al. 2021; Nsairat et al. 2022; Pattni et al. 2015).

2.2.4 Solvent Injection Method

The solvent injection method is a widely employed technique for the preparation of liposomes, offering versatility and control in nanocarrier design for biomedical and nanomedicine applications. This method involves dissolving lipid components in an organic solvent, followed by rapid injection of the lipid solution into an aqueous phase under vigorous stirring. As the organic solvent diffuses out, the lipids self-assemble into bilayer structures, forming liposomes. This technique allows for the encapsulation of various compounds, including hydrophilic and hydrophobic substances, within the liposome's aqueous core and lipid bilayers, respectively (Lombardo and Kiselev 2022). Furthermore, the solvent injection method provides the advantage of easy modulation of liposome size, composition, and surface properties, making it a versatile approach for drug delivery and studying interactions with different compounds, including polyphenols, in nanomedicine research (Šturm and Poklar Ulrih 2021). As such, the solvent injection method remains a valuable tool in the development of versatile nanocarriers with significant potential for biomedical applications (Fig. 5).

2.3 Characterization of Synthesized Liposomes

After the formation of liposomes through any of the above-mentioned methods, the liposomes are characterized for various properties to assess their stability, integrity and functionality. Characterization usually includes measurement of size, morphology, loading capacity and drug release. It also includes testing of the presence of different types of lipids, impurities and other contaminants in the liposomes. Different approaches can be adopted depending on the liposomal structure and

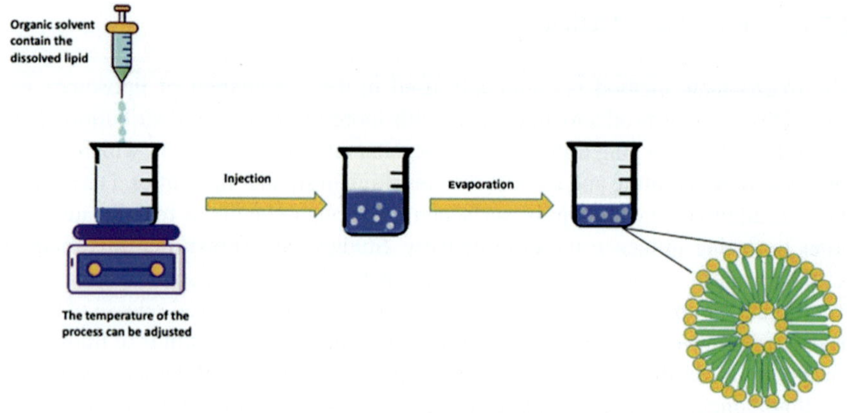

Fig. 5 Schematic representation of solvent injection method (Jaradat et al. 2021)

composition. These techniques involve physical methods such as measuring particle size and size distribution, determining zeta potential, and electron microscopy. Additionally, chemical methods such as chromatography, spectroscopy, and Nuclear Magnetic Resonance (NMR) spectroscopy can be used for characterization. Chemical characterization helps to understand the characteristics of the liposomal components and the entrapment and release of drugs. Moreover, stability, thermal, and photostability studies are necessary to assess the long-term stability and protection of the encapsulated drug. Studies have also shown that modified liposomes, such as surface-functionalized liposomes and those developed with strong adhesive polymers, can enhance the effectiveness of drug delivery. All of these techniques used in the development of liposomes have contributed to a better understanding of both the physical and chemical properties of liposomes needed for an optimal drug delivery system (Guimarães et al. 2021; Large et al. 2021; Pattni et al. 2015).

In conclusion, there are a variety of synthesis techniques that can be used to prepare liposomes. Each of these methods has its own advantages and disadvantages that must be taken into consideration prior to synthesis. Synthesis of liposomes is an important aspect of drug delivery systems as it can be used to finely tune the characteristics of the liposomes for improved bioavailability and efficacy of the drug.

3 Infectious Diseases

3.1 Overview of Liposomal Drug Delivery for Infectious Diseases

Liposomal DDS are one of the most innovative drug delivery mechanisms for treating infectious diseases. Infectious diseases are disorders caused by one or more organisms such as bacteria, viruses, parasites, and fungi, which can produce a wide range of health complications and even death. Common infectious diseases include measles, malaria, tuberculosis, HIV/AIDS, and influenza. Infectious diseases are one of the leading causes of mortality and morbidity worldwide (Ritchie et al. 2018). Liposomal DDS have demonstrated great potential for the treatment of infectious diseases. The control of infection requires the rapid and sustained delivery of antimicrobial agents to the site of infection, and liposomes offer a targeted approach to deliver the drug. In addition, liposomes are relatively safe and can reduce the potential side effects of the drug due to their ability to encapsulate the drug and reduce its systemic exposure. Liposomes can also act as an effective carrier for the delivery of antiviral drugs and provide sustained and targeted drug release. Additionally, liposomal formulations may provide protection to drugs from enzymatic degradation, thus increasing their stability and bioavailability. Furthermore, they can be used in combination with other DDS and enhance their therapeutic efficiency. It is recognized that widespread application of liposomal DDS in the treatment of infectious diseases necessitates a comprehensive understanding of the potential toxicity and biodistribution of liposomal formulations. Extensive research is being conducted to assess the safety and efficacy of these systems in clinical practice (Alhariri et al. 2013; Tiwari et al. 2012; Torchilin 2014).

A liposomal drug delivery system has increasingly been an appealing option for the treatment of infectious diseases because of its various advantages, in comparison to traditional therapies. Improved therapeutic index is one of the biggest advantages of using a liposomal drug delivery system for infectious diseases (Alhariri et al. 2013). When encapsulated within liposomes, drugs are able to evade a negative immune response, resulting in better efficacy and improved therapeutic index compared to traditional approaches. Targeted drug delivery is another advantage of using liposomes, as they can be designed in such a way that the drug selectively targets the affected organs and pathogenic microorganism to enhance bioavailability (Daraee et al. 2016; Torchilin 2014). This can lead to a higher level of drug potency within the targeted area and reduced toxic side effects in the host. The stability of drugs can also be improved by encapsulating them in liposomes. As the liposomal structure is a bio-friendly material, it helps to extend the life of the drug and protect it from external environmental factors such as pH, temperature and light. Moreover, it helps in protecting the drug from being degraded before it reaches its target site (Alhariri et al. 2013; Daraee et al. 2016; Goyal et al. 2005; Nisini et al. 2018; Torchilin 2014). All of these advantages of liposomal DDS can lead to improved treatments for infectious diseases, with better efficacy and fewer side effects on the body. Liposomes are lipid vesicles which carry drugs and can be used in disease control by targeting specific pathogens. However, there are significant challenges with using liposomal DDS that should be considered.

High cost is one major challenge with using liposomal DDS. Liposomes require a complex manufacturing process, with multiple steps and stages that are time-consuming and costly. This can limit their widespread use and prevent them from being an economically viable option for many countries and health systems. Furthermore, the cost of drugs in liposomal formulations are often much higher than traditional formulations, further limiting the ability to provide access to these drugs on a large scale. In addition to cost, liposomal DDS often have difficulty distinguishing between pathogenic and non-pathogenic organisms. To achieve a successful treatment, it is necessary to target the pathogenic microorganisms, while leaving the beneficial microbiome intact and unharmed. This specificity can be difficult to achieve with liposomes, as they interact with all organisms within their environment, regardless of their pathogenicity. Moreover, the effectiveness of the liposomal delivery can be diminished by the presence of serum proteins which may bind to, and inactivate the drug molecules. This further impairs the specificity and effectiveness of the delivery system (Alhariri et al. 2013; Nisini et al. 2018; Torchilin 2014).

3.2 Recent Technologies and Advances with Liposomal Drug Delivery

Recent advances in liposomal DDS allow for targeted and controlled release of drugs to specific sites of action. Liposomes are often used to deliver antibiotics, anti-cancer agents, vaccines, and other biological therapeutics. In addition, they can be loaded with drugs, nanoparticles, and other biological cargos to create multifunctional, stimuli-sensitive DDS. Recent technologies and advances with liposomal drug delivery have also been used in the development of nanoscale drugs for antimicrobial treatment of global infectious diseases.

3.2.1 Multifunctional, Stimulus-Sensitive Nanoparticulate Systems

Multifunctional, Stimulus-Sensitive Nanoparticulate Systems (MFSSNS) are a type of drug delivery system that uses nanoparticles to carry and release drugs in response to various external stimuli. These tiny particles can be incorporated into a variety of drug delivery methods, such as liposomes, nanoparticles, and micelles, which can be designed to respond to changes in pH, temperature, or stress. The resulting MFSSNS has applications in various medical treatments, including the delivery of antibiotics, antivirals, and other biologics. The MFSSNS can effectively target and penetrate drug resistant bacteria and viruses, offering enhanced drug efficacy and reducing side effects. In addition, they can be optimized for sustained or controlled release, which ensures consistent drug delivery over time. For example, Torchilin (2014) studied the use of an endothermic shell that uses pH to control the release of drugs. This shell is designed to dissolve under low pH conditions, releasing the drug at the desired rate. Additionally, this system can provide increased stability of drugs and reduce the amount of drug metabolized by the body. With these various advantages, MFSSNS offer promising treatments for a variety of infectious and disease-causing pathogens.

3.2.2 The Multirole of Liposomes in Therapy and Prevention of Infectious Diseases

Liposomes can be used in the therapy and prevention of infectious diseases. Nisini et al. (2018) studied the use of liposomes to prevent viral infection. They found that by coating the liposome with viral particles it was possible to reduce viral uptake and prevent the spread of the virus. Additionally, liposomes have also been studied for their potential use in the delivery of antibiotics, antivirals, and other antimicrobial agents for the treatment of bacterial, viral, and fungal infection (Alhariri et al. 2013). Furthermore, liposomes can be used as drug carriers to improve drug targeting and as adjuvants to enhance vaccine's immunogenicity. This action is possible due to their ability to stably encapsulate antigenic molecules and stimulate antigen-specific immune responses. They are also employed to help better identification of biomarkers and pathogens. This is important in the diagnosis and prognosis of infectious diseases. Moreover, recent research proves that liposomes can be used as prophylactic agents to prevent gastrointestinal or respiratory infections caused by bacteria or viruses.

3.2.3 Nanotechnology Approaches for Global Infectious Diseases

Nanotechnology has been studied for its potential use to combat global infectious diseases. Drug delivery systems are used in combination with other technologies, like nanotechnology, to create precise and targeted treatments. Nanoparticles provide enhanced cellular uptake, improved bioavailability, and improved drug biodistribution. Stimuli-sensitive nanoparticles offer targeted delivery, controlled and sustained release of drugs, and reduce cytotoxicity. Kirtane et al. (2021) studied the use of nanomedicines and nanotechnology to specifically target pathogens and deliver antimicrobial agents. Zong et al. (2022) studied the use of nanomaterials in the delivery of antibiotics for the treatment of bacterial infections. They found that these nanomaterials were able to increase the efficacy of drugs, reduce drug toxicity and side effects, and target bacterial cells more selectively. Additionally, nanotechnology can be used to develop point-of-care treatments to diagnose, monitor, and treat infection in resource-limited areas. Recent advances in nanotechnology have allowed for further development of novel drug delivery systems with improved efficiency and improved drug loading capacities. Different types of materials such as polymeric nanocarriers, silica nanoparticles, and gold nanoparticles have been used to develop new delivery systems for antimicrobial drugs (Kirtane et al. 2021). Furthermore, these systems are now much more cost-effective and eco-friendly. These advances have enabled the development of effective treatments for infectious diseases, making nano-drug delivery systems an important tool in the fight against drug-resistant diseases. With the aim of treating global infectious diseases, the development of efficient nanomaterials for drug delivery is a promising and viable approach for improved disease management.

Liposomes have become effective pharmaceutical DDS due to their flexibility, biodegradability, and stability. They are used to treat infections caused by bacteria, fungi, viruses, and parasites (Alhariri et al. 2013; Kirtane et al. 2021; Nisini et al. 2018). Liposomes help to target and enhance the permeation of drugs more effectively to the infected tissue due to their size, biodegradability, and the ability to carry a variety of molecules. These attributes help to achieve optimal therapeutic response by reducing the systemic side effects of drugs. However, challenges still exist for liposomal drug delivery for infectious diseases. These challenges include inefficient drug loading, decrease in liposome size, lack of targeting to the infectious site, and its decreased entrapment efficiency. It is also difficult to deliver drugs to the inaccessible tissue due to the liposome size and its tendency to bind to the plasma proteins. Furthermore, the drugs may be released too rapidly or too slowly due to the drug encapsulated in liposomes, which can affect their therapeutic efficacy. Recent advances in the field of liposomal drug delivery for infectious diseases include stimuli-responsive nanosystems for delivering drugs, such as pH-sensitive molecules, magnetic nanosystems, and temperature-sensitive molecules. The development of multifunctional liposomes has increased their drug encapsulation efficiency, which helps to increase the stability of liposomes in the bloodstream and target them more effectively to the infected site. Researchers have also employed other strategies such as membrane coating, double emulsion, and liposomal stabilization to increase the stability and drug loading capacity of liposomes. Nanodrugs have also been developed with antimicrobial properties, which have a higher efficiency and less toxicity compared to conventional drugs (Zhu et al. 2014). These nanodrugs have been used to successfully treat a wide range of infections, including tuberculosis and malaria. Additionally, drug-conjugated liposomes have been developed, which have improved drug loading efficiency and can target infection sites more directly. Overall, despite the benefits of liposomal DDS in treating infectious diseases, there are also considerable challenges associated with them. As such, developing improved delivery systems and strategies is necessary to overcome these challenges and make liposomal drug delivery a viable, cost-effective treatment option for infectious diseases.

4 Mycobacterial Infections

4.1 Overview of Mycobacterial Infections

Mycobacterial infectious diseases have become a major global health concern due to the emergence of drug-resistant variants of the bacteria (Ritchie et al. 2018). Mycobacterial infectious diseases are caused by different species of *Mycobacterium*, a type of bacteria. The common Mycobacterium-related diseases are tuberculosis (TB), nontuberculous mycobacterial (NTM) pulmonary disease, and other chronic lung infections including aspergillosis and cystic fibrosis. Tuberculosis is one of the most deadly infectious diseases and it can cause prolonged infections due to its slow multiplication rate (Pham et al. 2015; Pinheiro et al. 2011). Nontuberculous mycobacterial (NTM) pulmonary disease is caused by a variety of pathogens, such as *Mycobacterium avium*, which has been increasing in prevalence in recent years due to the emergence of antibiotic-resistant forms of the bacteria (Chalmers et al. 2021). Chronic lung infections including aspergillosis and cystic fibrosis are all caused by a range of pathogens, including *Mycobacterium* (Andrade et al. 2013; Goyal et al. 2005). As such, treatments for mycobacterial infectious diseases must address multiple pathogens and ensure that drug molecules reach their target with the minimum of side effects. This is where liposomal targeted drug delivery systems come into play, as they are capable of encapsulating and delivering a range of drug molecules to the lungs (Sercombe et al. 2015).

Current methods of treating mycobacterial infectious are wide-spread and diverse, due to the complex nature of mycobacterial infections. Traditional forms of treatments include combination chemotherapy and surgery, both of which employ a range of antimicrobial agents (Pham et al. 2015). As research progresses, there have been more specific and targeted treatments for mycobacterial infections. Currently available methods include combination chemotherapy and surgery, and more recently, specific DDS that include nanotechnologies, liposomes and pulmonary delivery systems. The potential limitation of current treatments involves the necessity for repeated treatments, and the potential for the emergence of drug-resistance strains, due to over-exposure to the drug (Andrade et al. 2013; Pham et al. 2015). For instance, research from Chalmers et al. (2021), has highlighted that repeated treatments and drug exposure can lead to the emergence of drug-resistance strains, thus decreasing the effectiveness of the drugs. Furthermore, current treatments can be lengthy and demanding, particularly the use of combination chemotherapy and pulmonary delivery systems. This can add a significant cost to the treatment, making it less accessible to certain patient demographics. Overall, the advantages of current treatments of mycobacterial infections primarily involve their efficacy and efficacy, as well as their ability to target a particular infection. However, the potential limitation of using these treatments involves the potential for the emergence of drug-resistance and the cost of repeated treatments (Andrade et al. 2013; Goyal et al. 2005; Pham et al. 2015; Pinheiro et al. 2011).

4.2 *Clinical Studies on Liposomal Drug Delivery for Mycobacterial Infectious*

Clinical studies on liposomal drug delivery for mycobacterial infectious mainly focus on two areas: pulmonary delivery and intravenous delivery. With regard to pulmonary delivery, Pham et al. (2015) found that liposomal drug delivery systems had great potential for TB treatment, with the use of right size of liposomes to ensure efficient delivery and retention of drugs to the targeted sites. Pinheiro et al. (2011) further studied the use of liposomal drug delivery systems for TB and suggested that liposomes of proper diameter can be used to achieve desirable drug targeting efficacy. In addition, Chalmers et al. (2021) conducted a liposomal drug delivery study to manage nontuberculous mycobacterial pulmonary disease and other chronic lung infections and found that liposomes have the potential to provide a sustained release of drugs in the lungs, which could potentially help to reduce the frequency of doses required. Furthermore, in terms of intravenous (IV) delivery, Goyal et al. (2005) and Pinheiro et al. (2011) studied the efficacy of liposomal drug delivery systems for *Mycobacterium tuberculosis*, such as a combination of Daunorubicin and Clarithromycin, and concluded that it may increase the efficacy and reduce the toxicity of the drug combination. Andrade et al. (2013) investigated the potential of using liposomal DDS to tackle drug resistance in mycobacterial infectious and suggested that liposomal DDS could have an important role in the development of new treatments for TB and other mycobacterial diseases. Thus, liposomal DDS could be used to overcome the transportation and access barriers to antibiotics in remote communities by delivering the drugs directly to the affected areas through a pulmonary inhalation route (Pham et al. 2015).

4.2.1 Challenges in Liposomal Targeted Drug Delivery for Mycobacterial Infectious

The main challenge of liposomal targeted drug delivery for mycobacterial infectious is related to achieving target specificity (Pham et al. 2015). The drug delivery systems must be able to selectively deliver the payloads to the targeted sites of action, such as the acetylcholine receptor in mycobacteria, so that the drugs can be fully absorbed and retain their intended pharmacological effects (Islas-Weinstein et al. 2021). Goyal et al. (2005) found that the size of liposomes used is critical for successful drug targeting and absorption, as small and medium liposomes may be quickly cleared from the body and offer poor targeting to mycobacteria, whereas large liposomes may be slow to degrade and unable to produce high drug accumulation at the target sites.

The potential of liposomal DDS is vast, and current research continues to focus on further improving their effectiveness in the treatment of mycobacterial infectious. Research efforts are needed to improve the targeted specificity of liposomes by further refining their size and composition for ideal drug targeting. Additionally, research should be directed to improve cost efficiency of liposomal DDS to make them more accessible to those who needed.

5 Anticancer Drug Delivery

5.1 *Overview of Anticancer Drug Delivery*

Cancer treatment is a challenging process due to the complexity of the disease, various types of cancer, and the range of cancer treatment options available. In addition, current treatments have a range of negative side effects, including damage to healthy cells and tissue (Gyanani et al. 2021; Kumari et al. 2016). Liposomal drug delivery systems offer a solution to this problem by enabling targeted drug delivery to cancerous cells and tissues. Targeted drug delivery has become an increasingly popular strategy for treating cancer in recent years (Saraf et al. 2020; Slingerland et al. 2012). The main premise is to deliver the therapeutic drugs specifically to the targeted cancer cells, thus reducing the risk of adverse side effects and increasing the efficacy of the drugs. A number of different types of targeted drug delivery strategies have been proposed and studied, including antibodies, nanotechnology-based drug delivery, and liposomal targeted drug delivery (Senapati et al. 2018; Zhong et al. 2021). Liposomal drug delivery systems have been developed as an attractive delivery vehicle for cancer drugs due to their high loading capacity, specificity, stability, and enhanced drug targeting. They have been used in the treatment of solid tumors, leukemia, and several other types of cancer. Additionally, they offer a number of advantages, such as increased drug stability, improved solubility, controlled release of the drug, and improved pharmacokinetic characteristics. Moreover, they are capable of delivering a wide range of drugs, including small molecules, peptides, antibodies, nucleic acids, and proteins (AlSawaftah et al. 2021; Gyanani et al. 2021; Kumari et al. 2016; Saraf et al. 2020). They can also be used to target multiple pathways with a single delivery system, allowing for the combination of different therapies and increased personalization of therapy.

5.2 *Examples of Anticancer Drugs Used in Targeted Liposomal Drug Delivery*

Clinical applications of Ligand-targeted liposomal anticancer drugs include doxorubicin, which is an anthracycline antibiotic used for the treatment of lymphomas, leukemias, and various solid tumors. It is one of the most important chemotherapeutics used in cancer treatment, with a maximum tolerated dose of 55–60 mg/m^2 body surface. Doxorubicin encapsulated in small unilamellar liposomes has been used to improve drug release, reduce systematic toxicity, prevent leakage and reduce cardiotoxicity associated with intravenous administration of the drug (Hofheinz et al. 2005; Rudokas et al. 2016; Slingerland et al. 2012; Tacar et al. 2013).

Another anticancer drug widely used in targeted liposomal drug delivery is paclitaxel, which is a natural product used in the treatment of a variety of malignancies including those of the breast, lung, ovary, and head and neck. Paclitaxel encapsulated in nanoliposomes has been used to reduce systemic toxicity, improve pharmacokinetics, and enhance drug delivery to cancer cells (Alavi and Hamidi 2019; Ravar et al. 2016). Such targeted liposomal drug delivery systems also enable higher drug concentrations to reach the target site, allowing for improved therapeutic efficacy. Together, liposomal-encapsulated doxorubicin and paclitaxel represent some of the most commonly used cancer drugs in targeted liposomal drug delivery systems. Cisplatin and Carboplatin are important platinum-based drugs that are used to treat certain type of cancer and are also available in liposomal formulations (Liu et al. 2013; Zhang et al. 2022). Epirubicin is an anthracycline-based chemotherapeutic agent used to treat breast cancer and is used in the form of liposomal formulations (Lao et al. 2013; Smith et al. 2010). Trastuzumab is a monoclonal antibody drug used to target HER2-positive tumors. Trastuzumab-liposomal formulations have been approved for use in the treatment of breast, ovarian and gastric cancer patients (Boekhout et al. 2011; Lao et al. 2013; Li and Li 2013).

Benefits of liposomal delivery of anticancer drugs are well-established and highly significant, and its effectiveness is attributable to improved bioavailability, increased stability, and enhanced tumor targeting (Bulbake et al. 2017; Daraee et al. 2016; Olusanya et al. 2018).

Improved Bioavailability: Liposomal delivery of anticancer drugs offer enhanced solubility in water, thus allowing higher plasma concentrations of the drug in the body which leads to more efficient drug absorption and bioavailability. For instance, the use of doxorubicin liposomes has resulted in increased bioavailability with improved biodistribution and lower hepatotoxicity than traditional doxorubicin components (Tacar et al. 2013).

Increased Stability: The outer membrane of the liposomes helps in providing stability to the drug in the serum resulting in improved therapeutic efficacy and reduced immunogenic activity (Gyanani et al. 2021). Liposomal drug delivery additionally improves the pharmacological longevity of drugs in the body, thereby reducing the number of treatments and toxicity (Slingerland et al. 2012).

Enhanced Tumor Targeting: The principles of enhanced drug accumulation in areas of inflammation, found through site-specific delivery of liposomal drugs, are known to be of great clinical value. Especially in cancer therapy, the active targeting of tumor cells via reactive ligands attached to the surface of liposomes allows high concentration of drug molecules in the tumor microenvironment. For example, cationic liposomes containing RGD peptides attached have allowed a greater concentration of drugs in tumor cells, as compared to the passive diffusion of common liposomes (Sapra and Allen 2003). Moreover, several other strategies such as UV-light triggered, heat-sensitive, and pH-sensitive liposome have also been utilized in cancer drug delivery (Daraee et al. 2016; Kumari et al. 2016).

5.3 Challenges of Liposomal Delivery of Anticancer Drugs

Compromised drug stability: In the targeted delivery of anticancer drugs through liposomes, there is the potential of compromising the stability of the beneficial drug within the system. This is mainly due to the complexity and varying physicochemical properties of the drug delivery system. In addition, the high drug loading capacity and the controlled drug release capability of the liposomes may lead to increased instability of the system (Gyanani et al. 2021). Liposomes as a drug delivery system consist of bilayer lipid membranes that enclose the drug molecules, which can lead to the breakdown of the liposomes due to hydrolysis of the lipid. This may consequently affect the physical stability of the liposome-encapsulated drug (Saraf et al. 2020).

High manufacturing cost: Liposomal drug delivery systems have certain advantages but also come at a high price. Liposomes are multi-component systems composed of numerous components such as phospholipids and other agents like surfactants, proteins, and dextrans. This in turn increases the complexity of their manufacturing process and makes them costly to produce. Moreover, the difficulties in reproducibility of the drug loading process and strict regulatory guidelines further add to the high manufacturing costs (Gyanani et al. 2021).

Inadequate tumor-tissue accumulation: An aspect of liposomal drug delivery system that needs to be addressed is adequate tumor accumulation of the therapeutic drugs. In order for the liposome-encapsulated drugs to react with their target cancer cell, they must remain in the physical space around the tumor. However, drug carriers such as liposomes can be easily cleared by the reticuloendothelial system, leading to a decrease in the tumor accumulation of the liposomes. This in turn may reduce the effectiveness of the therapeutic drugs, further compromising the anticancer drug delivery system (Daraee et al. 2016).

In conclusion, the use of liposomal targeted drug delivery of anticancer drugs (LTDD) can, without a doubt, significantly impact cancer treatments and provide promising results, specifically in regards to improved patient outcomes and a reduced number of side effects from chemotherapy. LTDD provides numerous advantages compared to conventional drug delivery systems, including an improved therapeutic index, enhanced drug stability, extended circulation in the bloodstream, accelerated cell delivery, and improved drug distribution in the target area. Consequently, researchers have been focused on exploring various liposomal formulations, ranging from nanoparticles, polymeric vesicles, and peptide nanocarriers to targeted drug delivery strategies.

5.4 *Future Directions of Research on Liposomal Targeted Drug Delivery of Anticancer Drugs*

Future directions of research on LTDD include further investigations of the effects of combination therapies and advanced active targeting strategies such as antibodies, peptides and ligands, which could potentially reduce cancer cell proliferation and increase the efficiency of LTDD. Additionally, combination strategies such as stimuli-sensitive lipid nanocarriers, polymersomes, siRNA liposomes, and lab-on-chip cancer therapy should be explored to further enhance the potential and maximize the advantages of LTDD. Furthermore, new biosensors, medical imaging technologies, and nanomaterial engineering are expected to assist in the development of more effective drug delivery platforms, which can be used to increase the efficacy of LTDD anticancer drugs.

6 Lung Specific Drug Delivery

The lungs are an important target for drug delivery due to their large internal surface area and high blood flow rates compared to other areas of the body. The four most common methods of delivering drugs to the lungs are dry-powder inhaler (DPI), inhalation solution, metered-dose inhaler (MDI), and nebulizer (Rudokas et al. 2016). These methods provide different advantages such as a rapid onset of action and a low rate of drug loss due to deposition in the upper airways. The therapeutic efficacy of drugs delivered directly to the lungs can be significantly enhanced in comparison to traditional systemic delivery methods, making local drug delivery to the lungs an attractive option for many drugs (Andrade et al. 2013; Loira-Pastoriza et al. 2014; Rudokas et al. 2016).

When delivering drugs to the lungs, there are a number of challenges that must be considered (Newman 2017). Firstly, lung tissue is highly impermeable and there are significant barriers to drug penetration, including an alveolar epithelium layer. Additionally, the distribution of drugs within the lungs is subject to deposition in different parts of the airways, and these factors can all affect the efficacy of the drug's action. Secondly, particle size should be carefully considered—particles that are too large may lead to obstruction of small airways or deposition in the throat, while particles that are too small may be exhaled and not have contact with the lungs. Finally, drugs must also be able to cross the pulmonary surfactant layer (inhalation solutions) or be stable and able to withstand changes in temperature and humidity (dry-powder inhalers) (He et al. 2022; Labiris and Dolovich 2003; Newman 2017). Targeted liposomal drug delivery is a promising approach for the delivery of drugs to the lungs. Using liposomes to encapsulate bioactive substances prevents adverse effects on the lungs caused by direct contact of drug particles with the mucus. Liposomal drug delivery systems can also improve the bioavailability of drugs, by increasing the bioactivity of the active ingredient, reducing systemic toxicity, increasing aerosol stability, and providing a sustained release of the drug. Furthermore, liposomes can be used to target drug delivery to specific regions of the lung, by incorporating specific molecules onto the liposomal surface, such as certain peptides or antibodies. This increases the drug's selectivity and specificity, and reduces off-target effects (Andrade et al. 2013; Antimisiaris et al. 2021; Sharma et al. 2019; Willis et al. 2012). Overall, targeted liposomal drug delivery systems have the potential to improve the efficacy of drugs delivered directly to the lungs and are a promising advance in local drug delivery.

Liposomal targeted drug delivery for lung delivery has been an effective way to deliver pharmaceuticals to the central compartment of the lung. Both in terms of its efficacy and safety, this approach has proven its potential in the treatment and prevention of various lung diseases. Below are few examples discussing the mechanisms of action of the liposomal targeted drug delivery system in lungs.

Inhaled Liposomal Drug Delivery: Inhaled liposomal drug delivery is a way of delivering therapeutic agents to the lungs through the use of drug-loaded liposomes. Inhaled liposomes are primarily used for the delivery of small molecules, such as antibiotics and anti-inflammatories, as well as larger therapeutic proteins and nucleic acid-based agents. Liposomes can be administered through an inhaler device, which aerosolizes them and allows them to penetrate deep into the respiratory system for targeted delivery to the lungs (Elhissi 2017; Gaspar et al. 2008; Rudokas et al. 2016). Tiwari et al. 2012 showed that inhalation of liposomes containing amphotericin B was effective in the treatment of pulmonary aspergillosis. Also, the study carried out by Willis et al. (2012) demonstrated the effects of inhaled liposomal epigallocatechin gallate in targeting pulmonary inflammation.

Intrapulmonary Drug Delivery: Intrapulmonary drug delivery involves the administration of therapeutic agents to the lungs via direct injection or implantation of drug-loaded liposomes. This method is used to localize the delivery of drugs to the lungs, provide a sustained release of drugs, and reduce systemic absorption and side effects (Lin et al. 2017; Shim et al. 2013). Andrade et al. (2013) demonstrated the effects of intrapulmonary delivery of liposomes containing an antifungal agent in mice. Rudokas et al. (2016) showed use of liposome delivery of siRNA molecules to the lungs for the treatment of lung cancer.

Local Surfactant Lipospheres: Local surfactant lipospheres involve the administration of therapeutic agents to the lungs by encapsulating the drugs in liposomes and combining them with surfactants for enhanced penetration and distribution of the drug to the lungs. This method has demonstrated to be useful for the delivery of proteins and peptides to the lungs, as well as nucleic acid-based agents (Singh et al. 2015; Willis et al. 2012). Antimisiaris et al. (2021) demonstrated the efficacy of local surfactant lipospheres in delivering different types of therapeutic agents for the treatment of bacterial diseases. The study carried out by Loira-Pastoriza et al. (2014) showed that local surfactant lipospheres were effective in the delivery of siRNA to the lungs.

Liposomal targeted drug delivery for lung delivery holds immense potential for treating various diseases. Liposomal drug delivery offers many advantages for targeted drug delivery in the lung, potentially improving both efficacy and safety of treatment. It has been found that liposomes are able to cross the barriers of the respiratory system and enter the deeper-lying parts of the respiratory tract, thereby enabling higher drug bioavailability in the lung (Rudokas et al. 2016). Furthermore, liposomal drug delivery systems can protect drugs from degradation in the environment and target specific cells in the lungs to increase the therapeutic efficacy of the drug (Tiwari et al. 2012). Furthermore, it allows drugs to be released over an extended period of time, reducing the dosing frequency and improving patient compliance (Loira-Pastoriza et al. 2014). Liposomal drug delivery systems also offer the advantage of reducing systemic toxicity caused by traditional routes of administration such as oral and intravenous (Antimisiaris et al. 2021). Despite offering many advantages for drug delivery for the lungs, there are also a number of drawbacks to using liposomal drug delivery systems. One disadvantage of liposomal drug delivery is the high cost associated with its production (Willis et al. 2012), as well as low stability of these systems and difficulty in controlling the drug release over time can lead to reduced efficacy (Andrade et al. 2013). Additionally, the size of liposomes has been shown to restrict their efficiency in targeting deeper-lying cells in the lungs (Sharma et al. 2019).

In conclusion, liposomes are an attractive way to carry out targeted drug delivery of drugs that require specific localization in order to be effective. In relation to the lungs, liposomes can be used to target areas of infection or diseases such as lung cancer directly, and they have the potential to increase the penetration of drugs and improve local drug delivery to the lungs, which presents an advantageous alternative to systemic drug administration. Given the complexity of lung anatomy, local drug delivery systems are becoming increasingly important to improve therapeutic efficacy while reducing risks of systemic toxicity. In the future, more research needs to be conducted on target drug delivery to the lungs so that it can become an effective delivery system for drugs, especially those used to treat respiratory diseases and malignant tumors. Improvements in the accuracy of liposomal targeting, development of new drug compounds with increased liposomal solubility, improved aerosol stability of liposomes, and increased drug loading are all areas of potential research. In contrast to conventional therapies, such as systemic drug delivery or intramuscular injection, liposomal-based drug delivery could provide beneficial therapeutic outcomes due to the improved local delivery to the lungs. Furthermore, new and innovative delivery systems are continuously being developed to further optimize and enhance drug delivery to the lungs. Thus, targeted drug delivery for the lungs is an important area of research that is set to continue in the future.

7 Future Prospects

The above chapter discusses the amount of progress in the field of liposomal drug delivery systems over the years, indicating a promising future outlook. New formulations, technologies, and techniques are being developed in order to enable more efficient and effective delivery of drugs. These developments, such as aptamers for targeted drug delivery, reveal that liposomal formulations open up avenues for more precise targeting and higher therapeutic effects. The increasing sophistication of drug delivery systems is enabling greater precision and therefore better efficacy of drugs, as improved drug targeting means a decrease in unwanted side effects. The field of liposome-based drug delivery is also continuing to grow and evolve. New lipid components, targeting molecules and innovative drug excipients are being developed which could enable more efficient drug release and improved pharmacokinetics.

Many recent advances in the field have focused on combining liposomes with nanomaterials, such as nanoparticles, cells, and peptides, as this opens up the possibility of targeted delivery. By combining liposomes with other materials it is possible to exploit certain properties, such as increased drug loading capacity, improved release profiles, as well as enhanced stability (Alavi and Hamidi 2019; Goyal et al. 2005; Kirtane et al. 2021; Lian and Ho 2001). It is expected that in the future, more intelligent delivery systems, such as those involving the use of stimuli-responsive systems, will be developed (AlSawaftah et al. 2021; Lee and Thompson 2017; Torchilin 2014). When designed correctly, these can provide controlled drug release and even allow for re-programmed dosages. This way, both the stability of the drug and its efficacy can be improved, making it a more attractive option for physicians. It is also expected that the development of smart and responsive materials will allow for better monodispersion of drug particles in the target area (Karimi et al. 2016; Patra et al. 2018). Further advancements in pharmaceutical manufacturing techniques will enable the production of liposomal drug delivery systems tailored to specific needs, increasing the chances of successful drug delivery. Thus, the future prospects of liposomal targeted drug delivery systems appear to be very promising.

8 Conclusion

In conclusion, with further developments in the field, this type of drug delivery technology could be applicable in a wide range of therapeutic areas, including skin disorders, eye diseases, systemic disorders, and even cancer treatments. Liposomal drug delivery systems could also be used across many different drug classes, including peptides, proteins, and anticancer treatments, as well as for combination therapies. Furthermore, with the emergence of biological macromolecules and nanotechnology, newer and enhanced strategies for targeted drug delivery can be better explored. Given the current developments in this area, it is likely that liposomal drug delivery systems will continue to revolutionize health care in the years to come.

References

Alavi M, Hamidi M (2019) Passive and active targeting in cancer therapy by liposomes and lipid nanoparticles. Drug Metab Personalized Therapy 34(1). https://doi.org/10.1515/dmpt-2018-0032

Alavi M, Karimi N, Safaei M (2017) Application of various types of liposomes in drug delivery systems. Adv Pharm Bull 7(1):3–9. https://doi.org/10.15171/apb.2017.002

Alhariri M, Azghani A, Omri A (2013) Liposomal antibiotics for the treatment of infectious diseases. Expert Opin Drug Deliv 10(11):1515–1532. https://doi.org/10.1517/17425247.2013.822860

AlSawaftah N, Pitt WG, Husseini GA (2021) Dual-targeting and stimuli-triggered liposomal drug delivery in cancer treatment. ACS Pharmacol Transl Sci 4(3):1028–1049. https://doi.org/10.1021/acsptsci.1c00066

Andrade F, Rafael D, Videira M, Ferreira D, Sosnik A, Sarmento B (2013) Nanotechnology and pulmonary delivery to overcome resistance in infectious diseases. Adv Drug Deliv Rev 65(13–14):1816–1827. https://doi.org/10.1016/j.addr.2013.07.020

Antimisiaris SG, Marazioti A, Kannavou M, Natsaridis E, Gkartziou F, Kogkos G, Mourtas S (2021) Overcoming barriers by local drug delivery with liposomes. Adv Drug Deliv Rev 174:53–86. https://doi.org/10.1016/j.addr.2021.01.019

Beltrán-Gracia E, López-Camacho A, Higuera-Ciapara I, Velázquez-Fernández JB, Vallejo-Cardona AA (2019) Nanomedicine review: clinical developments in liposomal applications. Cancer Nanotechnol 10(1):11. https://doi.org/10.1186/s12645-019-0055-y

Boekhout AH, Beijnen JH, Schellens JHM (2011) Trastuzumab. Oncologist 16(6):800–810. https://doi.org/10.1634/theoncologist.2010-0035

Bulbake U, Doppalapudi S, Kommineni N, Khan W (2017) Liposomal formulations in clinical use: an updated review. Pharmaceutics 9(2):12. https://doi.org/10.3390/pharmaceutics9020012

Chalmers JD, van Ingen J, van der Laan R, Herrmann J-L (2021) Liposomal drug delivery to manage nontuberculous mycobacterial pulmonary disease and other chronic lung infections. Eur Respir Rev Official J Eur Respir Soc 30(161):210010. https://doi.org/10.1183/16000617.0010-2021

Daraee H, Etemadi A, Kouhi M, Alimirzalu S, Akbarzadeh A (2016) Application of liposomes in medicine and drug delivery. Artif Cells, Nanomed Biotechnol 44(1):381–391. https://doi.org/10.3109/21691401.2014.953633

Deshmukh SK, Khan MA, Singh S, Singh AP (2021) Extracellular nanovesicles: from intercellular messengers to efficient drug delivery systems. ACS Omega 6(3):1773–1779. https://doi.org/10.1021/acsomega.0c05539

Dissanayake T, Sun X, Abbey L, Bandara N (2022) Recent advances in lipid-protein conjugate-based delivery systems in nutraceutical, drug, and gene delivery. Food Hydrocolloids Health 2:100054. https://doi.org/10.1016/j.fhfh.2022.100054

Durak S, Esmaeili Rad M, Alp Yetisgin A, Eda Sutova H, Kutlu O, Cetinel S, Zarrabi A (2020) Niosomal drug delivery systems for ocular disease—recent advances and future prospects. Nanomaterials 10(6):6. https://doi.org/10.3390/nano10061191

Elhissi A (2017) Liposomes for pulmonary drug delivery: the role of formulation and inhalation device design. Curr Pharm Des 23(3):362–372. https://doi.org/10.2174/13816128236661611116114732

Gaspar MM, Bakowsky U, Ehrhardt C (2008) Inhaled liposomes-current strategies and future challenges. J Biomed Nanotechnol 4(3):245–257. https://doi.org/10.1166/jbn.2008.334

Goyal P, Goyal K, Vijaya Kumar SG, Singh A, Katare OP, Mishra DN (2005) Liposomal drug delivery systems—clinical applications. Acta Pharm (zagreb, Croatia) 55(1):1–25

Guimarães D, Cavaco-Paulo A, Nogueira E (2021) Design of liposomes as drug delivery system for therapeutic applications. Int J Pharm 601:120571. https://doi.org/10.1016/j.ijpharm.2021.120571

Gyanani V, Haley JC, Goswami R (2021) Challenges of current anticancer treatment approaches with focus on liposomal drug delivery systems. Pharmaceuticals (basel, Switzerland) 14(9):835. https://doi.org/10.3390/ph14090835

He H, Lu Y, Qi J, Zhu Q, Chen Z, Wu W (2019) Adapting liposomes for oral drug delivery. Acta Pharm Sinica. B 9(1):36–48. https://doi.org/10.1016/j.apsb.2018.06.005

He S, Gui J, Xiong K, Chen M, Gao H, Fu Y (2022) A roadmap to pulmonary delivery strategies for the treatment of infectious lung diseases. J Nanobiotechnol 20(1):101. https://doi.org/10.1186/s12951-022-01307-x

Hofheinz R-D, Gnad-Vogt SU, Beyer U, Hochhaus A (2005) Liposomal encapsulated anti-cancer drugs. Anticancer Drugs 16(7):691–707. https://doi.org/10.1097/01.cad.0000167902.53039.5a

Huang X, Caddell R, Yu B, Xu S, Theobald B, Lee LJ, Lee RJ (2010) Ultrasound-enhanced microfluidic synthesis of liposomes. Anticancer Res 30(2):463–466

Ishak RAH, Mostafa NM, Kamel AO (2017) Stealth lipid polymer hybrid nanoparticles loaded with rutin for effective brain delivery—comparative study with the gold standard (Tween 80): optimization, characterization and biodistribution. Drug Deliv 24(1):1874–1890. https://doi.org/10.1080/10717544.2017.1410263

Islas-Weinstein L, Marquina-Castillo B, Mata-Espinosa D, Paredes-González IS, Chávez J, Balboa L, Marín Franco JL, Guerrero-Romero D, Barrios-Payan JA, Hernandez-Pando R (2021) The cholinergic system contributes to the immunopathological progression of experimental pulmonary tuberculosis. Front Immunol 11:581911. https://doi.org/10.3389/fimmu.2020.581911

Jaradat E, Weaver E, Meziane A, Lamprou D (2021) Microfluidics technology for the design and formulation of nanomedicines. Nanomaterials 11:3440. https://doi.org/10.3390/nano11123440

Karimi M, Ghasemi A, Zangabad PS, Rahighi R, Moosavi Basri SM, Mirshekari H, Amiri M, Pishabad ZS, Aslani A, Bozorgomid M, Ghosh D, Beyzavi A, Vaseghi A, Aref AR, Haghani L, Bahrami S, Hamblin MR (2016) Smart micro/nanoparticles in stimulus-responsive drug/gene delivery systems. Chem Soc Rev 45(5):1457–1501. https://doi.org/10.1039/c5cs00798d

Kim J-S (2016) Liposomal drug delivery system. J Pharm Investig 46(4):387–392. https://doi.org/10.1007/s40005-016-0260-1

Kirtane AR, Verma M, Karandikar P, Furin J, Langer R, Traverso G (2021) Nanotechnology approaches for global infectious diseases. Nat Nanotechnol 16(4):4. https://doi.org/10.1038/s41565-021-00866-8

Kulkarni JA, Witzigmann D, Thomson SB, Chen S, Leavitt BR, Cullis PR, van der Meel R (2021) The current landscape of nucleic acid therapeutics. Nat Nanotechnol 16(6):6. https://doi.org/10.1038/s41565-021-00898-0

Kumari P, Ghosh B, Biswas S (2016) Nanocarriers for cancer-targeted drug delivery. J Drug Target 24(3):179–191. https://doi.org/10.3109/1061186X.2015.1051049

Labiris NR, Dolovich MB (2003) Pulmonary drug delivery. Part I: physiological factors affecting therapeutic effectiveness of aerosolized medications. Br J Clin Pharmacol 56(6):588–599. https://doi.org/10.1046/j.1365-2125.2003.01892.x

Lamichhane TN, Raiker RS, Jay SM (2015) Exogenous DNA loading into extracellular vesicles via electroporation is size-dependent and enables limited gene delivery. Mol Pharm 12(10):3650–3657. https://doi.org/10.1021/acs.molpharmaceut.5b00364

Lamichhane N, Udayakumar TS, D'Souza WD, Simone CB, Raghavan SR, Polf J, Mahmood J (2018) Liposomes: clinical applications and potential for image-guided drug delivery. Molecules (basel, Switzerland) 23(2):288. https://doi.org/10.3390/molecules23020288

Lao J, Madani J, Puértolas T, Álvarez M, Hernández A, Pazo-Cid R, Artal Á, Antón Torres A (2013) Liposomal doxorubicin in the treatment of breast cancer patients: a review. J Drug Deliv 2013:456409. https://doi.org/10.1155/2013/456409

Large DE, Abdelmessih RG, Fink EA, Auguste DT (2021) Liposome composition in drug delivery design, synthesis, characterization, and clinical application. Adv Drug Deliv Rev 176:113851. https://doi.org/10.1016/j.addr.2021.113851

Le Meins J-F, Schatz C, Lecommandoux S, Sandre O (2013) Hybrid polymer/lipid vesicles: state of the art and future perspectives. Mater Today 16(10):397–402. https://doi.org/10.1016/j.mattod.2013.09.002

Lee Y, Thompson DH (2017) Stimuli-responsive liposomes for drug delivery. Wiley Interdisc Rev. Nanomed Nanobiotechnol 9(5). https://doi.org/10.1002/wnan.1450

Li SG, Li L (2013) Targeted therapy in HER2-positive breast cancer. Biomed Rep 1(4):499–505. https://doi.org/10.3892/br.2013.95

Li M, Du C, Guo N, Teng Y, Meng X, Sun H, Li S, Yu P, Galons H (2019) Composition design and medical application of liposomes. Eur J Med Chem 164:640–653. https://doi.org/10.1016/j.ejmech.2019.01.007

Lian T, Ho RJY (2001) Trends and developments in liposome drug delivery systems. J Pharm Sci 90(6):667–680. https://doi.org/10.1002/jps.1023

Lin C, Wong BCK, Chen H, Bian Z, Zhang G, Zhang X, Kashif Riaz M, Tyagi D, Lin G, Zhang Y, Wang J, Lu A, Yang Z (2017) Pulmonary delivery of triptolide-loaded liposomes decorated with anti-carbonic anhydrase IX antibody for lung cancer therapy. Sci Rep 7(1):1. https://doi.org/10.1038/s41598-017-00957-4

Liu D, He C, Wang AZ, Lin W (2013) Application of liposomal technologies for delivery of platinum analogs in oncology. Int J Nanomed 8:3309–3319. https://doi.org/10.2147/IJN.S38354

Liu P, Chen G, Zhang J (2022) A review of liposomes as a drug delivery system: current status of approved products, regulatory environments, and future perspectives. Molecules 27(4):1372. https://doi.org/10.3390/molecules27041372

Loira-Pastoriza C, Todoroff J, Vanbever R (2014) Delivery strategies for sustained drug release in the lungs. Adv Drug Deliv Rev 75:81–91. https://doi.org/10.1016/j.addr.2014.05.017

Lombardo D, Kiselev MA (2022) Methods of liposomes preparation: formation and control factors of versatile nanocarriers for biomedical and nanomedicine application. Pharmaceutics 14(3):543. https://doi.org/10.3390/pharmaceutics14030543

Luiz H, Oliveira Pinho J, Gaspar MM (2023) Advancing medicine with lipid-based nanosystems—the successful case of liposomes. Biomedicines 11(2):2. https://doi.org/10.3390/biomedicines11020435

Maurer N, Fenske DB, Cullis PR (2001) Developments in liposomal drug delivery systems. Expert Opin Biol Ther 1(6):923–947. https://doi.org/10.1517/14712598.1.6.923

Mishra GP, Bagui M, Tamboli V, Mitra AK (2011) Recent applications of liposomes in ophthalmic drug delivery. J Drug Deliv 2011:863734. https://doi.org/10.1155/2011/863734

Mukherjee A, Waters AK, Kalyan P, Achrol AS, Kesari S, Yenugonda VM (2019) Lipid–polymer hybrid nanoparticles as a next-generation drug delivery platform: State of the art, emerging technologies, and perspectives. Int J Nanomed 14:1937–1952. https://doi.org/10.2147/IJN.S198353

Newman SP (2017) Drug delivery to the lungs: challenges and opportunities. Ther Deliv 8(8):647–661. https://doi.org/10.4155/tde-2017-0037

Nisini R, Poerio N, Mariotti S, De Santis F, Fraziano M (2018) The multirole of liposomes in therapy and prevention of infectious diseases. Front Immunol 9:155. https://doi.org/10.3389/fimmu.2018.00155

Nsairat H, Khater D, Sayed U, Odeh F, Al Bawab A, Alshaer W (2022) Liposomes: structure, composition, types, and clinical applications. Heliyon 8(5):e09394. https://doi.org/10.1016/j.heliyon.2022.e09394

Olusanya TOB, Haj Ahmad RR, Ibegbu DM, Smith JR, Elkordy AA (2018) Liposomal drug delivery systems and anticancer drugs. Molecules (basel, Switzerland) 23(4):907. https://doi.org/10.3390/molecules23040907

Patra JK, Das G, Fraceto LF, Campos EVR, del Rodriguez-Torres MP, Acosta-Torres LS, Diaz-Torres LA, Grillo R, Swamy MK, Sharma S, Habtemariam S, Shin H-S (2018) Nano based drug delivery systems: Recent developments and future prospects. J Nanobiotechnol 16(1):71. https://doi.org/10.1186/s12951-018-0392-8

Pattni BS, Chupin VV, Torchilin VP (2015) New developments in liposomal drug delivery. Chem Rev 115(19):10938–10966. https://doi.org/10.1021/acs.chemrev.5b00046

Pham D-D, Fattal E, Tsapis N (2015) Pulmonary drug delivery systems for tuberculosis treatment. Int J Pharm 478(2):517–529. https://doi.org/10.1016/j.ijpharm.2014.12.009

Pierre MBR, dos Santos Miranda Costa I (2011) Liposomal systems as drug delivery vehicles for dermal and transdermal applications. Arch Dermatol Res 303(9):607–621. https://doi.org/10.1007/s00403-011-1166-4

Pinheiro M, Lúcio M, Lima JLFC, Reis S (2011) Liposomes as drug delivery systems for the treatment of TB. Nanomedicine (lond) 6(8):1413–1428. https://doi.org/10.2217/nnm.11.122

Ravar F, Saadat E, Gholami M, Dehghankelishadi P, Mahdavi M, Azami S, Dorkoosh FA (2016) Hyaluronic acid-coated liposomes for targeted delivery of paclitaxel, in-vitro characterization and in-vivo evaluation. J Controlled Release: Official J Controlled Release Soc 229:10–22. https://doi.org/10.1016/j.jconrel.2016.03.012

Richardson ES, Pitt WG, Woodbury DJ (2007) The role of cavitation in liposome formation. Biophys J 93(12):4100–4107. https://doi.org/10.1529/biophysj.107.104042

Ritchie H, Spooner F, Roser M (2018) Causes of death. Our world in data. https://ourworldindata.org/causes-of-death

Rudokas M, Najlah M, Alhnan MA, Elhissi A (2016) Liposome delivery systems for inhalation: a critical review highlighting formulation issues and anticancer applications. Med Principles Pract Int J Kuwait Univ Health Sci Centre 25 (Suppl 2):60–72. https://doi.org/10.1159/000445116

Sapra P, Allen TM (2003) Ligand-targeted liposomal anticancer drugs. Prog Lipid Res 42(5):439–462. https://doi.org/10.1016/s0163-7827(03)00032-8

Saraf S, Jain A, Tiwari A, Verma A, Panda PK, Jain SK (2020) Advances in liposomal drug delivery to cancer: an overview. J Drug Deliv Sci Technol 56:101549. https://doi.org/10.1016/j.jddst.2020.101549

Senapati S, Mahanta AK, Kumar S, Maiti P (2018) Controlled drug delivery vehicles for cancer treatment and their performance. Signal Transduct Target Ther 3:7. https://doi.org/10.1038/s41392-017-0004-3

Sercombe L, Veerati T, Moheimani F, Wu SY, Sood AK, Hua S (2015) Advances and challenges of liposome assisted drug delivery. Front Pharmacol 6:286. https://doi.org/10.3389/fphar.2015.00286

Sharma P, Mehta M, Dhanjal DS, Kaur S, Gupta G, Singh H, Thangavelu L, Rajeshkumar S, Tambuwala M, Bakshi HA, Chellappan DK, Dua K, Satija S (2019) Emerging trends in the novel drug delivery approaches for the treatment of lung cancer. Chem Biol Interact 309:108720. https://doi.org/10.1016/j.cbi.2019.06.033

Shim G, Choi H, Lee S, Choi J, Yu YH, Park D-E, Choi Y, Kim C-W, Oh Y-K (2013) Enhanced intrapulmonary delivery of anticancer siRNA for lung cancer therapy using cationic ethylphosphocholine-based nanolipoplexes. Mol Ther 21(4):816–824. https://doi.org/10.1038/mt.2013.10

Silva R, Ferreira H, Little C, Cavaco-Paulo A (2010) Effect of ultrasound parameters for unilamellar liposome preparation. Ultrason Sonochem 17(3):628–632. https://doi.org/10.1016/j.ultsonch.2009.10.010

Silva R, Ferreira H, Cavaco-Paulo A (2011) Sonoproduction of liposomes and protein particles as templates for delivery purposes. Biomacromol 12(10):3353–3368. https://doi.org/10.1021/bm200658b

Singh C, Koduri LVSK, Singh A, Suresh S (2015) Novel potential for optimization of antitubercular therapy: pulmonary delivery of rifampicin lipospheres. Asian J Pharm Sci 10(6):549–562. https://doi.org/10.1016/j.ajps.2015.08.003

Slingerland M, Guchelaar H-J, Gelderblom H (2012) Liposomal drug formulations in cancer therapy: 15 years along the road. Drug Discovery Today 17(3–4):160–166. https://doi.org/10.1016/j.drudis.2011.09.015

Smith LA, Cornelius VR, Plummer CJ, Levitt G, Verrill M, Canney P, Jones A (2010) Cardiotoxicity of anthracycline agents for the treatment of cancer: systematic review and meta-analysis of randomised controlled trials. BMC Cancer 10(1):337. https://doi.org/10.1186/1471-2407-10-337

Šturm L, Poklar Ulrih N (2021) Basic methods for preparation of liposomes and studying their interactions with different compounds, with the emphasis on polyphenols. Int J Mol Sci 22(12):6547. https://doi.org/10.3390/ijms22126547

Tacar O, Sriamornsak P, Dass CR (2013) Doxorubicin: An update on anticancer molecular action, toxicity and novel drug delivery systems. J Pharm Pharmacol 65(2):157–170. https://doi.org/10.1111/j.2042-7158.2012.01567.x

Tiwari G, Tiwari R, Sriwastawa B, Bhati L, Pandey S, Pandey P, Bannerjee SK (2012) Drug delivery systems: an updated review. Int J Pharm Invest 2(1):2–11. https://doi.org/10.4103/2230-973X.96920

Torchilin VP (2014) Multifunctional, stimuli-sensitive nanoparticulate systems for drug delivery. Nat Rev Drug Discov 13(11):11. https://doi.org/10.1038/nrd4333

Willis L, Hayes D, Mansour HM (2012) Therapeutic liposomal dry powder inhalation aerosols for targeted lung delivery. Lung 190(3):251–262. https://doi.org/10.1007/s00408-011-9360-x

Zhang C, Xu C, Gao X, Yao Q (2022) Platinum-based drugs for cancer therapy and anti-tumor strategies. Theranostics 12(5):2115–2132. https://doi.org/10.7150/thno.69424

Zhong L, Li Y, Xiong L, Wang W, Wu M, Yuan T, Yang W, Tian C, Miao Z, Wang T, Yang S (2021) Small molecules in targeted cancer therapy: advances, challenges, and future perspectives. Signal Transduction Targeted Therapy 6(1):1. https://doi.org/10.1038/s41392-021-00572-w

Zhu X, Radovic-Moreno AF, Wu J, Langer R, Shi J (2014) Nanomedicine in the management of microbial infection—overview and perspectives. Nano Today 9(4):478–498. https://doi.org/10.1016/j.nantod.2014.06.003

Zong T-X, Silveira AP, Morais JAV, Sampaio MC, Muehlmann LA, Zhang J, Jiang C-S, Liu S-K (2022) Recent advances in antimicrobial nano-drug delivery systems. Nanomaterials 12(11):11. https://doi.org/10.3390/nano12111855

Zununi Vahed S, Salehi R, Davaran S, Sharifi S (2017) Liposome-based drug co-delivery systems in cancer cells. Mater Sci Eng C Mater Biol Appl 71:1327–1341. https://doi.org/10.1016/j.msec.2016.11.073

Multifunctional Liposomes to Attain Targeting, Stimuli Sensitive Drug Release and Imaging Cancer

Vipin Kumar, Prerna Kewlani, Amita Singh, Sanjay, Anurag Kumar Gautam, and Vijayakumar Mahalingam Rajamanickam

Abstract Liposomes have been shown to be superior delivery systems for nutrients, drugs, vaccines, and bioactive substances. The concentric phospholipid bilayers that surround an aqueous core make up the spherical vesicles known as liposomes. Most of the molecules may be folded into the structure of these vesicular drug delivery systems during their fabrication, which is facilitated by techniques like the solvent injection process, thin film hydration, and reverse phase evaporation. Liposomal nanoformulations are a highly effective method for administering drugs using both active and passive targeting strategies. In order to actively target a specific cancer, liposomes can have their surface functionalized with specific targeting ligands such as protein molecules, sugars, and antibodies. Numerous recent kinds of literature were reported for the targeted delivery of small molecule drugs, genes and biological macromolecules using liposomal formulations. An emerging field of study investigates the use of stimuli-sensitive liposomes for the delivery of their payload. Recent developments have shown that stimuli-sensitive liposomes can be used to trigger drug release in response to both internal stimuli like acidic pH, abnormal enzyme concentrations, and temperature, and external stimuli like light waves, and magnetic

V. Kumar · P. Kewlani · A. Singh · Sanjay · A. K. Gautam · V. Mahalingam Rajamanickam (✉)
Department of Pharmaceutical Sciences, Babasaheb Bhimrao Ambedkar University (A Central University), Vidya Vihar, Rai Bareli Road, Lucknow 226025, India
e-mail: mrvceutics@gmail.com

V. Kumar
e-mail: vipinchoudhary1103@gmail.com

P. Kewlani
e-mail: prernakewlani18@gmail.com

A. Singh
e-mail: a.s.rathore2197@gmail.com

Sanjay
e-mail: atalsanjay18@gmail.com

A. K. Gautam
e-mail: kumargautamanurag@gmail.com

T. S. Santra and A. U. S. Shinde (eds.), *Advanced Drug Delivery*,
Studies in Mechanobiology, Tissue Engineering and Biomaterials 26,
https://doi.org/10.1007/978-981-99-6564-9_3

fields. Tumor imaging can also be done by incorporation of quantum dots or fluorescent dyes such as coumarin 6, indocyanine green, methylene blue, naphthalocyanine, etc. Multifunctional liposomes which combine, drug delivery, targeting, stimuli sensitiveness and imaging modalities are the cutting-edge innovations being carried out in this field. Scientists are interested in using liposomes for clinical applications in the treatment of cancer because of the special characteristics of liposomes such as being non-toxic, biodegradable, excellent biocompatibility, and non-immunogenicity. Liposomes that are undergoing clinical trials, and their novel innovations, such as cancer-targeting, imaging, and stimuli-sensitive tools, will soon be introduced to the market.

1 Introduction

The first liposomes were prepared in 1964 by Dr. Alec D. Bangham, a haematologist from the United Kingdom, and his colleagues from the Babraham Institute at the University of Cambridge (Bangham and Horne 1964). A spherical vesicle which is made up of many phospholipid molecules termed as "liposomes". Furthermore, the liposomes have one or more lipid bilayers that arrange around the internal aqueous core (Sebaaly et al. 2016). This structural arrangement of liposomes provides capacity to deliver the hydrophilic, hydrophobic and amphiphilic molecules. The hydrophilic molecules get entrapped into the internal aqueous core, hydrophobic molecules entrapped into the bilayer of lipid membrane and amphiphilic molecules present at the interface of water and lipid bilayer (Laouini et al. 2012). Liposomes are one of the important drug delivery systems to deliver the drug molecules because of their structural arrangement and biocompatibility. Additionally, the liposomes may improve the half-life of drug, increases efficiency and helpful for the targeted delivery of the drugs (Mathiyazhakan et al. 2018). Phospholipid molecules has amphiphilic characteristic similar to mammalian cell membrane allowing interaction between the cell membrane and the liposomes, as well as effectively enhance cellular uptake (He et al. 2019). Liposomes have many advantages that make it more specifically designed to transport drugs to the desired location. It possesses the ability to self-assemble and is capable of transporting big medicinal molecules. The physicochemical property of the liposomes can be improved to control biological characteristics (Sercombe et al. 2015). Liposomes improves therapeutic application of compounds by providing them stability and site-specific targeting (Ding 2006; Lu et al. 2013; Bozzuto and Molinari 2015). It protects the medication from enzymatic degradation, chemical and immunological inactivation, and prevents rapid plasma clearance, all of which contribute to an improvement in therapeutic activity. The drug entrapped inside the liposomes reduces its exposure to the healthy tissue and also minimize side effects compare the pristine form of the drug (Bozzuto and Molinari 2015).

In this chapter, we broadly focus on inventive ideas in methods of production, multifunctional liposomes to attain targeting, stimuli sensitive drug release and imaging cancer and their applications. Additionally, the formulations under clinical trials are also focused in this chapter. The purpose of this book chapter is to provide up-to-date information about liposomal nanoformulations and their applications, along with a general concept of the clinical research that is currently being conducted.

2 Structure of Liposomes

Depending on their design, liposomes are categorized according to the quantity of lipid bilayers (lamellae) assembly and the size of the vesicle (Fig. 1). Liposomes can be categorized as unilamellar vesicles (ULV, all size range), multilamellar vesicles (MLV, >500 nm), and multivesicular vesicles (MVV, >1000 nm) vesicles based on their lamellarity (Akbarzadeh et al. 2013; Emami et al. 2016). ULVs may be further classified into three different sizes: small unilamellar vesicles (SUVs), which range in size from 20 to 100 nm; large unilamellar vesicles (LUVs), which are larger than 100 nm; and giant unilamellar vesicles (GUVs), which are larger than 1000 nm. The presence of a single bilayer is what sets ULVs distinct, which allows for better hydrophilic chemical encapsulation. MLVs have two or more lipid bilayers that are concentric and structured in an onion-like shape, which is advantageous for encapsulating lipophilic substances. MVVs are made up of numerous small vesicles that are not concentrated together are entrapped inside a single lipid bilayer and are perfect for encapsulating large amounts of hydrophilic material (Emami et al. 2016). Multicompartment liposome (MCL) is another innovative vesicle-type construction. The MCL is a single delivery vehicle for combinatorial compounds that are composed of two distinct vesicle types fused together at a tight bilayer interface (Aljamal and Kostarelos 2007).

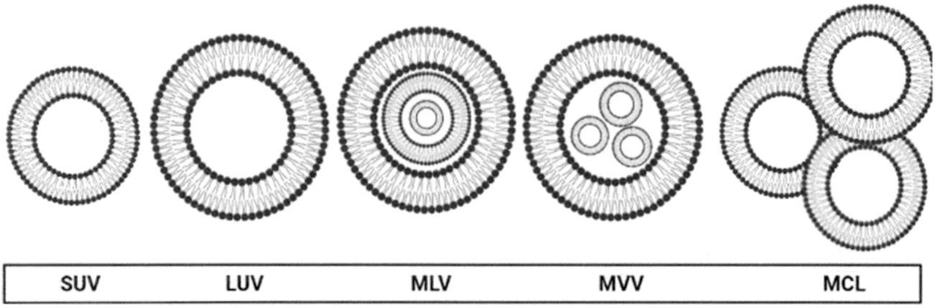

Fig. 1 Various categories of liposomes based on lamellarity

3 Liposomes in Cancer

Cancer is seen as being fatal disease in which the growth of cells become uncontrollable and disturbs the hemostatic conditions of the body (Srivastava et al. 2015). Cancer is responsible for the deaths of millions of individuals every year hence reported as one in six death causing disease (McGuire 2016). The important and effective treatment that is used for cancer is chemotherapy. However, there is a lack of specificity and sensitivity in chemotherapy (Schirrmacher 2019). Advanced research is being conducted in this area to discover new treatment for cancer therapy. As a result, nanomedicines, including viral nanoparticles (VNPs), quantum dots, polymer nanomaterials, and liposomes are some of the newer technologies that have been deployed in an effort to mitigate the adverse effects of some medications that is caused owing to the conventional chemotherapies (Rommasi and Esfandiari 2021). With respect to nanomedicine, liposomes have the benefit of being able to include both medications that are hydrophilic and those that are hydrophobic for effective anticancer therapy. This entrapment enhances the diversification in terms of specificity, bioavailability and biocompatibility properties (Sapra et al. 2005). Some of the examples of marketed liposomes for clinical applications are DaunoXome® and Caelyx® (Fanciullino and Ciccolini 2009). Further, liposomes have proven the potential in tissue regeneration and repairing, imaging and diagnosis purposes. The liposomes target the drug delivery of chemotherapeutics, immunotherapy, radiation, and gene therapy. The overall result to accomplish the use of liposomes is better drug efficacy, decreased side effects, enhanced physicochemical properties, efficiency to deliver macromolecules, and to bypass the resistance mechanism (Malam et al. 2009).

4 Methods for the Preparation of Liposomes

Traditional method for preparing liposomes such as thin film hydration technique (TFH), which is also known as the Bangham method, solvent injection, reverse phase evaporation, detergent removal methods and size reduction technique are discussed in detail.

4.1 Thin Film Hydration Process

This is the first documented manufacturing technique for the preparation of liposomes. In this method lipids are dissolved in an organic solvent like methanol, and/or chloroform, and then drying them out to produce a thin lipid layer in a round-bottom flask by the evaporation of the organic solvent under vacuum at temperatures between 45 and 60 °C. The liposomes are then formed by continuously swirling aqueous medium while a thin layer of lipids is being hydrated with it. The primary

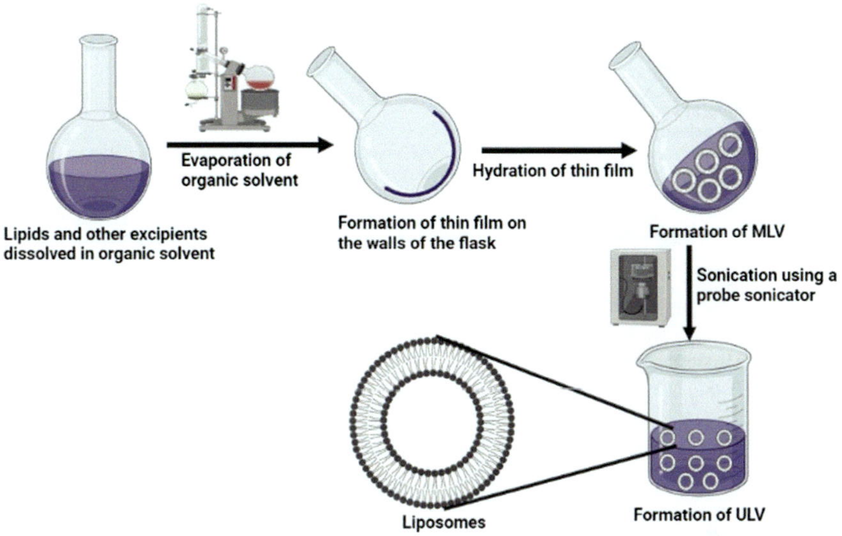

Fig. 2 Schematic depiction of the thin-film hydration process to prepare liposomes

drawbacks of this method are time-consuming, results in low entrapment efficiency, challenging to remove organic solvent completely in an appropriate manner, and require sterilisation (Bangham et al. 1967). Figure 2 depicts a graphical illustration of the steps involved in the process of hydrating thin films.

4.2 Solvent Injection Process

Liposomes can be prepared using the solvent injection method by dissolving lipids in an organic solvent like ethanol/ether and then introducing them into an aqueous phase (William et al. 2020). As a result of its simplicity, reliability, speed of implementation, and scalability, this method is one of the most used processes. Figure 3 depicts a diagrammatic representation of solvent injection process.

4.3 Reverse Phase Evaporation Method

The fabrication of liposomes may also be accomplished using the process of reverse phase evaporation. Thin film hydration technique is used in this procedure. This process involves dissolving a lipid mixture in an organic solvent, which is then evaporated off using a rotary evaporator to produce a thin film. An oil-in-water emulsion is created by dissolving the formed film in an organic solvent and then mixing it with an

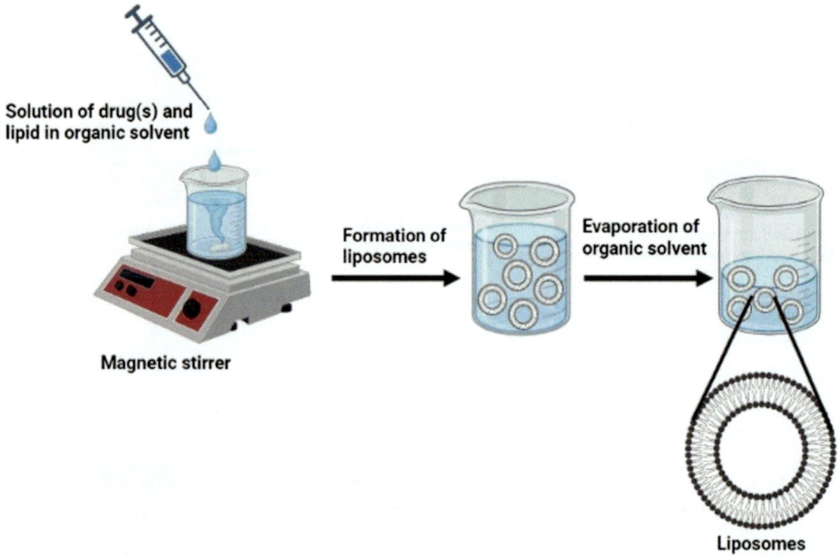

Fig. 3 Diagramatic representation of preparation of liposomes by solvent injection technique

aqueous phase (Pattni et al. 2015). In addition to the aforementioned steps, inverted micelles are formed by sonicating the mixture, creating a homogeneous emulsion; the organic solvent is then evaporated under low pressure, creating a viscous gel; and finally, liposomes are formed from the gel (Akbarzadeh et al. 2013). The reverse phase evaporation method has the advantage of being simple and permits a high encapsulation efficacy (Monteiro et al. 2014; Wagner and Vorauer-Uhl 2011). In addition, the procedure takes a long time, a lot of organic solvents is used, and sterilization is necessary (Antimisiaris and Kallinteri 2008). Figure 4 depicts a schematic representation of the reverse phase evaporation process.

4.4 Detergent Removal Method

Detergent removal is an additional method for producing liposomes in which lipids are solubilized by adding the appropriate detergent to an organic solvent at their critical micelle concentration to produce lipid micelles (Nkanga et al. 2019). When the detergent is removed, the phospholipid micelles gain an increased competitive edge and eventually self-assemble (Akbarzadeh et al. 2013; Emami et al. 2016; Aljamal and Kostarelos 2007; Bangham et al. 1967; William et al. 2020; Pattni et al. 2015). The pace and the amount to which detergent is eliminated, as well as the starting ratio of phospholipids to detergents, are the factors that may have an impact on the size and homogeneity of liposomes. The principal disadvantages of the detergent removal procedure include the possibility of contaminants in the final

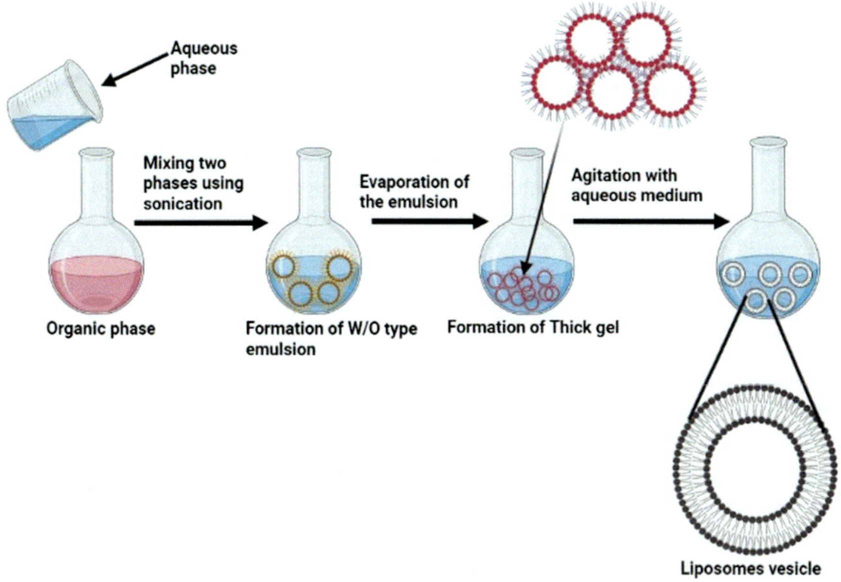

Fig. 4 Schematic representation of the fabrication of liposomes by the reverse-phase evaporation method

liposomal formulation, limited entrapment efficiency, and lengthy procedure (Meure et al. 2008). Figure 5 demonstrates the procedure of detergent removal technique.

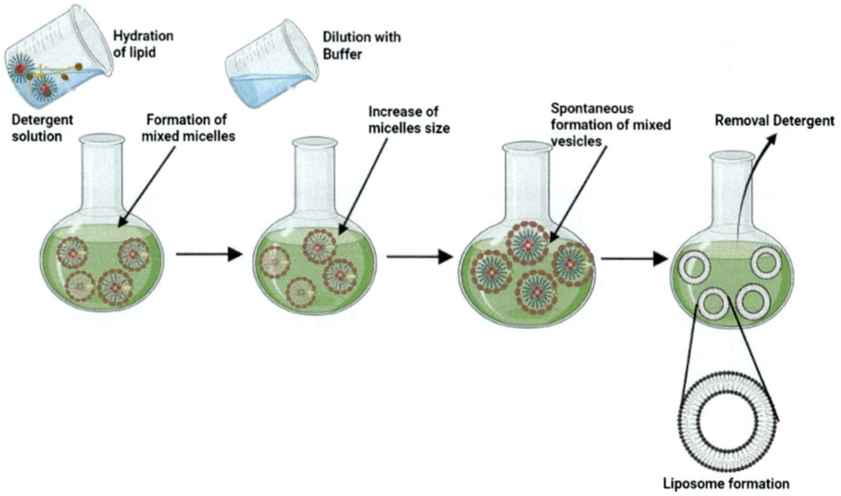

Fig. 5 Schematic portrayal of the detergent removal method for the preparation of liposomes

4.5 Size Reduction Method

Traditional methods of liposome formulation necessitate the use of size reduction techniques like sonication, homogenization, and extrusion (Kraft et al. 2014). Probe and bath sonication are the most common types of sonication (Akbarzadeh et al. 2013). Disadvantages of the sonication process include the potential for contamination from metal coming from the tip of the probe into the solution and the difficulty of providing the uniform energy generated by ultrasonic waves applied to a massive amount of liposomal suspension (scale-up). Additionally, phospholipid degradation is possible, which might potentially affect the drug molecule that is intended to be encapsulated (Batzri and Korn 1973; Tejera-Garcia et al. 2011). Homogenization techniques, on the other hand, use a high forced pressure to push liposomes through a small opening, resulting in a high velocity collision principle that shrinks the liposomes. Processes for reducing particle size such as microfluidization, high-pressure homogenization, and shear force-induced homogenization can be applied after homogenization (Wagner and Vorauer-Uhl 2011). The extrusion process is another method for decreasing liposome size. The liposomes go through a series of extrusion cycles in which they pass through a filter that has a predetermined pore size. This filter is commonly a polycarbonate membrane (Meure et al. 2008). Unlike homogenizers, this method uses significantly reduced pressure, as well as a significantly reduced volume of the liposomal solution (Kraft et al. 2014).

5 Marketed Liposomal Formulation

Clinical uses of liposomes have demonstrated considerable therapeutic advantages. However, the whole creation and manufacturing of liposomes, which entails manufacturing techniques, and regulatory permission by the appropriate authorities, together with intellectual property, limits the scope of their use (Saraf et al. 2020). Only a few liposomes have now been marketed, despite much research being done to produce compositions of liposomes that can be utilised in medicinal purposes (Moosavian et al. 2021). The first liposomal medication, Doxil®, was the first successful liposomal formulation the US market in 1995. Doxorubicin (DOX) hydrochloride is the drug component in Doxil® or Caelyx® in Europe which is delivered through intravenous injectable medication. Doxil® is a medication that is prescribed for the treatment of advanced Kaposi's sarcoma as well as ovarian cancer that is associated with AIDS (Bulbake et al. 2017). The pharmacokinetics of free DOX was shown to be significantly improved and the potential drug toxicities was reduced by the use of this liposomal formulation. The primary therapeutic areas that liposomal formulation products address is listed in Fig. 11 (Bulbake et al. 2017). Liposomes are reported in liquid form (suspension), solid (dry powder), and a semi-solid state (cream or gel). Liposomes can be parenterally or topically administered in vivo

(Laouini et al. 2012). It is crucial to emphasize that the majority of produced liposomal products are now undergoing various pre-clinical research projects and clinical trials. Clinical translation of liposomes needs sophisticated models and procedures. These models can help improve the therapeutic uses of liposomes by providing a forecast of their biosafety within the body (Saraf et al. 2020).

6 Dual or Multi Drug Loaded Liposomes

In comparison to single loading, the twofold incorporation of drug in dual drug loaded liposomes efficiently raises the drug: lipid ratio, which results in better action as well as extended impact (Bragagni et al. 2010). According to recent research, dual or multidrug loaded liposome based chemotherapy changes numerous target pathways for improved therapeutic efficiency and reduced toxicity. An example of combined drug delivery of 5-fluorouracil and apigenin in a single liposomal formulation demonstrates better synergistic study both in vitro and in vivo for potential treatments for colon cancer. These vesicles have increased cellular toxicity due to improved hemocompatibility and cytocompatibility. These liposomes had successfully internalized and enhanced angiogenesis inhibition, as well as decreased cell proliferation, and decreased cellular migration and invasion capabilities. The issue brought on by drug resistance and the harmful effects of 5-Fluorouracil are prevented by the liposomal version of the dual medicine 5-fluorouracil and Apigenin, loaded with both drugs. Consequently, it opens the door for a newer candidate with therapeutic uses (Sen et al. 2019). We report the development and evaluation of nanoparticles that are loaded with two different medications in order to successfully distribute carfilzomib and doxorubicin to cancer cells caused by multiple myeloma at the optimum ratio for their synergistic effects. This extra technique is necessary in order to successfully distribute these two drugs to cancer cells at the optimal ratio. According to the findings that we obtained, the dual drug-loaded liposomes exhibited synergy in vitro and were more effective at preventing tumour development in vivo than a combination of pristine drugs. This was the case even though both medications were contained inside the same liposome. This was done while lowering the systemic toxicity. Additionally, another study reported carfilzomib and doxorubicin-loaded liposomes for improved treatment effectiveness in multiple myeloma patients (Ashley et al. 2016).

Recently, Rolle et al. presented a dual-drug liposomal formulation that included disulfiram as the chemosensitizer and doxorubicin as the cytotoxic agent (Rolle et al. 2020). Meng et al. have demonstrated positive results in both in vitro and in vivo models findings using resveratrol (RES) and paclitaxel (PTX) co-encapsulated in a liposomal formulation (Meng et al. 2016). PTX with rapamycin co-loaded liposomes has been reported effectively in both vitro and in vivo settings (Eloy et al. 2016). Docetaxel, another extensively used taxane drug which is used in the treatment of breast cancer, has also been co-encapsulated into liposomes with thymoquinone (Odeh et al. 2019). The improved synergistic effectiveness of this dual-drug liposomal formulation against cells consistent with breast cancer was found. Jose et al. reported a liposomal formulation that delivers tamoxifen and imatinib simultaneously in yet another intriguing research (Jose et al. 2019). These two drugs had a synergistic effect on the growth rate of various breast cancer cell lines, which they slowed down by inhibiting the cell division process.

7 Prolongation of Systemic Circulation of Liposomes

In-depth research is being done on long-circulating liposomes right now, and they are used extensively in a variety of biomedical research settings, both in vitro and in vivo as well as in clinical applications. The flexibility of protective polymers is a significant property because it permits a minimal number of surface-grafted polymer molecules to form an impermeable barrier over the liposome surface. Long-circulating liposomes demonstrated dose-independent, log-linear kinetics, non-saturable, and higher bioavailability. To promote prolonged liposome circulation in vivo, a variety of techniques have been proposed, including covering the surface of liposomes coated with inert, biocompatible polymers such as polyethylene glycol (PEG) (Torchilin 2005). Generally, PEG is used as a polymer and the process is called PEGylation. These modifications demonstrated minimized dosing frequency decreased toxicity without reducing effectiveness (Veronese and Harris 2002). Long-chain PEG can have hydroxyl terminals that are connected to lipid anchors like DSPE (1,2-distearoyl-sn-glycero-3-phosphoethanolamine) and DOPE (1,2-dioleoyl-sn-glycero-3-phosphoethanolamine) and long chain PEG can range in length from 1 to 10 KD. It is possible to integrate these PEG chains into the liposomes either before or after they have been created (Yang et al. 2007). PEGylation not only helps to sterically stabilise the liposomes, but it also presents the opportunity for the attachment of a wide variety of targeting ligands, cell-penetrating peptides, stimuli-sensitive linkers, and other similar molecules (Aryasomayajula et al. 2017). Research on PEG liposomes is now focused on detaching PEG from liposomes in order to increase the likelihood that cells will take them up. After the accumulation of PEG-liposomes at the intended site, the enhanced permeability and retention (EPR) effect will come into play (Maeda et al. 2001), Under the influence of the regional pathological conditions, the PEG coating is stripped away (Zalipsky et al. 1999). Despite this, while PEG is still considered the gold standard for the steric protection of liposomes, research is ongoing to explore other polymers that could be useful in the development of long-circulating liposomes. Long-circulating liposomes are still being studied for their potential application in cancer chemotherapy, but they might also be used to deliver imaging agents and treating infections (Torchilin 2005) (Fig. 6).

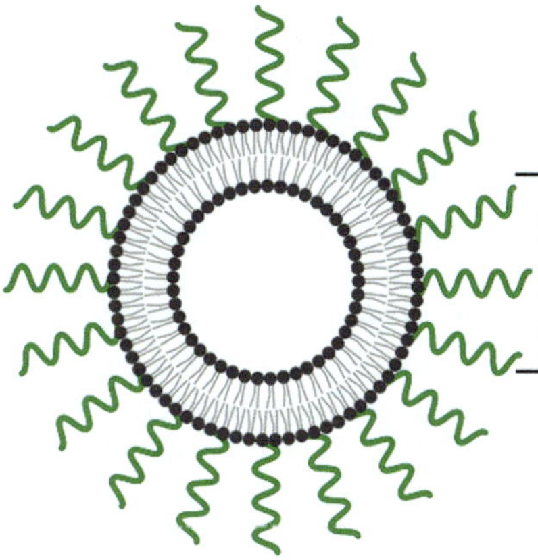

Fig. 6 Schematic representation a long-circulating liposome showing molecular arrangement of phopholipids and PEG

8 Strategies for Targeting Liposomes

The targeting strategies of the liposome formulation are attractive novel strategies in research and development. The most required functional feature of liposomes is the ability to target specific disease site. Therefore, focusing on specific areas prioritises the progression of the novel diagnostic instruments as well as the improvement pertaining to treatment efficacy. The tumour vasculature is the primary mediator of passive tissue targeting while on the surface of the liposomes, receptor-specific ligands that were intended for cell attachment are responsible for active tissue targeting (Lehner et al. 2013).

(a) Passive targeting of cancer

To date, oncology has relied primarily on passive targeting strategies due to the pathophysiological features of malignancies and their environments (Wicki et al. 2015). Liposomes are passively targeted to tissues or cells by molecularly driving them into the tumor interstitium through leaky tumor vasculature (Gogoi et al. 2016). Due to the increased permeability and retention (EPR) impact of the vasculature, non-targeted liposomes with diameters between 10 and 500 nm may concentrate preferentially on cancer and inflammatory tissues as a consequence of blood vessels that are unusually permeable to fluid and the absence of functioning lymphatics (Biswas and Torchilin 2014). For passive targeting to be effective, the liposomal formulation must be resistant to natural clearance pathways, include processes such as phagocytic uptake and the elimination of harmful substances by cells of the mononuclear phagocytic system (MPS) (Kraft et al. 2014). The modification of the surface of liposomes with PEG increases the circulation time, passive targeting techniques (Zylberberg and Matosevic 2016). This method also makes use of liposome characteristics, such charge, which can lead to more precise targeting of cancer cells. Cationic liposomes can serve as another illustration. Using electrostatic interactions, it was found that this liposomal form binds the phospholipid head groups (negatively charged) that are uniquely expressed on tumour endothelium cells (Byrne et al. 2008). As a result, alternative targeting strategies with cutting-edge functionality, including active targeting, have been explored and developed.

(b) Active targeting of cancer

In 1906, a visionary German scientist established the concept of active targeting by proposing a "magic bullet" to guide the delivery of specific medications to specific locations in the body (Lehner et al. 2013; Strebhardt and Ullrich 2008). Active targeting refers to the process by which a targeting ligand is attached to the surface of liposomes to improve the delivery efficiency of liposomal systems (Riaz et al. 2018). Active targeting has been used with a wide variety of ligands, including antibodies, nucleic acids (aptamers), peptides, entire proteins (transferrin), and even very small molecules such as vitamins (folic acid) (Wicki et al. 2015). Target ligand selection is based on a number of criteria, including the extent to which the target is over- or under-expressed, the efficiency with which the target cell absorbs the ligand-targeted formulation, and the level to which the target molecule is covered (Noble et al. 2014; Sawant and Torchilin 2012). There are primarily two approaches to functionalize liposomes. The first is to bind the desired targeted ligand to a lipid, then combine the ligand-containing lipid with the liposomes. Another strategy involves immediate functionalization of produced liposomes with the appropriate targeting ligand (Marqués-Gallego and Kroon 2014). The head group modified lipids combined with a PEG spacer that is functionalized at the end with amine, thiol, carboxylic acid, or maleimide groups are all feasible possibilities in this strategy (Conde et al. 2014) (Fig. 7; Table 1).

9 Stimuli Sensitive Drug Delivery by Liposomes

With the evolution of tumor cells and tissues, the nature of cells and its surrounding area also gets changed which can act as stimuli for a multifunctional targeting liposomes. These stimuli can be in the form of either pH change or temperature variations or overexpressed enzyme secretions such as matrix metallo-protease (MMPs) or can be any overexpressed receptors extended at the cell surface such as folate receptors, transferrin receptors, TNF receptors, interleukin receptors etc. Multifunctional liposomes were designed for a site specific and targeted delivery (Ishida et al. 2001; Trudel et al. 2003; Schafer and Buettner 2001; Gerweck and Seetharaman 1996; Kato et al. 2013). The release is triggered by 2 ways. (a) An intrinsic factor that trigger the drug release; (b) Extrinsic factor applied to the target tissues after the arrival of the liposomes (Needham et al. 2013). The stimuli-sensitive liposomes can be categorized in to (a) Internal stimuli sensitive liposomes and (b) External stimuli sensitive liposomes.

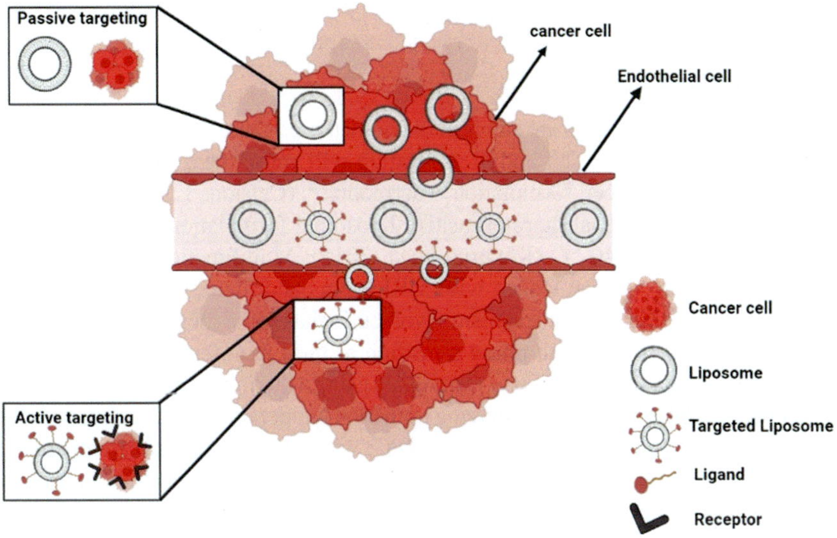

Fig. 7 Schematic depiction of passive and active targeting of liposomes into a tumour to enhance the therapeutic efficacy

Table 1 Functionalized liposomes with various targeting ligands for cancer therapy

Targeting ligand	Anticancer agent	Drug loading	Tumor treated	References
Folate	Doxorubicin	Ammonium sulphate gradient modified	Cervical cancer (HeLa cells) and Cervical adeno carcinoma (KBcells)	Pradhan et al. (2010)
Anti-EphA 10 antibody	MDRI-siRNA	Active loading	Multi-drug resistant breast tumor (MCF7/ ADR cells)	Zang et al. (2016)
RGD-peptide	Docetaxel	Passive loading	Cancer of the breast (MCF-7 cells)	Zuo et al. (2016)
Rhodamine-123	Paclitaxel	Passive loading	Cervical carcinoma	Biswas et al. (2011)
Anti EGFR antibody	Small interfering RNA (siRNA)	Active loading	Lung carcinoma	Lee et al. (2015)
AS1411aptamer	Gd-DTPA	Passive loading	Breast carcinoma	Zhang et al. (2015)
HER-2 antibody	Doxorubicin and hollow-gold nanospheres (HAuNS)	Ammonium sulfate Gradient	Cancer of the ovary (SKOV3 cells), cancer of the breast (BT474 cells)	Sudimack and Lee (2000)

(a) Internal stimuli sensitive liposomes

i. Thermo responsive liposomes

Thermosensitive liposomes are stable at the body temperature (37 °C), but a temperature that is greater than the body temperature will cause them to discharge their contents. The anticancer formulation Thermodox® (Celsion, Lawrenceville, NJ, USA) is a multifunctional thermosensitive liposomal formulation having the transition temperature at 40 °C (Kokuryo et al. 2015). Multifunctional thermosensitive liposomes can also be loaded with MRI agents and fluorescent dyes (Mabrey and Sturtevant 1976). The mechanism involved is that some thermosensitive lipids undergo gel-liquid crystalline phase transition at mild high temperatures than the body temperature. A liposomes was formulated using a thermosensitive lipid dipalmitoyl phosphatidylcholine (DPPC) having transition temperature 41 °C, along with the help of distearoyl phosphatidylcholine (DSPC) with a higher transition temperature of 54 °C in the molar ratio of DPPC:DSPC 3:1, to yield thermosensitive liposomes with transition temperature around 41 °C (Yatvin et al. 1978; Partanen et al. 2012).

A chronic method to disrupt a solid tumor using heat is magnetic resonance guided high intensity focused ultrasound (MR-HIFU) and therefore a slight increase in temperature is very useful for triggering the action of thermosensitive liposomes. A combination of low temperature sensitive liposomes (LTSLs) along with MR-HIFU was used over rabbits bearing VX2 tumors. When the temperature was maintained at 40.4 and 41.3 °C using ultrasound exposure in VX2 tumors there was a rapid drug release upon a controlled heating of tumor (Sullivan and Huang 1985). Antibody was also used as a targeting ligand for thermosensitive liposomes delivery. Sullivan and Huang in 1985 made a heat-sensitive liposomes by covalently bonding of anti-H2Kk antibody to a palmitic acid derivative with DPPC carrying carboxyfluorescein (Sorkin and Zastrow 2002).

ii. pH sensitive liposomes

Various reactions occurring in the tumor cells bring about the change in pH range mainly reducing the pH level. During anaerobic glycolysis, there is a lactate secretion which lowers the pH of tumor cells upto 4. This lower pH value is the identification marker for the site-specific delivery of liposomes by changing the bilayer membrane shape of liposomes (Needham et al. 2013; Soares et al. 2011). As the pH varies it provokes the permeability of the liposomal membrane by protonation/deprotonation of functional groups which characterizes the morphological changes of the lipid bilayers. For example, a natural phospholipid, 1,2-dioleoyl-sn-glycero-3-phosphoethanolamine (DOPE), acquires an inverted hexagonal phase II (HII phase) at low pH and a bilayer structure (Lα phase) at neutral pH to promote the membrane destabilization (Song et al. 2014). A multifunctional liposome was formulated which shows passive targeting by EPR effect, active mediated targeting by CD-44 receptors, which in return activates the drug release by lowering the pH in endosomal environment. This may increase the antitumor efficacy and inhibits systemic toxicity. Further, benzoporphyrin photosensitizer–encapsulated PEGylated liposomes functionalized with specific angiogenic endothelial cell peptide inhibits tumor growth as compared to conventional and nontargeted liposomes (Fouladi et al. 2017).

iii. Enzyme-sensitive liposomes

Enzyme plays a significant part in physiological as well as pathological conditions of the body. The concentration of intracellular or extracellular enzymes is always elevated by these pathological conditions. These enzyme stimuli nanoformulation go through structural changes to release at site specific delivery (Hu et al. 2012, 2014). The mechanism of action that work behind the enzyme-sensitive liposomes is that as the concentration of the enzyme increases in the pathological tissues it triggers to the sensitizing agent which in return increases the drug release at the targeted tissue (Hu et al. 2012, 2014; Quach et al. 2014a). The enzyme that triggers the liposomal action includes the extracellular enzymes phospholipase A2, matrix metalloproteinases, urokinase plasminogen activator, elastase, and prostate-specific antigen, and the intracellular enzyme cathepsin B (Hu et al. 2012, 2014; Quach et al. 2014a). Phospholipase A2 is mainly secreted in almost every type of cancer tissue, inflammatory disease, cardiovascular disease and immune disorders (Zuo et al. 2017) hence it can be used as the most fascinating trigger for enzyme responsive liposomes. The mechanism of the functioning of phospholipase sensitive liposomes involves the disruption of the integrity of lipid bilayer by enzyme-mediated phospholipids hydrolysis and dissociation of cleavable phospholipase A2-sensitive bond (Hu et al. 2012; Quach et al. 2014b; Yadav et al. 2014). MMPs work by degrading extracellular matrix through proteolysis. Pancreatic, colorectal, breast, and lung cancers are the primary sites where MMP-2 and MMP-9 are overexpressed (Zhu et al. 2012). MMPs liposomes are formulated using 2 strategies. Firstly, MMPs sensitive peptides are prepared attached to PEG. Then at the targeted area, peptides are disrupted for releasing PEG (Hu et al. 2012). An example of such MMP-responsive liposomes was reported by Zhu et al. Containing two different lipopolymers: mAb 2C5-PEG (3400)-MMP2 cleavable peptide-1,2-dioleoyl-sn-glycero-3-phosphoethanolamine (DOPE) and TATp-PEG(2000)-1,2-dioctadecanoyl-sn-glycero-3-phosphoethanolamine (DSPE). The former lipopolymer, having a longer chain of PEG, act as a protective shield over the latter lipopolymer having the TAT peptide. TAT provokes the endocytosis-mediated uptake of the liposomes. The targeting moiety for cancer cells is mAb 2C5. Breakage of the peptide (Gly-Pro-Leu-Gly-Ile-Ala-Gly-Gln) at the targeted site releases the long-chain PEG and diminishes its shielding effect. The hydrolysis causes the TAT peptide to be exposed and increases the absorption of liposomes (Romberg et al. 2008). Another lysosomal protease enzyme, cathepsin B is expressed in different cancer tissues of lung, brain and colon areas (Hu et al. 2012; Cavallo-Medved et al. 2011; Song et al. 2016). PEG was recently combined to a lipid with the use of an enzymatically-cleavable linker (glycine-phenylalanine-leucine-glycine). This can then result in the destruction of the peptide linkage, which can be triggered by the use of the enzyme cathepsin B. This leads to instability, endosomal rupture, and ultimately results in the the release of plasmid DNA into the cytoplasm (McCarley 2012).

iv. Redox responsive liposomes

The electron transfer reactions are the main phenomena to trigger the drug delivery using redoxsensitive liposomes. In order to ease phase transitions in the lipid system, this can be changed by modifying the amphiphile's charge or hydrophilicity, using chemical reducers, or removing cross-links (Aoyama and Nakaki 2015). A cysteine-containing tripeptide i.e. glutathione (GSH) is a potent intracellular reducing agent which functions in cellular growth and maintains cellular redox homeostasis (Marengo et al. 2016; Jhaveri et al. 2014). The increasing concentrations of GSH upto 100–1000 folds in the prominent intracellular areas such as cytosol, mitochondria and cell nucleus gives an honest candidate for tumor specific redox-sensitive liposomes because of the property of cleavable or reversible disulphide bond present in most of the reducing agent (Fleige et al. 2012; Sun et al. 2013; Yin et al. 2017). Another property to be stated for the good candidate to be modified as redox-sensitive liposomes includes the hydrophilic backbone, a lower rate of polymerization, a high-water solubility, a cationic nature, anti-angiogenesis and radical scavenging effectiveness, and cell adhesion qualities are all characteristics of this substance (Chen et al. 2017). An example for this includes a soyabean phosphatidylcholine, cholesterol and a redox-sensitizing cationic oligopeptide lipid nanosystem having a protonation system. The vesicles are modified by the disassembled within the presence of 10 mM GSH by monitoring size, zeta potential and morphological changes beneficiary in improving cellular uptake, lower surviving expression, efficient endolysosomal exchange, higher cellular degradation and overall synergistic in-vivo inhibitory effect on tumor growth and hence reduced pulmonary metatstasis of breast cancer (Chi et al. 2017). A brand-new detachable PEG linked to hyaluronic acid and cholesterol via a disulfide linkage, a ligand for CD44 were used for the preparation of cationic redox-sensitive liposomes which make them destabilized in reducing conditions and thus makes an excellent candidate developing an intracellular delivery method that is CD44-mediated in order to treat osteosarcoma in animal models (Fu et al. 2015).

The polyethylene (PE) polar headgroups, when paired with unsaturated acyl (dioleoyl) chains in the lipid, make it easier for cargo to be delivered by fusion processes, which in turn accelerates the transition from lamellar to hexagonal phase. A liposome formulation (DOPC/DOPE/SS 14;16.7:33.3:50 molar ratio) was prepared to give the maximal transfection efficiency (2.7-fold greater than that of the control, Lipofectamine 2000, against U87-MG tumour cell lines). This was accomplished by preparing a liposome formulation with a DOPC/DOPE/SS14; 16.7:33.3:50 molar ratio (Huang 2010). The combination of application potential of polyethylene glycol (PEG) for increasing blood circulation time and cell-penetrating peptides (CPPs) for enhancing cellular uptake was used in some studies. In 2015, Fu et al. formulated and developed TAT functionalized paclitaxel loaded liposomes with cleavable PEG using redox-stimuli disulfide linker. These liposomes are stable under physiological conditions such as increased blood circulation and provide a shielding effect to PEG. After reaching the location of the tumour, the cleavable bond was fractionized with the assistance of glutathione, an exogenous reducing agent. This finally resulted in the PEG shield being broken, which allowed TAT to become exposed and encouraged the cellular uptake of PTX-loaded liposomes. This can lead to an expanded tumour location both in vitro and in vivo (B16F1-bearing mice), which can result in a 69.4% increase in the rate of tumour blockage. In addition to this, these liposomes do not exhibit any signs of having a harmful effect on the important organs (Huang 2010).

(b) External stimuli sensitive liposomes

(i) **Ultrasound responsive liposomes**

Ultrasound imaging is made possible by liposomes that contain very small gas bubbles, when subjected to ultrasound radiation, create an echo sound. Additionally, by stimulating liposome structures with ultrasonic waves, It is possible for the medicine to be distributed to the location of the disease site (Han et al. 2015). Doxorubicin liposomes with a thermosensitive system that produces CO_2 bubbles significantly slowed the growth of breast tumors in MDA-MB-231 tumor-bearing mice as compared to plain thermosensitive doxorubicin liposomes that do not really produce any gas. Thermo-responsive liposomes that can rupture lipid bilayers by hydrating the dried lipid film with citrate buffer (300 mM, pH 4) have the ability to produce CO_2 bubbles. Because of a synergism between the burst drug release and the hyperthermia-induced generation of carbon dioxide, the anticancer activity of doxorubicin was significantly increased. For the purpose of monitoring the hyperthermia-induced generation of carbon dioxide, an ultrasonic imaging instrument was utilized (Marie et al. 2015).

(ii) **Magnetic-sensitive liposomes**

When the drug is fabricated inside the developed liposomes having the magnetic properties then the liposomes is said to be magneto-liposomes (Yoshida et al. 2012; Bolfarini et al. 2012; Qiu and An 2013; Guo et al. 2015; Hardiansyah et al. 2014) or magnetic stimuli liposomes (Yoshida et al. 2012; Nahar et al. 2014; Corato et al. 2015; Cuyper and Joniau 1988). Magneto-liposomes were first introduced by De Cuyper and Joniau in 1988 (Hervault and Thanh 2014). The strategies that are used in the formulation of magneto-liposomes are the preparation of hydrophilic nanoparticles that can be incorporated inside the aqueous core of liposomes, the lipid bilayer that is present on the liposomes that encapsulates hydrophobic nanoparticles and surface targeting with the magnetic nanoparticles (Langereis et al. 2013; Babincová et al. 2002). They work under a specific external guidance, targeted to the specific area of tumor cells which helps in drug release (Hardiansyah et al. 2014; Nahar et al. 2014; Tai et al. 2009; Kulshrestha et al. 2012) and inhibiting the growth of cancer cells (Bolfarini et al. 2012; Hardiansyah et al. 2014; Nahar et al. 2014; Peng et al. 2014; Kuo et al. 2016). High-frequency magnetic field (HFMF) is the system that is helpful in assisting the magnetic field that enhances the interaction between magnetic liposomes and HFMF (Nahar et al. 2014; Kuo et al. 2014; Hu et al. 2007; Liu et al. 2008; Park et al. 2000). Common metal and alloys that are used for magnetic nanoparticles are iron (Puntes et al. 1979), cobalt (Akbarzadeh et al. 2012), iron oxides such as maghemite (Fe_2O_3) and magnetite (Fe_3O_4). These metals are then coated with biocompatible polymers such as dextran, PEG that aid in the guidance to the target site of action under the influence of external magnetic field.

Magnetic stimuli-based liposomes having iron oxide (Fe_3O_4) is commonly used candidate for site specific stimuli-based targeting system due to its biocompatible nature and their physical properties (Peng et al. 2014; Nguyen et al. 2017). Magnetic iron oxide nanoparticles (MNPs), are also the good candidate for the imaging purpose in magnetic resonance imaging diagnosis. These iron oxides have their name as super-paramagnetic iron-oxide nanoparticles (SPIONS) due to their super-paramagnetic nature. The preparation of liposomes is a two step process i.e. first is to synthesize magnetic nanoparticles and secondly, encapsulation of magnetic nanoparticles into the liposomes by drug loading method.

Magnetic liposomes having attached ligands such as hyaluronic acid (Park et al. 2014; Bothun et al. 2011), folate (Xu et al. 2013), anti-αvβ3 antibody (Soenen et al. 2011), sugar moieties (Lin et al. 2016), and cell-penetrating peptides (CPPs) (Mattheolabakis et al. 2015) is a boon in the area of targeted drug delivery. Hyaluronic acid, mainly targets CD44 which is an overexpressed target in the areas of the colon, pancreas, breast, and ovarian (Visscher et al. 2011). In recent research, docetaxel that was incorporated inside the surface functionalized magneto-liposomes with hyaluronic acid, and citric acid coated magnetic nanoparticles were encapsulated inside the aqueous core, showed an accelerated drug release under near-IR laser radiation and superior cellular uptake in comparison to conventional docetaxel non-targeted liposomes (Park et al. 2014).

iii. **Light-responsive liposomes**

Light-sensitive liposomes works on the principle that when a certain wavelength of light is exposed to tissues and cells than the photosensitizing therapeutics agents present in the liposomes make them penetrate the tissues to inhibit the tumor growth. The wavelength that is most preferable for action lies in the region of near-infrared (NIR) regions (~700–1100 nm) because it can penetrate upto 1 cm deep inside the body. These candidates are the approved candidates for photodynamic therapy (PDT). PDT acts by light-mediated excitation which generates reactive oxygen species (ROS) to destruct and inhibit the unwanted cellular growth. The light responsive liposomal formulations have the challenges to meet vesicle retention stability or the entrapment of the therapeutic substance at the spot that is being addressed or the activation of liposomes in the penetrating tissue by the specific wavelength of light. The recently used example to work clinically under light sensitive liposomes is Visudyne with the photosensitizer element, verteporfin. The lipophilic chemical meta-tetra(hydroxyphenyl)chlorin is considered to be one of the most powerful photosensitizers now in therapeutic usage (mTHPC or Temo-porfin). An authorised formulation for the palliative treatment of advanced squamous cell carcinoma (SCC) of the head and neck is one that combines ethanol and propylene glycol with mTHPC. This formulation is known as Foscan®. In addition, there are two other liposomal formulations: Foslip®, which is made up of liposomes based on DPPC, and Fospeg®, which is a PEGylated liposome (Eloy et al. 2014; Morgan et al. 1992).

A light sensitive dye sulphonated aluminiumphthalocyanine (AlSPc) when formulated into liposomes and surface decorated with polyclonal sheep anti-mouse-Ig antibody, gets activated to induce singlet oxygen species when exposed to red light. These liposomes resulted to lysis of DW-BCL cells (Jain and Jain 2018). PSP (photo-sensitive peptides)/NGR (asparagine-glycine-arginine)-L, a dual modified liposomes that are formulated using emulsification method, having the concentrations of DSPE-PEG2000-psCPP and DSPE-PEG5000-NGR as 4% and 1% (molar ratio), respectively. These liposomes resulted in potent antitumor activity in the HT-1080 tumor model by the exposure of NIR illumination. This liposome showed effective in endosomal degradation and enhanced cell apoptosis when exposed to UV light (Zhao et al. 2016) (Fig. 8; Table 2).

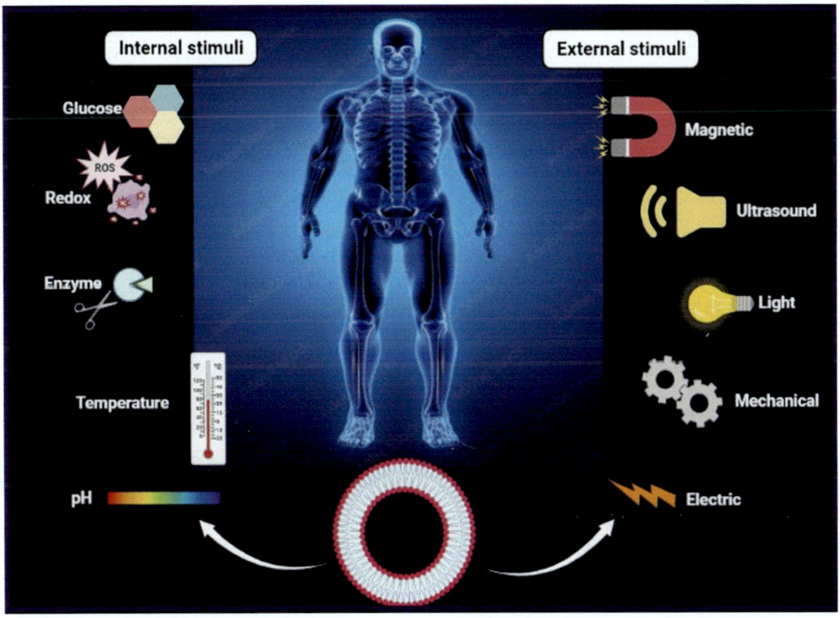

Fig. 8 An illustration of internal and external stimuli that triggers the liposomes

Table 2 Different stimuli sensitive liposomes in cancer drug delivery

Stimulus	Targeted cancer	Drug	Functionality	References
pH	Glioblastomas tumors	Doxorubicin	pH-sensitive loaded DOX liposomes activate at low pH thus targeting the tumor cells	Gu et al. (2023)
	Colon cancers	Docetaxel and pemetrexed	The combination of DTX and PMX improves the pharmacokinetics profile	Park et al. (2021)
	Lung cancer	Doxycycline and docetaxel	Exervating the synergistic effect with the reduction in tumor growth cells	Khan et al. (2022)
	Lung cancer	Thymoquinone	Enhances the drug solubility and aids in pulmonary drug delivery system	Zhang et al. (2021)
Thermo	Breast cancer	Resiquimod	Higher payload of drug concentrations	Xi et al. (2020)
	Lung cancer	Paclitaxel	Releases the drug at 40 °C	Yang et al. (2020)
	Gliomas cells	Lomustine	90% of drug release at 41 °C and therefore bioavailibility will be increased	Feng et al. (2019)
Redox	Orthotopic osteosarcoma	Doxorubicin	Bone and CD44-receptor dual targeted liposomes for the tumor suppression and increased survival rate	Chen et al. (2017)
	Breast cancer	Paclitaxel and survivin-siRNA	Exhibiting endosomal escape and the co-loaded drug delivery provides the synergistic cancer therapy	Dwivedi et al. (2020)
Magnetic	Pancreatic cancer	Doxorubicin	The release of drug by the stimulation of ultrasound magnetic parameters, requires low dose for inhibiting cancer cell growth by inducing cell proliferation	Lin et al. (2020)
	Liver cancer	Paclitaxel	Target overexpressed EGFR receptors with enhanced cytotoxicity against tumor cells	Maeda et al. (2000)

10 Imaging of Cancer Using Liposomes

In cancer studies, several alterations occur within or outside the tissues such as EPR and angiogenesis, which can be useful for targeted delivery system. As a result, there is an increase in the concentration of certain chemical agents such as prostaglandin, bradykinin, nitric oxide/peroxynitrite, vascular endothelial growth factor (VEGF), tumour necrosis factor, and other aspects of pathophysiology (Lukyanov and Torchilin 2004; Park 2002). An aid to these surface modified liposomes specially with PEG can be used to stay for a longer duration of time. Because of this, the only method to achieve targeted molecular imaging of medications is by the active targeting of molecules to the intended site, such as to the desired particular cell, sub-cell, molecule, receptor, protein, peptide, hormone, or enzyme linked with the sickness (Torchilin 2006).

This targeting strategy is favoured by conjugation of target specific ligand on liposomal surface, either through MRI, or through ultrasonication, etc. (Elbayoumi and Torchilin 2009; Dandamudi et al. 2009; Hu et al. 2009; Ren et al. 2008; Hamoudeh et al. 2008). Liposomes that can be used for nuclear imaging is formulated using conjugation of radionucleolus of 9 mTc, 67 Ga and 111 In (Patel 1990). The radionuclides used can be depending on the physicochemical property which can be changed through imaging functions. For instance, iodine is utilised for CT imaging, paramagnetic contrast agents such as gadolinium and superparamagnetic iron oxide are used for MRI imaging, and positron-emitting radionuclides such as 18F, 15O, and 13N can be used for PET and PET/CT imaging (Zhang et al. 2019). Methylene blue a dye can be used in diagnosing breast tumors with near-infrared (NIR) fluorescence imaging (Kosaka et al. 2009). Through NIR wavelength (650–900 nm) tissue penetration can be done which provides less autofluorescence from neighboring tissues (Silindir-Gunay et al. 2019; Kusano et al. 2008). An FDA approved indocyanine green (ICG) is another agent that uses a clinical near-infrared imaging, having higher safety margin, fewer undesirable effects, and a high signal-to-noise ratio in live tissue. It is the most common agent to be used in gastric cancer sentinel lymph node (SLN) imaging (Tajima et al. 2010; Tahara et al. 2013).

i. Quantum dots loaded liposomes

Liposomes combined with quantum dots aids in enhanced biocompatibility and reduce toxic effects of the incorporated theranostic agents. The incorporation of the quantum dots leads to establishment of the clear in vivo bioimaging of the tissue effected with tumors and the efficacy of different anti-tumor agents (Muthu et al. 2012; Fang et al. 2012; Bruun and Hille 2019; Muthu and Singh 2009). Quantum dots have the ability to be bonded covalently to the liposomal surface, or they can be integrated into the hydrophobic core of the liposome (Muthu et al. 2009; Sigot et al. 2010). In the recent studies, quantum dots are also proved to be the favourable imaging agents in comparison to the organic dyes, because of the advantageous of emitting high intensity signals and also more photostable (Pan et al. 2010; Weissleder and Pittet 2008). It is possible to increase the ability to diagnose cancer using liposomes that have been coated with TPGS and conjugated with folic acid. These liposomes are prepared for the co-delivery of quantum dots and docetaxel. Quantum dots used in this formulation are the model imaging agent for targeting folic acid conjugates (Fang et al. 2012).

ii. Fluorescent dye loaded liposomes

The imaging method that is utilised frequently is fluorescence imaging. This method can detect the position of biomolecules, gene expression, and enzyme activity in the tissues so that their effects may be studied (Al-Jamal et al. 2008). The integration of PEG-coated quantum dots into the hydrophilic core of DOPC liposomes was the first step in the development of a functionalized quantum dot liposome, which is abbreviated as f-QD-L. In comparison to unmodified QDs, the fluorescence imaging at the tumour cells demonstrates that these liposomes are significantly more effective. Moreover, the cellular uptake and tumour penetration capabilities of f-QD-L may be regulated by the surface charge density by utilising various liposomal combinations (Min et al. 2014).

Fluorescence imaging using near-infrared (NIR) light is also effective due to the low absorption qualities of NIR light and the scattering of photons that occur within the range of NIR light (Hilderbrand and Weissleder 2010). It also allows deeper penetration which provides clearer tissue imaging in compare to visible wavelengths (Eliseeva and Bünzli 2010). Er^{3+}-doped Y_2O_3 (Y_2O_3:Er^{3+}), which shows NIR fluorescence at 1550 nm under 980 nm illumination, is a possible candidate for NIR fluorescence imaging. After intravenous treatment, a superior candidate for fluorescence imaging in several organs was obtained when positively charged Y_2O_3:Er^{3+} nanoparticles were injected into negatively charged DPPG liposomes with surface decorating (Bünzli 2010; Soga et al. 2010; Ferguson et al. 2015).

iii. Iron oxide nanoparticles loaded liposomes

The superparamagnetic response of magnetic nanoparticles is essential to the operation of a recently developed imaging method known as magnetic particle imaging (MPI), produces higher sensitive imaging using both quantitative and tomographic methods at the point of distribution (Zhou et al. 2018; Gleich and Weizenecker 2005). MPI having the various applications in cancer imaging, stem cells distribution and clearance as well as the monitoring of the biodistribution of superparamagnetic iron oxide nanoparticles (Cuyper and Joniau 1988; Yu et al. 2017; Zheng et al. 2016; Keselman et al. 2017). Liposomes loaded with magnetic oxides known as magnetoliposomes, can be a better candidate for targeting theranostic agents along with imaging properties of iron oxides (Monnier et al. 2014; Maruyama et al. 2016).

A thermosensitive magneto-liposomes formulated using calcein or doxorubicin, to evaluate a potential MPI tracer. It shows the use of MPI to diagnose the magneto-liposomes and monitors the drug release. But these liposomes operate at a frequency of 400 Hz and drive field amplitude of 20 mT, while the commercially available works at the frequency of 25 kHz for the Bruker system and 45 kHz for the magnetic insight momentum imager at a lower amplitude of 16mT, which responds at the external magnetic fields (Murase et al. 2014; Bulte 2019; Zorko and Langel 2005) (Table 3).

Table 3 Cancer imaging by multifunctional liposomes

Type	Size (nm)	Active substance	Area	Targeting	Labeling	Imaging
MLV	100	–	Tumor imaging	Passive targeting	^{18}F-FDC	PET
MLV (pH sensitive liposomes)	124	–	Inflammation imaging	Passive targeting	^{99}Tc-HMPAO	SPECT, γ-Scintigraphy
SUV	70–200	Iodine	Tumor and inflammation imaging	Passive targeting	Iodine	CT
SUV	120	Doxorubicin	Kaposi sarcoma imaging and therapy	Active targeting with NCAM	Gd	MRI
MLV	150	–	Tumor imaging	Active targeting with mAb2C5	Gd	MRI
MLV	100–150	–	Tumor imaging	Active targeting with mAb2C5	^{111}In	γ-Scintigraphy
MLV	100–150	Iopromide	Tumor imaging	Active targeting with mAb2C5	^{99m}TC	SPECT/CT

NCAM-Neural cell adhesion molecule, SPECT-Single photon emission computerized tomography, ^{18}F-FDC-[^{18}F]Fluorodipalmitin, AT-Active targeting, PT-Passive targeting, Tc-HMPAO-Hexamethylpropylenamine-oxime, PET-Photon emission tomography, MRI-Magnetic resonance imaging

11 Cell Penetrating Liposomes

Liposomes are commonly utilised as drug delivery vehicles, and numerous liposome-based nanomedicines have received clinical approval. The plasma membrane of the cell is widely recognised to be a strong barrier to the passage of a variety of non-essential chemicals and medicines. Cell penetrating peptides (CPPs) are increasingly being used in order to boost the penetration of medicines and nanoformualtions across the cell membrane. To boost intracellular delivery effectiveness, cell penetrating peptides (CPPs) have been painted onto nanocarriers such as liposomes. CPPs are usually very short, their amino acid sequences are less than 40, amphipathic, and mostly positively charged (Frankel and Pabo 1988). The transactivator of transcription of human immunodeficiency virus (TAT peptide), which was isolated from the HIV virus in the 1980s, was the first CPP to be recognised (Bechara and Sagan 2013). Since then, several CPPs have been found and extensively employed for transporting medicinal molecules, peptides, proteins, oligonucleotides, nanoformulations, and other compounds across the cell membrane, including octa-arginine (R8), transportan, and Pep-1 (Gandhi et al. 2014).

12 Multifunctional Liposomes

Liposomes are effective lipid nanoparticle drug delivery systems that can be functionalized with noninvasive multimodality imaging agents. Each modality provides unique information and has synergistic advantages in therapeutic and diagnostic agents with different physicochemical properties. Hence, to make it easier many functional groups inside the confines of ligands to enhance the drug coming out of and act smartly inside as well as outside the liposomal conditions. These different attachments of functional groupings is possible to different areas of conventional liposomes on the basis of their chemical property, therapeutic activity of drugs, genes, molecular probes etc. either by mixing or adsorption or by covalent linking (Ying et al. 2010). A double targeting daunorubicin liposomes was formulated using p-amino-phenyl-a-D-mannopyranoside and transferrin, which boosts the activity of glucose transporters as well as transferrin receptors and hence enhance the transportation across blood brain barrier (Zhu et al. 2012; Vivo et al. 1991; Jefferies et al. 1984). The major properties of multifunctional liposomes are PEGylation of liposomes for passive targeting of the tumour using the EPR effect, active targeting of tumor, enhanced cellular internalization through TATp-endocytosis, stimuli sensitivity, and imaging (Veronese and Pasut 2005; Qiu et al. 2015).

The multifunctional approach is a prominent goal towards the tumor targeting and chemotherapy for gliomas by providing a hindrance towards multidrug resistance. Paclitaxel and chloroquine loaded dual functionalised liposomes is an effective treatment for paclitaxel-resistant carcinoma (Yang et al. 2016). CPPs and siRNA are incorporated inside magnetic hyperthermia dual responsive liposomes (siRNA-CPPsThermo-Magnetic Liposomes) for better physical and chemical characteristics, enhanced cellular uptake, functionalized endosomal escape and a promising gene silencing efficiency in MCF-7 cells in vitro. Further, during in-vivo studies siRNA-CPPs thermo-magnetic liposomes in the presence of magnetic field gives a significant targeted drug delivery, anticancer activity as well as gene silencing efficiency in a

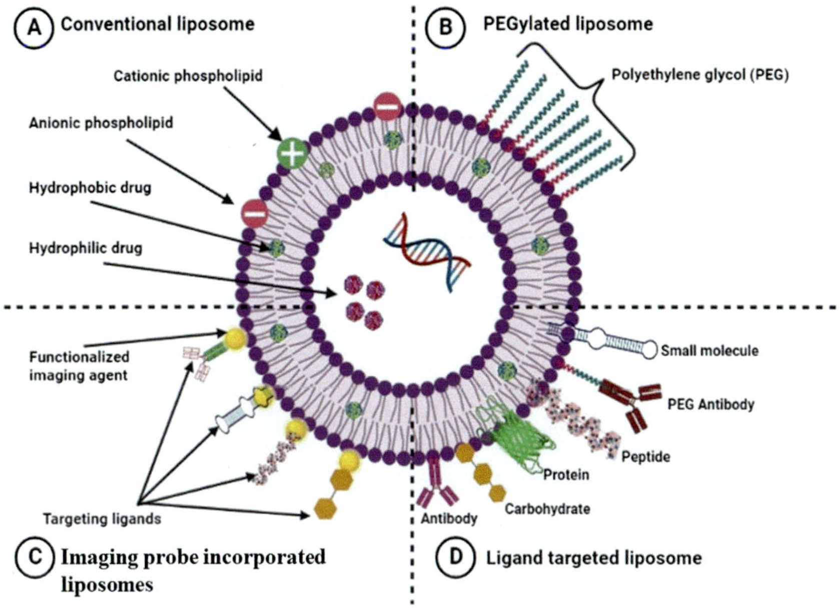

Fig. 9 Schematic illustration of multifuctional liposomes possessing multiple drug load, PEGylation, imaging probes and targeting ligands

MCF-7 xenograft murine model (Ta et al. 2014). A smart stimuli-sensitive multifunctional liposomes was developed which release DOX in high temperature condition that causes a significant decline in tumour growth that is significantly greater than that seen with traditional liposomes (Maurer et al. 2001). Figure 9 depicts a diagrammatic representation of multifuctional liposomes possessing multiple drug load, PEGylation, imaging probes and targeting ligands.

13 Clinical Trials Using Liposomal Formulations

Numerous new liposomal formulations have been developed for the treatment of wide variety of diseases, As a result of extensive clinical studies were conducted on lipid carriers. Figure 10 represents various liposomal formulations in different phases of clinical trials; the ongoing formulations of the current development phase and indication.

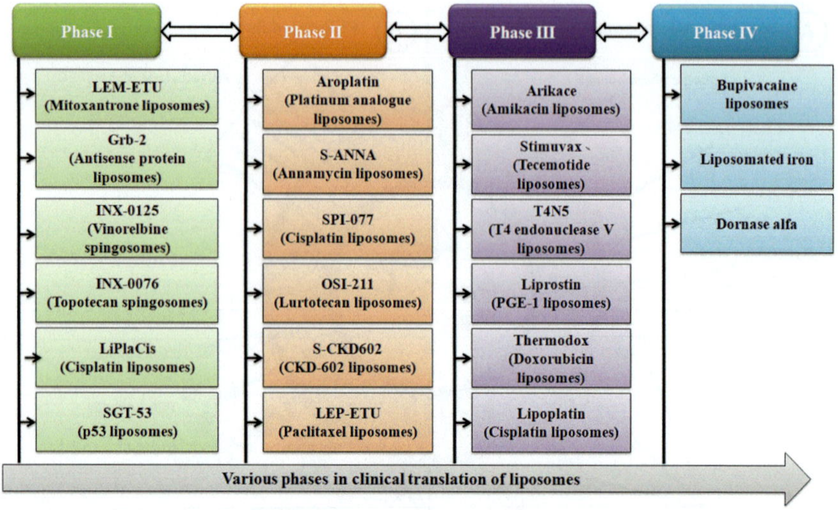

Fig. 10 Liposomal formulations in various stages of clinical trials

14 Applications of Liposomes in Healthcare

As a method for delivering a variety of medications, liposomes have shown encouraging outcomes. As a result, the thorough study of liposomes in medicine prompted researchers to create several liposomal formulations for the management and control of a variety of diseases in addition to a diverse assortment of therapeutic uses. Drugs are encapsulated inside liposomes, which improves their therapeutic impact owing to changes in both pharmacodynamics and pharmacokinetics (Bulbake et al. 2017). The key elements to design an effective liposomal formulation are the control of drug behaviour in vivo and the decrease of the adverse effects of the substance on the body. The therapy and identification of cancer are the primary uses of liposomes in clinical settings. The promise of liposomes for therapeutic uses is not restricted to the treatment of cancer. Liposomes are thought to be an incredibly versatile platform that may be used for different types of diseases. The current liposomal formulations on the market are discussed in this section (Fig. 11).

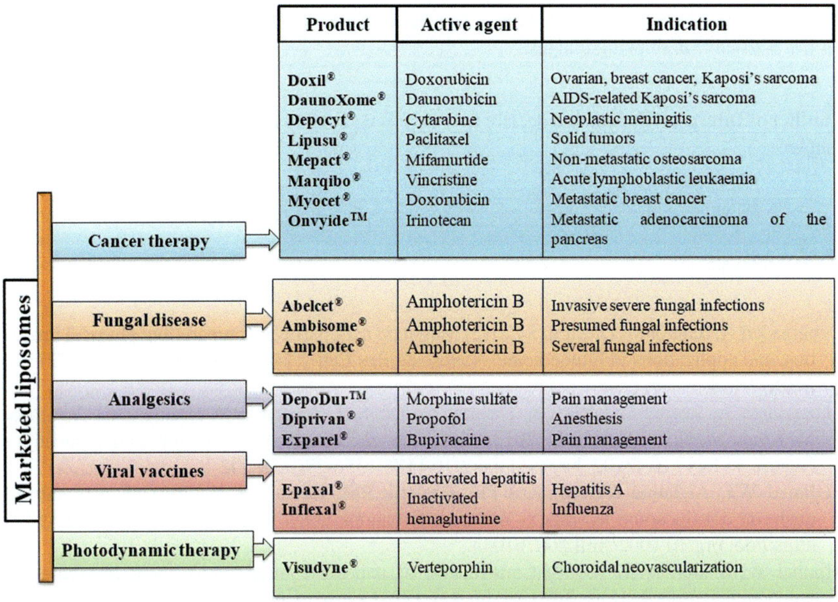

Fig. 11 Main therapeutic areas covered by liposomal formulations based marketed products

15 Conclusion

As a potential technique of transporting a wide range of medications into the body, liposomes have garnered a lot of interest. Because of the direct use of liposomes in medical practice, researchers are motivated to create novel liposomes for therapeutic applications, as well as, for the diagnosis purpose. This chapter demonstrated the structure, synthesis, multifunctionality, and stimulus responsiveness of liposomes using a variety of triggers, such as pH, temperature, light, enzyme, and redox potential, among others. It is feasible to assert that considerable work has to be done in the area of liposomal technology to get over the limits discussed in this analysis, despite the fact that there are certain liposomes certified for the market covering various health areas. Additionally, liposomes can support therapies with important performance, which will result in a more favourable result for therapeutic treatment, lower levels of toxicity and far fewer adverse consequences. Also, in terms of imaging technology it can be concluded in a broad aspect for diagnosis and theronaustic purpose. Other than this with different targeting ligands it provides a site specific delivery through which liposomes can be a preferable candidate for potent drugs.

Acknowledgements The principal investigator laboratory is supported by Science and Engineering Research Board (SERB) via Start-up Research Grant (SRG) [grant no. SRG/2021/002312] and the Empowerment and Equity Opportunities for Excellence in Science (EMEQ) grant [grant no. EEQ/2022/000218] during the scripting of the book chapter. The book chapter is related to the above research grants. The authors would like to acknowledge biorender (online imaging) service providers for the assistance in drawing images.

Conflict of Interest The authors declare that there is no conflict of interest in publishing the book chapter.

References

Akbarzadeh A, Samiei M, Davaran S (2012) Magnetic nanoparticles: preparation, physical properties, and applications in biomedicine. Nanoscale Res Lett 7:144. https://doi.org/10.1186/1556-276X-7-144

Akbarzadeh A, Rezaei-Sadabady R, Davaran S, Joo SW, Zarghami N, Hanifehpour Y, Samiei M, Kouhi M, Nejati-Koshki K (2013) Liposome: classification, preparation, and applications. Nanoscale Res Lett 8:102. https://doi.org/10.1186/1556-276X-8-102

Al-Jamal WT, Al-Jamal KT, Bomans PH, Frederik PM, Kostarelos K (2008) Functionalized-quantum-dot-liposome hybrids as multimodal nanoparticles for cancer. Small 4:1406–1415. https://doi.org/10.1002/smll.200701043

Aljamal W, Kostarelos K (2007) Construction of nanoscale multicompartment liposomes for combinatory drug delivery. Int J Pharm 331:182–185. https://doi.org/10.1016/j.ijpharm.2006.11.020

Antimisiaris SG, Kallinteri P, Fatouros DG (2008) Liposomes and drug delivery. In: Pharmaceutical manufacturing handbook. Wiley, Hoboken, NJ, pp 443–533

Aoyama K, Nakaki T (2015) Glutathione in cellular redox homeostasis: association with the excitatory amino acid carrier 1 (EAAC1). Molecules 20:8742–8758. https://doi.org/10.3390/molecules20058742

Aryasomayajula B, Salzano G, Torchilin VP (2017) Multifunctional liposomes, pp 41–61

Ashley JD, Quinlan CJ, Schroeder VA, Suckow MA, Pizzuti VJ, Kiziltepe T, Bilgicer B (2016) Dual carfilzomib and doxorubicin-loaded liposomal nanoparticles for synergistic efficacy in multiple myeloma. Mol Cancer Ther 15:1452–1459. https://doi.org/10.1158/1535-7163.MCT-15-0867

Babincová M, Čičmanec P, Altanerová V, Altaner Č, Babinec P (2002) AC-magnetic field controlled drug release from magnetoliposomes: design of a method for site-specific chemotherapy. Bioelectrochemistry 55:17–19. https://doi.org/10.1016/S1567-5394(01)00171-2

Bangham AD, Horne RW (1964) Negative staining of phospholipids and their structural modification by surface-active agents as observed in the electron microscope. J Mol Biol 8:660-IN10. https://doi.org/10.1016/S0022-2836(64)80115-7

Bangham AD, de Gier J, Greville GD (1967) Osmotic properties and water permeability of phospholipid liquid crystals. Chem Phys Lipids 1:225–246. https://doi.org/10.1016/0009-3084(67)90030-8

Batzri S, Korn ED (1973) Single bilayer liposomes prepared without sonication. Biochimica et Biophysica Acta (BBA) Biomembranes 298:1015–1019. https://doi.org/10.1016/0005-2736(73)90408-2

Bechara C, Sagan S (2013) Cell-penetrating peptides: 20 years later, where do we stand? FEBS Lett 587:1693–1702. https://doi.org/10.1016/j.febslet.2013.04.031

Biswas S, Torchilin VP (2014) Nanopreparations for organelle-specific delivery in cancer. Adv Drug Deliv Rev 66:26–41. https://doi.org/10.1016/j.addr.2013.11.004

Biswas S, Dodwadkar NS, Sawant RR, Koshkaryev A, Torchilin VP (2011) Surface modification of liposomes with rhodamine-123-conjugated polymer results in enhanced mitochondrial targeting. J Drug Target 19:552–561. https://doi.org/10.3109/1061186X.2010.536983

Bolfarini GC, Siqueira-Moura MP, Demets GJF, Morais PC, Tedesco AC (2012) In vitro evaluation of combined hyperthermia and photodynamic effects using magnetoliposomes loaded with cucurbituril zinc phthalocyanine complex on melanoma. J Photochem Photobiol B 115:1–4. https://doi.org/10.1016/j.jphotobiol.2012.05.009

Bothun GD, Lelis A, Chen Y, Scully K, Anderson LE, Stoner MA (2011) Multicomponent folate-targeted magnetoliposomes: design, characterization, and cellular uptake. Nanomedicine 7:797–805. https://doi.org/10.1016/j.nano.2011.02.007

Bozzuto G, Molinari A (2015) Liposomes as nanomedical devices. Int J Nanomed 975. https://doi.org/10.2147/IJN.S68861

Bragagni M, Maestrelli F, Mennini N, Ghelardini C, Mura P (2010) Liposomal formulations of prilocaine: effect of complexation with hydroxypropyl-ß-cyclodextrin on drug anesthetic efficacy. J Liposome Res 20:315–322. https://doi.org/10.3109/08982100903544169

Bruun K, Hille C (2019) Study on intracellular delivery of liposome encapsulated quantum dots using advanced fluorescence microscopy. Sci Rep 9:10504. https://doi.org/10.1038/s41598-019-46732-5

Bulbake U, Doppalapudi S, Kommineni N, Khan W (2017) Liposomal formulations in clinical use: an updated review. Pharmaceutics 9:12. https://doi.org/10.3390/pharmaceutics9020012

Bulte JWM (2019) Superparamagnetic iron oxides as MPI tracers: a primer and review of early applications. Adv Drug Deliv Rev 138:293–301. https://doi.org/10.1016/j.addr.2018.12.007

Bünzli J-CG (2010) Lanthanide luminescence for biomedical analyses and imaging. Chem Rev 110:2729–2755. https://doi.org/10.1021/cr900362e

Byrne JD, Betancourt T, Brannon-Peppas L (2008) Active targeting schemes for nanoparticle systems in cancer therapeutics. Adv Drug Deliv Rev 60:1615–1626. https://doi.org/10.1016/j.addr.2008.08.005

Cavallo-Medved D, Moin K, Sloane B (2011) Cathepsin B: basis sequence: mouse. AFCS Nat Mol Pages 2011

Chen X, Zhang Y, Tang C, Tian C, Sun Q, Su Z, Xue L, Yin Y, Ju C, Zhang C (2017) Co-delivery of paclitaxel and anti-survivin siRNA via redox-sensitive oligopeptide liposomes for the synergistic treatment of breast cancer and metastasis. Int J Pharm 529:102–115. https://doi.org/10.1016/j.ijpharm.2017.06.071

Chi Y, Yin X, Sun K, Feng S, Liu J, Chen D, Guo C, Wu Z (2017) Redox-sensitive and hyaluronic acid functionalized liposomes for cytoplasmic drug delivery to osteosarcoma in animal models. J Control Release 261:113–125. https://doi.org/10.1016/j.jconrel.2017.06.027

Conde J, Dias JT, GrazÃo V, Moros M, Baptista P, de la Fuente JM (2014) Revisiting 30 years of biofunctionalization and surface chemistry of inorganic nanoparticles for nanomedicine. Front Chem 2. https://doi.org/10.3389/fchem.2014.00048

Dandamudi S, Patil V, Fowle W, Khaw B-A, Campbell RB (2009) External magnet improves antitumor effect of vinblastine and the suppression of metastasis. Cancer Sci 100:1537–1543. https://doi.org/10.1111/j.1349-7006.2009.01201.x

de Cuyper M, Joniau M (1988) Magnetoliposomes. Eur Biophys J 15:311–319. https://doi.org/10.1007/BF00256482

de Visscher SAHJ, Kaščáková S, de Bruijn HS, van den Heuvel A, Amelink A, Sterenborg HJCM, Robinson DJ, Roodenburg JLN, Witjes MJH (2011) Fluorescence localization and kinetics of mTHPC and liposomal formulations of mTHPC in the window-chamber tumor model. Lasers Surg Med 43:528–536. https://doi.org/10.1002/lsm.21082

de Vivo DC, Trifiletti RR, Jacobson RI, Ronen GM, Behmand RA, Harik SI (1991) Defective glucose transport across the blood-brain barrier as a cause of persistent hypoglycorrhachia, seizures, and developmental delay. N Engl J Med 325:703–709. https://doi.org/10.1056/NEJM199109053251006

di Corato R, Béalle G, Kolosnjaj-Tabi J, Espinosa A, Clément O, Silva AKA, Ménager C, Wilhelm C (2015) Combining magnetic hyperthermia and photodynamic therapy for tumor ablation with photoresponsive magnetic liposomes. ACS Nano 9:2904–2916. https://doi.org/10.1021/nn506949t

Ding B-S (2006) Advanced drug delivery systems that target the vascular endothelium. Mol Interv 6:98–112. https://doi.org/10.1124/mi.6.2.7

du Nguyen V, Zheng S, Han J, Le VH, Park J-O, Park S (2017) Nanohybrid magnetic liposome functionalized with hyaluronic acid for enhanced cellular uptake and near-infrared-triggered drug release. Colloids Surf B Biointerfaces 154:104–114. https://doi.org/10.1016/j.colsurfb.2017.03.008

Dwivedi P, Kiran S, Han S, Dwivedi M, Khatik R, Fan R, Mangrio FA, Du K, Zhu Z, Yang C, Huang F, Ejaz A, Han R, Si T, Xu RX (2020) Magnetic targeting and ultrasound activation of liposome-microbubble conjugate for enhanced delivery of anticancer therapies. ACS Appl Mater Interfaces 12:23737–23751. https://doi.org/10.1021/acsami.0c05308

Elbayoumi TA, Torchilin VP (2009) Tumor-specific anti-nucleosome antibody improves therapeutic efficacy of doxorubicin-loaded long-circulating liposomes against primary and metastatic tumor in mice. Mol Pharm 6:246–254. https://doi.org/10.1021/mp8001528

Eliseeva S, Bünzli J-CG (2010) Lanthanide luminescence for functional materials and bio-sciences. Chem Soc Rev 39:189–227. https://doi.org/10.1039/B905604C

Eloy JO, Claro de Souza M, Petrilli R, Barcellos JPA, Lee RJ, Marchetti JM (2014) Liposomes as carriers of hydrophilic small molecule drugs: strategies to enhance encapsulation and delivery. Colloids Surf B Biointerfaces 123:345–363. https://doi.org/10.1016/j.colsurfb.2014.09.029

Eloy JO, Petrilli R, Topan JF, Antonio HMR, Barcellos JPA, Chesca DL, Serafini LN, Tiezzi DG, Lee RJ, Marchetti JM (2016) Co-loaded paclitaxel/rapamycin liposomes: development, characterization and in vitro and in vivo evaluation for breast cancer therapy. Colloids Surf B Biointerfaces 141:74–82. https://doi.org/10.1016/j.colsurfb.2016.01.032

Emami S, Azadmard-Damirchi S, Peighambardoust SH, Valizadeh H, Hesari J (2016) Liposomes as carrier vehicles for functional compounds in food sector. J Exp Nanosci 11:737–759. https://doi.org/10.1080/17458080.2016.1148273

Fanciullino R, Ciccolini J (2009) Liposome-encapsulated anticancer drugs: still waiting for the magic bullet? Curr Med Chem 16:4361–4373. https://doi.org/10.2174/092986709789712916

Fang J-Y, Wen C-J, Zhang, Al-Suwayeh, Yen T-C (2012) Theranostic liposomes loaded with quantum dots and apomorphine for brain targeting and bioimaging. Int J Nanomed 1599. https://doi.org/10.2147/IJN.S29369

Feng S, Wu Z-X, Zhao Z, Liu J, Sun K, Guo C, Wang H, Wu Z (2019) Engineering of bone- and CD44-dual-targeting redox-sensitive liposomes for the treatment of orthotopic osteosarcoma. ACS Appl Mater Interfaces 11:7357–7368. https://doi.org/10.1021/acsami.8b18820

Ferguson RM, Khandhar AP, Kemp SJ, Arami H, Saritas EU, Croft LR, Konkle J, Goodwill PW, Halkola A, Rahmer J, Borgert J, Conolly SM, Krishnan KM (2015) Magnetic particle imaging with tailored iron oxide nanoparticle tracers. IEEE Trans Med Imaging 34:1077–1084. https://doi.org/10.1109/TMI.2014.2375065

Fleige E, Quadir MA, Haag R (2012) Stimuli-responsive polymeric nanocarriers for the controlled transport of active compounds: concepts and applications. Adv Drug Deliv Rev 64:866–884. https://doi.org/10.1016/j.addr.2012.01.020

Fouladi F, Steffen KJ, Mallik S (2017) Enzyme-responsive liposomes for the delivery of anticancer drugs. Bioconjug Chem 28:857–868. https://doi.org/10.1021/acs.bioconjchem.6b00736

Frankel AD, Pabo CO (1988) Cellular uptake of the tat protein from human immunodeficiency virus. Cell 55:1189–1193. https://doi.org/10.1016/0092-8674(88)90263-2

Fu H, Shi K, Hu G, Yang Y, Kuang Q, Lu L, Zhang L, Chen W, Dong M, Chen Y, He Q (2015) Tumor-targeted paclitaxel delivery and enhanced penetration using TAT-decorated liposomes comprising redox-responsive poly(ethylene glycol). J Pharm Sci 104:1160–1173. https://doi.org/10.1002/jps.24291

Gandhi NS, Tekade RK, Chougule MB (2014) Nanocarrier mediated delivery of siRNA/miRNA in combination with chemotherapeutic agents for cancer therapy: current progress and advances. J Control Release 194:238–256. https://doi.org/10.1016/j.jconrel.2014.09.001

Gerweck LE, Seetharaman K (1996) Cellular pH gradient in tumor versus normal tissue: potential exploitation for the treatment of cancer. Cancer Res 56:1194–1198

Gleich B, Weizenecker J (2005) Tomographic imaging using the nonlinear response of magnetic particles. Nature 435:1214–1217. https://doi.org/10.1038/nature03808

Gogoi M, Kumar N, Patra S (2016) Multifunctional magnetic liposomes for cancer imaging and therapeutic applications. In: Nanoarchitectonics for smart delivery and drug targeting. Elsevier, pp 743–782

Gu Z, da Silva CG, Hao Y, Schomann T, Camps MGM, van der Maaden K, Liu Q, Ossendorp F, Cruz LJ (2023) Effective combination of liposome-targeted chemotherapy and PD-L1 blockade of murine colon cancer. J Control Release 353:490–506. https://doi.org/10.1016/j.jconrel.2022.11.049

Guo H, Chen W, Sun X, Liu Y-N, Li J, Wang J (2015) Theranosticmagnetoliposomes coated by carboxymethyl dextran with controlled release by low-frequency alternating magnetic field. Carbohydr Polym 118:209–217. https://doi.org/10.1016/j.carbpol.2014.10.076

Hamoudeh M, Kamleh MA, Diab R, Fessi H (2008) Radionuclides delivery systems for nuclear imaging and radiotherapy of cancer. Adv Drug Deliv Rev 60:1329–1346. https://doi.org/10.1016/j.addr.2008.04.013

Han HD, Jeon YW, Kwon HJ, Jeon HN, Byeon Y, Lee CO, Cho SH, Shin BC (2015) Therapeutic efficacy of doxorubicin delivery by a CO_2 generating liposomal platform in breast carcinoma. Acta Biomater 24:279–285. https://doi.org/10.1016/j.actbio.2015.06.019

Hardiansyah A, Huang L-Y, Yang M-C, Liu T-Y, Tsai S-C, Yang C-Y, Kuo C-Y, Chan T-Y, Zou H-M, Lian W-N, Lin C-H (2014) Magnetic liposomes for colorectal cancer cells therapy by high-frequency magnetic field treatment. Nanoscale Res Lett 9:497. https://doi.org/10.1186/1556-276X-9-497

He H, Lu Y, Qi J, Zhu Q, Chen Z, Wu W (2019) Adapting liposomes for oral drug delivery. Acta Pharm Sin B 9:36–48. https://doi.org/10.1016/j.apsb.2018.06.005

Hervault A, Thanh NTK (2014) Magnetic nanoparticle-based therapeutic agents for thermo-chemotherapy treatment of cancer. Nanoscale 6:11553–11573. https://doi.org/10.1039/C4NR03482A

Hilderbrand SA, Weissleder R (2010) Near-infrared fluorescence: application to in vivo molecular imaging. Curr Opin Chem Biol 14:71–79. https://doi.org/10.1016/j.cbpa.2009.09.029

Hu S-H, Liu T-Y, Liu D-M, Chen S-Y (2007) Controlled pulsatile drug release from a ferrogel by a high-frequency magnetic field. Macromolecules 40:6786–6788. https://doi.org/10.1021/ma0707584

Hu Y, Zhu J, Jiang Y, Hu B (2009) Ultrasound microbubble contrast agents: application to therapy for peripheral vascular disease. Adv Ther 26:425–434. https://doi.org/10.1007/s12325-009-0020-y

Hu J, Zhang G, Liu S (2012) Enzyme-responsive polymeric assemblies, nanoparticles and hydrogels. Chem Soc Rev 41:5933. https://doi.org/10.1039/c2cs35103j

Hu Q, Katti PS, Gu Z (2014) Enzyme-responsive nanomaterials for controlled drug delivery. Nanoscale 6:12273–12286. https://doi.org/10.1039/C4NR04249B

Huang S-L (2010) Ultrasound-responsive liposomes, pp 113–128

Ishida O, Maruyama K, Tanahashi H, Iwatsuru M, Sasaki K, Eriguchi M, Yanagie H (2001) Liposomes bearing polyethyleneglycol-coupled transferrin with intracellular targeting property to the solid tumors in vivo. Pharm Res 18:1042–1048. https://doi.org/10.1023/a:1010960900254

Jain A, Jain SK (2018) Stimuli-responsive smart liposomes in cancer targeting. Curr Drug Targets 19. https://doi.org/10.2174/1389450117666160208144143

Jefferies WA, Brandon MR, Hunt S, Williams AF, Gatter KC, Mason DY (1984) Transferrin receptor on endothelium of brain capillaries. Nature 312:162–163. https://doi.org/10.1038/312162a0

Jhaveri A, Deshpande P, Torchilin V (2014) Stimuli-sensitive nanopreparations for combination cancer therapy. J Control Release 190:352–370. https://doi.org/10.1016/j.jconrel.2014.05.002

Jose A, Ninave KM, Karnam S, Venuganti VVK (2019) Temperature-sensitive liposomes for co-delivery of tamoxifen and imatinib for synergistic breast cancer treatment. J Liposome Res 29:153–162. https://doi.org/10.1080/08982104.2018.1502315

Kato Y, Ozawa S, Miyamoto C, Maehata Y, Suzuki A, Maeda T, Baba Y (2013) Acidic extracellular microenvironment and cancer. Cancer Cell Int 13:89. https://doi.org/10.1186/1475-2867-13-89

Keselman P, Yu EY, Zhou XY, Goodwill PW, Chandrasekharan P, Ferguson RM, Khandhar AP, Kemp SJ, Krishnan KM, Zheng B, Conolly SM (2017) Tracking short-term biodistribution and long-term clearance of SPIO tracers in magnetic particle imaging. Phys Med Biol 62:3440–3453. https://doi.org/10.1088/1361-6560/aa5f48

Khan A, Alsahli MA, Aljasir MA, Maswadeh H, Mobark MA, Azam F, Allemailem KS, Alrumaihi F, Alhumaydhi FA, Almatroudi AA, AlSuhaymi N, Khan MA (2022) Experimental and theoretical insights on chemopreventive effect of the liposomal thymoquinone against benzo[a]pyrene-induced lung cancer in Swiss albino mice. J Inflamm Res 15:2263–2280. https://doi.org/10.2147/JIR.S358632

Kokuryo D, Nakashima S, Ozaki F, Yuba E, Chuang K-H, Aoshima S, Ishizaka Y, Saga T, Kono K, Aoki I (2015) Evaluation of thermo-triggered drug release in intramuscular-transplanted tumors using thermosensitive polymer-modified liposomes and MRI. Nanomedicine 11:229–238. https://doi.org/10.1016/j.nano.2014.09.001

Kosaka N, Ogawa M, Choyke PL, Kobayashi H (2009) Clinical implications of near-infrared fluorescence imaging in cancer. Future Oncol 5:1501–1511. https://doi.org/10.2217/fon.09.109

Kraft JC, Freeling JP, Ziyao W, Ho RJY (2014) Emerging research and clinical development trends of liposome and lipid nanoparticle drug delivery systems. J Pharm Sci 103:29–52. https://doi.org/10.1002/jps.23773

Kulshrestha P, Gogoi M, Bahadur D, Banerjee R (2012) In vitro application of paclitaxel loaded magnetoliposomes for combined chemotherapy and hyperthermia. Colloids Surf B Biointerfaces 96:1–7. https://doi.org/10.1016/j.colsurfb.2012.02.029

Kuo C-Y, Liu T-Y, Hardiansyah A, Lee C-F, Wang M-S, Chiu W-Y (2014) Self-assembly behaviors of thermal- and pH-sensitive magnetic nanocarriers for stimuli-triggered release. Nanoscale Res Lett 9:520. https://doi.org/10.1186/1556-276X-9-520

Kuo C-Y, Liu T-Y, Chan T-Y, Tsai S-C, Hardiansyah A, Huang L-Y, Yang M-C, Lu R-H, Jiang J-K, Yang C-Y, Lin C-H, Chiu W-Y (2016) Magnetically triggered nanovehicles for controlled drug release as a colorectal cancer therapy. Colloids Surf B Biointerfaces 140:567–573. https://doi.org/10.1016/j.colsurfb.2015.11.008

Kusano M, Tajima Y, Yamazaki K, Kato M, Watanabe M, Miwa M (2008) Sentinel node mapping guided by indocyanine green fluorescence imaging: a new method for sentinel node navigation surgery in gastrointestinal cancer. Dig Surg 25:103–108. https://doi.org/10.1159/000121905

Langereis S, Geelen T, Grüll H, Strijkers GJ, Nicolay K (2013) Paramagnetic liposomes for molecular MRI and MRI-guided drug delivery. NMR Biomed 26:728–744. https://doi.org/10.1002/nbm.2971

Laouini A, Jaafar-Maalej C, Limayem-Blouza I, Sfar S, Charcosset C, Fessi H (2012) Preparation, characterization and applications of liposomes: state of the art. J Colloid Sci Biotechnol 1:147–168. https://doi.org/10.1166/jcsb.2012.1020

Lee YK, Lee TS, Song IH, Jeong HY, Kang SJ, Kim MW, Ryu SH, Jung IH, Kim JS, Park YS (2015) Inhibition of pulmonary cancer progression by epidermal growth factor receptor-targeted transfection with Bcl-2 and survivin siRNAs. Cancer Gene Ther 22:335–343. https://doi.org/10.1038/cgt.2015.18

Lehner R, Wang X, Marsch S, Hunziker P (2013) Intelligent nanomaterials for medicine: carrier platforms and targeting strategies in the context of clinical application. Nanomedicine 9:742–757. https://doi.org/10.1016/j.nano.2013.01.012

Lin W, Xie X, Yang Y, Fu X, Liu H, Yang Y, Deng J (2016) Thermosensitive magnetic liposomes with doxorubicin cell-penetrating peptides conjugate for enhanced and targeted cancer therapy. Drug Deliv 23:3436–3443. https://doi.org/10.1080/10717544.2016.1189983

Lin Z, Ding J, Sun G, Li D, He S, Liang X, Huang X, Xie J (2020) Application of paclitaxel-loaded EGFR peptide-conjugated magnetic polymeric liposomes for liver cancer therapy. Curr Med Sci 40:145–154. https://doi.org/10.1007/s11596-020-2158-4

Liu T-Y, Hu S-H, Liu K-H, Shaiu R-S, Liu D-M, Chen S-Y (2008) Instantaneous drug delivery of magnetic/thermally sensitive nanospheres by a high-frequency magnetic field. Langmuir 24:13306–13311. https://doi.org/10.1021/la801451v

Lu R-M, Chen M-S, Chang D-K, Chiu C-Y, Lin W-C, Yan S-L, Wang Y-P, Kuo Y-S, Yeh C-Y, Lo A, Wu H-C (2013) Targeted drug delivery systems mediated by a novel peptide in breast cancer therapy and imaging. PLoS ONE 8:e66128. https://doi.org/10.1371/journal.pone.0066128

Lukyanov AN, Torchilin VP (2004) Micelles from lipid derivatives of water-soluble polymers as delivery systems for poorly soluble drugs. Adv Drug Deliv Rev 56:1273–1289. https://doi.org/10.1016/j.addr.2003.12.004

Mabrey S, Sturtevant JM (1976) Investigation of phase transitions of lipids and lipid mixtures by sensitivity differential scanning calorimetry. Proc Natl Acad Sci 73:3862–3866. https://doi.org/10.1073/pnas.73.11.3862

Maeda H, Wu J, Sawa T, Matsumura Y, Hori K (2000) Tumor vascular permeability and the EPR effect in macromolecular therapeutics: a review. J Control Release 65:271–284. https://doi.org/10.1016/s0168-3659(99)00248-5

Maeda H, Sawa T, Konno T (2001) Mechanism of tumor-targeted delivery of macromolecular drugs, including the EPR effect in solid tumor and clinical overview of the prototype polymeric drug SMANCS. J Control Release 74:47–61. https://doi.org/10.1016/S0168-3659(01)00309-1

Malam Y, Loizidou M, Seifalian AM (2009) Liposomes and nanoparticles: nanosized vehicles for drug delivery in cancer. Trends Pharmacol Sci 30:592–599. https://doi.org/10.1016/j.tips.2009.08.004

Marengo B, Nitti M, Furfaro AL, Colla R, de Ciucis C, Marinari UM, Pronzato MA, Traverso N, Domenicotti C (2016) Redox homeostasis and cellular antioxidant systems: crucial players in cancer growth and therapy. Oxid Med Cell Longev 2016:1–16. https://doi.org/10.1155/2016/6235641

Marie H, Lemaire L, Franconi F, Lajnef S, Frapart Y-M, Nicolas V, Frébourg G, Trichet M, Ménager C, Lesieur S (2015) Superparamagnetic liposomes for MRI monitoring and external magnetic field-induced selective targeting of malignant brain tumors. Adv Funct Mater 25:1258–1269. https://doi.org/10.1002/adfm.201402289

Marqués-Gallego P, de Kroon AIPM (2014) Ligation strategies for targeting liposomal nanocarriers. Biomed Res Int 2014:1–12. https://doi.org/10.1155/2014/129458

Maruyama S et al (2016) Development of magnetic nanocarriers based on thermosensitive liposomes and their visualization using magnetic particle imaging. Int J Nanomed Nanosurg 2. https://doi.org/10.16966/2470-3206.111. ISSN 2470-3206

Mathiyazhakan M, Wiraja C, Xu C (2018) A concise review of gold nanoparticles-based photo-responsive liposomes for controlled drug delivery. Nanomicro Lett 10:10. https://doi.org/10.1007/s40820-017-0166-0

Mattheolabakis G, Milane L, Singh A, Amiji MM (2015) Hyaluronic acid targeting of CD44 for cancer therapy: from receptor biology to nanomedicine. J Drug Target 23:605–618. https://doi.org/10.3109/1061186X.2015.1052072

Maurer N, Fenske DB, Cullis PR (2001) Developments in liposomal drug delivery systems. Expert Opin Biol Ther 1:923–947. https://doi.org/10.1517/14712598.1.6.923

McCarley RL (2012) Redox-responsive delivery systems. Annu Rev Anal Chem 5:391–411. https://doi.org/10.1146/annurev-anchem-062011-143157

McGuire S (2016) World cancer report 2014. World Health Organization, International Agency for Research on Cancer, WHO Press, Geneva, Switzerland, 2015. Adv Nutr 7:418–419. https://doi.org/10.3945/an.116.012211

Meng J, Guo F, Xu H, Liang W, Wang C, Yang X-D (2016) Combination therapy using co-encapsulated resveratrol and paclitaxel in liposomes for drug resistance reversal in breast cancer cells in vivo. Sci Rep 6:22390. https://doi.org/10.1038/srep22390

Meure LA, Foster NR, Dehghani F (2008) Conventional and dense gas techniques for the production of liposomes: a review. AAPS PharmSciTech 9:798. https://doi.org/10.1208/s12249-008-9097-x

Min Y, Li J, Liu F, Padmanabhan P, Yeow E, Xing B (2014) Recent advance of biological molecular imaging based on lanthanide-doped upconversion-luminescent nanomaterials. Nanomaterials 4:129–154. https://doi.org/10.3390/nano4010129

Monnier CA, Burnand D, Rothen-Rutishauser B, Lattuada M, Petri-Fink A (2014) Magnetoliposomes: opportunities and challenges. Eur J Nanomed 6. https://doi.org/10.1515/ejnm-2014-0042

Monteiro N, Martins A, Reis RL, Neves NM (2014) Liposomes in tissue engineering and regenerative medicine. J R Soc Interface 11:20140459. https://doi.org/10.1098/rsif.2014.0459

Moosavian SA, Bianconi V, Pirro M, Sahebkar A (2021) Challenges and pitfalls in the development of liposomal delivery systems for cancer therapy. Semin Cancer Biol 69:337–348. https://doi.org/10.1016/j.semcancer.2019.09.025

Morgan J, MacRobert A, Gray A, Huehns E (1992) Use of photosensitive, antibody directed liposomes to destroy target populations of cells in bone marrow: a potential purging method for autologous bone marrow transplantation. Br J Cancer 65:58–64. https://doi.org/10.1038/bjc.1992.11

Murase K, Hiratsuka S, Song R, Takeuchi Y (2014) Development of a system for magnetic particle imaging using neodymium magnets and gradiometer. Jpn J Appl Phys 53:067001. https://doi.org/10.7567/JJAP.53.067001

Muthu MS, Singh S (2009) Targeted nanomedicines: effective treatment modalities for cancer, AIDS and brain disorders. Nanomedicine 4:105–118. https://doi.org/10.2217/17435889.4.1.105

Muthu MS, Rajesh C, Mishra A, Singh S (2009) Stimulus-responsive targeted nanomicelles for effective cancer therapy. Nanomedicine 4:657–667. https://doi.org/10.2217/nnm.09.44

Muthu MS, Kulkarni SA, Raju A, Feng S-S (2012) Theranostic liposomes of TPGS coating for targeted co-delivery of docetaxel and quantum dots. Biomaterials 33:3494–3501. https://doi.org/10.1016/j.biomaterials.2012.01.036

Nahar K, Absar S, Patel B, Ahsan F (2014) Starch-coated magnetic liposomes as an inhalable carrier for accumulation of fasudil in the pulmonary vasculature. Int J Pharm 464:185–195. https://doi.org/10.1016/j.ijpharm.2014.01.007

Needham D, Park J-Y, Wright AM, Tong J (2013) Materials characterization of the low temperature sensitive liposome (LTSL): effects of the lipid composition (lysolipid and DSPE–PEG2000) on the thermal transition and release of doxorubicin. Faraday Discuss 161:515–534. https://doi.org/10.1039/C2FD20111A

Nkanga CI, Bapolisi AM, Okafor NI, Krause RWM (2019) General perception of liposomes: formation, manufacturing and applications. In: Liposomes—advances and perspectives. IntechOpen

Noble GT, Stefanick JF, Ashley JD, Kiziltepe T, Bilgicer B (2014) Ligand-targeted liposome design: challenges and fundamental considerations. Trends Biotechnol 32:32–45. https://doi.org/10.1016/j.tibtech.2013.09.007

Odeh F, Naffa R, Azzam H, Mahmoud IS, Alshaer W, al Bawab A, Ismail S (2019) Co-encapsulation of thymoquinone with docetaxel enhances the encapsulation efficiency into PEGylated liposomes and the chemosensitivity of MCF7 breast cancer cells to docetaxel. Heliyon 5:e02919. https://doi.org/10.1016/j.heliyon.2019.e02919

Pan J, Liu Y, Feng S-S (2010) Multifunctional nanoparticles of biodegradable copolymer blend for cancer diagnosis and treatment. Nanomedicine 5:347–360. https://doi.org/10.2217/nnm.10.13

Park YS (2002) Tumor-directed targeting of liposomes. Biosci Rep 22:267–281. https://doi.org/10.1023/a:1020190606757

Park S-J, Kim S, Lee S, Khim ZG, Char K, Hyeon T (2000) Synthesis and magnetic studies of uniform iron nanorods and nanospheres. J Am Chem Soc 122:8581–8582. https://doi.org/10.1021/ja001628c

Park J-H, Cho H-J, Yoon HY, Yoon I-S, Ko S-H, Shim J-S, Cho J-H, Park JH, Kim K, Kwon IC, Kim D-D (2014) Hyaluronic acid derivative-coated nanohybrid liposomes for cancer imaging and drug delivery. J Control Release 174:98–108. https://doi.org/10.1016/j.jconrel.2013.11.016

Park YI, Kwon S-H, Lee G, Motoyama K, Kim MW, Lin M, Niidome T, Choi JH, Lee R (2021) pH-sensitive multi-drug liposomes targeting folate receptor β for efficient treatment of non-small cell lung cancer. J Control Release 330:1–14. https://doi.org/10.1016/j.jconrel.2020.12.011

Partanen A, Yarmolenko PS, Viitala A, Appanaboyina S, Haemmerich D, Ranjan A, Jacobs G, Woods D, Enholm J, Wood BJ, Dreher MR (2012) Mild hyperthermia with magnetic resonance-guided high-intensity focused ultrasound for applications in drug delivery. Int J Hyperth 28:320–336. https://doi.org/10.3109/02656736.2012.680173

Patel HM (1990) Liposomes: a practical approach. FEBS Lett 275:242–243. https://doi.org/10.1016/0014-5793(90)81487-9

Pattni BS, Chupin V, Torchilin VP (2015) New developments in liposomal drug delivery. Chem Rev 115:10938–10966. https://doi.org/10.1021/acs.chemrev.5b00046

Peng Z, Wang C, Fang E, Lu X, Wang G, Tong Q (2014) Co-delivery of doxorubicin and SATB1 shRNA by thermosensitive magnetic cationic liposomes for gastric cancer therapy. PLoS ONE 9:e92924. https://doi.org/10.1371/journal.pone.0092924

Pradhan P, Giri J, Rieken F, Koch C, Mykhaylyk O, Döblinger M, Banerjee R, Bahadur D, Plank C (2010) Targeted temperature sensitive magnetic liposomes for thermo-chemotherapy. J Control Release 142:108–121. https://doi.org/10.1016/j.jconrel.2009.10.002

Puntes VF, Krishnan, Alivisatos AP KM (1979) Colloidal nanocrystal shape and size control: the case of cobalt. Science 291:2115–2117. https://doi.org/10.1126/science.1058495

Qiu D, An X (2013) Controllable release from magnetoliposomes by magnetic stimulation and thermal stimulation. Colloids Surf B Biointerfaces 104:326–329. https://doi.org/10.1016/j.colsurfb.2012.11.033

Qiu L, Gao M, Xu Y (2015) Enhanced combination therapy effect on paclitaxel-resistant carcinoma by chloroquine co-delivery via liposomes. Int J Nanomed 6615. https://doi.org/10.2147/IJN.S91463

Quach ND, Arnold RD, Cummings BS (2014a) Secretory phospholipase A2 enzymes as pharmacological targets for treatment of disease. BiochemPharmacol 90:338–348. https://doi.org/10.1016/j.bcp.2014.05.022

Quach ND, Mock JN, Scholpa NE, Eggert MW, Payré C, Lambeau G, Arnold RD, Cummings BS (2014b) Role of the phospholipase A2 receptor in liposome drug delivery in prostate cancer cells. Mol Pharm 11:3443–3451. https://doi.org/10.1021/mp500174p

Ren J-L, Wang Z-G, Zhang Y, Zheng Y-Y, Li X-S, Zhang Q-X, Wang Z-X, Xu C-S (2008) Transfection efficiency of TDL compound in HUVEC enhanced by ultrasound-targeted microbubble destruction. Ultrasound Med Biol 34:1857–1867. https://doi.org/10.1016/j.ultrasmedbio.2008.03.019

Riaz M, Riaz M, Zhang X, Lin C, Wong K, Chen X, Zhang G, Lu A, Yang Z (2018) Surface functionalization and targeting strategies of liposomes in solid tumor therapy: a review. Int J Mol Sci 19:195. https://doi.org/10.3390/ijms19010195

Rolle F, Bincoletto V, Gazzano E, Rolando B, Lollo G, Stella B, Riganti C, Arpicco S (2020) Coencapsulation of disulfiram and doxorubicin in liposomes strongly reverses multidrug resistance in breast cancer cells. Int J Pharm 580:119191. https://doi.org/10.1016/j.ijpharm.2020.119191

Romberg B, Flesch FM, Hennink WE, Storm G (2008) Enzyme-induced shedding of a poly(amino acid)-coating triggers contents release from dioleoyl phosphatidylethanolamine liposomes. Int J Pharm 355:108–113. https://doi.org/10.1016/j.ijpharm.2007.11.055

Rommasi F, Esfandiari N (2021) Liposomal nanomedicine: applications for drug delivery in cancer therapy. Nanoscale Res Lett 16:95. https://doi.org/10.1186/s11671-021-03553-8

Sapra P, Tyagi P, Allen T (2005) Ligand-targeted liposomes for cancer treatment. Curr Drug Deliv 2:369–381. https://doi.org/10.2174/156720105774370159

Saraf S, Jain A, Tiwari A, Verma A, Panda PK, Jain SK (2020) Advances in liposomal drug delivery to cancer: an overview. J Drug Deliv Sci Technol 56:101549. https://doi.org/10.1016/j.jddst.2020.101549

Sawant RR, Torchilin VP (2012) Challenges in development of targeted liposomal therapeutics. AAPS J 14:303–315. https://doi.org/10.1208/s12248-012-9330-0

Schafer FQ, Buettner GR (2001) Redox environment of the cell as viewed through the redox state of the glutathione disulfide/glutathione couple. Free Radic Biol Med 30:1191–1212. https://doi.org/10.1016/S0891-5849(01)00480-4

Schirrmacher V (2019) From chemotherapy to biological therapy: a review of novel concepts to reduce the side effects of systemic cancer treatment (review). Int J Oncol 54:407–419. https://doi.org/10.3892/ijo.2018.4661

Sebaaly C, Greige-Gerges H, Stainmesse S, Fessi H, Charcosset C (2016) Effect of composition, hydrogenation of phospholipids and lyophilization on the characteristics of eugenol-loaded liposomes prepared by ethanol injection method. Food Biosci 15:1–10. https://doi.org/10.1016/j.fbio.2016.04.005

Sen K, Banerjee S, Mandal M (2019) Dual drug loaded liposome bearing apigenin and 5-fluorouracil for synergistic therapeutic efficacy in colorectal cancer. Colloids Surf B Biointerfaces 180:9–22. https://doi.org/10.1016/j.colsurfb.2019.04.035

Sercombe L, Veerati T, Moheimani F, Wu SY, Sood AK, Hua S (2015) Advances and challenges of liposome assisted drug delivery. Front Pharmacol 6. https://doi.org/10.3389/fphar.2015.00286

Sigot V, Arndt-Jovin DJ, Jovin TM (2010) Targeted cellular delivery of quantum dots loaded on and in biotinylated liposomes. Bioconjug Chem 21:1465–1472. https://doi.org/10.1021/bc100054c

Silindir-Gunay M, Sarcan ET, Ozer AY (2019) Near-infrared imaging of diseases: a nanocarrier approach. Drug Dev Res. https://doi.org/10.1002/ddr.21532

Soares DCF, de Oliveira MC, dos Santos RG, Andrade MS, Vilela JMC, Cardoso VN, Ramaldes GA (2011) Liposomes radio labeled with 159Gd-DTPA-BMA: preparation, physicochemical characterization, release profile and in vitro cytotoxic evaluation. Eur J Pharm Sci 42:462–469. https://doi.org/10.1016/j.ejps.2011.01.010

Soenen SJH, Brisson AR, Jonckheere E, Nuytten N, Tan S, Himmelreich U, de Cuyper M (2011) The labeling of cationic iron oxide nanoparticle-resistant hepatocellular carcinoma cells using targeted magnetoliposomes. Biomaterials 32:1748–1758. https://doi.org/10.1016/j.biomaterials.2010.11.005

Soga K, Tokuzen K, Tsuji K, Yamano T, Hyodo H, Kishimoto H (2010) NIR bioimaging: development of liposome-encapsulated, rare-earth-doped Y_2O_3 nanoparticles as fluorescent probes. Eur J Inorg Chem 2010:2673–2677. https://doi.org/10.1002/ejic.201000201

Song S, Chen F, Qi H, Li F, Xin T, Xu J, Ye T, Sheng N, Yang X, Pan W (2014) Multifunctional tumor-targeting nanocarriers based on hyaluronic acid-mediated and pH-sensitive properties for efficient delivery of docetaxel. Pharm Res 31:1032–1045. https://doi.org/10.1007/s11095-013-1225-y

Song SJ, Lee S, Lee Y, Choi JS (2016) Enzyme-responsive destabilization of stabilized plasmid-lipid nanoparticles as an efficient gene delivery. Eur J Pharm Sci 91:20–30. https://doi.org/10.1016/j.ejps.2016.05.024

Sorkin A, von Zastrow M (2002) Signal transduction and endocytosis: close encounters of many kinds. Nat Rev Mol Cell Biol 3:600–614. https://doi.org/10.1038/nrm883

Srivastava SK, Bhardwaj A, Arora S, Tyagi N, Singh S, Andrews J, McClellan S, Wang B, Singh AP (2015) MicroRNA-345 induces apoptosis in pancreatic cancer cells through potentiation of caspase-dependent and -independent pathways. Br J Cancer 113:660–668. https://doi.org/10.1038/bjc.2015.252

Strebhardt K, Ullrich A (2008) Paul Ehrlich's magic bullet concept: 100 years of progress. Nat Rev Cancer 8:473–480. https://doi.org/10.1038/nrc2394

Sudimack J, Lee RJ (2000) Targeted drug delivery via the folate receptor. Adv Drug Deliv Rev 41:147–162. https://doi.org/10.1016/s0169-409x(99)00062-9

Sullivan SM, Huang L (1985) Preparation and characterization of heat-sensitive immunoliposomes. Biochimica et Biophysica Acta (BBA) Biomembranes 812:116–126. https://doi.org/10.1016/0005-2736(85)90528-0

Sun H, Meng F, Cheng R, Deng C, Zhong Z (2013) Reduction-sensitive degradable micellar nanoparticles as smart and intuitive delivery systems for cancer chemotherapy. Expert Opin Drug Deliv 10:1109–1122. https://doi.org/10.1517/17425247.2013.783009

Ta T, Bartolak-Suki E, Park E-J, Karrobi K, McDannold NJ, Porter TM (2014) Localized delivery of doxorubicin in vivo from polymer-modified thermosensitive liposomes with MR-guided focused

ultrasound-mediated heating. J Control Release 194:71–81. https://doi.org/10.1016/j.jconrel.2014.08.013
Tahara K, Fujimoto S, Fujii F, Tozuka Y, Jin T, Takeuchi H (2013) Quantum dot-loaded liposomes to evaluate the behavior of drug carriers after oral administration. J Pharm (Cairo) 2013:1–6. https://doi.org/10.1155/2013/848275
Tai L-A, Tsai P-J, Wang Y-C, Wang Y-J, Lo L-W, Yang C-S (2009) Thermosensitive liposomes entrapping iron oxide nanoparticles for controllable drug release. Nanotechnology 20:135101. https://doi.org/10.1088/0957-4484/20/13/135101
Tajima Y, Murakami M, Yamazaki K, Masuda Y, Kato M, Sato A, Goto S, Otsuka K, Kato T, Kusano M (2010) Sentinel node mapping guided by indocyanine green fluorescence imaging during laparoscopic surgery in gastric cancer. Ann Surg Oncol 17:1787–1793. https://doi.org/10.1245/s10434-010-0944-0
Tejera-Garcia R, Ranjan S, Zamotin V, Sood R, Kinnunen PKJ (2011) Making unilamellar liposomes using focused ultrasound. Langmuir 27:10088–10097. https://doi.org/10.1021/la201708x
Torchilin VP (2005) Recent advances with liposomes as pharmaceutical carriers. Nat Rev Drug Discov 4:145–160. https://doi.org/10.1038/nrd1632
Torchilin VP (2006) Multifunctional nanocarriers. Adv Drug Deliv Rev 58:1532–1555. https://doi.org/10.1016/j.addr.2006.09.009
Trudel D, Fradet Y, Meyer F, Harel F, Têtu B (2003) Significance of MMP-2 expression in prostate cancer: an immunohistochemical study. Cancer Res 63:8511–8515
Veronese FM, Harris JM (2002) Introduction and overview of peptide and protein pegylation. Adv Drug Deliv Rev 54:453–456. https://doi.org/10.1016/s0169-409x(02)00020-0
Veronese FM, Pasut G (2005) PEGylation, successful approach to drug delivery. Drug Discov Today 10:1451–1458. https://doi.org/10.1016/S1359-6446(05)03575-0
Wagner A, Vorauer-Uhl K (2011) Liposome technology for industrial purposes. J Drug Deliv 2011:1–9. https://doi.org/10.1155/2011/591325
Weissleder R, Pittet MJ (2008) Imaging in the era of molecular oncology. Nature 452:580–589. https://doi.org/10.1038/nature06917
Wicki A, Witzigmann D, Balasubramanian V, Huwyler J (2015) Nanomedicine in cancer therapy: challenges, opportunities, and clinical applications. J Control Release 200:138–157. https://doi.org/10.1016/j.jconrel.2014.12.030
William B, Noémie P, Brigitte E, Géraldine P (2020) Supercritical fluid methods: an alternative to conventional methods to prepare liposomes. Chem Eng J 383:123106. https://doi.org/10.1016/j.cej.2019.123106
Xi L, Li C, Wang Y, Gong Y, Su F, Li S (2020) Novel Thermosensitive polymer-modified liposomes as nano-carrier of hydrophobic antitumor drugs. J Pharm Sci 109:2544–2552. https://doi.org/10.1016/j.xphs.2020.05.006
Xu Y, Yan, Wu, Feng, Chen, Liu, Hao, Yang, Zhang, Lin, Liu (2013) Anti-αvβ3 antibody guided three-step pretargeting approach using magnetoliposomes for molecular magnetic resonance imaging of breast cancer angiogenesis. Int J Nanomed 245. https://doi.org/10.2147/IJN.S38678
Yadav L, Puri N, Rastogi V, Satpute P, Ahmad R, Kaur G (2014) Matrix metalloproteinases and cancer—roles in threat and therapy. Asian Pac J Cancer Prev 15:1085–1091. https://doi.org/10.7314/APJCP.2014.15.3.1085
Yang T, Cui F-D, Choi M-K, Cho J-W, Chung S-J, Shim C-K, Kim D-D (2007) Enhanced solubility and stability of PEGylated liposomal paclitaxel: in vitro and in vivo evaluation. Int J Pharm 338:317–326. https://doi.org/10.1016/j.ijpharm.2007.02.011
Yang Y, Xie X, Xu X, Xia X, Wang H, Li L, Dong W, Ma P, Yang Y, Liu Y, Mei X (2016) Thermal and magnetic dual-responsive liposomes with a cell-penetrating peptide-siRNA conjugate for enhanced and targeted cancer therapy. Colloids Surf B Biointerfaces 146:607–615. https://doi.org/10.1016/j.colsurfb.2016.07.002

Yang S, Zhu W, Wang Z, Xiao Y, Mao P, Qu L, Hu Y, Wang J (2020) Thermosensitive liposomes encapsulating anti-cancer agent lomustine, and contrast medium iohexol, for thermochemotherapy: preparation, characterization, and in vivo evaluation. J Nanosci Nanotechnol 20:6070–6076. https://doi.org/10.1166/jnn.2020.18590
Yatvin MB, Weinstein JN, Dennis, Blumenthal R WH (1978) Design of liposomes for enhanced local release of drugs by hyperthermia. Science 202:1290–1293. https://doi.org/10.1126/science.364652
Yin X, Chi Y, Guo C, Feng S, Liu J, Sun K, Wu Z (2017) Chitooligosaccharides modified reduction-sensitive liposomes: enhanced cytoplasmic drug delivery and osteosarcomas-tumor inhibition in animal models. Pharm Res 34:2172–2184. https://doi.org/10.1007/s11095-017-2225-0
Ying X, Wen H, Lu W-L, Du J, Guo J, Tian W, Men Y, Zhang Y, Li R-J, Yang T-Y, Shang D-W, Lou J-N, Zhang L-R, Zhang Q (2010) Dual-targeting daunorubicin liposomes improve the therapeutic efficacy of brain glioma in animals. J Control Release 141:183–192. https://doi.org/10.1016/j.jconrel.2009.09.020
Yoshida M, Sato M, Yamamoto Y, Maehara T, Naohara T, Aono H, Sugishita H, Sato K, Watanabe Y (2012) Tumor local chemohyperthermia using docetaxel-embedded magnetoliposomes: interaction of chemotherapy and hyperthermia. J Gastroenterol Hepatol 27:406–411. https://doi.org/10.1111/j.1440-1746.2011.06972.x
Yu EY, Bishop M, Zheng B, Ferguson RM, Khandhar AP, Kemp SJ, Krishnan KM, Goodwill PW, Conolly SM (2017) Magnetic particle imaging: a novel in vivo imaging platform for cancer detection. Nano Lett 17:1648–1654. https://doi.org/10.1021/acs.nanolett.6b04865
Zalipsky S, Qazen M, Walker JA, Mullah N, Quinn YP, Huang SK (1999) New detachable poly(ethylene glycol) conjugates: cysteine-cleavable lipopolymers regenerating natural phospholipid, diacyl phosphatidylethanolamine. Bioconjug Chem 10:703–707. https://doi.org/10.1021/bc990031n
Zang X, Ding H, Zhao X, Li X, Du Z, Hu H, Qiao M, Chen D, Deng Y, Zhao X (2016) Anti-EphA10 antibody-conjugated pH-sensitive liposomes for specific intracellular delivery of siRNA. Int J Nanomed 11:3951–3967. https://doi.org/10.2147/IJN.S107952
Zhang K, Liu M, Tong X, Sun N, Zhou L, Cao Y, Wang J, Zhang H, Pei R (2015) Aptamer-modified temperature-sensitive liposomal contrast agent for magnetic resonance imaging. Biomacromol 16:2618–2623. https://doi.org/10.1021/acs.biomac.5b00250
Zhang C, Jiang D, Huang B, Wang C, Zhao L, Xie X, Zhang Z, Wang K, Tian J, Luo Y (2019) Methylene blue-based near-infrared fluorescence imaging for breast cancer visualization in resected human tissues. Technol Cancer Res Treat 18:153303381989433. https://doi.org/10.1177/1533033819894331
Zhang H, Tang W-L, Kheirolomoom A, Fite BZ, Wu B, Lau K, Baikoghli M, Raie MN, Tumbale SK, Foiret J, Ingham ES, Mahakian LM, Tam SM, Cheng RH, Borowsky AD, Ferrara KW (2021) Development of thermosensitive resiquimod-loaded liposomes for enhanced cancer immunotherapy. J Control Release 330:1080–1094. https://doi.org/10.1016/j.jconrel.2020.11.013
Zhao Y, Ren W, Zhong T, Zhang S, Huang D, Guo Y, Yao X, Wang C, Zhang W-Q, Zhang X, Zhang Q (2016) Tumor-specific pH-responsive peptide-modified pH-sensitive liposomes containing doxorubicin for enhancing glioma targeting and anti-tumor activity. J Control Release 222:56–66. https://doi.org/10.1016/j.jconrel.2015.12.006
Zheng B, von See MP, Yu E, Gunel B, Lu K, Vazin T, Schaffer D, Goodwill PW, Conolly SM (2016) Quantitative magnetic particle imaging monitors the transplantation, biodistribution, and clearance of stem cells in vivo. Theranostics 6:291–301. https://doi.org/10.7150/thno.13728
Zhou XY, Tay ZW, Chandrasekharan P, Yu EY, Hensley DW, Orendorff R, Jeffris KE, Mai D, Zheng B, Goodwill PW, Conolly SM (2018) Magnetic particle imaging for radiation-free, sensitive and high-contrast vascular imaging and cell tracking. Curr Opin Chem Biol 45:131–138. https://doi.org/10.1016/j.cbpa.2018.04.014

Zhu L, Kate P, Torchilin VP (2012) Matrix metalloprotease 2-responsive multifunctional liposomal nanocarrier for enhanced tumor targeting. ACS Nano 6:3491–3498. https://doi.org/10.1021/nn300524f

Zorko M, Langel Ü (2005) Cell-penetrating peptides: mechanism and kinetics of cargo delivery. Adv Drug Deliv Rev 57:529–545. https://doi.org/10.1016/j.addr.2004.10.010

Zuo T, Guan Y, Chang M, Zhang F, Lu S, Wei T, Shao W, Lin G (2016) RGD(Arg-Gly-Asp) internalized docetaxel-loaded pH sensitive liposomes: preparation, characterization and antitumor efficacy in vivo and in vitro. Colloids Surf B Biointerfaces 147:90–99. https://doi.org/10.1016/j.colsurfb.2016.07.056

Zuo J, Tong L, Du L, Yang M, Jin Y (2017) Biomimetic nanoassemblies of 1-O-octodecyl-2-conjugated linoleoyl-sn-glycero-3-phosphatidyl gemcitabine with phospholipase A2-triggered degradation for the treatment of cancer. Colloids Surf B Biointerfaces 152:467–474. https://doi.org/10.1016/j.colsurfb.2017.02.001

Zylberberg C, Matosevic S (2016) Pharmaceutical liposomal drug delivery: a review of new delivery systems and a look at the regulatory landscape. Drug Deliv 23:3319–3329. https://doi.org/10.1080/10717544.2016.1177136

Electroporation-Based Drug Delivery

Souvik Pabi, Mohd. Kaleem Khan, Alexander Alexeev, Todd Sulchek, and Abhishek Raj

Abstract Tailored treatment of various diseases, including cancer, needs a proper design of patient-specific drugs, often requiring delivery of RNA, DNA, protein, genes and various drugs into single live cells with high viability and transfection efficiency. Rapid developments in microfluidics over the years have enabled the invention of various methods for drug delivery into cells. One of the most popular engineered techniques for cellular delivery is electroporation. It works on the principle of the cell membrane becoming permeable in response to a specific electrical pulse due to the reorganization of the structures within the cell or tissue. This technique is advantageous over other physical and chemical methods due to its easy and quick operation, higher transformation efficiency, and controllable and high throughput delivery. Due to its versatility, it is possible to perform bulk electroporation (BEP), single-cell electroporation (SCEP) and localized single-cell electroporation (LSCEP). SCEP is capable of withstanding a heterogeneous electrical field centred on a single adherent or suspended cell without impacting any nearby cells. In contrast, bulk electroporation can deliver drugs in a homogeneous electric field. On the other hand, in LSCEP, organelles and internal biochemical effects enable it to assess cell-to-cell variance accurately. This chapter presents a detailed discussion of the mechanism and various electroporation techniques, including BEP, SCEP and LSCEP. Further, the chapter is concluded with the future aspects of the electroporation technique.

S. Pabi · A. Raj (✉)
Bio-Microfluidics Laboratory, Department of Mechanical Engineering, Indian Institute of Technology Patna, Patna, Bihar 801106, India
e-mail: araj@iitp.ac.in

S. Pabi · Mohd. K. Khan
Sustainable Energy Research Laboratory (S.E.R.L.), Department of Mechanical Engineering, Indian Institute of Technology Patna, Patna, Bihar 801106, India

A. Alexeev · T. Sulchek
George W. Woodruff School of Mechanical Engineering, Georgia Institute of Technology, 801 Ferst Drive, Atlanta, GA 30332-0405, USA

T. Sulchek
Wallace H. Coulter Department of Biomedical Engineering, Georgia Institute of Technology, 313 Ferst Drive, Atlanta, GA 30332-0535, USA

T. S. Santra and A. U. S. Shinde (eds.), *Advanced Drug Delivery*,
Studies in Mechanobiology, Tissue Engineering and Biomaterials 26,
https://doi.org/10.1007/978-981-99-6564-9_4

1 Introduction

Cells communicate with each other and with their environment using molecules, which serve as carriers of information. DNA encodes the instructions for making proteins, that carry out the majority of the functions within the cell. Lipids form cell membranes that define the boundaries of the cell and organelle's and help to control the movement in and out of the cellular environment. New technologies have made it possible to introduce synthetic materials into cells, which can be used to manipulate their behavior and function. This is an essential step in understanding the complex workings of cells and developing new disease therapy (Stewart et al. 2018). New therapies that treat several diseases requires significant research on intercellular drug delivery in biomechanics and cell biology (Pecot et al. 2011; Phalon et al. 2010; Kar et al. 2018; Dey et al. 2020). The effective and secure transportation of various biological molecules inside cells, including large plasmids for gene expression and oligonucleotides like antisense DNA or interfering RNA for protein regulation, is crucial in cell biology research and helps understand mutagenesis and genetic modifications of microorganisms (Kar et al. 2018; Lam et al. 2015; Uprichard 2005). Figure 1 represents the applications of different molecular tools such as nucleic acids (DNA, mRNA, microRNA, and siRNA), inhibitors proteins, molecular probes, and nanodevices in biomedical science (such as cell-based therapies, regenerative medicine, biomolecule manufacture, diagnosis) (Shi et al. 2018).

Traditional methods of drug delivery involve the use of oral medications, injections, or topical ointments. However, these methods have limitations such as low bioavailability, poor penetration, and systemic toxicity. In recent years, electroporation-based drug delivery has emerged as a promising alternative that overcomes several limitations. The first instance of applying electrical sparks to human and animal skin to produce electroporation occurred in 1754, but it was not

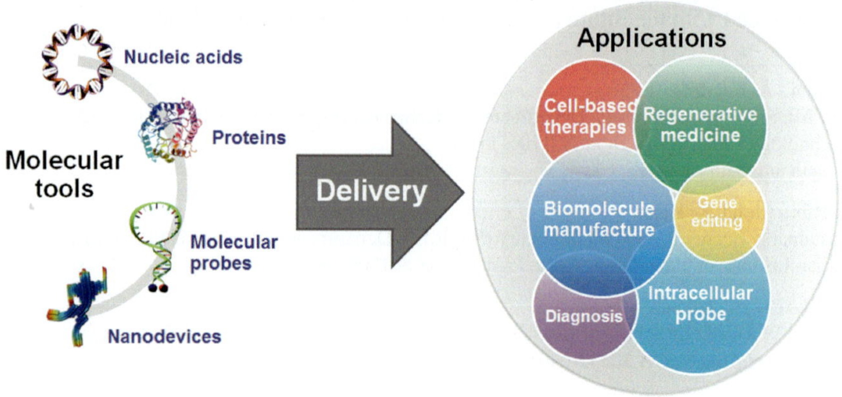

Fig. 1 Different molecular tools for intracellular delivery in biomedical science (Open access) (Shi et al. 2018)

recognized as such at the time (Kar et al. 2018). Subsequently, over the 18th and 19th centuries, numerous theoretical and experimental studies were conducted on various biological systems and their bilayer lipid membrane to better comprehend electricity's effects on biological materials (Hodgkin 1951; Yu and Peng 2017; Coster 1965; Pohl and Crane 1971; Hamilton and Sale 1967; Crowley 1973; Zimmermann et al. 1974). The term "*electroporation*" was first coined in the 1980's. Applying a powerful electric pulse on a cell or tissue causes its structures to reorganize, resulting in cell membrane permeability (Zimmermann et al. 1974; Santra et al. 2013a). The first successful gene transfer on murine cells using a specially designed electroporation chamber was accomplished in 1982 (Neumann et al. 1982). The brief administration of an electric pulse, alteration in the structure of the cell membrane due to the charging of the lipid bilayer, the creation of water-filled structures that can pierce the membrane (known as "aqueous pathways" or pores), and a rise in the mobility of molecules and ions are the key features of the electroporation based drug delivery system (Weaver 2000). Various techniques besides electroporation, such as micro precipitates, mechanoporation, photoporation, microinjection, thermoporation, sonoporation, magnetoporation, endocytosis, and liposome-mediated transport can be employed for gene transfer (Wells 2004; Chakrabarty et al. 2022; Capecchi 1980; Ewert et al. 2008; Fuller et al. 2006; Ohta et al. 2008). However, electroporation provides some advantages over other gene transfer methods. For instance, it is a quick and easy process with high reproducibility because electrical parameters can be easily controlled. Additionally, electroporation has a higher transformation efficiency compared to chemical transformations mediated by $CaCl_2$ and PEG, and the size of the pores can be regulated by adjusting the electrical pulse, which reduces the impact of cytosolic components.

Moreover, electroporation can facilitate DNA uptake into cells using lesser amounts of nucleic acid than other methods (Nickoloff 1995; Withers 1995; Prasanna and Panda 1997). Due to its versatility, it is possible to perform bulk electroporation (BEP), single cell electroporation (SCEP) and localized single-cell electroporation (LSCEP). Drug delivery through BEP can be carried out in a uniform electric field. In contrast, SCEP can introduce a non-uniform electric field that targets a single adherent or suspended cell without affecting its neighbouring cells. Currently, researchers are progressing research areas, such as LSCEP, as a fast and effective way of delivering drugs selectively and locally into a single cell from millions of cells (Santra et al. 2013a).

The primary focus of this review will be on intracellular delivery using different electroporation techniques in the drug delivery system. This review will present an overview of the history and mechanisms of electroporation, including the development of the traditional electroporation system and its principles, which are closely related to the difficulties and obstacles associated with drug delivery. Additionally, we will explore some of the most encouraging technological advancements in electroporation. Finally, we will discuss these advancements' potential clinical applications and future aspects.

2 Mechanisms

The cell membrane is primarily made up of a double layer of lipids approximately 5 nm thick. Its primary role is to create a boundary that prevents the movement of cellular components from the external environment (Shi et al. 2018; Neu and Krassowska 1999). The membrane structure is made up of phospholipid layers, where each layer has a polar head and hydrophobic hydrocarbon tails. These layers are organized so that the polar heads of one layer face the opposite direction of the polar heads of the other layer while the hydrophobic tails are positioned in the middle of the membrane (Kar et al. 2018; Neu and Krassowska 1999). The membrane functions as an electrical insulator in physiological conditions and has outstanding dielectric properties. It retains the electrical potential (~0.07 V) across the membrane by virtue of the significant disparity in ion concentration between the cytosol and extracellular fluid (Shi et al. 2018). Electroporation can successfully create pores in a cell membrane by applying a potential difference to the membrane. Once the voltage reaches a particular threshold, the likelihood of electroporation occurring in the cell membrane significantly rises. Consequently, the membrane experiences a powerful electrical force that creates a Maxwell stress causing the formation of pores (Crowley 1973; Lewis 2003; Zimmermann et al. 1976).

This force produces an electric field that alters the transmembrane potential (ΔV_m), or the difference in potential across the membrane. This transmembrane potential, $\Delta V_{m,}$ is given as (Shi et al. 2018):

$$\Delta V_m = -S \cdot r_C \cdot E \cdot \cos\theta \cdot (1 - \exp(-t/\tau)) \tag{1}$$

where, S is the cell's shape factor, r_C is the cell's radius, E is the external electric field, θ is the polar angle in the direction of the electric field, t is the time, and τ is the time constant associated with the cell membrane. For a steady-state condition ($\tau \ll t$) of spherical cells (shape factor, S = 1.5), the transmembrane potential, ΔV_m can be simplified as:

$$\Delta V_m = 1.5 \cdot r_C \cdot E \cdot \cos\theta \tag{2}$$

Researchers have extensively examined the connection between ΔV_m and pore formation using both experimental and theoretical methods. The findings show that pore development only happens in the membrane region when ΔV_m surpasses a particular threshold point, typically ranging from 0.2 to 1.0 V (Krassowska and Filev 2007; Kotnik et al. 2010). Debruin and Krassowska (1998) explored the formation and development of pores in 1998. They describe two kinds of pores: hydrophilic (conducting) and hydrophobic (non-conducting). They propose a model in which pores are initially hydrophobic, and are rapidly remodeled by changes of the lipid composition. A number of variables, including pore edge energy, membrane surface energy, steric repulsion of lipid heads, and transmembrane potential, affect the formation and stability of pores. The membrane becomes permeable when

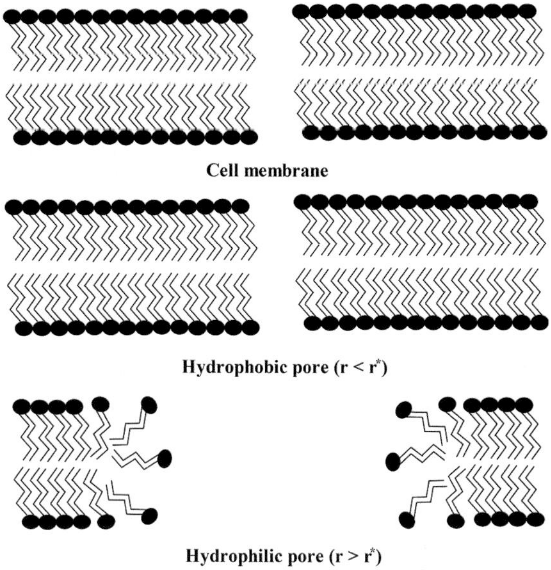

Fig. 2 Schematic representation of structural rearrangement of lipid bilayer membrane during electroporation (Kar et al. 2018) (redrawn)

hydrophobic pores with a radius larger than a specific value (r^*), around 0.5 nm, spontaneously change into long-lasting hydrophilic pores. The hydrophobic to hydrophilic transition size is correlated with the pore radius, r^* (Kar et al. 2018; Debruin and Krassowska 1998). This phenomenon of structural rearrangement of lipid bilayer membranes during electroporation is presented in Fig. 2 (Kar et al. 2018).

3 Bulk Electroporation (BEP)

The electroporation technique has extensively been used to transport various molecules into the intracellular area. Bulk electroporation (BEP) is a traditional technique for transporting molecules inside cells that involves parallel plate setups in a cuvette style. This method of BEP requires mixing the cell suspension and the molecules to be transported in a conductive buffer solution between two electrodes connected to a high-voltage electric source, as shown in Fig. 3 (Shi et al. 2018). A uniform electric source is applied to the cell suspension in a cuvette, causing rapid polarization resulting in mechanical deformation and redistribution of ionic charges across the membrane. The cell membrane is initially dielectric, but applying electric field pulses can increase membrane conductivity and cause hydrophobic pores to become hydrophilic pores (Santra et al. 2013a; Weaver and Chizmadzhev 1996). The electric field creates a roughly even distribution of electric potential across the suspension, but the cells experience different amounts of electrical stress depending on their position. As a result, the pores in the cells' plasma membranes are distributed

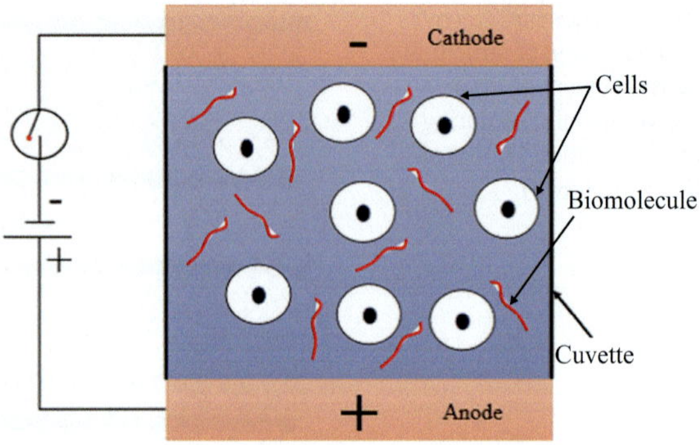

Fig. 3 Schematic representation of bulk electroporation (BEP) (redrawn with permission from Shi et al. 2018, Open Access)

unevenly, with more pores on cells facing the positive electrode. This effect is because of the negative charge on the cells, causing them to become more porous on the side of the positive side of the electrode. While on the depolarized side (negative electrode), the formation of the pores of cells is comparatively larger in size but less in number (Gehl 2003; Gabriel and Teissié 1999). Further, the threshold potential at which lipid bilayer breakdown ranges from 100 to 00 mV (Santra et al. 2013a), depending on lipid composition, and decreasing the electric pulse field can increase the breakdown voltage (Benz and Zimmermann 1980; Benz et al. 1979).

Bulk electroporation (BEP) can be reversible and irreversible. The "*reversible bulk electroporation*" exerts a strong external electric source on the cells. It can cause the cell membrane to become strongly polarized, significantly increasing membrane conductance and permeability and creating tiny openings called nanopores. However, once the external electric source is turned off, the membrane can come back to its normal insulating state from the conducting state (Zimmermann 1982). While in "*irreversible bulk electroporation*," a strong external electric source can make the membrane permanently permeable, resulting in cell death. The occurrence of irreversible electroporation was first observed in the mid-eighteenth century when a static electrical generator discharged onto the skin. The primary cause of the irreversible breakdown is due to a sudden decrease in the mean lifetime of the membrane as the voltage increases (Rubinsky 2007).

The use of bulk electroporation (BEP) in cell transfection has become prevalent in biomanufacturing due to its notable advantages. Some standard BEP devices for transfection or introducing foreign DNA or RNA molecules into cells are the Gene Pulser Xcell™, Nucleofector™, Neon® Transfection System, and MaxCyte STX® used in biomedical industries are presented in Fig. 4 (Shi et al. 2018; Brady et al.; Covello et al. 2014). The Gene Pulser Xcell™ uses electroporation, a process in which a sample of cells and foreign DNA or RNA is exposed to a high-voltage electrical pulse. The transient holes or openings in the cell membrane caused by the high-voltage pulse enable the entry of foreign genetic material (Shi et al. 2018; Covello et al. 2014; Technology N™ Manual). In Nucleofector™ the genetic material is delivered to the cells using a combination of electrical and chemical techniques. The cells are suspended in a mixture containing a specialised nucleofector solution

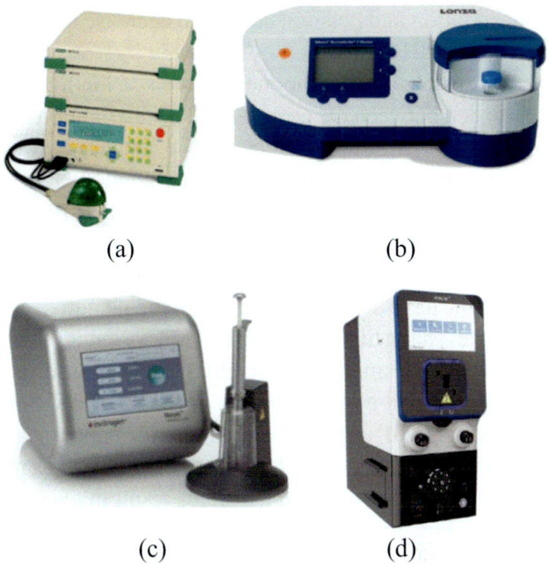

Fig. 4 Bulk electroporation (BEP) commercial devices (Open access) (Shi et al. 2018). **a** Gene Pulser Xcell™ (Open access) (Shi et al. 2018). **b** Nucleofector™ (Open access) (Shi et al. 2018). **c** Neon® Transfection system (Open access) (Shi et al. 2018) and **d** MaxCyte STX® (Open access) (Brady et al.)

and foreign DNA or RNA. The genetic material is then electroporated into the solution, which causes it to enter the cells (Shi et al. 2018; Manual, Instruction, "Gene Pulser XcellTM Electroporation system"; Duckert et al. 2021). The Neon® Transfection System also employs electroporation with a unique electrode configuration that enables a finer control over the electrical variables. The genetic material and cells are put into a disposable pipette tip after being suspended in a specific buffer. A strong voltage pulse is delivered while the tip is still placed into the electrode chamber, allowing the genetic material to enter the cells (Shi et al. 2018; Covello et al. 2014; Manual, Instruction, "Neon® Transfection Electroporation system). Flow electroporation is a distinct mechanism that is used by the MaxCyte STX® technology. The cells are sent through a small tube and exposed to an electric field with foreign DNA or RNA. The electric field momentarily damages the cell membrane allowing genetic material to enter the cells (Shi et al. 2018; Duckert et al. 2021; Manual 2022).

4 Single-Cell Electroporation (SCEP)

The strong electric field across the cells in bulk electroporation (BEP) can generate excess heat (Joule heating) due to resistance in the system. The BEP can result in low cell survival which can also lead to local pH fluctuation, electric field distortion, and irreversible electroporation because of excessive voltage (Kar et al. 2018). In contrast to the BEP, SCEP technique allows for precise transfection of a single cell with the ability to target specific cell numbers using low-voltage pulses that produce highly focused inhomogeneous electric fields. Moreover, SCEP methods allow cell concentration and position control to decrease reagent usage (Kar et al. 2018; Santra and Tseng 2016).

Figures 5a–c show the mechanisms of the SCEP in which an individual cell is positioned between the electrodes. The distribution of the electric field surrounding a single cell is illustrated in Fig. 5a. The distance between the electrodes is in microns, where a non-uniform electric field is applied. The electric potential requirements for the SCEP are lower than the BEP as the distance between the electrodes is kept small. The strength of the electric source is strongest at the central region of the cell (poles) and weakest at the ends (equator). As a result, different points on the cell membrane have different transmembrane potentials (ΔV_m). The central region of the cell (poles) has a high ΔV_m value, while the ends of the cell (equator) have a low ΔV_m value. This transmembrane potential (ΔV_m) causes pores to form at a higher density at the poles and a lower density at the equators. Further, in this process of SCEP, the electric field's strength affects the formation of pores on the cell membrane, as depicted in Fig. 5b, c. Higher electric fields lead to more pores forming at the poles of the cell, while lower electric fields result in fewer pores forming at the equator, as shown in Fig. 5b. Moreover, Fig. 5c demonstrates the resealing of the cell membrane (reversible process) upon withdrawing the electric field, allowing biomolecules to enter the cell successfully (Santra and Tseng 2013).

5 Localized Single-Cell Electroporation (LSCEP)

Over the years, researchers have been working on the concept of localized single-cell electroporation (LSCEP) (Jokilaakso et al. 2013; Santra et al. 2013b, 2014, 2020; Nawarathna et al. 2008; Chen et al. 2012). Recently, there has been a fast-growing interest in using SCEP research for biomedical applications. Nevertheless, to achieve targeted manipulation of specific organelles within cells, it is necessary to reduce the size of the electrodes to a nanoscale level. By using nanoelectrodes, it is possible to minimize toxicity and maintain high cell viability during electroporation experiments (Nawarathna et al. 2008). The nanoscale size of the electrodes allows it to intensify the electric field in the nanometer region. Single cells can be placed onto the nanoelectrode, and the cell membranes will only deform in the nanoscale area where the electrode is positioned. The remaining cell membrane area will remain unaffected, allowing drugs to be delivered from outside the cell to the inside, as shown in Fig. 6 (Santra and Tseng 2013; Nawarathna et al. 2008). Upon application of the electric field in the localized region of the cell, the pores open up, and molecules enter the cell. However, the cell membrane rescales the pores by withdrawing the electric field. Compared to BEP and SCEP processes, this novel technique requires low voltage input, improves transfection effectiveness, increases cellular viability, minimises toxicity, decreases sample volume, and lower Joule heating effects (Santra and Tseng 2013).

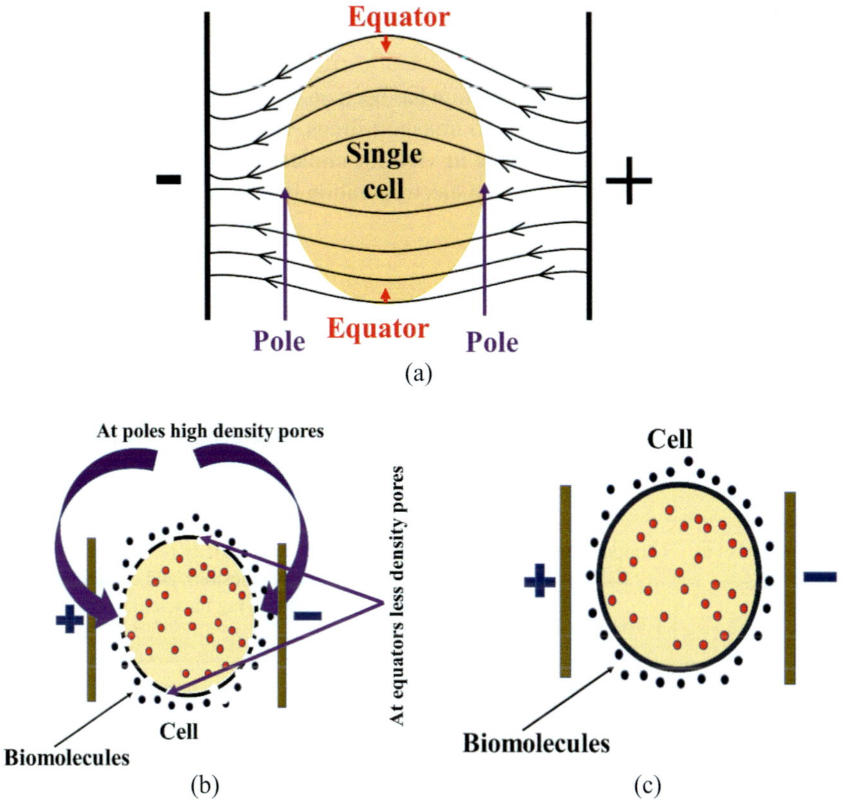

Fig. 5 **a** Distribution of electric field for SCEP. **b** Pore formation in SCEP (with electric field). **c** Regaining the cell structure upon withdrawing of the electric field (Santra and Tseng 2013) (redrawn with permission from Santra and Tseng (2013), Open Access)

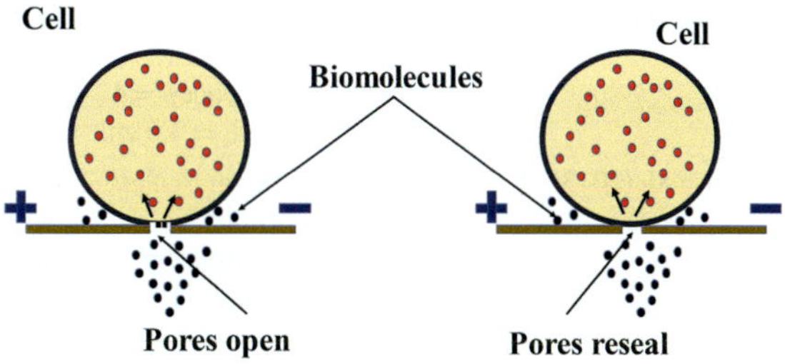

Fig. 6 Schematic representation of the localized single-cell electroporation (LSCEP) (redrawn with permission from Santra and Tseng (2013), Open Access)

6 Clinical Trials

Over the years, the use of electroporation has been on the rise. Recently, this method can be utilized in numerous ways to transport drugs, antibodies, oligonucleotides, RNA, DNA, proteins, and plasmids in vivo for clinical and biomedical purposes. Table 1 presents a description of the electroporation devices used in clinical trials (Shi et al. 2018).

Table 1 Modified table of electroporation medical devices in clinical trials (Open access) (Shi et al. 2018)

Author(s)	Electroporation device	Passage of delivery	General side effects
El-Kamary et al. (2012)	Easy Vax™	Intradermal: needle electrodes = 80 depth = 600 um	Burning sensation (few seconds) and tingling (<5 min)
Spanggaard et al. (2013)	Cliniporator	Intratumoral: needle electrodes = 2 rows of 4 linear needle electrodes depth = 4 mm between needle arrays	Mild local toxicity (erythema around the treated lesion), and contraction of the underlying muscle (transient brief)
Eriksson et al. (2013)	Dermavax™	Intradermal: 4–4-2 or 4–6-2 parallel row needle electrodes Depth = 2 mm	Momentary muscle fasciculation and Minor skin reaction (for a short span of time)
Wallace et al. (2009)	MedPulser	Intramuscular: needle electrodes = 4 depth = 1.5 cm	Injection pain (one minute)
Diehl et al. (2013)	CELLECTRA®	Intramuscular: needle electrodes = 5 depth = 18 mm Intradermal: needle electrodes = 3 depth = 3 mm	Injection pain (<10 min), involuntary muscle contraction after intradermal electroporation (one day), and local grade 1 erythema (maximum duration is for a day for a patient)
Vasan et al. (2011)	Trigrid	Intramuscular: needle electrodes = 4	Mild–moderate local pain (few minutes)

7 Pros and Cons of Electroporation

Electroporation occurs when cells are exposed to an external electric field with enough intensity to increase the permeability of the cells leading to intracellular drug delivery. Although this form of transfection has many benefits over other transfection techniques, there are some disadvantages that must be taken into account (Bolhassani et al. 2014).

Pros:

- This method is efficient for targeted delivery of molecules, including proteins, and DNA, to specific cells or tissues, allowing for more precise drug delivery (Bolhassani et al. 2014; Liu et al. 2022).
- Electroporation can achieve high transfection efficiency with minimal toxicity, allowing for more consistent experimental results.
- This method of drug delivery is a relatively fast and simple process that can be performed in a laboratory or clinical setting.

Cons:

- Electroporation requires specialized equipment and reagents, which can be costly and also requires significant technical expertise.
- The efficacy of electroporation-based drug delivery can vary depending on the cell type, the drug being delivered, and other factors, making it difficult to predict and optimize (Xi et al. 2021).
- This technique may not penetrate all cell types equally, and the effectiveness of the technique can be reduced in tissues with high electrical resistance.
- With increased electric field strength, the transmembrane potential increases, resulting in Joule heating. Joule heating in the channel which can destabilise the thermally liable samples, causing cell deaths and membrane disruption (Batista Napotnik et al. 2021; Garcia et al. 2016).
- Further, delivery through electroporation can be effective for unstructured proteins. Still, it may not be suitable for structured proteins due to the potential denaturation caused by the heat produced (joule heating) or reactive oxygen species during the process (Garcia et al. 2016).

This adverse effect may be minimized using the capillary electroporation technique. This method uses a tip-type capillary electroporator to keep efficiency high while simplifying cell transfection. The wire-type electrode-equipped capillary was developed to replace the conventional cuvette as the electroporation reaction chamber. Electroporation was quick, simple, and practical because the capillary module carrying the electrode was a disposable device. Capillary electroporation greatly enhanced the transfection rate to 70–80% compared to the conventional cuvette method by lowering the electrode's effective surface area (Kim et al. 2008).

8 Concluding Remarks and Future Aspect

The article reviews recent advancements in electroporation based drug delivery and highlights potential clinical applications. The use of electroporation based drug delivery systems to deliver therapeutic drugs to cells and tissues has shown a lot of promise. This method has been used to transport various medications, including proteins, nucleic acids, tiny compounds, and different types of cells, using specific pulse parameters. One of its key benefits is the ability to significantly improve cargo delivery into cells using electroporation-based drug delivery. This enables the delivery of therapeutic medication concentrations inside cells, which would otherwise be challenging or impossible to do using standard drug delivery techniques. Miniaturized electroporation is becoming more important in both scientific research and clinical trials as a promising non-viral method of transfection. By confining electroporation to micro- or nano-scale spaces, there are many advantages over traditional bulk electroporation, including faster uptake of molecules, more precise dosage control, and less cell disruption. Moreover, patients generally tolerate electroporation-based drug delivery devices well and are usually safe. They can deliver medications to specific body parts, lowering the danger of systemic toxicity.

The most recent electroporation platforms have excellent transfection capabilities. However, there are still many issues that need to be addressed. Numerous cells cannot yet be transfected using miniaturised electroporation techniques as effectively as traditional delivery methods. Additionally, while limiting unintended side effects, it is crucial to accurately target the transfected cells to the distinct disease locations in vivo. These approaches require a specialised level of instrument operating and microfabrication knowledge. Moreover, the increased demand for robust fabrication processes due to the micro/nanoscale of the miniaturised electroporation system provides a substantial hurdle for biomanufacturing. Owing to these problems, these innovative techniques have not been demonstrated to outperform conventional cuvette-style electroporation in clinical use.

Future studies on novel drug delivery strategies based on electroporation are anticipated to emphasise this approach more. Researchers are examining electroporation to transport pharmaceuticals to specific locations, such as the brain and spinal cord, where conventional drug delivery methods are occasionally ineffective. Moreover, the development of gene therapies may use electroporation-based medicine delivery, which can be used to deliver therapeutic genes to target cells. Biomedical research, encompassing fields like biomanufacturing, clinical trials for cancer immunotherapy, ex vivo cell-based gene therapy, and regenerative medicine, will surely be improved with improvements in electroporation technology and the creation of new molecular tools. Overall, electroporation-based drug delivery systems represent a significant technological advancement that has the potential to revolutionise the administration of drugs to cells and tissues. This technology is anticipated to contribute more to the invention of novel treatments with continuous research and development.

References

Batista Napotnik T, Polajžer T, Miklavčič D (2021) Cell death due to electroporation—a review. Bioelectrochemistry 141:107871. https://doi.org/10.1016/j.bioelechem.2021.107871

Benz R, Beckers F, Zimmermann U (1979) Reversible electrical breakdown of lipid bilayer membranes: a charge-pulse relaxation study, J Membrain Biol 48:181–204. https://doi.org/10.1007/BF01872858

Benz R, Zimmermann U (1980) 376-relaxation studies on cell membranes and lipid bilayers in the high electric field range. Bioelectrochem Bioenerg 7(4):723–739. https://doi.org/10.1016/0302-4598(80)80037-7

Bolhassani A, Khavari A, Oraf Z (2014) Electroporation—advantages and drawbacks for delivery of drug, gene and vaccine. In: Application of nanotechnology in drug delivery, pp 369–397

Brady J, Allen C, Shivakumar R, Maxcyte AV. Transfection of stem cells, primary cells and difficult-to-transfect cell lines: Jurkat, CHO, human skeletal muscle cells & primary neuronal cell transfection using a scalable, electroporation-based technology

Capecchi MR (1980) High efficiency transformation by direct microinjection of DNA into cultured mammalian cells. Cell 22(2):479–488. https://doi.org/10.1016/0092-8674(80)90358-X

Chakrabarty P, Gupta P, Illath K et al (2022) Microfluidic mechanoporation for cellular delivery and analysis. Mater Today Bio 13:100193. https://doi.org/10.1016/j.mtbio.2021.100193

Chen SC, Santra TS, Chang CJ et al (2012) Delivery of molecules into cells using localized single cell electroporation on ITO micro-electrode based transparent chip. Biomed Microdevices 14:811–817. https://doi.org/10.1007/s10544-012-9660-9

Coster HGL (1965) A quantitative analysis of the voltage-current relationships of fixed charge membranes and the associated property of "Punch-Through." Biophys J 5:669–686. https://doi.org/10.1016/S0006-3495(65)86745-5

Covello G, Siva K, Adami V, Denti MA (2014) An electroporation protocol for efficient DNA transfection in PC12 cells. Cytotechnology 66:543–553. https://doi.org/10.1007/s10616-013-9608-9

Crowley JM (1973) Electrical breakdown of bimolecular lipid membranes as an electromechanical instability. Biophys J 13:711–724. https://doi.org/10.1016/S0006-3495(73)86017-5

Debruin KA, Krassowska W (1998) Electroporation and shock-induced transmembrane potential in a cardiac fiber during defibrillation strength shocks. Ann Biomed Eng 26:584–596. https://doi.org/10.1114/1.101

Dey K, Kar S, Shinde P, Mohan L, Bajpai SK, Santra TS (2020) Microfluidic electroporation and applications. In: Microfluidics and bio-MEMS: devices and applications. Jenny Stanford Publisher Pte. Ltd., pp 343–400.

Diehl MC, Lee JC, Daniels SE et al (2013) Tolerability of intramuscular and intradermal delivery by CELLECTRA® adaptive constant current electroporation device in healthy volunteers. Hum Vaccin Immunother 9:2246–2252. https://doi.org/10.4161/hv.24702

Duckert B, Vinkx S, Braeken D, Fauvart M (2021) Single-cell transfection technologies for cell therapies and gene editing. J Control Release 330:963–975.https://doi.org/10.1016/j.jconrel.2020.10.068

El-Kamary SS, Billington M, Deitz S et al (2012) Safety and tolerability of the Easy Vax clinical epidermal electroporation system in healthy adults. Mol Ther 20:214–220. https://doi.org/10.1038/mt.2011.235

Eriksson F, Tötterman T, Maltais AK et al (2013) DNA vaccine coding for the rhesus prostate specific antigen delivered by intradermal electroporation in patients with relapsed prostate cancer. Vaccine 31:3843–3848. https://doi.org/10.1016/j.vaccine.2013.06.063

Ewert KK, Ahmad A, Bouxsein NF et al (2008) Non-viral gene delivery with cationic liposome-DNA complexes. Gene Therapy Protocols Methods Mol Biol 433:159–175. https://doi.org/10.1007/978-1-59745-237-3_10

Fuller DH, Loudon P, Schmaljohn C (2006) Preclinical and clinical progress of particle-mediated DNA vaccines for infectious diseases. Methods 40:86–97. https://doi.org/10.1016/j.ymeth.2006.05.022

Gabriel B, Teissié J (1999) Time courses of mammalian cell electropermeabilization observed by millisecond imaging of membrane property changes during the pulse. Biophys J 76:2158–2165. https://doi.org/10.1016/S0006-3495(99)77370-4

Garcia PA, Ge Z, Moran JL, Buie CR (2016) Microfluidic screening of electric fields for electroporation. Sci Rep. https://doi.org/10.1038/srep21238

Gehl J (2003) Electroporation: theory and methods, perspectives for drug delivery, gene therapy and research. Acta Physiol Scand 177:437–447. https://doi.org/10.1046/j.1365-201X.2003.01093.x

Hamilton WA, Sale AJH (1967) Effects of high electric fields on microorganisms II. mechanism of action of the lethal effect. Biochem Biophys Acta 148:789–800. https://doi.org/10.1016/0304-4165(67)90053-0

Hodgkin AL (1951) The ionic basis of electrical activity in nerve and muscle. Biol Rev 26(4):339–409. https://doi.org/10.1111/j.1469-185X.1951.tb01204.x

Jokilaakso N, Salm E, Chen A et al (2013) Ultra-localized single cell electroporation using silicon nanowires. Lab Chip 13:336–339. https://doi.org/10.1039/c2lc40837f

Kar S, Loganathan M, Dey K et al (2018) Single-cell electroporation: current trends, applications and future prospects. J Micromech Microeng 28(12):123002.https://doi.org/10.1088/1361-6439/aae5ae

Kim JA, Cho K, Shin MS et al (2008) A novel electroporation method using a capillary and wire-type electrode. Biosens Bioelectron 23:1353–1360. https://doi.org/10.1016/j.bios.2007.12.009

Kotnik T, Pucihar G, Miklavčič D (2010) Induced transmembrane voltage and its correlation with electroporation- mediated molecular transport. J Membr Biol 236:3–13. https://doi.org/10.1007/s00232-010-9279-9

Krassowska W, Filev PD (2007) Modeling electroporation in a single cell. Biophys J 92:404–417. https://doi.org/10.1529/biophysj.106.094235

Lam JKW, Chow MYT, Zhang Y, Leung SWS (2015) siRNA versus miRNA as therapeutics for gene silencing. Mol Ther Nucleic Acids 4:e252. https://doi.org/10.1038/mtna.2015.23

Lewis TJ (2003) A model for bilayer membrane electroporation based on resultant electromechanical stress. In: IEEE transactions on dielectrics and electrical insulation, pp 769–777. https://doi.org/10.1109/TDEI.2003.1237326

Liu F, Yang Z, Yao R et al (2022) Bulk electroporation for intracellular delivery directly driven by mechanical stimulus. ACS Nano 16:19363–19372. https://doi.org/10.1021/acsnano.2c08945

Manual, Instruction (2022) MaxCyte STX® electroporation system

Manual, Instruction. Neon® transfection electroporation system

Manual, Instruction. Gene pulser XcellTM electroporation system

Nawarathna D, Unal K, Wickramasinghe HK (2008) Localized electroporation and molecular delivery into single living cells by atomic force microscopy. Appl Phys Lett. https://doi.org/10.1063/1.2981568

Neu JC, Krassowska W (1999) Asymptotic model of electroporation. Phys Rev E 59:3471–3482. https://doi.org/10.1103/PhysRevE.59.3471

Neumann E, Schaefer-Ridder M, Wang Y, Hofschneider PH (1982) Gene transfer into mouse lyoma cells by electroporation in high electric fields. EMBO J 1:841–845. https://doi.org/10.1002/j.1460-2075.1982.tb01257.x

Nickoloff JA (1995) Plant cell electroporation and electrofusion protocols. Electroporation of Agrobacterium tumefaciensSpringer, New York, pp 63–72

Ohta S, Suzuki K, Ogino Y et al (2008) Gene transduction by sonoporation. Dev Growth Differ 50:517–520. https://doi.org/10.1111/j.1440-169X.2008.01026.x

Pecot CV, Calin GA, Coleman RL et al (2011) RNA interference in the clinic: challenges and future directions. Nat Rev Cancer 11:59–67. https://doi.org/10.1038/nrc2966

Phalon C, Rao DD, Nemunaitis J (2010) Potential use of RNA interference in cancer therapy. Expert Rev Mol Med 12. https://doi.org/10.1017/S1462399410001584

Pohl HA, Crane JS (1971) Dielectrophoresis of cells. Biophys J 11:711–727. https://doi.org/10.1016/S0006-3495(71)86249-5

Prasanna GL, Panda T (1997) Electroporation: basic principles, practical considerations and applications in molecular biology. Bioprocess Eng 261–264. https://doi.org/10.1007/s004490050319

Rubinsky B (2007) Irreversible electroporation in medicine. Technol Cancer Res Treat 6:255–259. https://doi.org/10.1177/153303460700600401

Santra TS, Tseng FG (2013) Recent trends on micro/nanofluidic single cell electroporation. Micromachines (basel) 4:333–356. https://doi.org/10.3390/mi4030333

Santra TS, Tseng F-G (2016) Electroporation for single cell analysis. Essentials of single-cell analysis. Springer-Verlag, Berlin Heidelberg, Germany, pp 55–84. https://doi.org/10.1007/978-3-662-49118-8_3

Santra T, Wang P-C, Gang F (2013a) Electroporation based drug delivery and its applications. Adv Micro/Nano Electromech Syst Fabrication Technol. https://doi.org/10.5772/55369

Santra TS, Wang PC, Chang HY, Tseng FG (2013b) Tuning nano electric field to affect restrictive membrane area on localized single cell nano-electroporation. Appl Phys Lett. https://doi.org/10.1063/1.4833535

Santra TS, Chang HY, Wang PC, Tseng FG (2014) Impact of pulse duration on localized single-cell nano-electroporation. Analyst 139:6249–6258. https://doi.org/10.1039/c4an01050g

Santra TS, Kar S, Chang HY, Tseng FG (2020) Nano-localized single-cell nano-electroporation. Lab Chip 20:4194–4204. https://doi.org/10.1039/d0lc00712a

Shi J, Ma Y, Zhu J et al (2018) A review on electroporation-based intracellular delivery. Molecules. https://doi.org/10.3390/molecules23113044

Spanggaard I, Snoj M, Cavalcanti A et al (2013) Gene electrotransfer of plasmid antiangiogenic metargidin peptide (AMEP) in disseminated melanoma: safety and efficacy results of a phase i first-in-man study. Hum Gene Ther Clin Dev 24:99–107. https://doi.org/10.1089/humc.2012.240

Stewart MP, Langer R, Jensen KF (2018) Intracellular delivery by membrane disruption: mechanisms, strategies, and concepts. Chem Rev 118:7409–7531. https://doi.org/10.1021/acs.chemrev.7b00678

Technology N^{TM} Manual, Instruction. Nucleofector electroporation system

Uprichard SL (2005) The therapeutic potential of RNA interference. FEBS Lett 579(26):5996–6007. https://doi.org/10.1016/j.febslet.2005.08.004

Vasan S, Hurley A, Schlesinger SJ et al (2011) In vivo electroporation enhances the immunogenicity of an HIV-1 DNA vaccine candidate in healthy volunteers. PLoS ONE. https://doi.org/10.1371/journal.pone.0019252

Wallace M, Evans B, Woods S et al (2009) Tolerability of two sequential electroporation treatments using MedPulser DNA delivery system (DDS) in healthy adults. Mol Ther 17:922–928. https://doi.org/10.1038/mt.2009.27

Weaver JC (2000) Electroporation of cells and tissues. IEEE Trans Plasma Sci 28:24–33. https://doi.org/10.1109/27.842820

Weaver JC, Chizmadzhev YA (1996) Theory of electroporation: a review. Bioelectrochem Bioenerg 41(2):135–160. https://doi.org/10.1016/S0302-4598(96)05062-3

Wells DJ (2004) Gene therapy progress and prospects: electroporation and other physical methods. Gene Ther 11(18):1363–1369. https://doi.org/10.1038/sj.gt.3302337

Withers HL (1995) Direct plasmid transfer between bacterial species and electrocuring. Electroporation Protocols Microorganisms Methods Mol Biol 47, Humana Press. https://doi.org/10.1385/0-89603-310-4:47

Xi XM, Chen-Meng XSJ, Lu R (2021) Drug loading techniques for exosome-based drug delivery systems. Pharmazie 76:61–67. https://doi.org/10.1691/ph.2021.0128

Yu C, Peng RY (2017) Biological effects and mechanisms of shortwave radiation: a review. Mil Med Res 4. https://doi.org/10.1186/s40779-017-0133-6

Zimmermann U (1982) Electric field-mediated fusion and related electrical phenomena. Biochim Biophys Acta 694(3):227–277. https://doi.org/10.1016/0304-4157(82)90007-7

Zimmermann U, Pilwat G, Riemann F (1974) Dielectric breakdown of cell membranes. Biophys J 14:881–899. https://doi.org/10.1016/S0006-3495(74)85956-4

Zimmermann U, Pilwat G, Beckers F, Riemann F (1976) Effects of external electrical fields on cell membranes, 3(1), 58–83. https://doi.org/10.1016/0302-4598(76)85007-6

Mechanoporation-Based Drug Delivery

Syed Ahsan Haider, Alexander Alexeev, Todd Sulchek, and Abhishek Raj

Abstract Cell-based therapy, genome editing and regenerative medicine are breakthrough technologies that require rapid and safe delivery of exogenous materials into huge populations of suspended cells. Different techniques have been devised to deliver the cargo of genetic modifying molecules to target cells that include, biochemical and physical techniques. This chapter presents a detailed discussion of various mechanoporation-based techniques for intracellular cargo delivery. In mechanoporation, the cell is deformed by mechanical forces, which creates momentary permeability. This temporary permeability facilitates diffusion or convection of macromolecules from the external fluid into the cell. With the use of these methods, a wide variety of molecules or compounds that can be dispersed in a solution can be delivered quickly and directly. Further, the chapter provides the historical background, operating principles, benefits, and drawbacks of mechanoporation processes. The chapter also discusses the advancement of new microfluidic and nanotechnological techniques along with their fabrication procedures that have enabled unprecedented levels of control over the membrane disruption process. Particular focus is placed on their applications, implementation challenges, and a discussion of potential future applications, if any. Finally, a detailed comparison of various mechanoporation-based drug delivery systems is presented. In future research on mechanoporation, there can be a focus on harnessing the distinct advantages of different intracellular delivery techniques to create multifunctional platforms capable of delivering biomolecules independently or concurrently, even under less severe operating conditions.

S. A. Haider · A. Raj (✉)
Department of Mechanical Engineering, Indian Institute of Technology Patna, Bihta 801106, Bihar, India
e-mail: araj@iitp.ac.in

A. Alexeev · T. Sulchek
George W. Woodruff School of Mechanical Engineering, Georgia Institute of Technology, 801 Ferst Drive, Atlanta, GA 30332-0405, USA

T. Sulchek
Wallace H. Coulter Department of Biomedical Engineering, Georgia Institute of Technology, 313 Ferst Drive, Atlanta, GA 30332-0535, USA

T. S. Santra and A. U. S. Shinde (eds.), *Advanced Drug Delivery*,
Studies in Mechanobiology, Tissue Engineering and Biomaterials 26,
https://doi.org/10.1007/978-981-99-6564-9_5

1 Introduction

A cell's interior is enclosed by the cell membrane, also referred to as the plasma membrane, which separates the cell's internal environment from the external environment and regulates the passage of materials inside and out of the cells. The cell membrane is made up of a double layer of phospholipids arranged in a bilayer with proteins, glycoproteins, and glycolipids embedded within it.

Cellular communication relies on molecules as messengers, much like electronic signals in computers. DNA encodes RNA and proteins, which are crucial for exchanging information within cells. Delivering materials and molecules into cells is essential to understand cell function, manipulating behaviour, and developing illness treatments (Stewart et al. 2018). The uptake of exogenous molecules by the cells, such as drugs, genes, and proteins, is restricted by the plasma membrane, which functions as a barrier. While several natural mechanisms allow molecules to enter the cells and get secreted through the cell membrane (e.g., passive diffusion, endocytosis, exocytosis, active and co-transport), many synthetic macromolecules require alternative ways to enter the cell interior (Meacham et al. 2018).

To advance biomedical applications, it is necessary to efficiently deliver impermeable cargo molecules across the cell membrane, such as nucleic acids, antibodies, and nanomaterials, such as quantum dots (Lukacs et al. 2000; Belting et al. 2005; Lechardeur et al. 2005; Vaughan et al. 2006). Conventional approaches to intracellular cargo delivery may utilise viral vectors and chemical transfections, which incur many limitations (Meacham et al. 2018). For example, chemical transfection using cationic lipids may disrupt cell membranes and cytotoxicity. Its effectiveness is limited by cell types and culture conditions, resulting in inconsistent outcomes. Conversely, viral vector methods are hindered by reduced efficacy due to targeting difficulties for specific cells, limited cargo capacity, and increased manufacturing complexity, as well as undesired host cell immune response. Physical approaches, such as mechanoporation, offer advantages to allow the uptake of extracelleluar cargo molecules without the need for further modifications of the molecules to be delivered or the needing the use of viral vectors (Meacham et al. 2014; Williams et al. 1999; Stewart et al. 2016). Mechanoporation platforms based on micro/nanofluidics additionally offer precise control over various parameters that affect delivery, and that manipulates cellular behaviour at the single-cell level. As a result, mechanoporation has become a promising method for both intracellular delivery and analysis, which could have potential uses in therapeutics, diagnostics, and disease analysis.

This book chapter provides a comprehensive overview of mechanoporation techniques, including the mechanisms underlying the techniques, different methods used, various tools and materials required, and the advantages and limitations of these techniques.

2 Basic Mechanism

The mechanisms of mechanoporation can vary based on the specific technique employed, as well as the magnitude of the force applied and the type of cell involved. The basic principle uses mechanical forces to modify cells to accept various substances, such as small molecules or drugs by disrupting the cell membrane. (Chakrabarty et al. 2023). This results in the formation of transient pores in the membrane, allowing the molecules to enter the cell. Unlike carrier-based methods, mechanoporation is not limited to specific combinations of cargo and carriers and can deliver almost any type of cargo that can be dispersed in a solution (Kaladharan et al. 2021). There are two main methods of mechanoporation: active penetration and permeabilisation.

Active penetration methods require a conduit or vehicle to penetrate the cell membrane and create a path for exogenous molecules to enter. The vehicle could be solid, such as a needle, or liquid, like a micro jet of fluid that causes a localised rupture in the cell. In contrast, permeabilisation strategies temporarily increase the permeability of the cell membrane to allow extracellular cargo to enter. The membrane is called permeable when the generated transient pores are both large enough and long-lasting enough to permit the passage of the cargo. The critical parameters to be controlled in both these methods are the size and duration of the membrane disruptions.

Applying mechanical force to cells can take several forms, such as ultrasound waves or pressure, resulting in stress on the plasma membrane and the generation of openings. These openings facilitate molecule entry to the cell but it can also lead to cell resource loss. Different molecules, including DNA and proteins, can be delivered into the cell through these pores. After delivery, the plasma membrane reseals to restore its integrity and prevent leakage of intracellular contents, as excessive or prolonged poration can lead to cell damage or death. The speed and efficiency of resealing can depend on various factors, such as the magnitude and duration of the mechanical stress, the type of cell being mechanoporated, and the presence of any additional factors or substances that may affect membrane repair.

There are various factors that influence the phenomenon of mechanoporation. These factors include the type of deformation experienced by cells, the duration, frequency and rate of deformation, and the dimensions of the pores that are formed. The type of deformation caused by the forces applied is critical to the results in mechanoporation studies. For example, in experiments where cells pass through excessively constricted channels, it has been observed that the cell's nuclear membrane can rupture, leading to DNA damage (Raab et al. 2016; Harada et al. 2014; Mayr et al. 2002). The threshold for compression damage varies among different cells, and further research is needed to clearly articulate the factors that lead to prolonged cell damage.

Similarly, there is a threshold for tension-induced lysis, where cells may die if their plasma membrane is stretched beyond its limit. The cell can, however, increase its surface area in response to the stretching to relieve the excess tension generated. The cell's ability to ease membrane tension is limited by its reservoirs, such as caveolae, filopodia, endocytic pits, microvilli, and loose membrane folds, from where lipids are pulled to increase the surface area in response to external effects (Clark et al. 2014; Sinha et al. 2011). However, global and local limits exist to using resources from these reservoirs, determining how much the cell can stretch before rupture occurs. Although, the excess surface area of the cell membrane is believed to be between two and ten times the observable surface area of the cell (Clark et al. 2014), in reality, cell membranes can typically endure mechanical area strains of up to 3% before losing integrity (Bloom and Evans 1991).

The speed and rate at which force is applied also play a crucial role in determining membrane disruption behaviour. For example, in cases where nanowires pierce cells, the piercing velocity has been shown to have a more significant impact than the final deformation magnitude caused by the nanowires. Slow piercing rates allow lipids to flow from the reservoirs to compensate for local tension increase. In contrast, high deformation rates do not provide enough time for the cell to bring in resources to the target site, leading to the energetically favoured generation of pores or holes in the membrane (Marmottant et al. 2008).

Another crucial aspect in determining the success of mechanoporation methods is how the cell membrane repairs itself after being disrupted by external factors. After a membrane breach, cells respond by rapidly repairing the membrane to avoid losing resources that risk cell death. The speed of this repair response can determine whether the cell will heal properly or undergo permanent changes in morphology that may affect its viability. The size of the pores created heavily influences this repair response. While larger pores (>0.2 μm) can facilitate the transfection of larger cargo volumes, they can also cause significant trauma to cells and are quickly detected and repaired. On the other hand, minor disruptions, such those caused by electroporation or residual pore-forming toxins, can last for extended periods of time and deplete the cell's resources over time (Jimenez et al. 2014; Bhakdi et al. 1993; Weaver 2000). Therefore, it is important to carefully consider pore size in mechanoporation methods to ensure that the cell achieves its initial homeostatic conditions without compromising its potency or causing morphological changes after plasma membrane recovery.

3 Different Methods of Mechanoporation

There are several methods used in mechanoporation to introduce molecules into cells. Some common methods include (Fig. 1):

Microinjection: Microinjection is a classic method of delivering molecules into cells, involving the use of a micropipette tip of small diameter (0.1–5 μm) to penetrate the plasma membrane and introduce a solution with the desired cargo and was the first intracellular method to be used for cargo delivery (Korzh and Strähle 2002). The initial application of this method was documented in the cloning of bacteria,

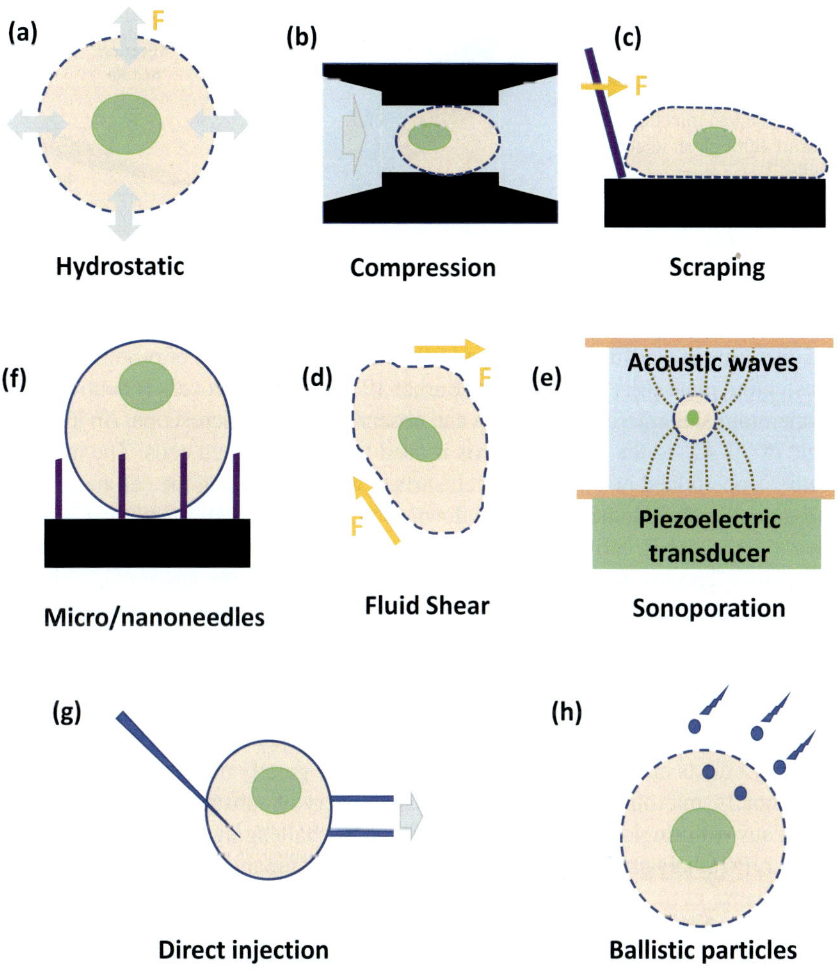

Fig. 1 The above image displays various methods used for intracellular delivery. These include permeabilisation-based techniques such as **a** hydrostatic, **b** cell compression, **c** cell scraping, **d** fluid shear-based, **e** sonoporation, and active penetration methods like **f** micro/nanoneedle arrays, **g** microinjection, and **h** ballistic particles (biolistic). It should be highlighted that sonoporation primarily relies on permeabilisation for transfection. However, in the event of bubble collapse, the primary method of cargo transfer can be active penetration, as it causes localised membrane rupture that facilitates cargo transfer. Similarly, in some instances, irradiation of acoustic waves can cause a bubble to directly impact the membrane, resulting in membrane disruption through a direct penetrating bubble (redrawn with permission from Stewart et al. (2018), copyright 2020, American Chemical Society)

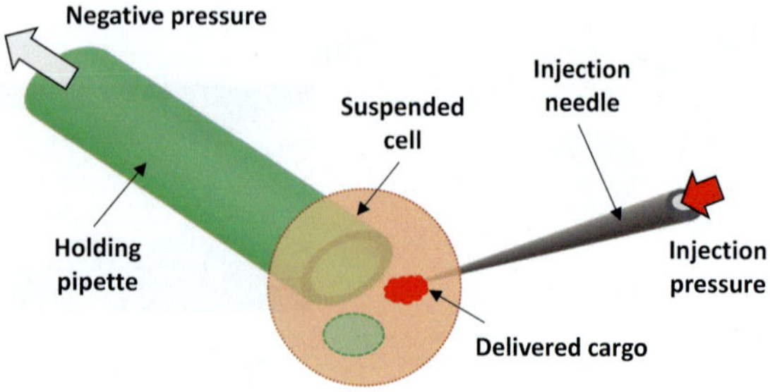

Fig. 2 The microinjection of suspended cells uses a holding needle to immobilise the cell with negative pressure and an injection pressure through an injection needle

and over time, it was further developed for influencing the nuclei of protozoa and for transfecting plant cells with bacteria (Barber 1911a, b). The process is controlled by an operator using micromanipulators and observed under a microscope. An injection angle of 30–45° to the horizontal axis is used to inject attached cells. The injection needle is positioned just above the cell surface without piercing the cell membrane. Subsequently, the needle is lowered down to penetrate the cell membrane, and the injection process is carried out using a microinjector under control. The method can also be employed to transfect suspended cells where the suspended cells are immobilised using a holding pipette that uses a suction pressure to immobilise the cell, and injection is achieved using positive pressure (as shown in Fig. 2) (Zhang and Yu 2008; Munish and Santra 2016).

As a physical delivery method, microinjection offers several advantages for well-controlled experiments. As the delivered material serves as the only independent variable, any effects brought about by the substance are greatly magnified and isolated. Additionally, microinjection enables precise delivery of substances into the cytosol or nucleus with single cell precision, which can be challenging to achieve with other methods (Muthaiyan Shanmugam and Manoj 2022; Kumar et al. 2018; Shinde et al. 2020; Santra et al. 2020a). To differentiate between the injected and non-injected cells, a membrane impermeable marker dye or a fluorescence-conjugated protein can be injected into the cells along with the target molecule, enabling the targeted cells to be easily identified (Zhang and Yu 2008).

Microinjection is also a time-consuming and labour-intensive technique with a slow delivery rate, limiting its use to low-throughput processes such as in vitro fertilisation and transgenic animal creation (Mehier-Humbert and Guy 2005). For applications requiring the processing of many cells, it is not commonly used due to its demanding nature compared to other techniques.

Micro/nanoneedle arrays: Microneedle diameters are typically 1–300 μm, with lengths up to 1 mm. Microneedles are a simple and pain-free solution for transdermic delivery of a diverse range of molecules (Maaden et al. 2012; Harshan et al. 2020). Similar to microneedles, nanoneedles come in a range of shapes and sizes and can be manufactured using a variety of materials, with the selection of dimensions depending on the particular application being targeted. Their diameters are typically in the range of 10–200 nm, with a length 1–5 μm (He et al. 2020). Such minute dimensions result in minimal membrane disruption during penetration, minimising cell damage. However, recent studies suggest that even with such marginal dimensions, nanoneedle arrays can alter the growth and proliferation rates of the cells. In longer nanowires, this effect on cell division rate after transfection is even more pronounced (McKnight et al. 2004; Persson et al. 2013).

Micro/nanoneedle arrays that are intricately fabricated, high aspect ratio nanostructures with diameters in the range of nanometers while lengths in the range of micrometres are utilised for mechanoporation by delivering molecules into cells through direct penetration (Maaden et al. 2012; He et al. 2020; Adamo and Jensen 2008; Chakrabarty et al. 2022). In this process, needles are produced on a disposable microfluidic device, and cells are injected into them, akin to microinjection techniques (Liu and Sun 2009) (Fig. 3a). For improved efficiency, the devices utilise massively parallel injector arrays (Zhang et al. 2012) (Fig. 3b, c). A combination of channel geometry and the applied pressure is used to immobilise the cells to ensure accuracy and appropriate penetration through the needles (Dixit et al. 2020; Park et al. 2016). Following the injection, cells are freed from the capture sites via flow and pressure conditions changes.

Shear-based methods: Applying mechanical stresses in confined flow geometries induces transient pore formation in cell membranes, utilising shear-based methods. Fluid shear forces have the potential to disturb lipid bilayers in multiple ways. When water flow in parallel to a plasma membrane surface at a high speed, it can induce a tilt in the lipid heads towards the direction of shear, leading to buckling instabilities and, eventually, the breach of the bilayer (Hanasaki et al. 2010). Alternatively, a microjet of water molecules can pierce a membrane perpendicularly, like a mechanical object (Yuan et al. 2015). Shear-based devices can vary in their shape and size. For instance, cone plate shearing devices can adjust the flow shear rate by changing the rotational speed of the cone (Hallow et al. 2007) (Fig. 4). Simpler flow shear devices, with geometries of microchannels that have diameters comparable to cell size use fluid velocity profiles to generate fluid shear forces (Davis et al. 2005). Shear-based membrane disruption is less invasive than that caused by solid contact. However, managing the shear forces generated by the fluid can be a challenging task. Applying regulated pressure-driven flow and constriction channels provides a means to impart a shear force on cells. Parameters critical to this process include the magnitude and duration of the introduced shear stress (Zarnitsyn et al. 2008). This technique has the potential to be universal, as it has shown broad applicability to the many types of cells and transported cargo (Sharei et al. 2013a). However, due to the dimensions, cells must pass single-file resulting in a low yield that can be increased using parallel arrays of flow constrictions in microchannels or orifice plates. This approach has facilitated cargo delivery such as proteins, siRNA, and quantum dots into primary and stem cells at rates of up to 1×10^5 cells/s (Hallow et al. 2007).

Biolistics (Penetrating/ballistic particle delivery): In 1987, the first use of biolistic delivery was reported for DNA transfection in plants (Klein et al. 1992). It later became popular for transfecting microbes that were challenging to transfect by other

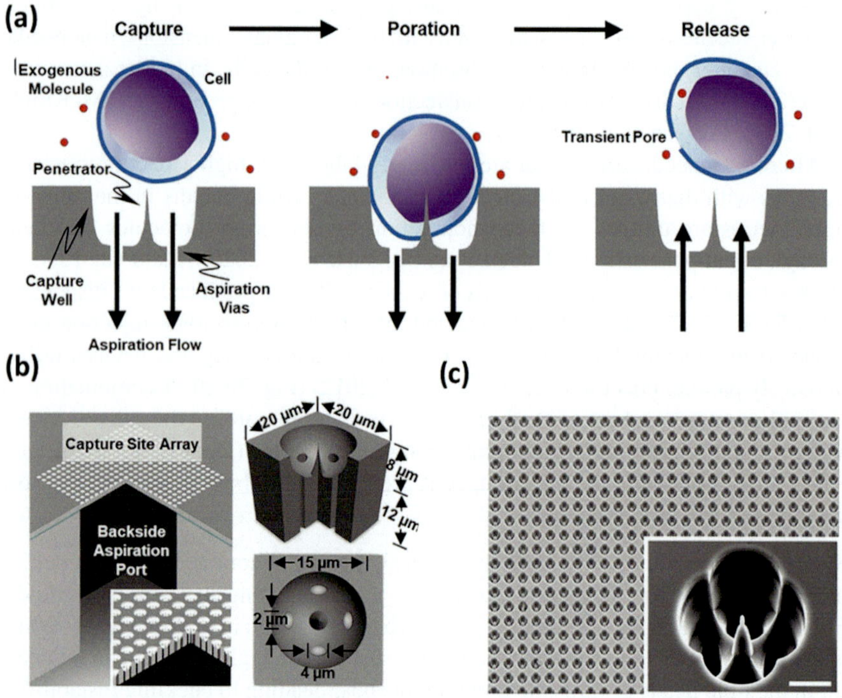

Fig. 3 **a** The process involves capturing cells using negative aspiration flow, creating a pore in the cell membrane through impingement on a penetrator, and releasing the cells by reversing the flow. Intracellular delivery occurs via diffusive influx of extracellular materials through the opening created in the cell membrane. **b** The design schematics display the key features of the device with one fourth of the section removed in isometric views for better visualisation. The actual device comprises a 100 × 100 array of capture sites. Image credits go to the source. **c** A scanning electron micrograph displays a a small section of the device array, with a higher magnification image of a single capture site shown in the inset. The scale bar equals 5 μm (reprinted with permission from Dixit et al. 2020, copyright 2020, American Chemical Society)

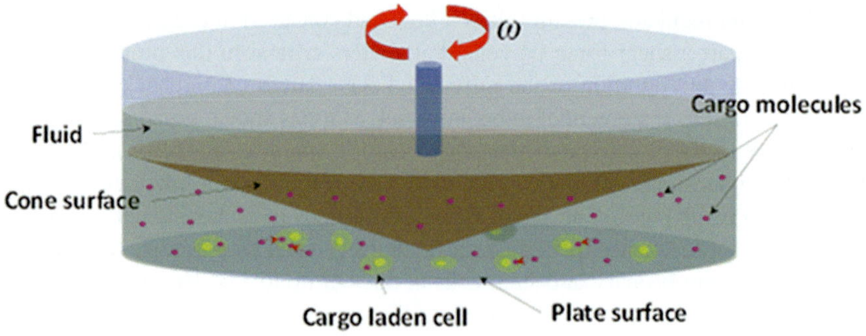

Fig. 4 Schematic of a cone plate shearing device. A servo motor propels the cone, and the shear within the fluid can be adjusted by modifying the cone's rotational speed, ω

methods. The process involves launching particles coated with desired material at high speeds into the cell. The method is classified as a membrane disruption due to the active force required to puncture the cell membrane (Stewart et al. 2018).

To carry out biolistics, tiny heavy metal particles (usually gold or tungsten) are covered with the desired cargo and mixed in a solution that is then put onto the surface of a projectile. The particles are then accelerated using a gas shock that can be generated through methods like a chemical explosion, high-voltage electric spark or helium discharge (Klein and Fitzpatrick-Mcelligott 1993; O'Brien and Lummis 2006a). Finally, the projectile is suddenly stopped (such as by a mesh), which releases the microparticles at great speeds towards the target cells or tissues.

Biolistic delivery can be used to transfect cells and tissues that are challenging to transfect through other means, including neurons, organotypic brain slices, microorganisms, and skin or muscle inoculation for vaccination (O'Brien and Lummis 2006b; Klimaschewski et al. 2002; Usachev et al. 2000; Mcallister 2000; Wellmann et al. 1999; Kikkert 1993; Khan 2002). Projectile bombardment is effective, even with the transfection of a small percentage of cells, making it suitable for immunisation. However a balance is needed to be achieved between power, size, amount of bombarding particles and cell survival rate for intracellular delivery to skin cells. The trend is towards smaller, less damaging projectiles and attempts to improve consistency in cell treatment could lead to beneficial outcomes if limitations related to cargo and cell type suitability and uniformity of cell treatment can be solved.

Cell deformation: This method utilises a robust and cost-effective microfluidic platform to induce rapid cell deformation that leads to high strain. There are several methods to induce cell deformation. One method uses narrow pores to induce multistage deformation of cells as they pass through (Sharei et al. 2014). Membrane poration is achieved by combining hydrodynamic forces in microchannels and gradual squeezing with contact-based shearing in microchannels with a width smaller than the cell diameter leading to membrane shear disruption. After the transient pores open in the membrane, cargo molecules diffuse into cell interior untill the pores reseal and membrane recovers. Figure 5 depicts the mechanoporation process used for loading suspended cells with cargo. The factors that control cell deformation and delivery include transit duration through the constriction, determined by the applied pressure and resulting flow rate of the cell-biomolecule suspension, and associated non-dimensional characteristics such as Reynolds, Weber and Bond number among others (Zhang et al. 2018). Cell stiffness and diameter can also play a significant role in determining travel duration (Adamo et al. 2012). Other essential delivery characteristics include the dimensions of the channel sections and number of the constriction channel in series, and concentration of cargo molecules in the solution. High transfection efficiency requires a higher concentration of deliverable material, as the delivery process is mainly diffusion-dominated. Additionally, the size and diffusivity of the cargo molecule also affect transfection efficiency (Sharei et al. 2013b; Williams et al. 1999; Uvizl et al. 2021). Commercial microfluidic platforms have been developed using this method of cell transfection, and they have found various applications, including the delivery of quantum dots and cargo to primary immune cells

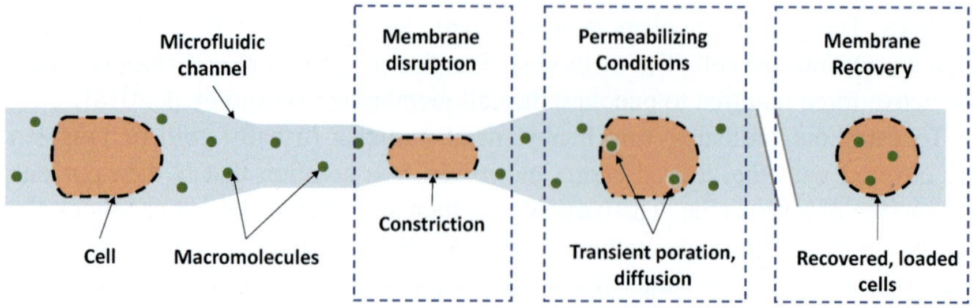

Fig. 5 Microfluidic cell compression involves the rapid deformation of cell shape during the passage, leading to the disruption of cell membranes. Redrawn from Uvizl et al. (2021) (open access)

(Sharei et al. 2013b). Compared to shear-induced and direct penetration methods, this approach achieves high transfection rates among cells with diverse mechanical characteristics without compromising their survival and proliferation rates.

An example of an approach for cargo delivery through abrupt cell compression is the use of chevron ridge-based microfluidic devices for Volume Exchange for Convective Transfer (VECT) (Fig. 6), In this approach, rapid cell compression induce membrane poration and cell volume reduction. Cell recovery drives an influx of the medium with suspended cargo molecules into the cell interior (Liu et al. 2018). VECT has shown promising results in delivering a cargo of varying sizes into different cell lines, including natural killer (NK) cells, T cells, HeLa cells, adipose-derived stem cells and hematopoietic stem and progenitor cells (HSPCs) (Kiru et al. 2022; Nejadnik et al. 2020; Sciolino et al. 2022; Loo et al. 2021; Jung et al. 2022). Notably, post-delivery, cells loaded with cargo exhibit excellent recovery and continue to grow exponentially, with minimal changes in cell morphology (Sciolino et al. 2022; Loo et al. 2021). This can be attributed to the fact that VECT does not subject cells to significant prolonged stress during cargo delivery. The convective transport results in a wide range of molecule size being delivered through the delivery of cargo that was influenced by the dimensions and velocity of the cell in the microchannel, with the effectiveness of the process was determined by the number of ridges and the gap distance between them in the microchannel (Liu et al. 2020). Moreover, for methods that allow wider channels, throughput can be increased.

Hydroporation: Although cell squeezing techniques can achieve high throughput rates, they are often impeded by device failure due to channel obstruction. To tackle this problem, multiple research groups have created microfluidic devices that utilise hydrodynamic forces for controlling the flow of cells and fluids, known as hydroporation (Adamo et al. 2013; Deng et al. 2018). In this method, single cells are poked, elongated, or compressed (or a combination thereof) in a microchannel with the help of generated hydrodynamic forces, resulting in transient pores for material delivery through diffusion or fluid convection (Fig. 7). (Jarrell et al. 2019) demonstrated an alternative approach to affect cells by utilising hydrodynamic forces, specifically

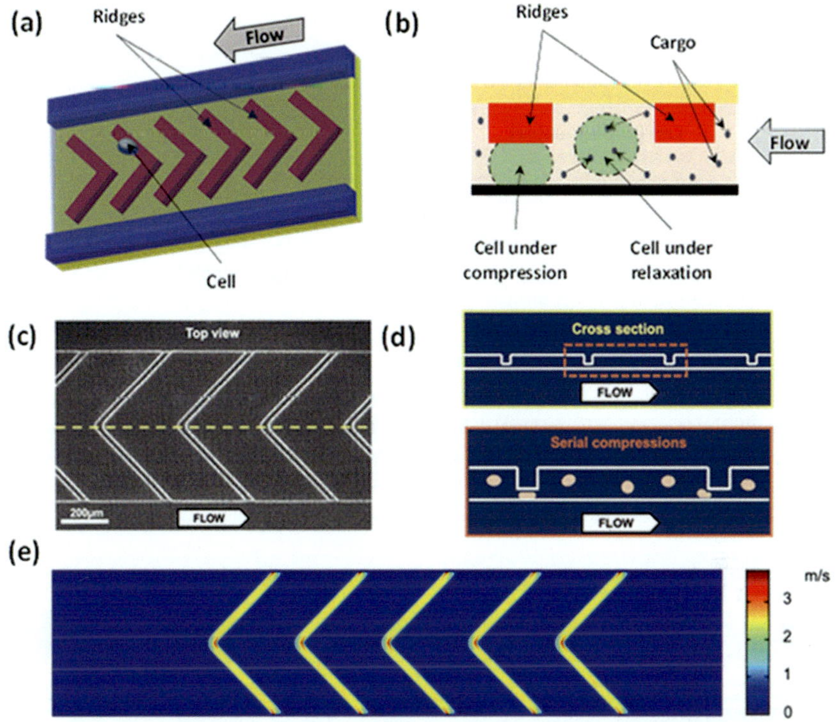

Fig. 6 Isometric diagram of the microfluidic device utilised in volume exchange for convective transfection (VECT). The figure shows: **a** the arrangement of ridges in the microfluidic channel, **b** a cross-sectional view of the microchannel exhibiting the uptake of cargo molecules by cells from the surrounding medium, **c** a top view of the VECT device, **d** a compressed cell undergoing serial compression cycle in the device, and **e** a velocity profile of the fluid from fluid simulations (open access (Loo et al. 2021))

through the creation of vortices in microchannels (Fig. 8). The resulting mechanical forces were then applied to the cell membrane. By utilising a technique called microfluidic vortex shedding (μVS), the group successfully delivered mRNA into primary human T lymphocytes, achieving both high cell recovery rate and transfection efficiency. The (μVS) method was later employed by Indee Labs in their commercialised product, the Hydropore™, which enables swift delivery of diverse cargos to immune cells within seconds (Morshedi Rad et al. 2021). The device has demonstrated high transfection rates and excellent cell viability.

Compared to constriction channels, hydroporation methods present a lower risk of device clogging and cell lysis. These approaches offer several benefits, including their straightforward design, use of cost-effective equipment, and ability to efficiently deliver diverse biomolecules to various cell types at high throughput rates.

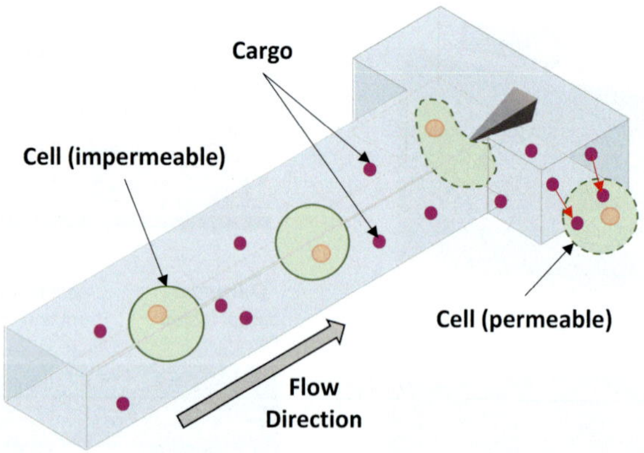

Fig. 7 Inertial hydroporation for material delivery. The interaction between the cells and the channel geometry causes the cells to become permeable, enabling the transfer of extracellular nanocargo into the cells (redrawn with permission from Deng et al. (2018), copyright 2018, American Chemical Society)

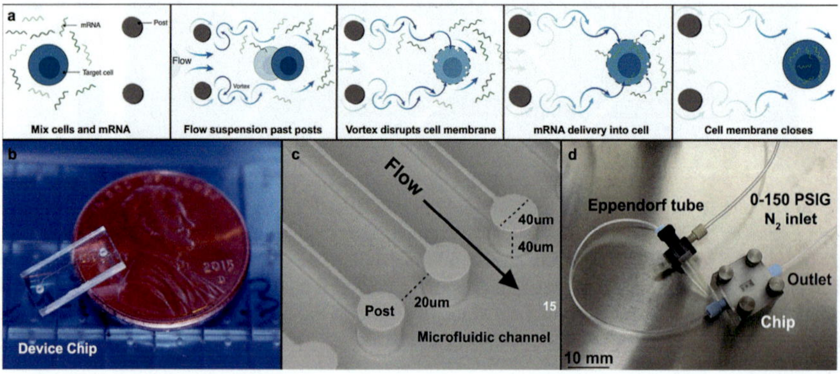

Fig. 8 **a** An overall process overview, **b** a microfluidic chip of a specific size, **c** a schematic that depicts the flow regions and poles needed to generate vortices, and **d** a pneumatic unit used to regulate the flow rate within the microfluidic chip (open access (Jarrell et al. 2019))

Scrape/scratch loading: The scrape loading method involves using a scraping tool to both remove and permeabilise cells, making it effective in delivering various substances like proteins, antibodies, peptides, and dyes. It can also be used for transfecting plasmids (Frankel and Pabo 1988; Bernat et al. 1990; Adler et al. 1996; Fechheimer et al. 1987). Another similar method is scratch loading, which involves dragging a sharp object across a layer of cells to cause membrane damage, allowing for the delivery of dextrans, nucleotides, and quantum dots to nearby cells (Legenzov et al. 2015; Emerson et al. 2014; Schermelleh et al. 2000). One advantage of scratch loading is that cells can be imaged immediately while remaining adherent, although it has a lower throughput than scrape loading. These techniques have been demonstrated to effectively deliver a range of materials to cells.

Sonoporation: Sonoporation is a method of delivering substances into cells by using pressure waves generated in the ultrasound frequency range, typically between 20 kHz to GHz. The technique was developed in the mid-1980s to permeabilise cultured cells using ultrasound. This was done by applying three half-second ultrasound pulses to a plastic tube containing cell suspensions. This rudimentary approach was used to load dextran and proteins into amoebae (Fechheimer et al. 1987; Emerson et al. 2014; Fechheimer and Taylor 1987; Furukawa et al. 1988).

In a more recent development, microbubbles are used to transfect cells to specific areas in the body. Microbubbles are small particles filled with gas and surrounded by a stabilising shell of lipids, proteins, or polymers (Dijkmans et al. 2004). When a frequency close to the resonance frequency of the bubbles is applied, the bubbles generate acoustic waves through oscillations.. This can render the plasma membrane porous, which facilitates the cellular uptake of drug molecules present in the extracellular fluids (Morgan et al. 2000; Dayton et al. 2002). This enhanced cellular uptake is attributed to the increased porosity of the cell membrane due to the interaction with these oscillating bubbles in the ultrasound field (Deng et al. 2018). Several mechanisms have been proposed to explain the formation of pores in the cell membrane through sonoporation. These mechanisms include (Kotopoulis et al. 2014): (A) Push, where a microbubble in its expansion phase touches the cell membrane and pushes it apart, potentially causing pore formation. (B) Pull, where the cell membrane is drawn towards the contracting microbubble, trying to fill the space left by a contracting microbubble, potentially compromising the integrity of the cell membrane. (C) Jetting, where the asymmetric collapse of a bubble creates a conical-shaped protuberance through the bubble that is directed towards a boundary, creating a fluid jet that disrupts the cell membrane. (D) Streaming, where fluid streaming around oscillating microbubbles attached to cell membranes generates enough enough fluid shear forces in the surrounding liquids, that it could rupture the membrane. (E) Translation, where microbubbles covered in lipids may directly pass through the plasma membrane due to radiation forces, potentially losing part of their lipid shell during the process. Figure 9 shows the different effects of microbubble cavitation when exposed to ultrasound.

4 Device Fabrication

Some mechanoporation techniques require simple equipment, such as scrape/scratch loading, where the user scratches the cell adhered to a wall with the help of needles, rubber spatula or glass beads to make the cell porous momentarily. Glass beads can also be used for this purpose. For this, the adherent cells and glass beads are put in a flask and shaken (McNeil and Warder 1988). The collision of glass beads with cells

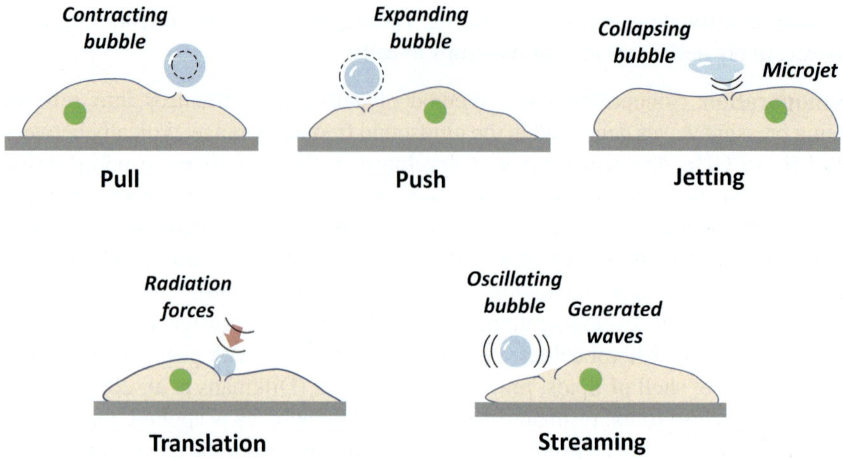

Fig. 9 Generation of sonoporation. Redrawn with permission from Rich et al. (2022), copyright 2021, John Wiley and Sons

can render the cells permeable, but there is a potential risk of dislodging adherent cells.

Another cheap and straight forward technique is using a simple syringe for shear-based mechanoporation techniques. To transiently permeabilise cells using this method, cell suspensions are mixed with high cargo concentration and passed back and forth through fine gauge syringe needles (Clarke and McNeil 1992). Other shear-induced mechanoporation may generate fluid shear forces by driving cell suspension through narrow constrictions. For instance, cone and plate viscometers can produce and control hydrodynamically applied shear stress over cell monolayers (Laplaca et al. 1997). The method of μVS to render the plasma membrane porous via fluid shear has also been leveraged in developing a commercial microfluidic device that achieves shear-induced transfection of a wide range of cells by regulating the flow rate (Morshedi Rad et al. 2021). While manual flow control in shear based mechanoporation method offers precision and reproducibility, combining this strategy with other techniques, such as sonoporation, can further enhance its efficacy.

However, many mechanoporation methods do require specialised tools. For example, sonoporation requires ultrasonic transducers to generate high-frequency sound waves that create temporary pores in the cell membrane. In vitro studies on ultrasound-mediated membrane, poration have generally been conducted using custom experimental setups consisting of commercial electronics, custom or commercial ultrasound transducers, custom sample chambers, and micropositioning systems. Alternatively, some implementations have employed commercial therapeutic ultrasound probes submerged directly into cells cultured in multiwell plates (Duvshani-Eshet et al. 2006; Karshafian et al. 2009; Schlicher et al. 2010). The addition of echo contrast microbubbles (such as Albumex) has significantly improved transfection efficiency, suggesting that acoustic cavitation is likely the underlying mechanism in the sonoporation process (Bao et al. 1997). However, the use of sonoporation for intracellular delivery has mainly been limited to the ultrasound community despite the availability of commercial systems. The diffraction limit problem limits Sub-MHz ultrasound technology, but researchers are developing miniaturised ultrasound point sources with sub-millimetre confinement using fibre-based optoacoustic emitters (Li et al. 2009).

Similar to sonoporation, biolistics also requires specialised tools. Biolistic accelerators come in various types and utilise different methods to accelerate particles. One such device, commercially available for a long time, is the Biolistic PDS-1000/He, which is used for transforming cells and operates at a pressure range of 450–2200 psi (Kikkert 1993). This device used helium pressure and vacuum to accelerate particles coated with recombinant DNA into target cells. The Helios gene gun, manufactured by BioRad Laboratories in Hercules, CA, USA, and the Accell gene gun, developed by Agracetus, Inc. in Middleton, WI, USA, are among the other gene gun devices that are available commercially (Wang et al. 2004). The sample is stored in chambers with a maintained vacuum, as this determines the drag experienced by the particles. Particles of sub-micron diameters of gold, platinum, tungsten or iridium are commonly used (Harrier and Millam 2001). Gold particles have advantages such as uniformity and non-toxicity to cells and do not attack DNA bound to cells catalytically, unlike tungsten particles. However, they are costly and unstable in sterile solutions, leading to irreversible agglomeration. In recent times, there has been a growing trend towards smaller, less damaging particle alternatives (Yu and Peng 2017; O'Brien and Lummis 2011).

Microinjection is another technique that requires dedicated equipment. One of the traditional methods for microinjection involves using an inverted microscope to provide sufficient distance for the injection needles, a light source for illumination, micromanipulators for needle positioning, microinjectors for applying positive pressure to inject materials or negative pressure to hold suspended cells, and an injection chamber for maintaining a cell survival environment (Zhang and Yu 2008). For microinjection systems, low-cost implementation and affordable, disposable parts are appealing characteristics. To achieve high throughput while maintaining single-cell scale control, Zhang et al. (2012) suggested an ultra-high throughput microneedle platform that includes an array of capture sites with silicon penetrators, either hollow or solid, for inserting macromolecules into cells. Microfluidics was also used to maintain high throughput while preserving single-cell scale control, making jet injection more commonly used in vitro. Adamo and Jensen (2008) developed a prototype microfluidics-based jet injection system that could treat 500–1000 cells using polydimethylsiloxane (PDMS) to produce master moulds via soft lithography.

PDMS-based microfluidic devices fabricated through soft lithography are widely used in mechanoporation techniques such as hydroporation and cell squeezing. Photolithography using SU8 moulds has been employed to fabricate PDMS microfluidic devices with dimensions compatible with cell sizes (Sciolino et al. 2022). For cell squeezing devices, PDMS microfluidic channel moulds made through photolithography are designed to match the dimensions of cells being processed, with multichannel designs incorporating self-cleaning properties to reduce clogging and device failure (Loo et al. 2021; Jung et al. 2022). PDMS has been a popular material for prototyping mechanoporation devices in research settings due to its flexibility, transparency, and biocompatibility, making it suitable for proof-of-concept studies and initial device testing (Sia and Whitesides 2003).

However, for more robust and clinically translatable mechanoporation devices that can be used by a wider range of users and are suitable for clinical use, a transition to other materials such as silicon, microinjection-moulded plastic, and other product materials may be necessary. Silicon and silica-based materials like glass and quartz offer advantages in terms of well-established fabrication techniques, commercial availability, excellent optical and electroosmotic properties, and integration with other electronic components for multifunctional devices (Park et al. 2016; Alrifaiy et al. 2012; Zhuang et al. 2006). Silicon microfabrication techniques such as photolithography and etching allow for the precise and reproducible fabrication of microscale structures with scalability (Li et al. 2019) (Fig. 10), making it suitable not only for microfluidic devices themselves but also for making master moulds used in the fabrication of other materials. Additionally, these materials have excellent mechanical properties, resulting in more durable and reliable mechanoporation devices (Alrifaiy et al. 2012; Hong et al. 2010). Unlike PDMS devices, silicon and plastic devices do not deform under high pressure applied to generate flows, and they also have low air permeability and stable surface zeta potential (Lee et al. 2003; Sollier et al. 2011; Zhao and Yang 2009; Sze et al. 2003).

Microinjection-moulded plastic is another attractive class of materials for mechanoporation devices due to its scalability, geometrical quality, and cost-effectiveness (Alrifaiy et al. 2012). Recent research has shown the use of thermoplastics and thermoset polymers, such as polyethylene terephthalate (PET), polypropylene (PP), Polymethyl methacrylate (PMMA), and cyclic olefin copolymer (COC), in mechanoporation devices. These materials offer unique advantages, such as improved mechanical strength, chemical resistance, and biocompatibility. For example, mechanoporation devices fabricated from thermoplastics like PET, PP, COC and PMMA have shown improved durability and stability compared to PDMS, with reduced cost and ease of manufacturing, making them suitable for long-term use in clinical settings (Alrifaiy et al. 2012; Zhuang et al. 2006; Hu and Chen 2018; Doronin et al. 2021; Li et al. 2012; Lee and Kim 2022; Maddah 2016; Mark et al. 2010). PMMA is cheap, widely available, and chemically inert in neutral aqueous solutions, showing resistance to hydrolysis (Hong et al. 2010). PP and COC have high chemical resistance and transparency, making them reliable materials for fabricating devices used in biochemical reactions (Keller et al. 2016; Sun et al. 2019; Shirani et al. 2017). Polystyrene is another thermoplastic that has gained interest due to its easy functionalisation and surface modification and its compatibility with high-throughput equipment (Li et al. 2012; Johnson et al. 2013; Anderson et al. 2013). It can be easily fabricated using laser ablation, a cost-effective alternative to traditional polymer fabrication techniques such as hot embossing, injection moulding, and photolithography (Li et al. 2012; Johnson et al. 2013; Anderson et al. 2013). Polystyrene's

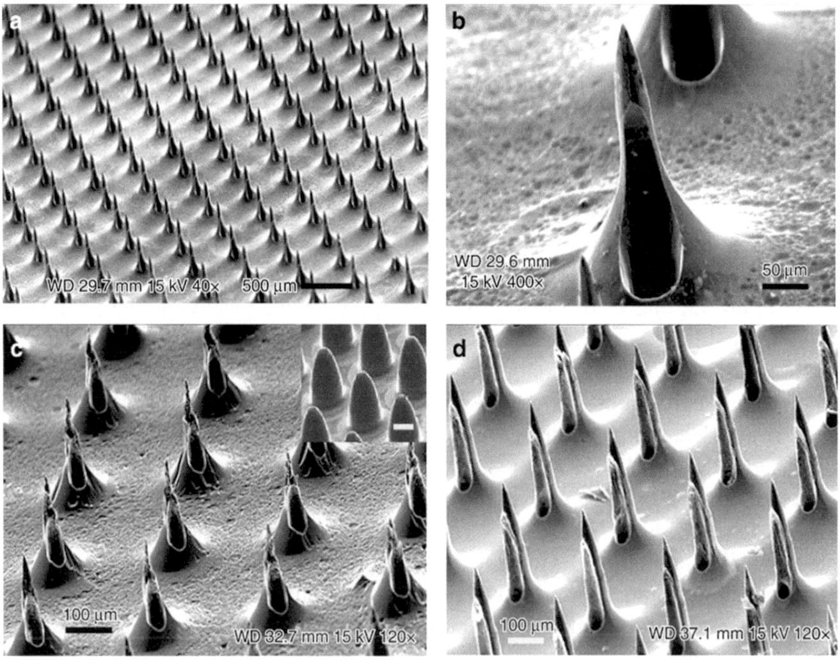

Fig. 10 The sharpening process of silicon hollow microneedles array generated by **a** wet etching with **b** the zoomed-in image of the microneedles, **c** plasma etched microneedle arrays and **d** arrays fabricated using a combination of plasma and wet etching (open access (Li et al. 2019))

strength and rigidity make it suitable for integration into high throughput equipment, and its compatibility with cell culture makes it suitable for on-chip cell culture compared to PDMS.

Furthermore, thermoset polymers such as epoxy resins have been used to create robust and biocompatible mechanoporation devices with low molecule absorption, making them suitable for chemically demanding applications (Lee and Kim 2022). They also offer lower air permeability and higher resistance to elastic deformation caused by high pressure in microchannels compared to other materials. However, their fabrication times may be longer compared to other devices.

Ceramics are another potential material for mechanoporation devices, as they exhibit minimal swelling upon solvent absorption, unlike PDMS (Toepke and Beebe 2006; Wang et al. 2012), and also demonstrate high resistance to reactive reagents that may adversely affect PDMS walls (Lee and Kim 2022).

In addition to materials, fabrication methods are also important design parameters for producing microfluidic devices. Generally, a material can be used for microfluidic devices with multiple fabrication methods, allowing researchers to select a fabrication method appropriate for the lab environment. For example, thermoplastic-based microfluidic devices can be fabricated using various fabrication techniques, such as hot embossing, injection moulding, and micro-milling. Hot embossing and injection moulding can create complex parts with very high precision and reproducibility, which can be employed in the mass production of mechanoporation devices as the moulds used in these techniques can fabricate multiple devices without degradation (Alrifaiy et al. 2012; Heckele and Schomburg 2004). However, hot embossing and injection moulding are rarely used in microdevices fabricated using inorganic materials. And while micro-milling enables the microchannel fabrication from harder materials with a cutting tool (Chen et al. 2014; Vázquez et al. 2010) and is more widely available in workshops and universities compared to other fabrication methods (Alrifaiy et al. 2012), the process is also associated with the surface roughness generated from the milling marks of the tool, rendering the device unsuitable for operations where optical transparency is required (Guckenberger et al. 2015).

Laser ablation is another commonly used method for microdevice fabrication (Mansour et al. 2022). It is a rapid process that does not require dedicated chemical reagents or cleanroom facilities and can be used with a wide range of materials such as PDMS, PMMA (Hong et al. 2010), PET (Hu and Chen 2018), and ceramics (Umroh et al. 2020). However, cost is one disadvantage.

In conclusion, while PDMS has traditionally been used as a prototyping material in mechanoporation devices, a transition to other materials such as silicon, microinjection-moulded plastic, and other product materials may be necessary to develop more reliable, scalable, and clinically viable mechanoporation devices for various biomedical applications. Silicon offers precise fabrication techniques and integration with other components, while injection-moulded plastic offers scalability and cost-effectiveness. Other product materials like thermoplastics and thermoset polymers provide improved mechanical properties and biocompatibility. These material transitions can enable the development of mechanoporation devices suitable for clinical translation and can be used by a wider range of users. The fabrication process for these devices will depend on the specific device and method being used for mechanoporation.

5 Advantages of Mechanoporation

Mechanoporation has several advantages over other methods of introducing molecules into cells, including:

Payload flexibility: Mechanoporation is a flexible technique that allows for the introduction of various types of molecules, including small molecules, nucleic acids, and proteins, into cells. Due to its versatility, it is an appealing method for researchers and scientists in various fields, including gene therapy, drug discovery, and basic research. The different techniques of mechanoporation can be used to deliver various particles, ranging from small nanoparticles to micron-sized beads. Mechanoporation techniques like microinjection allow for the direct injection of molecules into the cytoplasm of cells, which can enhance the efficiency of cargo delivery. Furthermore, mechanoporation can be combined with other techniques like microfluidics to enable high-throughput delivery processes. Overall, mechanoporation's versatility makes it a valuable tool for various applications in different applications, such as biotechnology, medicine, as well as basic research.

Minimal effect on target cell morphology: Mechanoporation is a technique that is generally regarded as non-toxic, as it does not require the use of foreign carriers or chemicals. Instead, it relies on mechanical forces to create temporary openings in the plasma membrane, facilitating the uptake of exogenous molecules. As this approach operates by exerting physical forces on the plasma membrane, it does not require an external carrier or an electric source, significantly reducing the risk of cell toxicity and enhancing cell viability (Morgan et al. 2000). Based on the data obtained from the long-term recovery of various cell lines, it has been demonstrated that cells subjected to mechanoporation-based methods exhibited normal morphology and metabolism in contrast to electroporated cells. Furthermore, mechanoporation-transfected cells were found to recover post-delivery and retain the capacity for exponential growth, similar to non-transfected cells, whereas the growth rate decreased in electroporated cells (Liu et al. 2018, 2020).

However, some mechanoporation methods may still involve a certain degree of physical manipulation of the cell or tissue, such as stretching or compression, which can be considered minimally invasive. Nevertheless, compared to other intrusive techniques such as viral transfection or chemical transfection, mechanoporation is generally considered less invasive and has reduced lingering effects on the transfected cells, making it an attractive option for various applications.

Precise control: Mechanoporation techniques, like microinjection, offer a way to achieve accurate control over both injection location and dosage volume. Researchers have achieved control over volume to the extent of femtoliters to attoliters using various methods, as demonstrated in Knoblauch et al.'s (1999) study utilising a Galinstan expansion syringe. The exceptional precision of this method enables even the delivering of a single chloroplast into a plant cell without causing any harm to other cellular components.

Low cost and simplicity: Mechanoporation can be a relatively straightforward and cost-efficient method of introducing molecules into cells, particularly compared to more complex techniques like viral vectors, electroporation or optoporation/photoporation techniques (Shinde et al. 2021, 2023; Kar et al. 2020; Gupta et al. 2021; Mohan et al. 2021a, b, c; Dey et al. 2020; Santra and Tseng 2016).

The cost-effectiveness of mechanoporation methods depends on the specific method used and the context of the application. In general, some mechanoporation techniques, such as biolistics and microinjection-based methods, may require expensive equipment and may not be cost-effective for routine use in a laboratory setting.

In contrast, alternative mechanoporation techniques like cell squeezing may offer greater cost-effectiveness, as they require less expensive equipment and reagents. This approach is characterised by simplicity, as it does not necessitate small moving parts, making it a cost-effective strategy for applying mechanical force to cell membranes. This characteristic has been utilised for intracellular delivery, targeting both the external plasma membrane and internal endosomal membrane systems (Stewart et al. 2018). In these applications, only a flow rate and pressure controller are necessary. If cost is more important than achieving high cell viability, scrape and scratch loading techniques can be used as convenient substitutes (Tzavelas et al. 2004).

In addition to the initial cost of equipment and reagents, the cost-effectiveness of mechanoporation methods also depends on the efficiency and reliability of the method. If a mechanoporation method can achieve high efficiency and reproducibility with minimal cellular damage, it may be considered more cost-effective in the long run, as it can reduce the need for repeated experiments and increase the yield of the desired product.

Overall, the cost-effectiveness of mechanoporation methods is dependent on various aspects, including the specific method used, the application's context, and the method's efficiency and reliability.

6 Drawbacks of Mechanoporation

Lower transfection efficiency: Mechanoporation can sometimes result in low efficiency in delivering molecules into cells, which can be a drawback of this technique. Low efficiency can be due to several factors, including the type of cells being targeted, the specific mechanoporation method used, and the properties of the molecules being delivered.

For example, some cell types may be more challenging to mechanoporate than others due to differences in their cell membrane properties. Different cell lines show different transfection efficiency for various mechanoporation techniques (Kang et al. 2020). Additionally, some mechanoporation methods may be more efficient than others for specific types of cells or molecules. Reported data for biolistics (ballistic particles) have also shown that during the arbitrary spraying of particles across sample cells, the particles were unable to breach the nuclear membrane of each cell to deliver the DNA cargo to prostate cancer cell lines (Zhang et al. 2002), resulting in low efficiency (30–40%), while for larger cells like myotubes, the reported efficiency was higher (20–70%) (Antolik et al. 2003). Likewise, the mechanism of nanoneedle penetration also required the optimisation of nanoneedle arrays for a particular cell type (Shalek et al. 2012). Constriction-based mechanoporation also has to balance efficiency with potential cell damage. With the increase in pressure, the efficiency increases, but with adverse effects on the cells, causing damage to the cells (Williams et al. 1999).

The properties of the molecules being delivered can also affect mechanoporation efficiency. For example, the size and charge of the molecules can impact their ability to pass through the temporary pores in the cell membrane created by the mechanoporation method. As a general guideline, it is recommended to use particle sizes that are no larger than one-tenth of the size of the target cells (Sanford et al. 1993).

However, some strategies can be used to improve the efficiency of mechanoporation, such as optimising the parameters of the mechanical force used, such as pressure, or using a combination of mechanoporation with other methods of delivery to improve overall efficiency. Additionally, advancements in technology and the development of new mechanoporation techniques are improving the efficiency of this method, making it a promising option for gene and molecule delivery.

Potential for cell damage: Applying mechanical forces in mechanoporation can harm cells, especially those that are fragile or sensitive, resulting in decreased cell viability and impacting downstream processes. The fluid shear forces can disrupt the lipid bilayers forming the cell membrane in various ways. For example, if the flow is parallel to the membrane surface, it can cause buckling instabilities or bilayer rupture (Hanasaki et al. 2010), and if the flow is perpendicular, it can pierce the membrane (Yuan et al. 2015). Biolistic delivery has the major drawback of causing damage to cells due to high-velocity particles, especially when the particle diameter is large relative to the cell size. Biolistic techniques utilise durable and dense heavy metal particles that effectively maintain the momentum to breach the plasma membrane. However, this comes at the expense of decreased cell viability (Sanford et al. 1993). In such cases, nanoparticles have been reported to provide a better survival rate (O'Brien and Lummis 2011).

Mechanoporation can also cause cell damage or stress, especially when excessive force is applied or extended, leading to changes in intracellular structure and function, such as the cytoskeleton, nucleus, and genomic DNA (Raab et al. 2016). In such experiments, the nucleus's stiffness and size are critical factors in determining the passage during mechanoporation (Versaevel et al. 2013; Rowat et al. 2013). Similarly, in sonoporation, an increased efficiency in cargo delivery is associated with increased levels of cell death (Meacham et al. 2018).

However, it is important to note that not all mechanoporation methods or conditions result in significant cell damage. It is possible to minimise cell damage by optimising the parameters of the mechanical force used in mechanoporation, such as the intensity, duration, and frequency of the force. The choice of method can also impact the level of cell damage. For example, due to high precision, microinjection may cause less cell damage than some other methods of mechanoporation (Meacham et al. 2018).

Overall, while some level of cell damage may occur during mechanoporation, it is possible to minimise this by using appropriate parameters and selecting the most suitable method for the specific cell type and application.

Limited throughput: Mechanoporation can be time-consuming and labour-intensive, especially when targeting individual cells. This can limit the throughput of the technique and make it challenging to scale up for high-throughput applications, which puts it at a disadvantage to other gene or molecule delivery methods, such as viral transduction or chemical transfection. This is because the process typically requires treating cells individually or in small batches, which can be time-consuming and limit the number of cells treated at once.

Microinjection is a widely used technique in mechanoporation that offers efficient control over the delivered substance and the injection site within a cell while maintaining high cell viability. However, it has a low throughput. The low throughput of microinjection arises from several factors, including the need for prior immobilisation of cells and the manual positioning of the cargo-laden needle onto the cell for injection. Adherent cells can be immobilised using the substrate itself, while suspended cells may require methods such as micropipette suction to hold them in a position (Muthaiyan Shanmugam and Manoj 2022; Chakrabarty et al. 2022). This process demands skilled technicians for precise needle placement onto the cells. To overcome these limitations, various strategies can be employed, such as incorporating automation or robotics for precise process control or utilising microfluidic platforms where the cells approach the cargo needles instead of the other way around (Muthaiyan Shanmugam and Manoj 2022; Chakrabarty et al. 2022; Permana et al. 2016; Chi et al. 2020; Xu 2018).

Despite its lower throughput compared to other methods, mechanoporation is still a valuable tool for gene and molecule delivery, particularly in applications where cell viability and function are critical. With advancements in technology and the development of new mechanoporation techniques, the throughput of this method is likely to improve in the future.

Limited control over pore size: Mechanoporation can create openings in the cell membrane of varying sizes, which can hinder the precision of the technique and make it challenging to regulate the uptake of molecules and target specific cells or subcellular regions. Pores formed through fluid shear are especially difficult to control. Similarly, sonoporation has been found to create pore sizes ranging from nanometers to micrometres due to the limited control over the magnitude and mode of fluid shear phenomena (Stewart et al. 2018).

The variability in pore size can result in several consequences. For instance, if the pores are too small, it may be difficult for foreign molecules to enter the cell. Conversely, if the pores are too large, the cell membrane may be damaged, leading to unwanted effects or cell death.

Therefore, while it is possible to exert some level of control over pore size in mechanoporation, achieving precise and uniform control over the pore size is challenging. Other methods, such as electroporation or chemical treatment, may be better suited for creating pores of a specific size (Dey et al. 2020; Santra and Tseng 2016; Santra et al. 2020b; Kar et al. 2018). The different advantages and limitations of the mechanoporation techniques discussed above are summarised in Table 1.

Researchers are continuously working to enhance the precision and dependability of mechanoporation, including developing new tools and techniques to regulate the magnitude and duration of the mechanical force applied to the cell membrane.

Despite these limitations, mechanoporation remains a promising tool for a wide range of applications in fields such as gene therapy, drug delivery, and basic research in cell biology. Current research efforts are dedicated to enhancing the efficiency and specificity of the technique while also addressing its inherent limitations.

Table 1 The table provides a concise overview of the pros and cons of various mechanoporation techniques as documented in the literature

Methods	Advantages	Disadvantages	Source
Microinjection	Cargo can be delivered to adhered as well as suspended cells The volume of cargo can be controlled Highly accurate High cell viability	Low throughput Skilled technician required	Stewart et al. (2018), Mehier-Humbert and Guy (2005), Xu (2018)
Micro/Nanoneedle arrays	Enables high throughput capabilities Applicable for diverse cargo types	Low efficiency Transfected cells show an altered growth rate	McKnight et al. (2004), Zhang et al. (2012, 2014), Suzuki et al. (2021)
Shear based	Simple fabrication process Less invasive than membrane disruption methods	Low efficiency Difficult to control	Hallow et al. (2007)
Biolistics	Useful in delivery to superficial tissue High throughput It has the potential to effectively transfect neurons, which is a challenging task with other conventional methods	Damage to cellular structures Bulk delivery Many parameters need to be optimised	Mcallister (2000), Zhang et al. (2014), Uchida et al. (2009)
Cell squeezing	Achieves high throughput rates An efficient method of cargo delivery Good cell viability post-treatment with no adverse effect on cell growth rate Cell morphology does not get affected	Faster speeds used for delivering larger molecules have an adverse effect on cell viability Clogging of the devices might happen It can cause nuclear ruptures in cells if the constriction is too small	Chakrabarty et al. (2022), Sharei et al. (2013a), Sciolino et al. (2022), Loo et al. (2021), Fechheimer et al. (1987)

(continued)

Table 1 (continued)

Methods	Advantages	Disadvantages	Source
Hydroporation	Cheap Remarkably high throughput capabilities High transfection efficiency Maintains high cell viability during the process User-friendly and straightforward to operate Decreased probability of clogging	Limited to suspended cell types The delivery process is non-uniform Occasional device clogging	Deng et al. (2018), Kang et al. (2020), Chakrabarty et al. (2022)
Scrape/scratch loading	Economical Cell remains adherent for immediate analysis	Low cell viability Low throughput	Stewart et al. (2018), Fechheimer et al. (1987)
Sonoporation	The hydrodynamic effect used as a delivery mechanism Independent of the complex fabrication process High efficiency	Single-cell delivery is not possible The creation of transient pores cannot be regulated closely Excessive cell deaths due to uncontrolled cavitation	Stewart et al. (2018), Dijkmans et al. (2004), Kotopoulis et al. (2014), Shi et al. (2020), Yang et al. (2020)

7 Conclusions

Mechanoporation is a powerful technique that uses mechanical force to create transient pores in cell membranes, with potential applications in drug delivery, gene therapy, and tissue engineering. While controlling pore size precisely may be challenging, mechanoporation offers advantages over other methods due to its non-invasive and gentle nature, not requiring exogenous agents. Mechanoporation has the potential to democratise access to advanced biomedical technologies, benefiting fundamental research, drug discovery, and gene therapy.

Mechanoporation can enable personalised medicine by delivering therapeutic agents directly into cells with controlled mechanical force, allowing targeted therapies tailored to individual patients. This can lead to improved treatment outcomes and reduced side effects. Mechanoporation can also drive innovation in medicine by making advanced technologies more accessible to a wider range of users, stimulating the development of novel applications and approaches in gene editing, regenerative medicine, cancer immunotherapy, and other areas.

Furthermore, mechanoporation can enhance education and training in the biomedical sciences by facilitating hands-on learning experiences, fostering critical thinking, and promoting interdisciplinary collaboration. This can empower the next Generation of scientists and clinicians with the skills and knowledge needed to advance the field of medicine.

To summarise, mechanoporation has the potential to revolutionise biomedical technology, leading to better healthcare outcomes and more widespread access to advanced medical procedures. However, this is still a nascent area of study, and further research could focus on improving control over the shape and size of pores created, as well as gaining deeper insights into the mechanisms of various mechanoporation techniques to achieve better control over pore size and duration. By doing so, this could lead to improved cell transfection rates and cell viability, which could enable testing with a broader range of cargos delivered to cells. Additionally, standardising reported data, with a focus on cell viability and its behaviour after transfection, including proliferation rates and the ability to perform various functions with minimal change, is necessary. Therefore, continued research has the potential to produce new techniques and applications that enhance mechanoporation's efficiency, specificity, and safety across a range of uses.

Conflict of Interest Todd Sulchek and Alexander Alexeev are inventors of patents that have been licensed to a company developing cell engineering related products and equity holders of CellFE, Inc. Their conflicts of interest have been disclosed to and are managed by the Georgia Institute of Technology Office of Research Integrity Assurance.

References

Adamo A, Jensen KF (2008) Microfluidic based single cell microinjection. Lab Chip 8:1258–1261. https://doi.org/10.1039/b803212b

Adamo A, Sharei A, Adamo L et al (2012) Microfluidics-based assessment of cell deformability. Anal Chem 84:6438–6443. https://doi.org/10.1021/ac300264v

Adamo A, Roushdy O, Dokov R et al (2013) Microfluidic jet injection for delivering macromolecules into cells. J Micromech Microeng 23. https://doi.org/10.1088/0960-1317/23/3/035026

Adler V, Pincus MR, Polotskaya A et al (1996) Activation of c-Jun-NH2-kinase by UV irradiation is dependent on p21(ras). J Biol Chem 271:23304–23309. https://doi.org/10.1074/jbc.271.38.23304

Alrifaiy A, Lindahl OA, Ramser K (2012) Polymer-based microfluidic devices for pharmacy, biology and tissue engineering. Polymers (Basel) 4:1349–1398. https://doi.org/10.3390/polym4031349

Anderson KB, Halpin ST, Johnson AS et al (2013) Integration of multiple components in polystyrene-based microfluidic devices Part II: Cellular analysis. Analyst 138:137–143. https://doi.org/10.1039/c2an36171j

Antolik C, De Deyne PG, Bloch RJ (2003) Biolistic transfection of cultured myotubes. Sci STKE 2003:1–14. https://doi.org/10.1126/scisignal.1922003pl11

Bao S, Thrall BD, Miller DL (1997) Transfection of a reporter plasmid into cultured cells by sonoporation in vitro. Ultrasound Med Biol 23:953–959. https://doi.org/10.1016/S0301-5629(97)00025-2

Barber MA (1911a) The effect on the protoplasm of nitella of various chemical substances and of microorganisms introduced into the cavity of the living cell. J Infect Dis 9:117–129. https://doi.org/10.1093/infdis/9.2.117
Barber MA (1911b) A technic for the inoculation of bacteria and other substances into living cells. J Infect Dis 8:348–360. https://doi.org/10.1093/infdis/8.3.348
Belting M, Sandgren S, Wittrup A (2005) Nuclear delivery of macromolecules: barriers and carriers. Adv Drug Deliv Rev 57:505–527. https://doi.org/10.1016/j.addr.2004.10.004
Bernat RL, Borisy GG, Rothfield NF, Earnshaw WC (1990) Injection of anticentromere antibodies in interphase disrupts events required for chromosome movement at mitosis. J Cell Biol 111:1519–1533. https://doi.org/10.1083/jcb.111.4.1519
Bhakdi S, Weller U, Walev I et al (1993) A guide to the use of pore-forming toxins for controlled permeabilization of cell membranes. Med Microbiol Immunol 182:167–175. https://doi.org/10.1007/BF00219946
Bloom M, Evans E (1991) Physical properties of the fluid lipid-bilayer component of cell membranes: a perspective. Q Rev Biophys 24:293–397. https://doi.org/10.1017/S0033583500003735
Chakrabarty P, Gupta P, Illath K et al (2022) Microfluidic mechanoporation for cellular delivery and analysis. Mater Today Bio 13. https://doi.org/10.1016/j.mtbio.2021.100193
Chakrabarty P, Illath K, Kar S et al (2023) Combinatorial physical methods for cellular therapy: towards the future of cellular analysis? J Control Release 353:1084–1095. https://doi.org/10.1016/j.jconrel.2022.12.038
Chen PC, Pan CW, Lee WC, Li KM (2014) An experimental study of micromilling parameters to manufacture microchannels on a PMMA substrate. Int J Adv Manuf Technol 71:1623–1630. https://doi.org/10.1007/s00170-013-5555-z
Chi Z, Xu Q, Zhu L (2020) A review of recent advances in robotic cell microinjection. IEEE Access 8:8520–8532. https://doi.org/10.1109/ACCESS.2020.2964305
Clarke MSF, McNeil PL (1992) Syringe loading introduces macromolecules into living mammalian cell cytosol. J Cell Sci 102:533–541. https://doi.org/10.1242/jcs.102.3.533
Clark AG, Wartlick O, Salbreux G, Paluch EK (2014) Stresses at the cell surface during animal cell morphogenesis. Curr Biol 24:R484–R494. https://doi.org/10.1016/j.cub.2014.03.059
Davis SP, Martanto W, Allen MG, Prausnitz MR (2005) Hollow metal microneedles for insulin delivery to diabetic rats. IEEE Trans Biomed Eng 52:909–915. https://doi.org/10.1109/TBME.2005.845240
Dayton PA, Allen JS, Ferrara KW (2002) The magnitude of radiation force on ultrasound contrast agents. J Acoust Soc Am 112:2183–2192. https://doi.org/10.1121/1.1509428
Deng Y, Kizer M, Rada M et al (2018) Intracellular delivery of nanomaterials via an inertial microfluidic cell hydroporator. Nano Lett 18:2705–2710. https://doi.org/10.1021/acs.nanolett.8b00704
Dey K, Kar S, Shinde P et al (2020) Microfluidic electroporation and applications. In: Microfluidics and bio-MEMS: devices and applications. Jenny Stanford Publisher Pte. Ltd., pp 343–400
Dijkmans PA, Juffermans LJM, Musters RJP et al (2004) Microbubbles and ultrasound: from diagnosis to therapy. Eur J Echocardiogr 5:245–256. https://doi.org/10.1016/j.euje.2004.02.001
Dixit HG, Starr R, Dundon ML et al (2020) Massively-parallelized, deterministic mechanoporation for intracellular delivery. Nano Lett 20:860–867. https://doi.org/10.1021/acs.nanolett.9b03175
Doronin FA, Rudyak YV, Rytikov GO et al (2021) 3D-printed planar microfluidic device on oxyfluorinated PET-substrate. Polym Test 99:107209. https://doi.org/10.1016/j.polymertesting.2021.107209
Duvshani-Eshet M, Baruch L, Kesselman E et al (2006) Therapeutic ultrasound-mediated DNA to cell and nucleus: bioeffects revealed by confocal and atomic force microscopy. Gene Ther 13:163–172. https://doi.org/10.1038/sj.gt.3302642
Emerson NT, Hsia CH, Rafalska-Metcalf IU, Yang H (2014) Mechanodelivery of nanoparticles to the cytoplasm of living cells. Nanoscale 6:4538–4543. https://doi.org/10.1039/c3nr06468a

Fechheimer M, Taylor DL (1987) Introduction of exogenous molecules into the cytoplasm of dictyostelium discoideum amoebae by controlled sonication. Methods Cell Biol 28:179–190. https://doi.org/10.1016/S0091-679X(08)61644-1

Fechheimer M, Boylan JF, Parker S et al (1987) Transfection of mammalian cells with plasmid DNA by scrape loading and sonication loading. Proc Natl Acad Sci USA 84:8463–8467. https://doi.org/10.1073/pnas.84.23.8463

Frankel AD, Pabo CO (1988) Cellular uptake of the tat protein from human immunodeficiency virus. Cell 55:1189–1193. https://doi.org/10.1016/0092-8674(88)90263-2

Furukawa R, Wampler JE, Fechheimer M (1988) Measurement of the cytoplasmic pH of dictyostelium discoideum using a low light level microspectrofluorometer. J Cell Biol 107:2541–2549. https://doi.org/10.1083/jcb.107.6.2541

Guckenberger DJ, De Groot TE, Wan AMD et al (2015) Micromilling: a method for ultra-rapid prototyping of plastic microfluidic devices. Lab Chip 15:2364–2378. https://doi.org/10.1039/c5lc00234f

Gupta P, Kar S, Kumar A et al (2021) Pulsed laser assisted high-throughput intracellular delivery in hanging drop based three dimensional cancer spheroids. Analyst 146:4756–4766. https://doi.org/10.1039/d0an02432e

Hallow DM, Seeger RA, Kamaev PP, Prado GR, LaPlaca MC (2007) Shear-induced intracellular loading of cells with molecules by controlled microfluidics. Biotechnol Bioeng 99:503–505. https://doi.org/10.1002/bit

Hanasaki I, Walther JH, Kawano S, Koumoutsakos P (2010) Coarse-grained molecular dynamics simulations of shear-induced instabilities of lipid bilayer membranes in water. Phys Rev E Stat Nonlinear Soft Matter Phys 82:1–6. https://doi.org/10.1103/PhysRevE.82.051602

Harada T, Swift J, Irianto J et al (2014) Nuclear lamin stiffness is a barrier to 3D migration, but softness can limit survival. J Cell Biol 204:669–682. https://doi.org/10.1083/jcb.201308029

Harrier LA, Millam S (2001) Biolistic transformation of arbuscular mycorrhizal fungi: progress and perspectives. Appl Biochem Biotechnol Part B Mol Biotechnol 18:25–33. https://doi.org/10.1385/MB:18:1:25

Harshan, Gupta P et al (2020) Microneedles: current trends and applications. In: Microfluidics and bio-MEMS: devices and applications. Jenny Stanford Publisher Pte. Ltd., pp 275–324

He G, Hu N, Xu AM et al (2020) Nanoneedle platforms: the many ways to pierce the cell membrane. Adv Funct Mater 30:1–30. https://doi.org/10.1002/adfm.201909890

Heckele M, Schomburg WK (2004) Review on micro molding of thermoplastic polymers. J Micromech Microeng 14. https://doi.org/10.1088/0960-1317/14/3/R01

Hong TF, Ju WJ, Wu MC et al (2010) Rapid prototyping of PMMA microfluidic chips utilizing a CO_2 laser. Microfluid Nanofluidics 9:1125–1133. https://doi.org/10.1007/s10404-010-0633-0

Hu ZL, Chen XY (2018) Fabrication of polyethylene terephthalate microfluidic chip using CO_2 laser system. Int Polym Process 33:106–109. https://doi.org/10.3139/217.3447

Jarrell JA, Twite AA, Lau KHWJ et al (2019) Intracellular delivery of mRNA to human primary T cells with microfluidic vortex shedding. Sci Rep 9:1–11. https://doi.org/10.1038/s41598-019-40147-y

Jimenez AJ, Maiuri P, Lafaurie-Janvore J et al (2014) ESCRT machinery is required for plasma membrane repair. Science (80–) 343. https://doi.org/10.1126/science.1247136

Johnson AS, Anderson KB, Halpin ST et al (2013) Integration of multiple components in polystyrene-based microfluidic devices Part I: Fabrication and characterization. Analyst 138:129–136. https://doi.org/10.1039/c2an36168j

Jung KO, Theruvath AJ, Nejadnik H et al (2022) Mechanoporation enables rapid and efficient radiolabeling of stem cells for PET imaging. Sci Rep 12:1–10. https://doi.org/10.1038/s41598-022-06938-6

Kaladharan K, Kumar A, Gupta P et al (2021) Microfluidic based physical approaches towards single-cell intracellular delivery and analysis

Kang G, Carlson DW, Kang TH et al (2020) Intracellular nanomaterial delivery via spiral hydroporation. ACS Nano 14:3048–3058. https://doi.org/10.1021/acsnano.9b07930

Kar S, Loganathan M, Dey K et al (2018) Single-cell electroporation: current trends, applications and future prospects. J Micromech Microeng 28. https://doi.org/10.1088/1361-6439/aae5ae
Kar S, Shinde P, Nagai M, Santra TS (2020) Optical manipulation of cells. In: Microfluidics and bio-MEMS: devices and applications. Jenny Stanford Publisher Pte. Ltd., pp 49–94
Karshafian R, Bevan PD, Williams R et al (2009) Sonoporation by ultrasound-activated microbubble contrast agents: effect of acoustic exposure parameters on cell membrane permeability and cell viability. Ultrasound Med Biol 35:847–860. https://doi.org/10.1016/j.ultrasmedbio.2008.10.013
Keller N, Nargang TM, Runck M et al (2016) Tacky cyclic olefin copolymer: a biocompatible bonding technique for the fabrication of microfluidic channels in COC. Lab Chip 16:1561–1564. https://doi.org/10.1039/c5lc01498k
Khan Y (2002) The gene gun: current application in cutaneous gene therapy. J Pak Assoc Dermatol 12:167–170
Kikkert JR (1993) The Biolistic® PDS-1000/He device. Plant Cell Tissue Organ Cult 33:221–226. https://doi.org/10.1007/BF02319005
Kiru L, Zlitni A, Tousley AM et al (2022) In vivo imaging of nanoparticle-labeled CAR T cells. Proc Natl Acad Sci USA 119. https://doi.org/10.1073/pnas.2102363119
Klein TM, Fitzpatrick-Mcelligott S (1993) Particle bombardment: a universal approach for gene transfer to cells and tissues. Curr Opin Biotechnol 4:583–590. https://doi.org/10.1016/0958-1669(93)90081-7
Klein RM, Wolf ED, Wu R, Sanford JC (1992) High-velocity microprojectiles for delivering nucleic acids into living cells. 1987. Biotechnology 24:384–386
Klimaschewski L, Nindl W, Pimpl M et al (2002) Biolistic transfection and morphological analysis of cultured sympathetic neurons. J Neurosci Methods 113:63–71. https://doi.org/10.1016/S0165-0270(01)00473-3
Knoblauch M, Hibberd JM, Gray JC, Van Bel AJE (1999) A galinstan expansion femtosyringe for microinjection of eukaryotic organelles and prokaryotes. Nat Biotechnol 17:906–909. https://doi.org/10.1038/12902
Korzh V, Strähle U (2002) Marshall Barber and the century of microinjection: from cloning of bacteria to cloning of everything. Differentiation 70:221–226. https://doi.org/10.1046/j.1432-0436.2002.700601.x
Kotopoulis S, Delalande A, Popa M et al (2014) Sonoporation-enhanced chemotherapy significantly reduces primary tumour burden in an orthotopic pancreatic cancer xenograft. Mol Imaging Biol 16:53–62. https://doi.org/10.1007/s11307-013-0672-5
Kumar A, Mohan L, Shinde P et al (2018) Mechanoporation: towards single cell approaches.pdf. In: Handbook of single cell technologies. Springer, pp 1–29
Laplaca MC, Lee VMY, Thibault LE (1997) An in vitro model of traumatic neuronal injury: loading rate-dependent changes in acute cytosolic calcium and lactate dehydrogenase release. J Neurotrauma 14:355–368. https://doi.org/10.1089/neu.1997.14.355
Lechardeur D, Verkman AS, Lukacs GL (2005) Intracellular routing of plasmid DNA during non-viral gene transfer. Adv Drug Deliv Rev 57:755–767. https://doi.org/10.1016/j.addr.2004.12.008
Lee J, Kim M (2022) Polymeric microfluidic devices fabricated using epoxy resin for chemically demanding and day-long experiments. Biosensors 12. https://doi.org/10.3390/bios12100838
Lee JN, Park C, Whitesides GM (2003) Solvent compatibility of poly(dimethylsiloxane)-based microfluidic devices. Anal Chem 75:6544–6554. https://doi.org/10.1021/ac0346712
Legenzov EA, Dirda NDA, Hagen BM, Kao JPY (2015) Synthesis and characterization of 8-ocarboxymethylpyranine (CM-Pyranine) as a bright, violet-emitting, fluid-phase fluorescent marker in cell biology. PLoS ONE 10:1–18. https://doi.org/10.1371/journal.pone.0133518
Li YS, Davidson E, Reid CN, McHale AP (2009) Optimising ultrasound-mediated gene transfer (sonoporation) in vitro and prolonged expression of a transgene in vivo: potential applications for gene therapy of cancer. Cancer Lett 273:62–69. https://doi.org/10.1016/j.canlet.2008.07.030
Li H, Fan Y, Kodzius R, Foulds IG (2012) Fabrication of polystyrene microfluidic devices using a pulsed CO_2 laser system. Microsyst Technol 18:373–379. https://doi.org/10.1007/s00542-011-1410-z

Li Y, Zhang H, Yang R et al (2019) Fabrication of sharp silicon hollow microneedles by deep-reactive ion etching towards minimally invasive diagnostics. https://doi.org/10.1038/s41378-019-0077-y
Liu X, Sun Y (2009) Microfabricated glass devices for rapid single cell immobilization in mouse zygote microinjection. Biomed Microdevices 11:1169–1174. https://doi.org/10.1007/s10544-009-9333-5
Liu A, Islam M, Stone N et al (2018) Microfluidic generation of transient cell volume exchange for convectively driven intracellular delivery of large macromolecules. Mater Today 21:703–712. https://doi.org/10.1016/j.mattod.2018.03.002
Liu A, Yu T, Young K et al (2020) Cell mechanical and physiological behavior in the regime of rapid mechanical compressions that lead to cell volume change. Small 16:1–11. https://doi.org/10.1002/smll.201903857
Loo J, Sicher I, Goff A et al (2021) Microfluidic transfection of mRNA into human primary lymphocytes and hematopoietic stem and progenitor cells using ultra-fast physical deformations. Sci Rep 11:1–11. https://doi.org/10.1038/s41598-021-00893-4
Lukacs GL, Haggie P, Seksek O et al (2000) Size-dependent DNA mobility in cytoplasm and nucleus. J Biol Chem 275:1625–1629. https://doi.org/10.1074/jbc.275.3.1625
Maddah HA (2016) Polypropylene as a promising plastic: a review. Am J Polym Sci 6:1–11. https://doi.org/10.5923/j.ajps.20160601.01
Mansour H, Soliman EA, El-Bab AMF, Abdel-Mawgood AL (2022) Development of epoxy resin-based microfluidic devices using CO_2 laser ablation for DNA amplification point-of-care (POC) applications. Int J Adv Manuf Technol 120:4355–4372. https://doi.org/10.1007/s00170-022-08992-w
Mark D, Haeberle S, Roth G et al (2010) Microfluidic lab-on-a-chip platforms: requirements, characteristics and applications. Chem Soc Rev 39:1153–1182. https://doi.org/10.1039/b820557b
Marmottant P, Biben T, Hilgenfeldt S (2008) Deformation and rupture of lipid vesicles in the strong shear flow generated by ultrasound-driven microbubbles. Proc R Soc A Math Phys Eng Sci 464:1781–1800. https://doi.org/10.1098/rspa.2007.0362
Mayr M, Hu Y, Hainaut H, Xu Q (2002) Mechanical stress-induced DNA damage and rac-p38MAPK signal pathways mediate p53-dependent apoptosis in vascular smooth muscle cells. FASEB J 16:1423–1425. https://doi.org/10.1096/fj.02-0042fje
Mcallister AK (2000) Biolistic transfection of neurons materials recipes notes and remarks references introduction. Plasmid 1–13
McKnight TE, Melechko AV, Hensley DK et al (2004) Tracking gene expression after DNA delivery using spatially indexed nanofiber arrays. Nano Lett 4:1213–1219. https://doi.org/10.1021/nl049504b
McNeil PL, Warder E (1988) Glass beads load macromolecules into living cells. J Cell Sci 88:669–678. https://doi.org/10.1242/jcs.88.5.669
Meacham JM, Durvasula K, Degertekin FL, Fedorov AG (2014) Physical methods for intracellular delivery: practical aspects from laboratory use to industrial-scale processing. J Lab Autom 19:1–18. https://doi.org/10.1177/2211068213494388
Meacham JM, Durvasula K, Degertekin FL, Fedorov AG (2018) Enhanced intracellular delivery via coordinated acoustically driven shear mechanoporation and electrophoretic insertion. Sci Rep 8:1–10. https://doi.org/10.1038/s41598-018-22042-0
Mehier-Humbert S, Guy RH (2005) Physical methods for gene transfer: Improving the kinetics of gene delivery into cells. Adv Drug Deliv Rev 57:733–753. https://doi.org/10.1016/j.addr.2004.12.007
Mohan L, Kar S, Nagai M, Santra TS (2021a) Electrochemical fabrication of TiO_2 micro-flowers for an efficient intracellular delivery using nanosecond light pulse. Mater Chem Phys 267:124604. https://doi.org/10.1016/j.matchemphys.2021.124604

Mohan L, Kar S, Mahapatra PS et al (2021b) Fabrication of TiO_2 microspikes for highly efficient intracellular delivery by pulse laser-assisted photoporation. RSC Adv 11:9336–9348. https://doi.org/10.1039/d0ra09785c

Mohan L, Kar S, Hattori R et al (2021c) Can titanium oxide nanotubes facilitate intracellular delivery by laser-assisted photoporation? Appl Surf Sci 543:148815. https://doi.org/10.1016/j.apsusc.2020.148815

Morgan KE, Allen JS, Dayton PA et al (2000) Experimental and theoretical evaluation of microbubble behavior: effect of transmitted phase and bubble size. IEEE Trans Ultrason Ferroelectr Freq Control 47:1494–1509. https://doi.org/10.1109/58.883539

Morshedi Rad D, Alsadat Rad M, Razavi Bazaz S et al (2021) A comprehensive review on intracellular delivery. Adv Mater 33:1–36. https://doi.org/10.1002/adma.202005363

Munish MS, Santra TS (2016) Microinjection for single cell analysis. Essential of single-cell analysis. Springer, Berlin Heidelberg, pp 85–130

Muthaiyan Shanmugam M, Manoj H (2022) Microinjection for single-cell analysis and therapy

Nejadnik H, Jung KO, Theruvath AJ et al (2020) Instant labeling of therapeutic cells for multimodality imaging. Theranostics 10:6024–6034. https://doi.org/10.7150/thno.39554

O'Brien JA, Lummis SCR (2006a) Biolistic transfection of neuronal cultures using a hand-held gene gun. Nat Protoc 1:977–981. https://doi.org/10.1038/nprot.2006.145

O'Brien JA, Lummis SCR (2006b) Diolistic labeling of neuronal cultures and intact tissue using a hand-held gene gun. Nat Protoc 1:1517–1521. https://doi.org/10.1038/nprot.2006.258

O'Brien JA, Lummis SCR (2011) Nano-biolistics: a method of biolistic transfection of cells and tissues using a gene gun with novel nanometer-sized projectiles. BMC Biotechnol 11:2–7. https://doi.org/10.1186/1472-6750-11-66

Park S, Choi SO, Paik S, Joon et al (2016) Intracellular delivery of molecules using microfabricated nanoneedle arrays. Biomed Microdevices 18:1–13. https://doi.org/10.1007/s10544-016-0038-2

Permana S, Grant E, Walker GM, Yoder JA (2016) A review of automated microinjection systems for single cells in the embryogenesis stage. IEEE/ASME Trans Mechatron 21:2391–2404. https://doi.org/10.1109/TMECH.2016.2574871

Persson H, Købler C, Mølhave K et al (2013) Fibroblasts cultured on nanowires exhibit low motility, impaired cell division, and DNA damage. Small 9:4006–4016. https://doi.org/10.1002/smll.201300644

Raab M, Gentili M, De Belly H et al (2016) ESCRT III repairs nuclear envelope ruptures during cell migration to limit DNA damage and cell death. Science (80–) 352:359–362. https://doi.org/10.1126/science.aad7611

Rich J, Tian Z, Huang TJ (2022) Sonoporation: past, present, and future. Adv Mater Technol 7:1–17. https://doi.org/10.1002/admt.202100885

Rowat AC, Jaalouk DE, Zwerger M et al (2013) Nuclear envelope composition determines the ability of neutrophil-type cells to passage through micron-scale constrictions. J Biol Chem 288:8610–8618. https://doi.org/10.1074/jbc.M112.441535

Sanford JC, Smith FD, Russell JA (1993) Optimizing the biolistic process for different biological applications. Methods Enzymol 217:483–509. https://doi.org/10.1016/0076-6879(93)17086-K

Santra TS, Kar S, Chen TC et al (2020a) Near-infrared nanosecond-pulsed laser-activated highly efficient intracellular delivery mediated by nano-corrugated mushroom-shaped gold-coated polystyrene nanoparticles. Nanoscale 12:12057–12067. https://doi.org/10.1039/d0nr01792b

Santra TS, Kar S, Chang HY, Tseng FG (2020b) Nano-localized single-cell nano-electroporation. Lab Chip 20:4194–4204. https://doi.org/10.1039/d0lc00712a

Santra TS, Tseng F-G (2016) Electroporation for single cell analysis.pdf. In: Santra T, Tseng F-G (eds) Essentials of Single-cell analysis. Springer, Berlin Heidelberg, pp 55–84

Schermelleh L, Solovei I, Zink D, Cremer T (2000) Two-color fluorescence labeling of early and mid-to-late replicating chromatin in living cells. Chromosom Res 103:77–80. 10.1023/A

Schlicher RK, Hutcheson JD, Radhakrishna H et al (2010) Changes in cell morphology due to plasma membrane wounding by acoustic cavitation. Ultrasound Med Biol 36:677–692. https://doi.org/10.1016/j.ultrasmedbio.2010.01.010

Sciolino N, Liu A, Breindel L et al (2022) Microfluidics delivery of DARPP-32 into HeLa cells maintains viability for in-cell NMR spectroscopy. Commun Biol 5:1–8. https://doi.org/10.1038/s42003-022-03412-x

Shalek AK, Gaublomme JT, Wang L et al (2012) Nanowire-mediated delivery enables functional interrogation of primary immune cells: application to the analysis of chronic lymphocytic leukemia. Nano Lett 12:6498–6504. https://doi.org/10.1021/nl3042917

Sharei A, Zoldan J, Adamo A et al (2013a) A vector-free microfluidic platform for intracellular delivery. Proc Natl Acad Sci USA 110:2082–2087. https://doi.org/10.1073/pnas.1218705110

Sharei A, Cho N, Mao S et al (2013b) Cell squeezing as a robust, microfluidic intracellular delivery platform. J Vis Exp 1–7. https://doi.org/10.3791/50980

Sharei A, Poceviciute R, Jackson EL et al (2014) Plasma membrane recovery kinetics of a microfluidic intracellular delivery platform. Integr Biol (United Kingdom) 6:470–475. https://doi.org/10.1039/c3ib40215k

Shi L, Jiang Y, Zhang Y et al (2020) A fiber optoacoustic emitter with controlled ultrasound frequency for cell membrane sonoporation at submillimeter spatial resolution. Photoacoustics 20:100208. https://doi.org/10.1016/j.pacs.2020.100208

Shinde P, Kar S, Loganathan M et al (2020) Infrared pulse laser-activated highly efficient intracellular delivery using titanium microdish device. ACS Biomater Sci Eng 6:5645–5652. https://doi.org/10.1021/acsbiomaterials.0c00785

Shinde A, Kar S, Nagai M et al (2021) Light induced cellular delivery and analysis. In: Handbook of single cell technologies. Springer Nature, pp 1–30

Shinde A, Shinde P, Kar S et al (2023) Metallic micro-ring device for highly efficient large cargo delivery in mammalian cells using infrared light pulses. Lab Chip. https://doi.org/10.1039/d2lc00899h

Shirani E, Razmjou A, Tavassoli H et al (2017) Strategically designing a pumpless microfluidic device on an "inert" polypropylene substrate with potential application in biosensing and diagnostics. Langmuir 33:5565–5576. https://doi.org/10.1021/acs.langmuir.7b00537

Sia SK, Whitesides GM (2003) Microfluidic devices fabricated in poly(dimethylsiloxane) for biological studies. Electrophoresis 24:3563–3576. https://doi.org/10.1002/elps.200305584

Sinha B, Köster D, Ruez R et al (2011) Cells respond to mechanical stress by rapid disassembly of caveolae. Cell 144:402–413. https://doi.org/10.1016/j.cell.2010.12.031

Sollier E, Murray C, Maoddi P, Di Carlo D (2011) Rapid prototyping polymers for microfluidic devices and high pressure injections. Lab Chip 11:3752–3765. https://doi.org/10.1039/c1lc20514e

Stewart MP, Sharei A, Ding X et al (2016) In vitro and ex vivo strategies for intracellular delivery. Nature 538:183–192. https://doi.org/10.1038/nature19764

Stewart MP, Langer R, Jensen KF (2018) Intracellular delivery by membrane disruption: mechanisms, strategies, and concepts. Chem Rev 118:7409–7531. https://doi.org/10.1021/acs.chemrev.7b00678

Sun H, Chan CW, Wang Y et al (2019) Reliable and reusable whole polypropylene plastic microfluidic devices for a rapid, low-cost antimicrobial susceptibility test. Lab Chip 19:2915–2924. https://doi.org/10.1039/c9lc00502a

Suzuki T, Takao H, Suzuki T et al (2021) Proposal of hematocrit-based non-Newtonian viscosity model and its significance in intracranial aneurysm blood flow simulation. J Nonnewton Fluid Mech 290:104511. https://doi.org/10.1016/j.jnnfm.2021.104511

Sze A, Erickson D, Ren L, Li D (2003) Zeta-potential measurement using the Smoluchowski equation and the slope of the current-time relationship in electroosmotic flow. J Colloid Interface Sci 261:402–410. https://doi.org/10.1016/S0021-9797(03)00142-5

Toepke MW, Beebe DJ (2006) PDMS absorption of small molecules and consequences in microfluidic applications. Lab Chip 6:1484–1486. https://doi.org/10.1039/b612140c

Tzavelas C, Bildirici L, Rickwood D (2004) Factors that affect the efficiency of cell transfection by immunoporation. Anal Biochem 328:219–224. https://doi.org/10.1016/j.ab.2004.01.036

Uchida M, Li XW, Mertens P, Alpar HO (2009) Transfection by particle bombardment: delivery of plasmid DNA into mammalian cells using gene gun. Biochim Biophys Acta Gen Subj 1790:754–764. https://doi.org/10.1016/j.bbagen.2009.05.013

Umroh B, Ginting A, Rahman MNA (2020) CO_2 laser machining on alumina ceramic: a review. IOP Conf Ser Mater Sci Eng 1003.https://doi.org/10.1088/1757-899X/1003/1/012131

Usachev YM, Khammanivong A, Campbell C, Thayer SA (2000) Particle-mediated gene transfer to rat neurons in primary culture. Pflugers Arch Eur J Physiol 439:730–738. https://doi.org/10.1007/s004240050999

Uvizl A, Goswami R, Gandhi SD et al (2021) Efficient and gentle delivery of molecules into cells with different elasticity via progressive mechanoporation. Lab Chip 21:2437–2452. https://doi.org/10.1039/d0lc01224f

Van Der Maaden K, Jiskoot W, Bouwstra J (2012) Microneedle technologies for (trans)dermal drug and vaccine delivery. J Control Release 161:645–655. https://doi.org/10.1016/j.jconrel.2012.01.042

Vaughan E, DeGiulio J, Dean D (2006) Intracellular trafficking of plasmids for gene therapy: mechanisms of cytoplasmic movement and nuclear import. Curr Gene Ther 6:671–681. https://doi.org/10.2174/156652306779010688

Vázquez E, Rodríguez CA, Elías-Zúñiga A, Ciurana J (2010) An experimental analysis of process parameters to manufacture metallic micro-channels by micro-milling. Int J Adv Manuf Technol 51:945–955. https://doi.org/10.1007/s00170-010-2685-4

Versaevel M, Riaz M, Grevesse T, Gabriele S (2013) Cell confinement: putting the squeeze on the nucleus. Soft Matter 9:6665–6676. https://doi.org/10.1039/c3sm00147d

Wang S, Heilman D, Liu F et al (2004) A DNA vaccine producing LcrV antigen in oligomers is effective in protecting mice from lethal mucosal challenge of plague. Vaccine 22:3348–3357. https://doi.org/10.1016/j.vaccine.2004.02.036

Wang JD, Douville NJ, Takayama S, Elsayed M (2012) Quantitative analysis of molecular absorption into PDMS microfluidic channels. Ann Biomed Eng 40:1862–1873. https://doi.org/10.1007/s10439-012-0562-z

Weaver JC (2000) Electroporation of cells and tissues. IEEE Trans Plasma Sci 28:24–33. https://doi.org/10.1109/27.842820

Wellmann H, Kaltschmidt B, Kaltschmidt C (1999) Optimized protocol for biolistic transfection of brain slices and dissociated cultured neurons with a hand-held gene gun. J Neurosci Methods 92:55–64. https://doi.org/10.1016/S0165-0270(99)00094-1

Williams AR, Bao S, Miller DL (1999) Filtroporation: A simple, reliable technique for transfection and macromolecular loading of cells in suspension. Biotechnol Bioeng 65(3):341–346. https://doi.org/10.1002/(SICI)1097-0290(19991105)65:3%3c341::AID-BIT12%3e3.0.CO;2-I

Xu Q (2018) Review of microinjection systems

Yang Y, Li Q, Guo X et al (2020) Mechanisms underlying sonoporation: interaction between microbubbles and cells. Ultrason Sonochem 67:105096. https://doi.org/10.1016/j.ultsonch.2020.105096

Yu C, Peng RY (2017) Biological effects and mechanisms of shortwave radiation: a review. Mil Med Res 4

Yuan F, Yang C, Zhong P (2015) Cell membrane deformation and bioeffects produced by tandem bubble-induced jetting flow. Proc Natl Acad Sci USA 112:E7039–E7047. https://doi.org/10.1073/pnas.1518679112

Zarnitsyn VG, Meacham JM, Varady MJ et al (2008) Electrosonic ejector microarray for drug and gene delivery. Biomed Microdevices 10:299–308. https://doi.org/10.1007/s10544-007-9137-4

Zhang Y, Yu LC (2008) Microinjection as a tool of mechanical delivery. Curr Opin Biotechnol 19:506–510. https://doi.org/10.1016/j.copbio.2008.07.005

Zhang S, Gu J, Yang NS et al (2002) Relative promoter strengths in four human prostate cancer cell lines evaluated by particle bombardment-mediated gene transfer. Prostate 51:286–292. https://doi.org/10.1002/pros.10081

Zhang Y, Ballas CB, Rao MP (2012) Towards ultrahigh throughput microinjection: MEMS-based massively-parallelized mechanoporation. Proc Annu Int Conf IEEE Eng Med Biol Soc EMBS 594–597.https://doi.org/10.1109/EMBC.2012.6346001
Zhang D, Das DB, Rielly CD (2014) Potential of microneedle-assisted micro-particle delivery by gene guns: a review. Drug Deliv 21:571–587. https://doi.org/10.3109/10717544.2013.864345
Zhang Z, Xu J, Drapaca C (2018) Particle squeezing in narrow confinements. Microfluid Nanofluidics 22:1–26. https://doi.org/10.1007/s10404-018-2129-2
Zhao C, Yang C (2009) Analysis of induced-charge electro-osmotic flow in a microchannel embedded with polarizable dielectric blocks. Phys Rev E Stat Nonlinear Soft Matter Phys 80:1–11. https://doi.org/10.1103/PhysRevE.80.046312
Zhuang G, Jin Q, Liu J et al (2006) A low temperature bonding of quartz microfluidic chip for serum lipoproteins analysis. Biomed Microdevices 8:255–261. https://doi.org/10.1007/s10544-006-9142-z

Thermoporation Based Drug Delivery Systems

Rohit and Abhishek Raj

Abstract Delivery of drugs into the cells needs the opening of pores of the cellular membrane which eases the diffusion or, convection of drugs into the cell. Temperature has been found to affect the characteristics of cell membrane. Lipid bilayer dissociates at temperature higher than 37 °C as a result of the increased kinetic energy of the constituent molecules. The bonding between tail parts of lipid bilayer weakens at higher temperatures, which reduces the resistance to flow across cell membrane for biomolecules or, drugs. Use of heat for the generation of tiny openings in stratum corneum is called Thermoporation or Microporation. This chapter presents a detailed discussion on mechanism, applications and limitations of thermoporation and photothermal techniques for drug delivery. Various strategies for thermoporation are elaborated and explained. Further, various techniques of photothermal process of cell transfection such as use of excited photosensitizers, near-infrared radiation and various nanoparticles are described. At the end, membrane abnormalities caused by rapid temperature fluctuations are discussed, which is one of the limitations of thermoporation and photoporation based system for drug delivery. This chapter will be advantageous to the readers interested in cell transfection or cell lysis using the photothermal methods.

1 Introduction

The skin is without a doubt the biggest organ of the human body and makes up about 15 percent of the body's mass. Its surface area ranges from 1.50 to 2.00 m^2, depending on the gender and the anatomical region (Menon 2002; Waller and Maibach 2005). The skin is considered to be a vital protective layer and has a significant homeostatic function (Parhi et al. 2012). The epidermis, dermis, and subcutaneous layers make up its several layers, which are essentially its three main layers. The appearance and

Rohit · A. Raj (✉)
Department of Mechanical Engineering, Indian Institute of Technology Patna, Bihta 801106, Bihar, India
e-mail: araj@iitp.ac.in

T. S. Santra and A. U. S. Shinde (eds.), *Advanced Drug Delivery*,
Studies in Mechanobiology, Tissue Engineering and Biomaterials 26,
https://doi.org/10.1007/978-981-99-6564-9_6

function of each of these layers differs from the others. Owing to its enormously accessible wide surface area, the skin continues to be the focus of significant study attention (Parhi et al. 2012; Shahzad et al. 2013). The stratum corneum, or outermost layer of skin, is a significant obstacle, making it difficult for many medications to be administered via the skin. In order to achieve therapeutic success, the skin's natural barrier function need be changed in such a way that allows the drug to be delivered at a rate fast enough to reach the target plasma. Transdermal Drug Delivery (TDD) refers to drug administration through the skin into the bloodstream. When compared to oral drug delivery, TDD has a number of potential benefits, including avoiding preliminary metabolism and the harsh surroundings of the stomach, minimising inter and intra-patient differences, increasing patient comfort, promoting controlled and consistent medication delivery, delivering medications with decreased biological half-lives and limited therapeutic indices, allowing for self-administration, and being simple to stop using in the event of side effects (Anjos et al. 2007; Mills and Cross 2006; Jayaprakash et al. 2017; Jawale et al. 2017). Despite the foregoing advantages, TDD was only explored after the Food and Drug Administration (FDA) in 1979 approved the first transdermal patch, Transderm Scope® (Alza Corporation, USA). TDD was previously limited to easier drug delivery methods such as gels and ointments with a small number of active components (e.g. nitroglycerine and estradiol). The transdermal form of drug delivery, however, didn't become widely accepted by patients until 1991, when nicotine patches for quitting smoking were released.

Thermoporation (TP), also known as "microporation," uses heat to create minute pores in the stratum corneum (Ivanov et al. 1999). By briefly ablating the stratum corneum without considerably increasing the temperature or harming the deeper tissues, exposing the skin to brief, increased temperature pulses can lead to structural disturbance and loss of stratum corneum (Lee et al. 2011). Similar to how microneedles create such channels by carving a path through the skin, thermal ablation can also produce micron-sized microchannels in the skin, but it does so by vaporising and decomposing stratum corneum in a specific area rather than by cutting the stratum corneum as microneedles do. Thermal ablation offers non-invasive medication distribution through skin-created microchannels, making it a favourable approach for increasing the permeability of the SC. The size of these microchannels is still suitably small enough for avoiding side effects like bleeding, inflammation, infection, and discomfort. This method also has more control over the physiological and physical effects on the skin. Moreover, it provides efficient delivery of high Molecular weight substances like proteins and peptides (Park et al. 2008). The benefits of thermal ablation over other techniques are as follows (Lee et al. 2011).

1. Vaccines, macromolecules, and tiny compounds can all be delivered using this method.
2. Thermal ablation equipment does not need to be sterilised, but microneedles do.
3. The solute molecules do not need to carry charges across the skin as they do in iontophoresis.

4. Thermal ablation does not require any changes to the formulation, unlike chemical penetration enhancement, which requires these changes in order to be incorporated.
5. There are many methods for performing thermal ablation, including using chemicals, Radio frequency, and laser.

2 Mechanism of Thermoporation

The technique known as thermoporation or microporation allows active compounds to be more permeable through the skin and into the bloodstream by opening up aqueous channels across the SC. In this technique a number of metallic filaments are kept in touch with the surface of the skin for a short period of time. These filaments heat up as an electric current flow through them, which led to the SC vaporising and decomposing in a precise location (Fig. 1).

Microchannels started to emerge on the skin's surface as a result. Following that, using transdermal formulations like vaccines, gels, creams, or patches will boost the penetration of infused medications (Herwadkar and Banga 2012; Lakshmanan et al. 2014). The benefit of microporation devices is that it can considerably minimise the risk of spreading diseases transmitted by blood because of the use of disposable and sterilizable metal filaments (Garg et al. 2007). The effects of high temperature on model drug calcein permeability across the human cadaver epidermis were thoroughly investigated by Park et al. (2008). A Franz diffusion cell was used to perform the ex vivo experiment with a duration range of 100 ms to 5 s, and a temperature range of 100–315 °C. Skin permeability was shown to be strongly influenced by temperature, but not by the duration. In a different work, Paranjape et al. (2003) developed an integrated patch from human graft skin samples using a polydimethylsiloxane (PDMS) patch and micro-heaters for transdermal glucose administration. They stated that to produce micropores in the SC, an average temperature of 130 °C is required to be maintained for more than 33 ms. These newly created micropores could transport biomolecules like glucose across the skin.

Thermoporation is a method that can be used to transfer drugs directly into cells. Temperature has the ability to increase the permeability of plasma membrane.

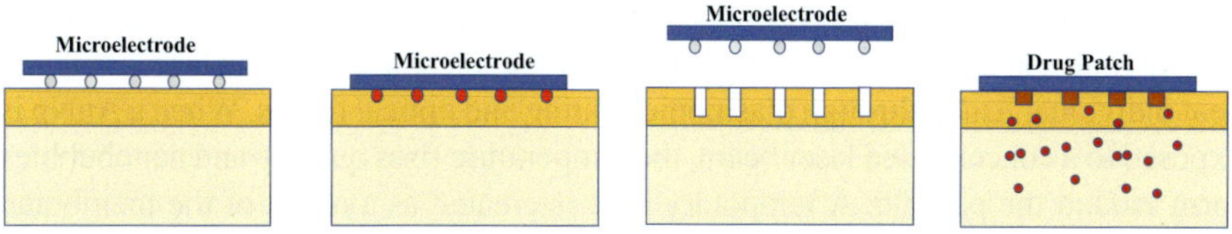

Fig. 1 Steps involved in drug delivery using thermoporation approach. **a** Placement of microelectrodes on skin, **b** skin was ablated as a result of heat produced by microelectrodes pressed over it, **c** microelectrodes are removed after the formation of micropores and **d** to distribute the encapsulated medicine, a drug patch is placed over the micropores

In the thermally assisted (Thermoporation) delivery method, cells either undergo several cooling-heating cycles or undergo a bulk supraphysiological heating scenario (above 37 °C) that results in membrane rupture and phospholipid bilayer dissociation (Jacobson and Papahadjopoulos). Since the early 1980s, thermal shock has been a frequently used method for bacterial transformation to introduce plasmid DNA into capable bacteria. Competent bacteria are incubated at 0 °C, then subjected to a heat pulse at 42 °C for a brief duration of time, followed by cooling on ice, exposing them to a series of thermal shocks (Hanahan 1983). The membrane potential is dramatically reduced by these cooling-heating cycles, which also lead to momentary membrane ruptures that increase the membrane's permeability to foreign DNA. As a result, this technique has been utilised to amplify and isolate DNA plasmids from bacteria in the log phase, which is a crucial step for gene editing and cloning applications (Das et al. 2017).

Another technique for thermally assisted delivery involves heating the cellular membranes to a temperature above the physiological range. As a result, forces preserving the integrity of the membrane are overpowered by the constituent molecules' kinetic energy, causing spontaneous disruptions in the plasma membrane (He et al. 2006). These stochastic thermally generated flaws are directly influenced by variables like pH, hydrostatic pressure, temperature fluctuations, and ion concentration.

3 Applications of Thermoporation

Recent years have seen an increase in interest in the thermal membrane disruption caused by heating single cells for the purposes of cell poration (Li et al. 2015), gene expression (Ginet et al. 2011), nano surgery (Vogel et al. 2005; Lachaine et al. 2016), cell migration (Zhu et al. 2012), cell fusing (Bahadori et al. 2017), and the creation of neural spikes (Lavoie-Cardinal et al. 2016; Ermakova et al. 2017).

3.1 Use of Thermal Membrane Disruption for Cell Poration

Li et al. (2015) used an entirely optical method to introduce 80 nm gold nanoparticle (AuNp) into a live mammalian cell. This technique solely uses focused laser energy to heat individual AuNp through plasmonic heating and optical forces. When a AuNp is exposed to a concentrated laser beam, the temperature rises quickly and nanobubbles form around the particle. A temporary hole is created as a result of the membrane rupturing as the bubble expands and contracts. The optical force quickly drives the nanoparticle into the cell after that. It was discovered that the creation of nanobubbles helped to push the particles over the membrane barrier and into the cells. Figure 2 shows the steps involved in the process.

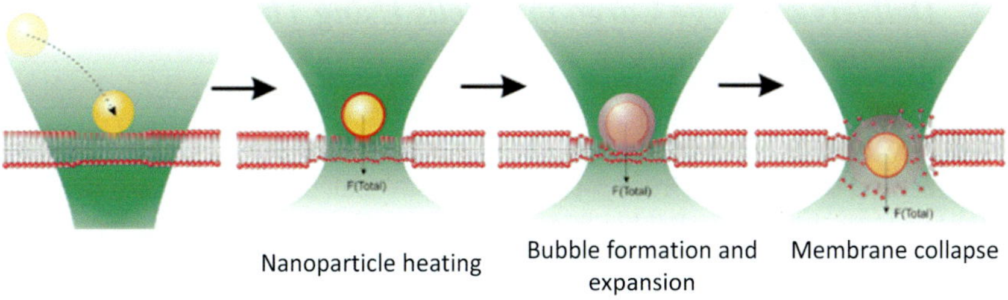

Fig. 2 Illustration of the injection procedure: a focused laser beam exposure to a gold particle causes a rapid rise in temperature and the creation of nanobubbles around the particle. A temporary hole is created as the bubble expands and contracts, rupturing the membrane. The optical force then immediately forces the nanoparticle into the cell (Reprinted with permission Li et al. (2015). Copyright 2015 American Chemical Society.)

3.2 *Use of Thermal Membrane Disruption for Cell Fusion*

Bahadori et al. (2017) showed how to optically regulate the fusion of two selected cells or one cell and one vesicle. AuNps with plasmonic properties that are optically trapped between two target cells (or a vesicle containing a cell) are used to mediate fusion (refer to Fig. 3). The two cytoplasms and two cell membranes are completely mixed during this hot-particle-mediated fusion, creating a new hybrid cell with an undamaged cellular membranes and enzymatic activity after fusion. A syncytium with two nuclei and a cytoplasm made up of the cytoplasms of the two initial cells is produced by this fusion process. After fusion, the new cell maintains its viability and a normal metabolism for at least 4 h. The technique enables single-cell focused drug delivery and has significant promise for cellular control and design.

Using optical traps, two cells of interest are chosen and positioned adjacent to one another. The fusion process is catalysed by the heat created when a AuNp (Diameter = 150 nm) diffuses into the optical trap situated at the contact zone between the two cells. When cells fully fuse, their cytoplasmic components and cell membranes combine.

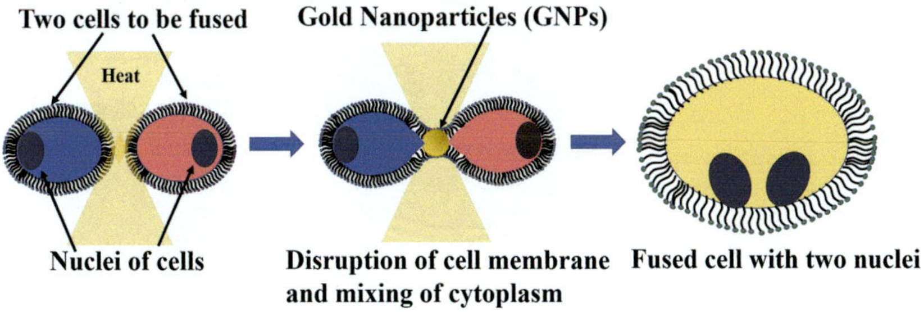

Fig. 3 Fusion of cells caused by optically heated AuNPs (redrawn)

3.3 *Delivery of Large Cargo into Cells Using Thermal Membrane Disruption*

Wu et al. (2011) demonstrated a photothermal nanoblade that uses a metallic nanostructure to absorb energy from brief laser pulses and transform it into a highly localised explosive vapour bubble that quickly ruptures a cell membrane by causing high-velocity fluid flows and temporary shear stresses. The metallic structural arrangement, duration of laser pulse, and energy all affect the cavitation bubble pattern. The photothermal nanoblade has the potential to transport huge cargo that is currently impractical to transfer into mammalian cells, such as chromosomes, organelles, and intracellular infections, because they exceed the size restrictions of existing delivery techniques.

An exterior Titanium thin film is applied to a glass micropipette. The Titanium heats up quickly through heat conduction after being excited by a nanosecond laser pulse, coupled with a thin aqueous layer around it. The contacting cell membrane is locally cut by an explosive vapour nanobubble that swells and deflates in less than one microsecond in time with the delivery of the microcapillary contents driven by pressure (refer to Fig. 4).

The ease of usage of photothermal nanoblade is another advantage. The laser pulse energy and Titanium coating govern membrane cutting, thus all that is required to complete delivery is for the operator to maintain a delicate exposure of the micropipette tip to the cell membrane.

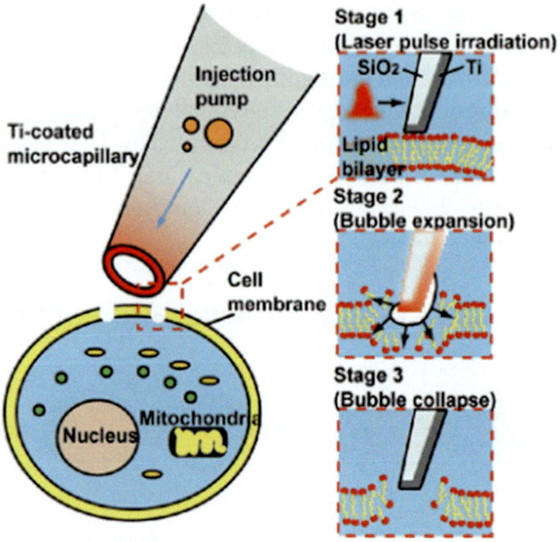

Fig. 4 Different stages involved in cargo delivery into the cell (Reprinted with permission Wu et al. (2011). Copyright 2011 American Chemical Society.)

4 Photothermal Therapy for Drug Delivery

Photothermal therapy makes use of the Photosynthesizers (PS) excitation, which causes it to release heat-based vibrational energy. With the electrons containing highest energy in the highest occupied molecular orbital (HOMO), the PS molecule possesses a steady electronic structure in the singlet state (St. Denis et al. 2011). An electron in the HOMO is excited to the lowest unoccupied molecular orbital (LUMO) after absorbing a photon of light with the correct wavelength according to its absorption spectrum, which causes the PS to attain the unstable and transient excited singlet state. Photothermal therapy kills undesired cells or tissues by using a variety of nanoparticles and near-infrared radiation (NIR) as agents. Many things need to be taken into account for effectiveness when lasers are utilised as the activation source (Jori and Spikes 1990):

1. Tissue Thickness
2. Tissue Sensitization
3. Thermal Relaxation Time
4. Tissue Absorption Coefficient
5. Time Under the Laser
6. Thermal Relaxation Time.

Photothermal sensitizers for Photothermal Therapy include metallo derivatives of porphyrins and porphyrinoid chemicals, azo dyes, and derivatives of triphenylmethane (Jori and Spikes 1990). Pulsed irradiation of either indigenous chromophores added or dyes can cause photothermal damage to tissues. The advantage of dyes is their substantial absorption at wavelengths $\geq$600 nm, which allows them to deeply penetrate biological tissues. The photoexcited absorption of chromophore coefficient and its thermal relaxation period determine the spatial confinement of photothermal process. Also, since the beginning of the decade, the usage of lasers in PTT has demonstrated promising outcomes for the treatment of cancer. In the study done by Chen et al., indocyanine was used as the PTT agent, and murine mammary tumours were treated using an 808 nm laser (Chen et al. 1996). Breast tumours (Gu et al. 2007), Lewis lung carcinoma in mice (Liu et al. 2008), localised prostate cancer (Lindner et al. 2009), murine glioma model (Day et al. 2011), Ehrlich carcinoma tumours (Khlebtsov et al. 2012), and gene therapy (Everts et al. 2006) have all been successfully treated with PTT. PTT can also be used for large cargo delivery into the cells (Shinde et al. 2020, 2023; Mohan et al. 2021a; Gupta et al. 2021; Santra et al. 2020) (Fig. 5).

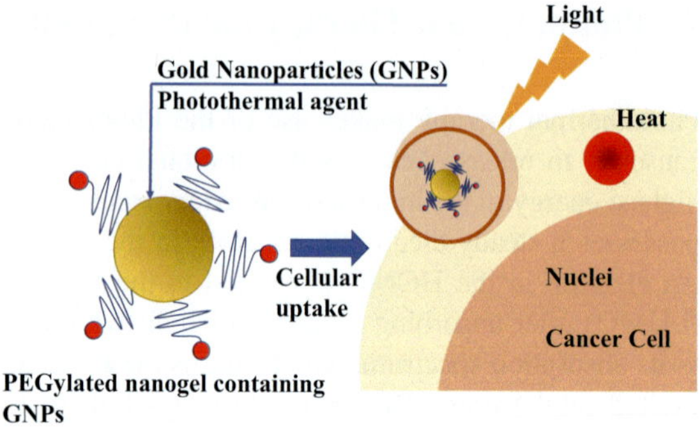

Fig. 5 Cells absorbs AuNps contained in PEGylated nanogels. Cancer cells are eliminated by irradiation, including increased temperature and light

4.1 Photoporation Assisted Intracellular Delivery Using Titanium Oxide Nanotubes

Mohan et al. (2021a), in their study developed a technique for efficient intracellular delivery assisted by nanosecond pulse laser using Titanium oxide nanotubes (Mohan et al. 2021b, c). Dextran and PI dye were successfully delivered into HeLa cells using TNT. The following were listed as prospective delivery operating principles:

(a) **Photochemical induced photoporation**: In the near field enhancement zone, photochemical initiated photoporation occurs when lasers interact with nanotubes. Reactive oxygen species (ROS) are created when water molecules are ionized (Kalies et al. 2014) resulting in decreased hydrophobicity of the lipid bilayer.

(b) **Nanobubbles induced photoporation**: The environment becomes heated as a result of the temperature of the nanotubes. As a result, the temperature of nanotubes can rise quickly in a matter of seconds, which leads to the formation of nanobubbles when water evaporates nearby nanotubes (refer to Fig. 6). These ensuing nanobubbles of water vapour are described as thermal-mediated nanobubbles.

(c) **Photoporation by heating of plasma membrane**: The heat transfer from the nanotubes that are adsorbed to the cell membrane following laser irradiation causes cell membrane rupture (Shinde et al. 2020; Hatef and Meunier 2015; Pustovalov et al. 2008). The thermal denaturation of glycoproteins causes transitory hydrophilic holes to open in the cell membrane, or the lipid bilayer may undergo a local phase change (Urban et al. 2009). In addition to the heat, plasma formation due to SPR also may creates the nanobubbles. Near field enhancement at the edges of the nanotubes may caused by SPR. In this area, due to the multiphoton ionization, plasma forms. The formed plasma cooled by collision with existing water molecules. Consequently, an increase in pressure and temperature may leads to the formation of nanobubbles near the nanotubes.

(d) **Laser interaction with plasmonic nanotubes**: Localized surface plasmon resonance (LSPR) may enhance the optical absorption of quasi-metallic nanotube

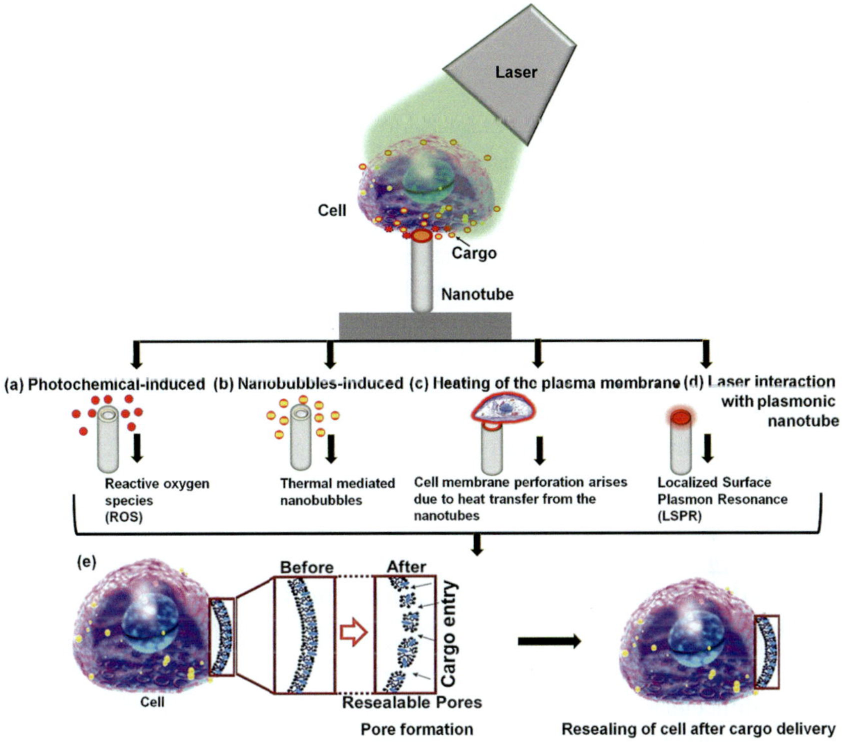

Fig. 6 Different mechanism of cargo delivery into cells on Titanium oxide nanotubes (Reprinted with permission Mohan et al. (2021a). Copyright 2020 Elsevier.)

oxides (Shinde et al. 2020; Santra et al. 2020; Li et al. 2019). This absorption may have been triggered by the LSPR effect brought about by the excitation of free electrons in the presence of light. This LSPR might make it possible to increase the magnetic field. A higher concentration of free electrons, on the other hand, produces excellent photothermal transformation features. Because the Ti_3O_5 LSPR wavelength (551 nm) is close to the wavelength of incident light (532 nm) used for excitation, the powerful resonance is simple to occur. In each cell membrane-nanostructure contact, the resonantly excited optical hotspots produce local heat that may result in the development of plasmonic nano-bubbles (PNBs). The nanostructure-cell membrane interfaces may experience constant fluid flow as a result of these PNBs' fast growth, coalescence, and collapse (Shinde et al. 2020; Santra et al. 2020; Boulais et al. 2013).

4.2 *Cargo Delivery in Mammalian Cells Using Infrared Light Pulses*

Using a titanium micro-ring (TMR) device driven by an infrared light pulse, Shinde et al. (2023) achieved highly efficient, uniform parallel intracellular transport of tiny to very big biomolecules into different cell types. Each micro-ring in a TMR array device (2 cm by 2 cm) has an outside diameter of 10 mm, an inner diameter of 3 mm, and an interspacing of 10 mm. The cell membrane is broken by photothermal bubbles produced by cavitation after the TMR device is subjected to an infrared (1050 nm) pulse laser. Biomolecules are then softly injected into the cells by a simple diffusion process. A variety of small to very large biomolecules, including dextran (3 kDa), small interfering RNA (13.3 kDa), enhanced green fluorescent protein expression plasmidDNA (6.2 kb), and galactosidase enzyme (465 kDa), were successfully delivered into human cervical (SiHa), mouse fibroblast (L929), and mouse neural crest-derived (N2a) cancer cells by this TMR device.

5 Disadvantages of Thermoporation and Photothermal Therapy

Due of difficulties with unspecificity, off-target damage, and challenges with spatiotemporal control of temperature exposure, this delivery technique has not been frequently utilised for delivery purposes in animal cells despite its relative simplicity (Boheim et al. 1980).

Here are some of the disadvantages of thermoporation:

1. Limited cell types: Molecules can be introduced into certain types of cells via thermoporation, such as mammalian cells and bacteria, but it may not work as well for other types of cells (Fan et al. 2012). Certain cell types may not respond to this treatment since the procedure depends on the cell membrane's ability to become more permeable to molecules when exposed to heat.
2. Non-specific: Thermoporation has the potential to have off-target effects by introducing molecules into both target and non-target cells. This lack of specificity may even result in cell death by causing unintended modifications to cellular functions (Boheim et al. 1980; Antonov et al. 1980).
3. Inconsistent results: Even when the identical methodology is followed, thermoporation effectiveness can change from experiment to experiment. Differences in the characteristics of the cells being employed, the size and concentration of the molecules being injected, and the precise temperature and length of the heat pulse all contribute to the unpredictability.
4. Limited throughput: Thermoporation is a relatively slow and labor-intensive technique, making it unsuitable for high-throughput applications. For example, it may not be possible to utilize thermoporation to introduce molecules into a huge number of cells in a short period.

5. Risk of damaging cells: The heat required for thermoporation can potentially damage the cells being treated. While low temperatures and short pulses can minimize the risk of cell damage, the risk cannot be completely eliminated (Robert et al. 2018).
6. Limited penetration depth: Thermoporation is not suitable for applications requiring deeper penetration into the cells (Ermakova et al. 2017).

In summary, while thermoporation has several advantages as a technique for introducing molecules into cells, it also has several disadvantages that limit its use in certain applications. Researchers must weigh the benefits and drawbacks of thermoporation carefully and choose the most appropriate technique for their specific needs.

6 Future Applications

Thermoporation and Photothermal therapy has a huge range of potential usage in various fields, including medicine, biotechnology, and environmental science.

Thermoporation has already demonstrated potential in the field of medicine for drug delivery. Drugs can be delivered directly into the cell by making tiny pores in the cell membrane, removing the need for uptake through a carrier system. This might improve the effectiveness of medications while lowering their adverse effects. Also, the delivery of large molecules like DNA into cells by thermoporation has significance for gene therapy and other genetic treatments. In the future, targeted thermal energy could be utilised to produce pores solely in particular cells or tissues, allowing for the targeted administration of drugs.

In addition to drug delivery, thermoporation has potential applications in cancer treatment. Since cancer cells are more vulnerable to high temperatures than healthy cells, heat can be employed to destroy specific cancer cells. Thermoporation could be used to target cancer treatment by heating up particular cancer cells. The use of conventional chemotherapy and radiation therapy, which can have serious adverse effects, might be lessened as a result.

Furthermore, thermoporation is used in biotechnology. Large molecules can be introduced into cells for use in bioprocessing by employing heat to make pores in the cell membrane. This has ramifications for the complicated fermentation procedures now used to manufacture biologics like proteins and antibodies. Biologics could be produced more efficiently, with lower costs and higher yields, by employing thermoporation to transport the required genetic material directly into cells.

Thermoporation has potential uses in bioremediation in the field of environmental research. By employing thermoporation to introduce genetic material into bacteria, some contaminants, like oil or heavy metals, could be more efficiently degraded by the bacteria. For environmental cleanup initiatives, particularly in areas plagued by industrial pollution, this may have substantial ramifications.

Overall, prospective applications of thermoporation span a wide range of fields, including biotechnology, environmental research, and medicine. The method has already demonstrated promise in the delivery of drugs and the treatment of cancer, and its application to bioprocessing and bioremediation could completely transform both industries. The potential impact of thermoporation will certainly increase as new uses for the technology come to light as it develops.

7 Conclusion

In conclusion, the transfer of molecules into cells has become more effective owing to the development of thermoporation. In comparison to conventional delivery techniques, it has a number of benefits, including the ability to be targeted and its non-invasiveness. The transport of several substances, including medicines, genes, and proteins, both in vitro and in vivo, has been accomplished with the help of thermoporation.

The adaptability of thermoporation is one of its main benefits. Several different compounds, including some that are typically challenging to transport using conventional techniques, can be delivered using it. Additionally, because it may be applied to various cell types and tissues, it is a helpful tool for a wide range of applications.

However, while the potential of thermoporation is clear, there are still challenges that need to be addressed. For example, the precise mechanism of thermoporation is not yet fully understood, and there is a need for more efficient and controlled methods of heating. In order to reduce damage to cells and tissues, it is also necessary to optimise the delivery conditions.

The potential of thermoporation as a method for delivering molecules into cells is obvious despite these difficulties. As more research is done and new techniques are developed, thermoporation is anticipated to become a more effective tool in the fields of drug delivery and gene therapy.

References

Antonov VF, Petrov VV, Molnar AA et al (1980) The appearance of single-ion channels in unmodified lipid bilayer membranes at the phase transition temperature. Nature 283:585–586. https://doi.org/10.1038/283585a0

Bahadori A, Oddershede LB, Bendix PM (2017) Hot-nanoparticle-mediated fusion of selected cells. Nano Res 10:2034–2045. https://doi.org/10.1007/s12274-016-1392-3

Boheim G, Hanke W, Eiblt H (1980) Lipid phase transition in planar bilayer membrane and its effect on carrier-and pore-mediated ion transport (fluctuation phenomena/mixed-chain lipids/ionophores). https://doi.org/10.1073/pnas.77.6.3403

Boulais E, Lachaine R, Hatef A, Meunier M (2013) Plasmonics for pulsed-laser cell nanosurgery: fundamentals and applications. J Photochem Photobiol C 17:26–49. https://doi.org/10.1016/j.jphotochemrev.2013.06.001

Chen WR, Adams RL, Higgins AK et al (1996) Photothermal effects on murine mammary tumors using indocyanine green and an 808-nm diode laser: an in vivo efficacy study. Cancer Lett (Elsevier). https://doi.org/10.1016/S0304-3835(06)80028-5

Das M, Raythata H, Chatterjee S (2017) Bacterial transformation: what? why? how? and when? Annu Res Rev Biol 16. https://doi.org/10.9734/ARRB/2017/35872

Day ES, Thompson PA, Zhang L et al (2011) Nanoshell-mediated photothermal therapy improves survival in a murine glioma model. J Neurooncol 104:55–63. https://doi.org/10.1007/s11060-010-0470-8

dos Anjos JLV, de Neto DS, Alonso A (2007) Effects of 1,8-cineole on the dynamics of lipids and proteins of stratum corneum. Int J Pharm 345:81–87. https://doi.org/10.1016/j.ijpharm.2007.05.041

Ermakova YG, Lanin AA, Fedotov IV et al (2017) Thermogenetic neurostimulation with single-cell resolution. Nat Commun 8. https://doi.org/10.1038/ncomms15362

Everts M, Saini V, Leddon JL et al (2006) Covalently linked Au nanoparticles to a viral vector: potential for combined photothermal and gene cancer therapy. Nano Lett 6:587–591. https://doi.org/10.1021/nl0500555

Fan Z, Shelton M, Singh AK et al (2012) Multifunctional plasmonic shell-magnetic core nanoparticles for targeted diagnostics, isolation, and photothermal destruction of tumor cells. ACS Nano 6:1065–1073. https://doi.org/10.1021/nn2045246

Garg S, Hoelscher M, Belser JA et al (2007) Needle-free skin patch delivery of a vaccine for a potentially pandemic influenza virus provides protection against lethal challenge in mice. Clin Vaccine Immunol 14:926–928. https://doi.org/10.1128/CVI.00450-06

Ginet P, Montagne K, Akiyama S et al (2011) Towards single cell heat shock response by accurate control on thermal confinement with an on-chip microwire electrode. Lab Chip 11:1513–1520. https://doi.org/10.1039/c0lc00701c

Gu Y, Chen WR, Xia M et al (2007) Effect of photothermal therapy on breast tumor vascular contents: noninvasive monitoring by near-infrared spectroscopy. Photochem Photobiol 81:1002–1009. https://doi.org/10.1111/j.1751-1097.2005.tb01475.x

Gupta P, Kar S, Kumar A et al (2021) Pulsed laser assisted high-throughput intracellular delivery in hanging drop based three dimensional cancer spheroids. Analyst 146:4756–4766. https://doi.org/10.1039/d0an02432e

Hanahan D (1983) Studies on transformation of Escherichia coli with plasmids. J Mol Biol 166(4):557–580. https://doi.org/10.1016/S0022-2836(83)80284-8

Hatef A, Meunier M (2015) Plasma-mediated photothermal effects in ultrafast laser irradiation of gold nanoparticle dimers in water. Opt Express 23:1967. https://doi.org/10.1364/oe.23.001967

He X, Amin AA, Fowler A, Toner M (2006) Thermally induced introduction of trehalose into primary rat hepatocytes. Cell Preserv Technol 4(3):178–187. https://doi.org/10.1089/cpt.2006.4.178

Herwadkar A, Banga AK (2012) Peptide and protein transdermal drug delivery. Drug Discov Today: Technol 9(2):e147–e154. https://doi.org/10.1016/j.ddtec.2011.11.007

Ivanov IT, Todorova R, Zlatanov I (1999) Spectrofluorometric and microcalorimetric study of the thermal poration relevant to the mechanism of thermohaemolysis. Int J Hyperth 15:29–43. https://doi.org/10.1080/026567399285837

Jacobson K, Papahadjopoulos D (1975) Phase transitions and phase separations in phospholipid membranes induced by changes in temperature, pH, and concentration of bivalent cations. Biochemistry 14(1):152–161

Jawale N, Bhangale C, Chaudhari M, Deshmukh TA (2017) Physical approach to transdermal drug delivery: a review. J Drug Deliv Ther 7. https://doi.org/10.22270/jddt.v7i3.1435

Jayaprakash R, Hameed J, Anupriya (2017) An overview of transdermal delivery system. Asian J Pharm Clin Res 10:36–40. https://doi.org/10.22159/ajpcr.2017.v10i10.19909

Jori G, Spikes JD (1990) Photothermal sensitizers: possible use in tumor therapy. J Photochem Photobiol B: Biol 6(1–2):93–101. https://doi.org/10.1016/1011-1344(90)85078-B

Kalies S, Birr T, Heinemann D et al (2014) Enhancement of extracellular molecule uptake in plasmonic laser perforation. J Biophotonics 7:474–482. https://doi.org/10.1002/jbio.201200200

Khlebtsov BN, Panfilova EV, Terentyuk GS et al (2012) Plasmonic nanopowders for photothermal therapy of tumors. Langmuir 28:8994–9002. https://doi.org/10.1021/la300022k

Lachaine R, Boutopoulos C, Lajoie PY et al (2016) Rational design of plasmonic nanoparticles for enhanced cavitation and cell perforation. Nano Lett 16:3187–3194. https://doi.org/10.1021/acs.nanolett.6b00562

Lakshmanan S, Gupta GK, Avci P et al (2014) Physical energy for drug delivery; poration, concentration and activation. Adv Drug Deliv Rev 71:98–114

Lavoie-Cardinal F, Salesse C, Bergeron É et al (2016) Gold nanoparticle-assisted all optical localized stimulation and monitoring of Ca^{2+} signaling in neurons. Sci Rep 6. https://doi.org/10.1038/srep20619

Lee JW, Gadiraju P, Park JH et al (2011) Microsecond thermal ablation of skin for transdermal drug delivery. J Control Release 154:58–68. https://doi.org/10.1016/j.jconrel.2011.05.003

Li M, Lohmüller T, Feldmann J (2015) Optical injection of gold nanoparticles into living cells. Nano Lett 15:770–775. https://doi.org/10.1021/nl504497m

Li Y, Bai H, Zhai J et al (2019) Alternative to noble metal substrates: metallic and plasmonic Ti_3O_5 hierarchical microspheres for surface enhanced Raman spectroscopy. Anal Chem 91:4496–4503. https://doi.org/10.1021/acs.analchem.8b05282

Lindner U, Weersink RA, Haider MA et al (2009) Image guided photothermal focal therapy for localized prostate cancer: phase I trial. J Urol 182:1371–1377. https://doi.org/10.1016/j.juro.2009.06.035

Liu H, Chen D, Tang F et al (2008) Photothermal therapy of Lewis lung carcinoma in mice using gold nanoshells on carboxylated polystyrene spheres. Nanotechnology 19. https://doi.org/10.1088/0957-4484/19/45/455101

Menon GK (2002) New insights into skin structure: scratching the surface

Mills PC, Cross SE (2006) Transdermal drug delivery: basic principles for the veterinarian. Vet J 172:218–233. https://doi.org/10.1016/j.tvjl.2005.09.006

Mohan L, Kar S, Ren Hattori et al (2021a) Can titanium oxide nanotubes facilitate intracellular delivery by laser-assisted photoporation? Appl Surf Sci 543. https://doi.org/10.1016/j.apsusc.2020.148815

Mohan L, Kar S, Nagai M, Santra TS (2021b) Electrochemical fabrication of TiO_2 micro-flowers for an efficient intracellular delivery using nanosecond light pulse. Mater Chem Phys 267. https://doi.org/10.1016/j.matchemphys.2021.124604

Mohan L, Kar S, Mahapatra PS et al (2021c) Fabrication of TiO_2 microspikes for highly efficient intracellular delivery by pulse laser-assisted photoporation. RSC Adv 11:9336–9348. https://doi.org/10.1039/d0ra09785c

Paranjape M, Garra J, Brida S et al (2003) A PDMS dermal patch for non-intrusive transdermal glucose sensing. Sens Actuators A Phys 104:195–204. https://doi.org/10.1016/S0924-4247(03)00049-9

Parhi R, Suresh P, Mondal S, Kumar PM (2012) Novel penetration enhancers for skin applications: a review

Park JH, Lee JW, Kim YC, Prausnitz MR (2008) The effect of heat on skin permeability. Int J Pharm 359:94–103. https://doi.org/10.1016/j.ijpharm.2008.03.032

Pustovalov VK, Smetannikov AS, Zharov VP (2008) Photothermal and accompanied phenomena of selective nanophotothermolysis with gold nanoparticles and laser pulses. Laser Phys Lett 5:775–792

Robert HML, Savatier J, Vial S et al (2018) Photothermal control of heat-shock protein expression at the single cell level. Small 14. https://doi.org/10.1002/smll.201801910

Santra TS, Kar S, Chen TC et al (2020) Near-infrared nanosecond-pulsed laser-activated highly efficient intracellular delivery mediated by nano-corrugated mushroom-shaped gold-coated polystyrene nanoparticles. Nanoscale 12:12057–12067. https://doi.org/10.1039/d0nr01792b

Shahzad Y, Afreen U, Nisar Hussain Shah S, Hussain T (2013) Applying response surface methodology to optimize nimesulide permeation from topical formulation. Pharm Dev Technol 18:1391–1398. https://doi.org/10.3109/10837450.2012.723721

Shinde P, Kar S, Loganathan M et al (2020) Infrared pulse laser-activated highly efficient intracellular delivery using titanium microdish device. ACS Biomater Sci Eng 6:5645–5652. https://doi.org/10.1021/acsbiomaterials.0c00785

Shinde A, Shinde P, Kar S et al (2023) Metallic micro-ring device for highly efficient large cargo delivery in mammalian cells using infrared light pulses. Lab Chip. https://doi.org/10.1039/d21c00899h

St. Denis TG, Dai T, Izikson L et al (2011) All you need is light, antimicrobial photoinactivation as an evolving and emerging discovery strategy against infectious disease. Virulence 2:509–520

Urban AS, Fedoruk M, Horton MR et al (2009) Controlled nanometric phase transitions of phospholipid membranes by plasmonic heating of single gold nanoparticles. Nano Lett 9:2903–2908. https://doi.org/10.1021/nl901201h

Vogel A, Noack J, Hüttman G, Paltauf G (2005) Mechanisms of femtosecond laser nanosurgery of cells and tissues. Appl Phys B 81:1015–1047

Waller JM, Maibach HI (2005) Age and skin structure and function, a quantitative approach (I): blood flow, pH, thickness, and ultrasound echogenicity. Skin Res Technol 11:221–235

Wu TH, Teslaa T, Kalim S et al (2011) Photothermal nanoblade for large cargo delivery into mammalian cells. Anal Chem 83:1321–1327. https://doi.org/10.1021/ac102532w

Zhu M, Baffou G, Meyerbröker N, Polleux J (2012) Micropatterning thermoplasmonic gold nanoarrays to manipulate cell adhesion. ACS Nano 6:7227–7233. https://doi.org/10.1021/nn302329c

Microinjection-Based Drug Delivery

Hima Manoj and Muniesh Muthaiyan Shanmugam

Abstract Microinjection is a gold standard technique to understand physiological and pathological processes at single-cell level; as well as has applications in delivery of therapeutic drugs for treatment. Needle microinjection involves physical delivery of foreign cargoes (such as biomolecules, microorganisms, therapeutic agents, etc.) using a microneedle which can directly penetrate a single cell. Direct physical delivery overcomes the structural barriers in the cells resulting in 100% transfection efficiency. Obtaining technical expertise in the microinjection provides superior advantage of 100% success rate by causing minimal damages to the cells. Earlier, microinjection was utilized only for research to understand several biological processes; however, recently the technology was applied for clinical treatment to directly benefit humankind in a process called intracytoplasmic sperm injection (ICSI). Parallelly, microneedles were developed for efficient drug delivery transdermally with less pain and discomfort. Development in the field of Bio-MEMS (MicroElectroMechanical Systems) resulted in wide varieties of microneedles utilizing varied materials resulting in unique properties. Apart from delivery of drugs, microneedles were also developed for biospecimen collection and metabolite monitoring. This book chapter provides an overall understanding of basics of microinjection, devices utilized in microinjection procedure, development of microneedles as well as its applications and comprehensively lists the technique's advantages and disadvantages.

Both the authors contributed equally.

H. Manoj (✉)
Department of Engineering Design, Indian Institute of Technology Madras, Chennai, India
e-mail: himabiomems2402@gmail.com

M. Muthaiyan Shanmugam
Department of Life Science, National Tsing Hua University, Hsinchu, Taiwan

Present Address:
Buck Institute for Research on Aging, Novato, CA, USA

T. S. Santra and A. U. S. Shinde (eds.), *Advanced Drug Delivery*,
Studies in Mechanobiology, Tissue Engineering and Biomaterials 26,
https://doi.org/10.1007/978-981-99-6564-9_7

1 Introduction

It is necessary to deliver foreign substance (such as ions, RNA, DNA, protein, antibodies, microorganisms, therapeutic agents, larger biomolecules, sugar molecules, cellular organelles, cytoplasmic contents, radioactive molecules, nanoparticles, sperm, pronuclei, oligonucleotide, metabolites, etc.) into a single-cell or a population of cells to understand the cell autonomous physiological and therapeutic responses. There are several methods of transfection of host cells that have been developed. Following are few methodologies of transfection (Meacham et al. 2014; Muthaiyan Shanmugam and Manoj 2022; Shanmugam and Santra 2016):

1. Viral-mediated transfection (also called as transduction)—Viruses such as Retrovirus, Adenovirus, Adeno-associated virus, Herpes virus, Baculovirus, etc. can be used to deliver foreign biomolecules (either RNA or DNA) to host cell (Kay et al. 2001; Robbins and Ghivizzani 1998; Thomas et al. 2003; Zhang and Godbey 2006).
2. Chemical mediated transfection—Several chemicals that can penetrate cellular barriers, such as cell wall and plasma/cell membrane, are successfully used to deliver foreign agents into host cell. Examples include calcium phosphate, DEAE-dextran, lipid/liposome-mediated delivery, etc. (Freitag and Wagner 2021; Mintzer and Simanek 2009).
3. Physical methods of delivery—Here, a physical force is utilized to deliver the foreign agents into host cells. Following are few physical methods of delivery:

   (a) Microinjection—Microinjection involves the use of a microneedle to penetrate the host cell to deliver the cargoes.
   (b) Bombardment—Nano/Micro scale projectiles coated with cargo are used to bombard the host cells with velocities sufficient to penetrate and deliver the cargoes (Kimura et al. 2023; Schweinsberg 2013; Yang et al. 1990).
   (c) Magnetofection—This technique utilizes magnetic field to accelerate the transfection of cargoes coated on magnetic nanoparticles (Scherer et al. 2002; Sizikov et al. 2021).
   (d) Electroporation—Electroporation involves generation of membrane pores using electric fields to facilitate the diffusion of cargo from the surrounding media into the host cell (Kar et al. 2018; Dey 2020; Santra et al. 2013, 2014, 2020a; Santra and Tseng 2013, 2016; Subhra et al. 2013).
   (e) Mechanoporation—Here, membrane pores are created by imparting mechanical stress or physical forces. The transient pores allow for translocation of foreign cargoes into the cell (Chakrabarty et al. 2022, 2023; Kaladharan et al. 2021; Santra and Tseng 2022; Tseng and Santra 2016).
   (f) Sonoporation—Here, ultrasound is used for generation of pores on the cell membrane to facilitate cargo delivery (Belling et al. 2020; Du et al. 2022).

(g) Optoinjection—Optoinjection utilizes laser irradiation to create membrane pores to facilitate the diffusion of foreign cargo into the host cell (Gupta et al. 2021; Jiang 2016; Mohan et al. 2022; Mohan et al. 2021a, b, c; Santra et al. 2020b; Shinde et al. 2023; Shinde et al. 2020).

This chapter gives an overview of understanding on microinjection technique and its clinical uses; however, details about above mentioned other techniques can be obtained in the other chapters of this book.

2 Brief History of Microinjection

Félix Dujardin, a French biologist, introduced the concept of microsurgery on a single cell in 1835 (Zhang and LeBlanc 2002) and nearly after half-century, M. A. Barber in between 1904–1911 performed first successful microinjection/microsurgery of introduction of foreign bacteria and other substances into eukaryotic living cells (Barber 1904, 1911). M. A. Barber designed and developed an elegant microinjection setup as shown in Fig. 1. He used a compound microscope to visualize the cells, a microneedle to penetrate the host cell and a simple microinjector system with hot/cold water-mercury to inject the foreign substance into host cell, where the temperature-controlled expansion and contraction of mercury is utilized to load the microneedle as well as inject the cargoes into the host cell (Barber 1911). Following which several foreign substances were microinjected into different unicellular host cells to understand various aspects of biology such as immunity, host cell defense, organismal development, fertilization, cell autonomous behavior, signaling pathways, etc. (Barber 1911; Capecchi 1980; Gurdon et al. 1971; Kimura and Yanagimachi 1995; Knowles 1974; Maller and Koontz 1981; Palermo 1992; Shanmugam and Santra 2016). Further, the equipments used for microinjection was improved to increase efficiency as well as the rate of injection.

The evolution of microneedles into microelectrodes/micropipettes was achieved by 1919 followed by development of patch-clamp techniques (another technique that developed parallel to microinjection) to study electrophysiology by 1970 (Neher and Sakmann 1976; Pratt and Eisenberger 1919). During the latter half of twentieth century several biomolecules were injected in frog egg, mouse egg as well as in cultured cell lines to demonstrate signaling pathways as well as the high efficiency of transfection (Birchmeier et al. 1985; Capecchi 1980; Gebauer et al. 1994; Heidemann and Kirschner 1975). Development of transgenic (genetic modification by insertion of genetic materials) lines during the late 1960s and 1980s marked an important milestone where microinjection was used to deliver either nucleic acid for recombination into genome or mutated embryonic stem cells (Gordon and Ruddle 1981; Lin 1966). Later, microinjection of either sperm (intracytoplasmic sperm injection—ICSI) or male pronuclei into egg mimicking fertilization, termed in vitro fertilization (IVF) was developed into a successful treatment strategy for infertility in humans (Palermo 1992). The success in IVF was followed by cloning of whole organisms

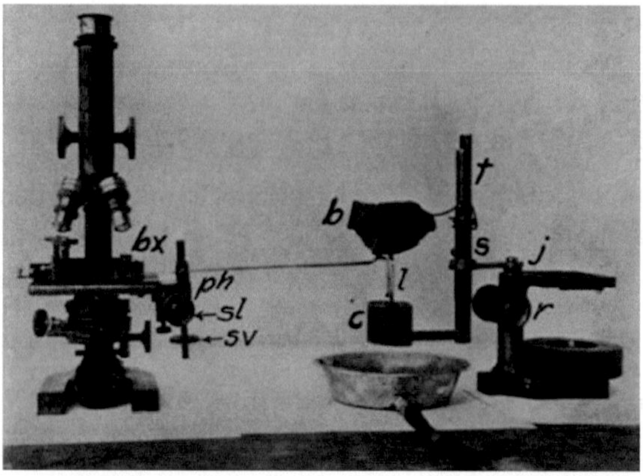

Fig. 1 Microinjection equipments used by M. A. Barber for first successful injection procedure. Towards the left is a compound microscope containing micropipette/microneedle holder '*ph*' as well as '*sl*' and '*sv*' micromanipulator for movement of the microneedle. While '*s*' attaches '*t*' to a dissection microscope or stand where '*j*' and '*r*' moves '*t*' to desired position as well as brings 'b' (hot water) and 'c' (cold water) to '*l*' which is at room temperature to regulate the contraction and expansion of mercury for loading and injection of cargoes, respectively. Reprinted from Barber (1911) by permission of Oxford University Press

by enucleation of oocyte using micropipette (microinjection technique) followed by somatic cell nuclear transfer (SCNT) of enucleated eggs marked another milestone where microinjection techniques played very important role, which includes cloning of the famous sheep 'Dolly the sheep' (Wilmut et al. 1997, Keefer et al. 2015).

3 Basic Mechanism of Microinjection

It is necessary to have several basic instruments to manipulate at micron scale either to inject or to remove foreign cargoes into or retrieve from a single cell. However, recent technological development gave several expensive instruments to increase the efficacy as well as to perform the procedure with less fatigue. Following are essential equipments to carry out a microinjection procedure:

1. Microscope,
2. Micromanipulator,
3. Injection chamber,
4. Microinjector,
5. Antivibration table,
6. Pressure pump,
7. Needle puller, and
8. Microneedle.

This chapter briefly describes the instrumentation below.

Microscopes: One of the essential equipment for injection of foreign cargoes into a single cell is microscope that helps to magnify single cell as well as microneedle. An inverted microscope provides sufficient workplace to accommodate the micropipette and microneedle comparted to upright microscope, because the space above the microscope stage in upright microscope is occupied by objective lens.

Micromanipulator: This device holds and positions the microneedle in desired place and in desired angle as well as within the microscope's field of view to perform the microinjection. In few instances two micromanipulators can be used, one to hold the microneedle (that injects the foreign cargo) and another to hold the micropipette or holding pipette (that can be used to immobilize either a suspension cell or oocyte, etc.). A micromanipulator allows controlled movement of microneedle in three different axes (x, y and z axes, Fig. 3) and holds the needle at an appropriate angle for successful penetration of the host cell (Figs. 2 and 3).

Injection chamber: Certain host cell types are very sensitive to changes in environmental conditions such as temperature, CO_2 concentrations, O_2 concentrations, humidity, etc. The injection chamber covers the stage of the microscope and the

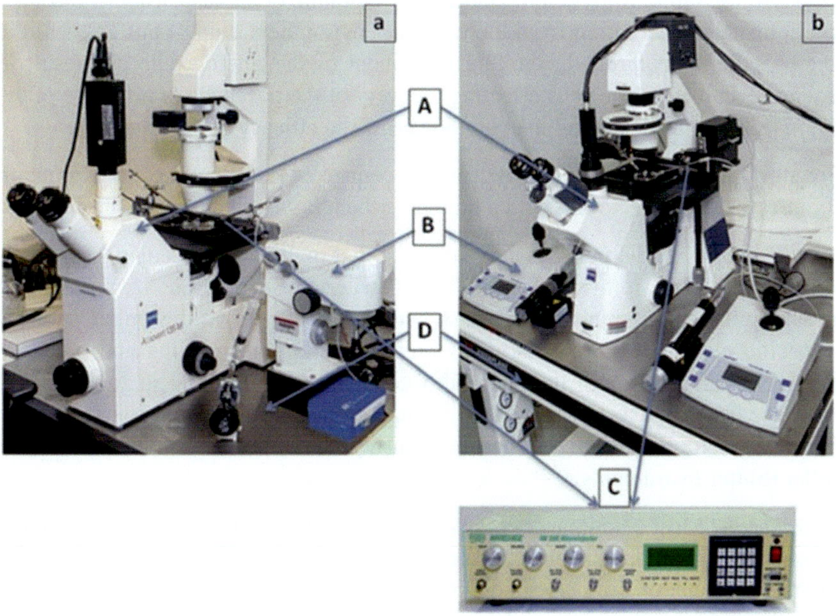

Fig. 2 Image of a microinjection setup indicating different instruments. **a** Microscope, **b** micromanipulators, **c** microinjector and **d** antivibration table. Copyright © Springer Science + Business Media, LLC, part of Springer Nature. Figure reprinted from Pu et al. (2019) with permission from Springer Nature

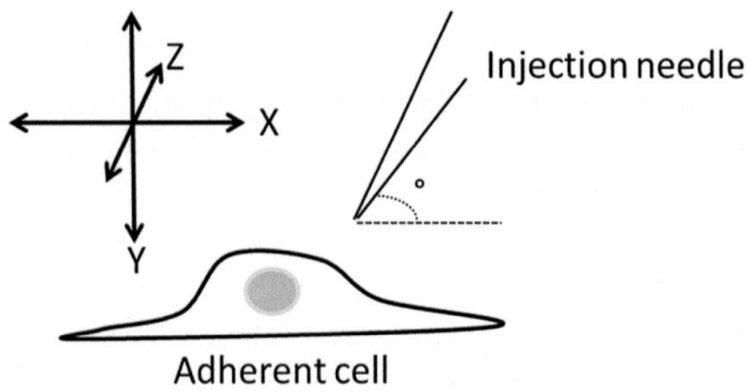

Fig. 3 Pictorial representation of microinjection needle and axes of movement in respect to an adherent cell. Copyright © Springer-Verlag Berlin Heidelberg. Figure reprinted from Shanmugam and Santra (2016) with permission from Springer

micromanipulator holding the microneedles to provide a suitable environment for the survival of the host cell during injection procedure.

Microinjector: Microinjector is a device that controls the pressure of injection and the duration of injection of the foreign cargo into the host cell once the microneedle is penetrated. It also maintains a compensation pressure to prevent the retrieval of cellular content into the microneedle because of capillary force. The microinjector is connected to a pressure pump and regulates the pressure that is available for microinjection.

Antivibration table: Antivibration table with hydraulic or air suspension is essential to eliminate any vibration from the external environment being transferred to the microneedle. Uncontrolled vibrating movement of the microneedle can result in extensive damage to the cells effecting recovery of the host cells or breakage of the microneedle or unsuccessful injection lowering the efficacy of the procedure.

Pressure pump: A pressure pump is generally connected to a microinjector. This pump compresses air to create and hold a pressure which is supplied to the microinjector.

Needle puller: Needle puller is an instrument that makes microneedles from capillaries. Capillaries are available in different materials such as quartz, borosilicate glass capillaries, aluminosilicate glass capillaries, etc. A needle puller subjects a specific region of the capillaries to precise heating and cooling and pulls the capillary apart from both the ends with defined speed resulting in formation of microneedles (Miller et al. 2002). Two examples of needle puller instruments include PMP102 micropipette puller (MicroData Instruments *Inc.*) and P-97 Micropipette puller (Sutter Instruments).

Microneedle: Most commonly borosilicate glass capillaries are used for microinjection. Although the dimensions of the capillaries vary, we recommend starting with an outer diameter of 1 mm, an inner diameter of 0.5–0.75 mm and ~10 cm in length. Once the microneedle is prepared using needle puller the inner diameter of the injection needle ranges between 3 and 5 μm and the holding pipette ranges between 10 and 20 μm. Commercial microneedles are also available, examples include Femtotips I and Femtotips II. However, recent developments in bio-MEMS (MicroElectroMechanical Systems) brought several types of micro/nano needles as well as array of microneedles for drug delivery in humans whose details can be found below.

3.1 *Types of Host Cells*

Several eukaryotic cells have been microinjected successfully to study various aspects of biology such as immunity, signal transduction, development, cancer, etc. Adherent cell, suspension cell, oocytes, cultured tissue, skin tissue and cancerous tissue are popular host cell or tissue type for delivery of foreign cargoes.

Adherent cells—These are cultured cells mostly utilized in research to understand various aspects of biology. At a given time they are immobile and adhered to the growth surface, thus do not require a holding pipette for injection (Fig. 3).

Suspension cells—These cells do not require any attachment or growth surface to survive and divide. Suspension cells float around the media, thus requiring a holding pipette to immobilize the cells for injection (Fig. 7).

Oocytes—Microinjection of oocytes with sperm marked a notable milestone in the treatment of infertility. Oocytes do not require attachment surface, thus require a holding pipette for injection of single sperm (Sciorio and Esteves 2022) (Fig. 7).

Cultured tissue—A single cell in the cultured tissue can be injected with foreign cargoes, however the disadvantage is that only cell on the outer layer of the tissue can be injected without damaging the tissue itself (Wong et al. 2014).

Skin tissue—Recently microneedles are developed for delivery of drugs, therapeutic agents, vaccine, etc. transdermally to bypass skin barriers resulting in minimal damage to the skin tissue as well as minimal pain (Avcil and Çelik 2021).

Cancerous tissue—Few testing for utilization of microneedles to selectively eliminate cancerous tissue or vaccinate against cancer by administration of therapeutic agents are under study (Cole et al. 2018; Zhi et al. 2021).

3.2 *Basic Injection Procedure*

Simple understanding on the procedure involved in microinjection of single cell is provided below.

Injection cargoes—It is important to prepare the foreign cargoes that are being injected in appropriate buffer which is suitable for the host cell cytoplasmic environment. Proper preparation of the cargoes ensures minimal damage to the host cell as well as ensures acceptance and further processing of cargoes by host cell.

Host cell—As explained above there are different types of host cells which are subjected to microinjection procedure, and they should be grown in proper condition ensuring their survival before and during injection as well as recovery after injection. Preparation of specific host cell require specific conditions which can be found elsewhere appropriately.

Microneedle—Microneedle for injection of most eukaryotic host cell should have an injection diameter of 3–5 μm. The capillaries can be pulled in needle-puller to obtain microneedles with specific dimensions which can be loaded with foreign cargoes. Buffers containing nucleic acids, protein, etc. can be loaded using a pipette behind the microneedle using narrow pipette tip (e.g. GELoader tips, Eppendorf) and single sperm can be loaded into the microneedle by negative succession while visualizing using a microscope. In a few instants a freshly prepared microneedle tip is blocked and needs to be broken to allow for injection of foreign cargoes. Care is to be taken to balance the microneedle tip size as too small internal diameter will result in frequent blockage and a bigger diameter will damage the host cell perturbing the recovery efficiency. Holding pipette, if needed, should also be prepared accordingly. Thus, prepared microneedles and holding pipette should be attached to holders in the micromanipulator at defined angle to penetrate the host cell (Figs. 3 and 7).

Microinjection—During microinjection place the microneedles in the field of view in the microscope above the focal plane as well as focus the host cells. A combination of 10X and 40X objective can be used for microinjection procedure. At appropriate magnification, move the microneedle above a single host cell and gently lower the needle to penetrate the host cell. Once penetrated, use microinjector to inject specific volume of cargoes (commonly around 5–40 picolitres) and gently pull the microneedle back using micromanipulator. Repeat this to inject as many adherent cells as possible that are required to be transfected. In case of suspension cell or oocyte use the holding pipette to immobilize the host cell at the focal plane followed by injection from the opposite side as shown in Fig. 7.

3.3 Microfluidic, Semi-automated and Automated Systems for Microinjection

Microinjection can be performed only on a single cell at a given time, thus making the rate of injection very slower. To increase the rate of microinjection several microfluidic devices, semi-automated and automated systems have been developed. Few of the systems are discussed below:

Microfluidic systems: Microfluidic devices containing fluidic channels, at micron scale, create constant flow of individual cells which can be injected with cargo. The design of microfluidic devices is varied and limited only by the creativity of the engineers (Fig. 4). Unlike traditional microinjection where the cells are immobilized with mobile microneedle injecting the cargoes, in most microfluidic devices the cells are mobilized, and the microneedle is made stationary. A device developed by Adamo and Jensen let the cells flow through microchannel where it is captured by a microneedle for injection and after injection the cells are collected in a reservoir

(Adamo and Jensen 2008). Instead of microneedle a jet of injection fluid was generated by piezoelectric membrane to penetrate the floating cell to deliver the cargoes in another device developed by Adamo et al. (2013). Another device has an array of microneedles where the cells are captured via negative aspiration flow leading to penetration of single cell for cargo injection and released after injection (Zhang et al. 2012). These microfluidic devices increased the rate of injection from a few hundred cells to ~3500 cells per hour.

Semi-automated systems: Step by step process in a manual microinjection is explained above and it can be noticed that the rate is slower. To increase the rate of injection semi-automated and automated systems were developed. In semi-automated systems

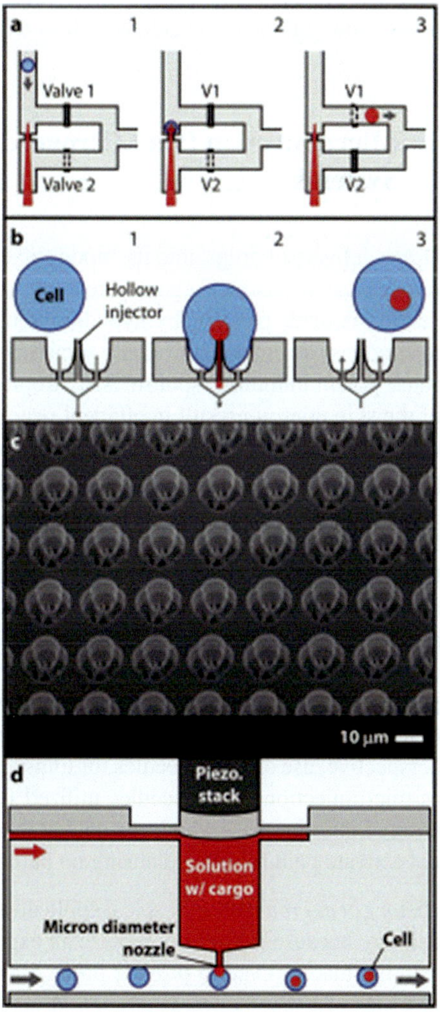

Fig. 4 Microfluidic devices for high-throughput microinjection. **a** Microfluidic device designed by Adamo and Jensen (2008) showing different stages of injection procedure where 1 represent the entry of cell, 2 represent penetration of the host cell by immobilized microneedle and 3 represent release and flow towards collection reservoir. **b** Concept of microinjection designed by Zhang et al. (2012) showing different stages of injection as mentioned above. **c** SEM image of array of microneedles designed by Zhang et al. (2012). **d** Concept designed by Adamo et al. (2013) where the injection is carried by piezoelectric element into the host cell as it flows through the microfluidic device. Copyright ©2014 Society for Laboratory Automation and Screening. Published by Elsevier Inc. Reprinted from Meacham et al. (2014). This is free to use content under the terms and conditions of the Creative Commons Attribution—NonCommercial-NoDerivs (CC BY-NC-ND 4.0) license

the injection paths for microinjection are calibrated once, as all the adherent cells exist in monolayer. Thus, a technician must move the needle to a new position where untransfected cell is located and a press of a button will lower the microneedle to inject the cargoes followed by returning of the needle to its original position (Viigipuu and Kallio 2004).

Automated systems: Further development in robotic systems is used to identify the coordinates of the cells that need to be injected apart from calibration of injection path resulted in development of fully automated system. In automated system, the technician must load samples and define the coordinates as well as the injection path which will be followed by injection of several hundreds of cells as defined in the system (Pepperkok et al. 1991; Truong et al. 2012; Wang et al. 2007).

These systems increased the rate of microinjection and helped to generate a larger population of transfected cell as well as reduced the fatigue to the technicians.

4 Microinjection (via Microneedles) as Drug Delivery Systems

Direct delivery of drugs, into the blood stream (or quick absorption into blood stream) for delivery to distinct organ system, as well as biomolecules as therapeutic intervention is made possible by hypodermic needle/syringe injection; however, needle injection is extremely painful and sometime damages the area of injection as well as cause potent immune response. Also, tropical administration of drugs on the surface of the skin does not result in efficient penetration throughout the local area because skin is resistance for several materials as well as protects from toxic chemicals and pathogens. Skin has three layers, the outer layer is stratum corneum, followed by epidermis and the final dermis. One alternative to reduce the overall discomfort is by introducing microneedles which can deliver the drugs transdermally to be up taken into the blood stream as well as increase the efficiency of tropical administration, respectively. The two major factors increasing the efficiency of drug deliveries are (1) micromechanical disruption of the skin and (2) release of drug within the epidermis layer across the stratum corneum. Microneedles used in experimental procedures have evolved into injection systems that can deliver therapeutic agents transdermally as well as collect samples and continuously monitor glucose. In once perspective, use of microneedles for transdermal delivery of drugs can be considered as microinjection. Microneedles utilized for transdermal drug delivery contain an array of microneedles which can deliver therapeutic agents transdermally thereby do not activate pain receptors causing no pain or negligible pain.

History of microneedles: Tropical application of drugs is the simplest method of drug delivery because it does not require an expert, non-invasive and painless. However, skin has also evolved to provide protection against hazardous external factors, thus provides resistance towards efficient drug delivery (Smith 1919; Scheuplein 1967).

Microneedles for efficient drug delivery via skin were conceptualized in 1970s followed by significant innovation by late 1990s which resulted in development of silicon microneedle for delivery of calcein through human skin (Henry et al. 1998). Later changes in permeability of skin upon utilization of microneedles for various cargoes or drugs such as insulin, bovine serum albumin, latex nanoparticles, etc. was evaluated. Mathematical models demonstrated that simple diffusion delivers drug after injection using microneedles. In the last decade, innovations in microneedle technology exploded resulting in unique designs using varieties of materials for efficient delivery of drugs (Manoj et al. 2020; McAllister et al. 2003; Prausnitz 2004; Waghule et al. 2019).

Microneedle materials: Microneedle array (MNA), several microneedles organized in a tight defined space (Fig. 5), also enhances drug delivery. Microneedles are 150–1500 μm in length and 50–250 μm in width with a tip size of 1–25 μm. They are made of wide variety of materials such as metal, silicon, glass, sugars that can dissolve into the blood stream after dispensing the therapeutic agent upon injection (such as polysaccharides, mannitol, maltose, sucrose, etc.), polymers, ceramic, etc. (Bhatnagar et al. 2017; Prausnitz and Langer 2008; Takeuchi and Kim 2018; Waghule et al. 2019).

Types of microneedles: Current development in material science and engineering has brought us several types of microneedles for transdermal delivery of drugs. Following are few unique designs of microneedles:

Solid microneedle—These needles are solid and commonly made of either metals or silicon and they appear in variety of shapes as shown in Fig. 5. Solid microneedles create pores in the skin in micrometer range which can be used to deliver therapeutic agents. Therapeutic agents are generally applied over the skin after poking with the solid microneedles. Thus, this method of drug administration is called "poke and patch approach" (Avcil and Çelik 2021). Silicon microneedles are fabricated by etching technique of silicon. Apart from delivery of drugs, an array of micromechanical structures was used to deliver foreign genes in the nematode worm model as early as 1995 (Hashmi 1995).

Coated microneedle—These are solid microneedles that are coated with appropriate therapeutic agents on their surface. These microneedles utilize "coat and poke approach" of drug administration, which allows for effective delivery when the formulations are uniformly layered or coated and stable. However, the therapeutic agents should be water-soluble and allow layer-by-layer coating on the surface of the microneedle (Avcil and Çelik 2021).

Hollow microneedle—Hollow microneedle contains microchannels inside the needles which connect a reservoir (for drug storage) to a hollow tip for drug delivery (Mukerjee et al. 2004), as shown in Figs. 5 and 6. They work like conventional hypodermic needles; however, does not reach the dermis layers of the skin to activate the pain receptors. Hollow microneedles are more advantageous over solid microneedles because hollow microneedles facilitate pressure-driven injection of drugs into the skin; thus follow "poke and flow approach" of drug administration (Avcil and Çelik 2021; Chaudhri et al. 2010; Chua et al. 2013; Kim and Lee 2006). Apart from

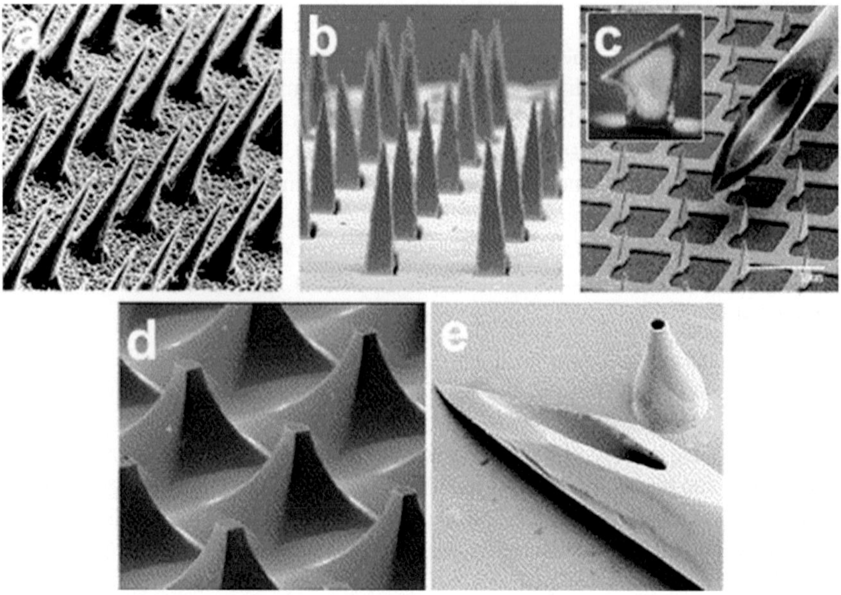

Fig. 5 SEM images of array of microneedles. **a** Solid microneedles (150 μm tall) etched from a silicon wafer. **b** Solid microneedles (1000 μm tall) laser-cut from a stainless-steel sheet. **c** Solid microneedle ("micro projection array", 330 μm tall) acid-etched from a titanium sheet, in comparison with the hypodermal syringe needle, were coated with protein antigen for vaccine delivery. **d** Solid microneedles ("micro enhancer array", 200 μm tall) chemically etched from a silicon wafer were dipped in plasmid DNA solution for vaccine delivery in vivo. **e** Hollow microneedles (500 μm tall) formed by electrodeposition of metal onto a polymer mold, in comparison with the hypodermal syringe needle. Copyright © 2003 Elsevier B.V. Figure reprinted from Prausnitz (2004) with permission from Elsevier

injection of drugs into the biological system, hollow microneedle array can also be used for extraction of biological fluid samples for various analysis (Mukerjee et al. 2004). Wide varieties of materials are used for fabrication of hollow microneedles beside silicon such as SU8 polymers, clay-reinforced polyimide, metal electroplated polymer MNA (Chaudhri et al. 2010; Kim and Lee 2006; Wang et al. 2013). Hollow microneedles were successfully tested to deliver insulin and dermal polio vaccination (Lee et al. 2017; van der Maaden et al. 2014). Further, hybrid additive manufacturing approach (combining digital light processing 3D printing with ex situ direct laser writing) resulted in development of new class of MNAs for fluidic microinjection and was tested for effective distribution of surrogate fluids and nanoparticle suspensions directly into brains (Sarker et al. 2023). Although hollow microneedles demonstrate several advantages, they possess the following disadvantages such as they are inherently weaker than the solid microneedles, the necessity of bore opening reduced the tip sharpness and the fluid flow can be slower or obstructed upon blockage of the dense dermal tissue (Martanto et al. 2006).

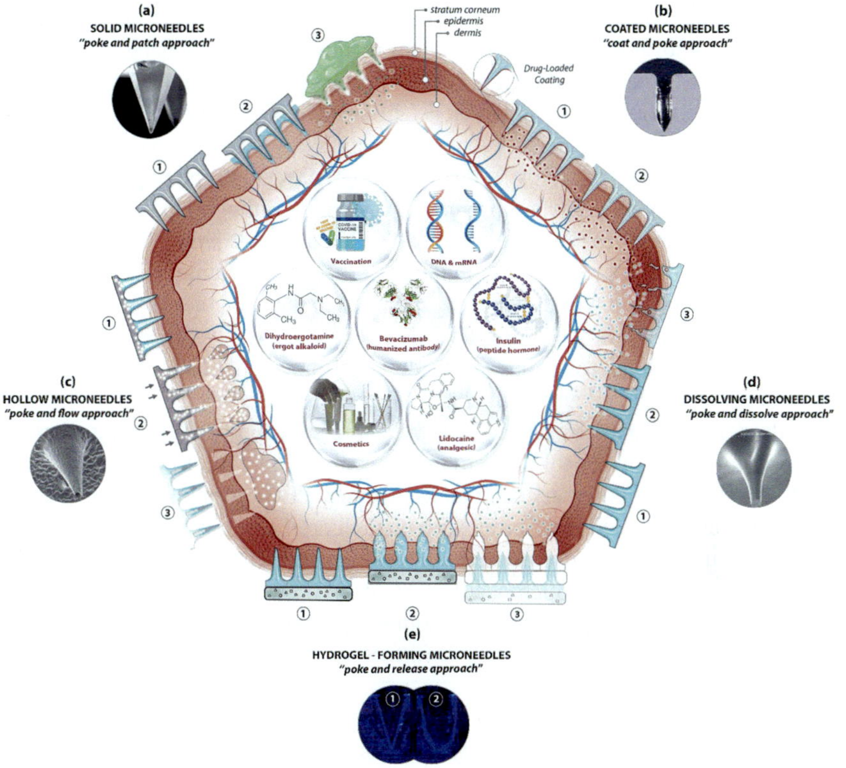

Fig. 6 Pictorial diagram of microneedle-based drug delivery approaches. Cartoon of **a** solid, **b** coated, **c** hollow, **d** dissolving and **e** hydrogel-forming microneedles as well as representative microscopic images of microneedles. Step-by-step process of drug delivery approaches are indicated from 1 to 3. Copyright © 2021 by the authors. Licensee MDPI, Basel, Switzerland. Figure reprinted from Avcil and Çelik (2021). This is free to use content under the terms and conditions of the Creative Commons Attribution (CC BY) license

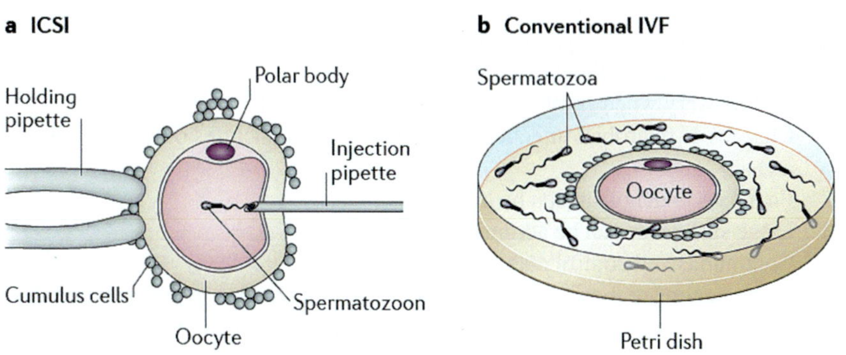

Fig. 7 Pictorial representation of assisted reproductive techniques (ART). **a** Intracytoplasmic sperm injection (ICSI). **b** In vitro fertilization. Figure reprinted from Esteves et al. (2018) with permission from Springer Nature

Polymer microneedles—These microneedles are fabricated using polymers such as polylactic acid, polymethylmethacrylate, maltose, etc. (Janphuang et al. 2018; Prausnitz 2004). These needles are utilized for superior mechanical strength and designed with profound tapering which is fabricated using inclined ultraviolet exposure technology. Apart from UV source, deep X-ray exposure is also used to fabricate microneedle structures, known as lithography, electroplating and molding (LIGA—Lithographie, Galvanoformung, Abformung-lithography, electroplating and molding) techniques (Pérennès et al. 2006; Yoon et al. 2006).

Dissolvable microneedles—Dissolving microneedles containing the therapeutic agent in a reservoir and will dissolve once applied to the skin, there by releasing the drugs for absorption via the skin following "poke and dissolve approach" of drug administration. The microneedles are made up of biocompatible or biodegradable low-cost polymers (Avcil and Çelik 2021; Waghule et al. 2019).

Hydrogel-forming microneedles—"Poke and release approach" of drug administration is utilized by hydrogel forming microneedles. These microneedles are made up of soft swellable polymers such as poloxamer, PEG-crosslinked poly (methyl vinil-co-maleic acid) and silk fibroin with phenylboronic acid/acrylamide to absorb the interstitial fluid upon insertion into the skin resulting in delivery of therapeutic agents. Hydrogel-forming microneedles are also utilized for detection of glucose as well as insulin delivery and lithium monitoring (Avcil and Çelik 2021; Caffarel-Salvador et al. 2015; Chen et al. 2019; Eltayib et al. 2016; Sivaraman and Banga 2017).

Use of microneedles for drug delivery has several advantages as well as accompanied by few disadvantages (Donnelly and Douroumis 2015; Gill et al. 2008; Gupta et al. 2011; Li et al. 2017; Park et al. 2005; Prausnitz 2004, 2017; Sebastien Henry 1988; van der Maaden et al. 2012).

Advantages of Microneedles:

1. Ease of administration.
2. Delivering drug with minimal discomfort and best patient compliance.
3. Specific skin area can be targeted for the administration of drug with reduced dosage than normal hypodermic injection.
4. Good tolerance without oedema and erythema.
5. Fast delivery when combining microneedles with other technologies.
6. Chances of microbial infection are reducing as it punctures only the epidermis area.
7. First pass metabolism can be avoided.
8. Large size drugs can be administered.
9. Continuous and controllable release.
10. Faster healing at the site of application when compared to hypodermic needles.
11. Polymer microneedles have the advantages like biocompatible and biodegradable in nature.
12. Reduction or elimination of sharps waste.
13. Lack of pain and needle phobia.

Disadvantage of Microneedles:

1. Microneedle delivery can be affected by the external factors like hydration of skin, thickness of skin, etc.
2. There is a chance of break of tip of microneedle and will remain in skin even after the removal of patch.
3. The microneedle administration can be carried out by expert trained person only.
4. Local inflammation during the repetitive injection.
5. If the microneedle is not placed properly, it will cause the drug escape in different angles under the skin with different concentration.
6. The large and hydrophilic compounds could not be delivered using microneedles.

4.1 Microinjection of Sperm for Treatment of Infertility

Couple's inability to conceive in a period of one year is defined as infertility; which is also associated with anxiety, depression, and psychological stress (Cousineau and Domar 2007; Inhorn and Patrizio 2015; Rooney and Domar 2018). Assisted reproductive techniques (ART) are used to clinically treat infertility and in vitro fertilization (IVF), where oocyte and sperm are incubated in appropriate laboratory condition for fertilization followed by implantation (Fig. 7b), plays a major role.

Intracytoplasmic sperm injection (ICSI) (Fig. 7a) is a modification of IVF, where a single spermatozoon is microinjected into an oocyte resulting in fertilization and was introduced by Dr. Gianpiero Palermo in 1992 (Palermo 1992). ICSI was first identified after an accidental injection of sperm into oocyte during sub zonal insemination procedure and later developed into ICSI with adaptation of standard protocols (Jain and Gupta 2007; O'Neill et al. 2018; Palermo et al. 2017). Although IVF as well as ICSI is constantly used for treatment of infertility, it is necessary to acknowledge the risks on offspring's health associated with the procedure such as preterm birth, chromosomal and congenital abnormalities, childhood cancer, psychological and neurological disorders, etc. (Cavoretto et al. 2018; Davies et al. 2017; Esteves et al. 2018; Hargreave et al. 2013; Zheng et al. 2018).

4.2 Gene Therapy Using Microinjection and Microneedles

Microinjection can be successfully used for transfection of cells on the surface of the tissue as well as skin surfaces. Further application of microinjection into internal organs is technically difficult, in which case other methods of transfection or transduction must be applied. Ex vivo genetic manipulation by gene therapy utilizing microinjection techniques had been demonstrated on isolated human blood stem cells and mesenchymal stem cells which can be developed into gene therapy treatment strategy in future (Davis et al. 2000; Tsulaia et al. 2003). Furthermore, there are several exciting recent developments in use of microneedles for cutaneous gene therapy to destroy skin cancer cells, to induce immunity (vaccination) as well as for wide variety of skin disease (Coulman et al. 2006; McCaffrey et al. 2015; Zhi et al. 2021; Zhu et al. 2022). Further, utilization of lyophilization to increase the loading of genetic material into nanoparticles in dissolvable microneedles has been successfully demonstrated in preclinical cervical cancer model (Cole et al. 2018). However, extensive studies are required to successfully translate gene therapy using microneedles to benefit humankind.

5 Advantages and Disadvantages of Microinjection

Every technique invented up to date comes with both advantages and disadvantages. As such, microinjection has several unique advantages which guarantee the continued use of microinjection technique both in research as well as in clinical treatment. Following are few notable advantages of microinjection:

1. With sufficient technical expertise, the success rate of 100% can be achieved in microinjection with minimal damage for quick recovery of the injected cells.
2. Localized injection of therapeutic substance for treatment without causing systemic toxicity. E.g. treatment of cancerous tissue without damaging or with minimal damage to the surrounding normal tissue.
3. Microneedles developed for localized drug delivery cause minimal damage and stress with efficient delivery of therapeutic agent.
4. Microinjection is the best choice for transfection of cells and tissues which are sensitive or impossible to either chemical-mediated or viral-mediated transfection.
5. This technique can deliver cargoes to specific subcellular location such as nucleus or cytoplasm. DNA in the cytoplasm has a very less half-life before DNA degradation; thus, microinjection of DNA at nucleus is essential for successful gene expression. Similarly, appropriate subcellular locations must be chosen for injection of mRNA and proteins depending on the functional goal to increase the success of the experiment.
6. Microinjection can delivery mitochondria containing non-mutated DNA into host cells to treat disease caused by defects in mitochondrial DNA which follows non-mendelian inheritance pattern (Kagawa et al. 2001).
7. Large cargoes and biomolecules which cannot be delivered by both chemical-mediated transfection and viral-mediated transfection can be delivered using microinjection technique.
8. Multiple cargo types can be delivered during a single microinjection procedure. E.g. mixture of nucleic acid and proteins can be delivered to the nucleus of a single cell.

9. Microinjection is popularly used in the generation of transgenic model organisms such as *C. elegans*, *Drosophila*, mice, etc. for various research (Evans 2006; Muthaiyan Shanmugam 2016; Ringrose 2009).
10. Cell autonomous behavior caused by a particular signaling pathways can be demonstrated using microinjection of necessary proteins involved in the signaling.

Although microinjection techniques have several unique advantages, they also have a few disadvantages. Following are disadvantages of the microinjection techniques:

1. The number of cells that can be transfected at a given time is very low, i.e., very low rate of transfection, in a population of cells. Better transfection rate can be obtained with extensive practice. Recently, semi-automated and robotic methods of needle injection have been developed to increase the rate as well as the efficiency of injection.
2. Expensive microinjection equipment is necessary to perform microinjections.
3. Extensive practice and technical expertise (such as microneedle preparation, understanding of sample cell type, identification of clogged microneedle, understanding depth perception to prevent needle breaks and to prevent extensive tissue damage, etc.) is necessary for proper application of the methodology as well as maintenance of the equipments.
4. Single cell microinjection can be performed only on the outer layer of cells in a tissue as injection of inner cells can cause extensive tissue damage.
5. Material used for preparation of microneedles can cause allergic reactions.
6. Precise amount or rate of drug delivery is hard to maintain while using transdermal microneedles for treatment. Microneedle materials can profoundly affect the rate of delivery and metabolism of drug.

Understanding the pros and cons of a specific technique is necessary for informed selection of research methodology as well as treatment strategies. It is very important to note that microinjection is a gold standard technique for specific experiment, such as development of transgenic lines, and microneedles for minimal invasive transdermal delivery of drugs. Also, its unique advantages ensured continued use of microinjection techniques in research and clinical practice.

6 Conclusions

Microinjection is an irreplaceable gold standard technique for certain applications such as research applications where other methods of transfection cannot be used, transgenic line generations, intracytoplasmic sperm injection (ICSI), etc. Further recent development in bioengineering microneedle arrays has demonstrated its potential to be a painless alternative for painful hypodermal injections for delivery of therapeutic agents as well as sample collection, continuous metabolite/glucose monitoring and gene therapy. With the development of genome editing tools, such as

CRISPR/Cas9, microinjection becomes an inevitable technique in research. Further fusion of bioengineering with artificial intelligent will give rise to technologies using microneedles for continuous monitoring of cellular metabolites and biomolecules as well as automated injection of therapeutic agents for several life-threatening illnesses. With improving industrialization, commercialization, and distribution networks; it is certain that use of microneedles (with affordable cost) for safe, painless, self-administration of therapeutic agents is not far away.

References

Adamo A, Jensen KF (2008) Microfluidic based single cell microinjection. Lab Chip 8(8):1258. https://doi.org/10.1039/b803212b

Adamo A, Roushdy O, Dokov R, Sharei A, Jensen KF (2013) Microfluidic jet injection for delivering macromolecules into cells. J Micromech Microeng 23(3):035026. https://doi.org/10.1088/0960-1317/23/3/035026

Avcil M, Çelik A (2021) Microneedles in drug delivery: progress and challenges. Micromachines 12(11):1321. https://doi.org/10.3390/mi12111321

Barber MA (1904) A new method of isolating microorganisms. Annu Rev Physiol 4(487)

Barber MA (1911) A technic for the inoculation of bacteria and other substances into living cells. J Infect Dis 8(3):348–360. https://doi.org/10.1093/infdis/8.3.348

Belling JN, Heidenreich LK, Tian Z, Mendoza AM, Chiou T-T, Gong Y, Chen NY, Young TD, Wattanatorn N, Park JH, Scarabelli L, Chiang N, Takahashi J, Young SG, Stieg AZ, De Oliveira S, Huang TJ, Weiss PS, Jonas SJ (2020) Acoustofluidic sonoporation for gene delivery to human hematopoietic stem and progenitor cells. Proc Natl Acad Sci 117(20):10976–10982. https://doi.org/10.1073/pnas.1917125117

Bhatnagar S, Dave K, Venuganti VVK (2017) Microneedles in the clinic. J Control Release 260:164–182. https://doi.org/10.1016/j.jconrel.2017.05.029

Birchmeier C, Broek D, Wigler M (1985) RAS proteins can induce meiosis in Xenopus oocytes. Cell 43(3):615–621. https://doi.org/10.1016/0092-8674(85)90233-8

Briggs R, King TJ (1952) Transplantation of living nuclei from blastula cells into enucleated frogs' eggs. Proc Natl Acad Sci 38(5):455–463. https://doi.org/10.1073/pnas.38.5.455

Caffarel-Salvador E, Brady AJ, Eltayib E, Meng T, Alonso-Vicente A, Gonzalez-Vazquez P, Torrisi BM, Vicente-Perez EM, Mooney K, Jones DS, Bell SEJ, McCoy CP, McCarthy HO, McElnay JC, Donnelly RF (2015) Hydrogel-forming microneedle arrays allow detection of drugs and glucose in vivo: potential for use in diagnosis and therapeutic drug monitoring. PLoS ONE 10(12):e0145644. https://doi.org/10.1371/journal.pone.0145644

Capecchi MR (1980) High efficiency transformation by direct microinjection of DNA into cultured mammalian cells. Cell 22(2):479–488. https://doi.org/10.1016/0092-8674(80)90358-X

Cavoretto P, Candiani M, Giorgione V, Inversetti A, Abu-Saba MM, Tiberio F, Sigismondi C, Farina A (2018) Risk of spontaneous preterm birth in singleton pregnancies conceived after IVF/ICSI treatment: meta-analysis of cohort studies. Ultrasound Obstet Gynecol 51(1):43–53. https://doi.org/10.1002/uog.18930

Chakrabarty P, Gupta P, Illath K, Kar S, Nagai M, Tseng F-G, Santra TS (2022) Microfluidic mechanoporation for cellular delivery and analysis. Mater Today Bio 13:100193. https://doi.org/10.1016/j.mtbio.2021.100193

Chakrabarty P, Illath K, Kar S, Nagai M, Santra TS (2023) Combinatorial physical methods for cellular therapy: towards the future of cellular analysis? J Control Release 353:1084–1095. https://doi.org/10.1016/j.jconrel.2022.12.038

Chaudhri BP, Ceyssens F, De Moor P, Van Hoof C, Puers R (2010) A high aspect ratio SU-8 fabrication technique for hollow microneedles for transdermal drug delivery and blood extraction. J Micromech Microeng 20(6):064006. https://doi.org/10.1088/0960-1317/20/6/064006

Chen S, Matsumoto H, Moro-oka Y, Tanaka M, Miyahara Y, Suganami T, Matsumoto A (2019) Smart microneedle fabricated with silk fibroin combined semi-interpenetrating network hydrogel for glucose-responsive insulin delivery. ACS Biomater Sci Eng 5(11):5781–5789. https://doi.org/10.1021/acsbiomaterials.9b00532

Chua B, Desai SP, Tierney MJ, Tamada JA, Jina AN (2013) Effect of microneedles shape on skin penetration and minimally invasive continuous glucose monitoring in vivo. Sens Actuators A 203:373–381. https://doi.org/10.1016/j.sna.2013.09.026

Cole G, Ali AA, McCrudden CM, McBride JW, McCaffrey J, Robson T, Kett VL, Dunne NJ, Donnelly RF, McCarthy HO (2018) DNA vaccination for cervical cancer: strategic optimisation of RALA mediated gene delivery from a biodegradable microneedle system. Eur J Pharm Biopharm 127:288–297. https://doi.org/10.1016/j.ejpb.2018.02.029

Coulman S, Allender C, Birchall J (2006) Microneedles and other physical methods for overcoming the stratum corneum barrier for cutaneous gene therapy. Crit Rev Ther Drug Carrier Syst 23(3):205–258. https://doi.org/10.1615/CritRevTherDrugCarrierSyst.v23.i3.20

Cousineau TM, Domar AD (2007) Psychological impact of infertility. Best Pract Res Clin Obstet Gynaecol 21(2):293–308. https://doi.org/10.1016/j.bpobgyn.2006.12.003

Davies M, Rumbold A, Marino J, Willson K, Giles L, Whitrow M, Scheil W, Moran L, Thompson J, Lane M, Moore V (2017) Maternal factors and the risk of birth defects after IVF and ICSI: a whole of population cohort study. BJOG: Int J Obstet Gynaecol 124(10):1537–1544. https://doi.org/10.1111/1471-0528.14365

Davis BR, Yannariello-Brown J, Prokopishyn NL, Luo Z, Smith MR, Wang J, Carsrud ND, Brown DB (2000) Glass needle-mediated microinjection of macromolecules and transgenes into primary human blood stem/progenitor cells. Blood 95(2):437–444

Dey K (2020) Microfluidic electroporation and applications. In: Microfluidics and bio-MEMS, pp 343–400

Donnelly R, Douroumis D (2015) Microneedles for drug and vaccine delivery and patient monitoring. Drug Deliv Transl Res 5(4):311–312. https://doi.org/10.1007/s13346-015-0250-2

Du M, Li Y, Chen Z (2022) Sonoporation-mediated gene transfection: a novel direction for cell reprogramming in vivo. Front Bioeng Biotechnol 9. https://doi.org/10.3389/fbioe.2021.803055

Eltayib E, Brady AJ, Caffarel-Salvador E, Gonzalez-Vazquez P, Zaid Alkilani A, McCarthy HO, McElnay JC, Donnelly RF (2016) Hydrogel-forming microneedle arrays: potential for use in minimally-invasive lithium monitoring. Eur J Pharm Biopharm 102:123–131. https://doi.org/10.1016/j.ejpb.2016.03.009

Esteves SC, Roque M, Bedoschi G, Haahr T, Humaidan P (2018) Intracytoplasmic sperm injection for male infertility and consequences for offspring. Nat Rev Urol 15(9):535–562. https://doi.org/10.1038/s41585-018-0051-8

Evans T (2006) Transformation and microinjection. WormBook. https://doi.org/10.1895/wormbook.1.108.1

Freitag F, Wagner E (2021) Optimizing synthetic nucleic acid and protein nanocarriers: the chemical evolution approach. Adv Drug Deliv Rev 168:30–54. https://doi.org/10.1016/j.addr.2020.03.005

Gebauer F, Xu W, Cooper GM, Richter JD (1994) Translational control by cytoplasmic polyadenylation of c-mos mRNA is necessary for oocyte maturation in the mouse. EMBO J 13(23):5712–5720. https://doi.org/10.1002/j.1460-2075.1994.tb06909.x

Gill HS, Denson DD, Burris BA, Prausnitz MR (2008) Effect of microneedle design on pain in human volunteers. Clin J Pain 24(7):585–594. https://doi.org/10.1097/AJP.0b013e31816778f9

Gordon JW, Ruddle FH (1981) Integration and stable germ line transmission of genes injected into mouse pronuclei. Science 214(4526):1244–1246. https://doi.org/10.1126/science.6272397

Gupta J, Gill HS, Andrews SN, Prausnitz MR (2011) Kinetics of skin resealing after insertion of microneedles in human subjects. J Control Release 154(2):148–155. https://doi.org/10.1016/j.jconrel.2011.05.021

Gupta P, Kar S, Kumar A, Tseng F-G, Pradhan S, Mahapatra PS, Santra TS (2021) Pulsed laser assisted high-throughput intracellular delivery in hanging drop based three dimensional cancer spheroids. Analyst 146(15):4756–4766. https://doi.org/10.1039/D0AN02432E

Gurdon JB, Elsdale TR, Fischberg M (1958) Sexually mature individuals of Xenopus laevis from the transplantation of single somatic nuclei. Nature 182(4627):64–65. https://doi.org/10.1038/182064a0

Gurdon JB, Lane CD, Woodland HR, Marbaix G (1971) Use of frog eggs and oocytes for the study of messenger RNA and its translation in living cells. Nature 233(5316):177–182. https://doi.org/10.1038/233177a0

Hargreave M, Jensen A, Toender A, Andersen KK, Kjaer SK (2013) Fertility treatment and childhood cancer risk: a systematic meta-analysis. Fertil Steril 100(1):150–161. https://doi.org/10.1016/j.fertnstert.2013.03.017

Hashmi S (1995) Genetic transformation of nematodes using arrays of micromechanical piercing structures. Biotechiques 19(5):766–770

Heidemann SR, Kirschner MW (1975) Aster formation in eggs of Xenopus laevis. Induction by isolated basal bodies. J Cell Biol 67(1):105–117. https://doi.org/10.1083/jcb.67.1.105

Henry S, McAllister DV, Allen MG, Prausnitz MR (1998) Microfabricated microneedles: a novel approach to transdermal drug delivery. J Pharm Sci 87(8):922–925. https://doi.org/10.1021/js980042+

Inhorn MC, Patrizio P (2015) Infertility around the globe: new thinking on gender, reproductive technologies and global movements in the 21st century. Hum Reprod Update 21(4):411–426. https://doi.org/10.1093/humupd/dmv016

Jain T, Gupta RS (2007) Trends in the use of intracytoplasmic sperm injection in the United States. N Engl J Med 357(3):251–257. https://doi.org/10.1056/NEJMsa070707

Janphuang P, Laebua M, Sriphung C, Taweewat P, Sirichalarmkul A, Sukjantha K, Promsawat N, Leuasoongnoen P, Suphachiaraphan S, Phimol K, Isarakorn D, Tancharakorn S (2018) Polymer based microneedle patch fabricated using microinjection moulding. MATEC Web Conf 192:01039. https://doi.org/10.1051/matecconf/201819201039

Jiang H (ed) (2016) Optical MEMS for chemical analysis and biomedicine. Institution of Engineering and Technology. https://doi.org/10.1049/PBCS025E

Kagawa Y, Inoki Y, Endo H (2001) Gene therapy by mitochondrial transfer. Adv Drug Deliv Rev 49(1–2):107–119. https://doi.org/10.1016/S0169-409X(01)00129-6

Kaladharan K, Kumar A, Gupta P, Illath K, Santra T, Tseng F-G (2021) Microfluidic based physical approaches towards single-cell intracellular delivery and analysis. Micromachines 12(6):631. https://doi.org/10.3390/mi12060631

Kar S, Loganathan M, Dey K, Shinde P, Chang H-Y, Nagai M, Santra TS (2018) Single-cell electroporation: current trends, applications and future prospects. J Micromech Microeng 28(12):123002. https://doi.org/10.1088/1361-6439/aae5ae

Kay MA, Glorioso JC, Naldini L (2001) Viral vectors for gene therapy: the art of turning infectious agents into vehicles of therapeutics. Nat Med 7(1):33–40. https://doi.org/10.1038/83324

Keefer CL (2015) Artificial cloning of domestic animals. PNAS 112(29):8874–8878. https://doi.org/10.1073/pnas.1501718112

Kim K, Lee J-B (2006) High aspect ratio tapered hollow metallic microneedle arrays with microfluidic interconnector. Microsyst Technol 13(3–4):231–235. https://doi.org/10.1007/s00542-006-0221-0

Kimura Y, Yanagimachi R (1995) Intracytoplasmic sperm injection in the mouse1. Biol Reprod 52(4):709–720. https://doi.org/10.1095/biolreprod52.4.709

Kimura M, Endo A, Nagira Y, Yoshizumi T (2023) Particle bombardment-assisted peptide-mediated gene transfer for highly efficient transient assay. BMC Res Notes 16(1):46. https://doi.org/10.1186/s13104-023-06320-3

Knowles JKC (1974) An improved microinjection technique in Paramecium aurelia. Exp Cell Res 88(1):79–87. https://doi.org/10.1016/0014-4827(74)90620-X

Lee J, Kim D-H, Lee KJ, Seo IH, Park SH, Jang EH, Park Y, Youn Y-N, Ryu W (2017) Transfer-molded wrappable microneedle meshes for perivascular drug delivery. J Control Release 268:237–246. https://doi.org/10.1016/j.jconrel.2017.10.007
Li J, Zeng M, Shan H, Tong C (2017) Microneedle patches as drug and vaccine delivery platform. Curr Med Chem 24(22). https://doi.org/10.2174/0929867324666170526124053
Lin TP (1966) Microinjection of mouse eggs. Science 151(3708):333–337. https://doi.org/10.1126/science.151.3708.333
Maller JL, Koontz JW (1981) A study of the induction of cell division in amphibian oocytes by insulin. Dev Biol 85(2):309–316. https://doi.org/10.1016/0012-1606(81)90262-1
Manoj H, Gupta P, Mohan L, Nagai M, Wankhar S, Santra TS (2020) Microneedles: current trends and applications. In: Microfluidics and bio-MEMS. Jenny Stanford Publishing, pp 275–342. https://doi.org/10.1201/9781003014935-7
Martanto W, Moore JS, Kashlan O, Kamath R, Wang PM, O'Neal JM, Prausnitz MR (2006) Microinfusion using hollow microneedles. Pharm Res 23(1):104–113. https://doi.org/10.1007/s11095-005-8498-8
McAllister DV, Wang PM, Davis SP, Park J-H, Canatella PJ, Allen MG, Prausnitz MR (2003) Microfabricated needles for transdermal delivery of macromolecules and nanoparticles: fabrication methods and transport studies. Proc Natl Acad Sci 100(24):13755–13760. https://doi.org/10.1073/pnas.2331316100
McCaffrey J, Donnelly RF, McCarthy HO (2015) Microneedles: an innovative platform for gene delivery. Drug Deliv Transl Res 5(4):424–437. https://doi.org/10.1007/s13346-015-0243-1
Meacham JM, Durvasula K, Degertekin FL, Fedorov AG (2014) Physical methods for intracellular delivery: practical aspects from laboratory use to industrial-scale processing. SLAS Technol 19(1):1–18. https://doi.org/10.1177/2211068213494388
Miller DFB, Holtzman SL, Kaufman TC (2002) Customized microinjection glass capillary needles for P-element transformations in Drosophila melanogaster. Biotechniques 33(2):366–375. https://doi.org/10.2144/02332rr03
Mintzer MA, Simanek EE (2009) Nonviral vectors for gene delivery. Chem Rev 109(2):259–302. https://doi.org/10.1021/cr800409e
Mohan L, Kar S, Mahapatra PS, Nagai M, Santra TS (2021a) Fabrication of TiO_2 microspikes for highly efficient intracellular delivery by pulse laser-assisted photoporation. RSC Adv 11(16):9336–9348. https://doi.org/10.1039/D0RA09785C
Mohan L, Kar S, Nagai M, Santra TS (2021b) Electrochemical fabrication of TiO_2 micro-flowers for an efficient intracellular delivery using nanosecond light pulse. Mater Chem Phys 267:124604. https://doi.org/10.1016/j.matchemphys.2021.124604
Mohan L, Kar S, Hattori R, Ishii-Teshima M, Bera P, Roy S, Santra TS, Shibata T, Nagai M (2021c) Can titanium oxide nanotubes facilitate intracellular delivery by laser-assisted photoporation? Appl Surf Sci 543:148815. https://doi.org/10.1016/j.apsusc.2020.148815
Mohan L, Hattori R, Zhang H, Matsumura Y, Santra TS, Shibata T, Ryu S, Nagai M (2022) Effect of size and interparticle distance of nanoparticles on the formation of bubbles induced by nanosecond laser. Surf Interfaces 30:101820. https://doi.org/10.1016/j.surfin.2022.101820
Mukerjee EV, Collins SD, Isseroff RR, Smith RL (2004) Microneedle array for transdermal biological fluid extraction and in situ analysis. Sens Actuators A 114(2–3):267–275. https://doi.org/10.1016/j.sna.2003.11.008
Muthaiyan Shanmugam M (2016) Improved methods of microinjection in Caenorhabditis elegans: automation and microfluidic systems. JSM Biotechnol Bioeng 3(5):1072
Muthaiyan Shanmugam M, Manoj H (2022) Microinjection for single-cell analysis and therapy. In: Handbook of single-cell technologies. Springer Singapore, pp 81–107. https://doi.org/10.1007/978-981-10-8953-4_32
Neher E, Sakmann B (1976) Single-channel currents recorded from membrane of denervated frog muscle fibres. Nature 260(5554):799–802. https://doi.org/10.1038/260799a0
O'Neill CL, Chow S, Rosenwaks Z, Palermo GD (2018) Development of ICSI. Reproduction 156(1):F51–F58. https://doi.org/10.1530/REP-18-0011

Palermo G (1992) Pregnancies after intracytoplasmic injection of single spermatozoon into an oocyte. Lancet 340(8810):17–18. https://doi.org/10.1016/0140-6736(92)92425-F

Palermo GD, O'Neill CL, Chow S, Cheung S, Parrella A, Pereira N, Rosenwaks Z (2017) Intracytoplasmic sperm injection: state of the art in humans. Reproduction 154(6):F93–F110. https://doi.org/10.1530/REP-17-0374

Park J-H, Allen MG, Prausnitz MR (2005) Biodegradable polymer microneedles: fabrication, mechanics and transdermal drug delivery. J Control Release 104(1):51–66. https://doi.org/10.1016/j.jconrel.2005.02.002

Pepperkok R, Schneider C, Philipson L, Ansorgel W (1991) Single cell assay with an automated capillary microinjection system. Cytotechnology 5(S1):93–98. https://doi.org/10.1007/BF00736821

Pérennès F, Marmiroli B, Matteucci M, Tormen M, Vaccari L, Fabrizio ED (2006) Sharp beveled tip hollow microneedle arrays fabricated by LIGA and 3D soft lithography with polyvinyl alcohol. J Micromech Microeng 16(3):473–479. https://doi.org/10.1088/0960-1317/16/3/001

Pratt FH, Eisenberger JP (1919) The quantal phenomena in muscle: methods, with further evidence of the all-or-none principle for the skeletal fiber. Am J Physiol-Leg Content 49(1):1–54. https://doi.org/10.1152/ajplegacy.1919.49.1.1

Prausnitz MR (2004) Microneedles for transdermal drug delivery. Adv Drug Deliv Rev 56(5):581–587. https://doi.org/10.1016/j.addr.2003.10.023

Prausnitz MR (2017) Engineering microneedle patches for vaccination and drug delivery to skin. Annu Rev Chem Biomol Eng 8(1):177–200. https://doi.org/10.1146/annurev-chembioeng-060816-101514

Prausnitz MR, Langer R (2008) Transdermal drug delivery. Nat Biotechnol 26(11):1261–1268. https://doi.org/10.1038/nbt.1504

Pu X, Young AP, Kubisch HM (2019) Production of transgenic mice by pronuclear microinjection, pp 17–41. https://doi.org/10.1007/978-1-4939-8831-0_2

Ringrose L (2009) Transgenesis in Drosophila melanogaster, pp 3–19. https://doi.org/10.1007/978-1-60327-019-9_1

Robbins PD, Ghivizzani SC (1998) Viral vectors for gene therapy. Pharmacol Ther 80(1):35–47. https://doi.org/10.1016/S0163-7258(98)00020-5

Rooney KL, Domar AD (2018) The relationship between stress and infertility. Dialog Clin Neurosci 20(1):41–47. https://doi.org/10.31887/DCNS.2018.20.1/klrooney

Santra T, Tseng F (2013) Recent trends on micro/nanofluidic single cell electroporation. Micromachines 4(3):333–356. https://doi.org/10.3390/mi4030333

Santra TS, Tseng F-G (2016) Electroporation for single-cell analysis, pp 55–83. https://doi.org/10.1007/978-3-662-49118-8_3

Santra TS, Tseng F-G (eds) (2022) Handbook of single-cell technologies. Springer Singapore. https://doi.org/10.1007/978-981-10-8953-4

Santra TS, Wang P-C, Chang H-Y, Tseng F-G (2013) Tuning nano electric field to affect restrictive membrane area on localized single cell nano-electroporation. Appl Phys Lett 103(23). https://doi.org/10.1063/1.4833535

Santra TS, Chang H-Y, Wang P-C, Tseng F-G (2014) Impact of pulse duration on localized single-cell nano-electroporation. Analyst 139(23):6249–6258. https://doi.org/10.1039/C4AN01050G

Santra TS, Kar S, Chang H-Y, Tseng F-G (2020a) Nano-localized single-cell nano-electroporation. Lab Chip 20(22):4194–4204. https://doi.org/10.1039/D0LC00712A

Santra TS, Kar S, Chen T-C, Chen C-W, Borana J, Lee M-C, Tseng F-G (2020b) Near-infrared nanosecond-pulsed laser-activated highly efficient intracellular delivery mediated by nano-corrugated mushroom-shaped gold-coated polystyrene nanoparticles. Nanoscale 12(22):12057–12067. https://doi.org/10.1039/D0NR01792B

Sarker S, Colton A, Wen Z, Xu X, Erdi M, Jones A, Kofinas P, Tubaldi E, Walczak P, Janowski M, Liang Y, Sochol RD (2023) 3D-printed microinjection needle arrays via a hybrid DLP-direct laser writing strategy. Adv Mater Technol 8(5):2201641. https://doi.org/10.1002/admt.202201641

Scherer F, Anton M, Schillinger U, Henke J, Bergemann C, Krüger A, Gänsbacher B, Plank C (2002) Magnetofection: enhancing and targeting gene delivery by magnetic force in vitro and in vivo. Gene Ther 9(2):102–109. https://doi.org/10.1038/sj.gt.3301624

Scheuplein RJ (1967) Mechanism of percutaneous absorption. J Investig Dermatol 48(1):79–88. https://doi.org/10.1038/jid.1967.11

Schweinsberg PJ (2013) C. elegans gene transformation by microparticle bombardment. WormBook 1–10. https://doi.org/10.1895/wormbook.1.166.1

Sciorio R, Esteves SC (2022) Contemporary use of ICSI and epigenetic risks to future generations. J Clin Med 11(8):2135. https://doi.org/10.3390/jcm11082135

Sebastien Henry D (1988) Microfabricated microneedles: a novel approach to transdermal drug delivery. J Pharm Sci 87(8):922–925

Shanmugam MM, Santra TS (2016) Microinjection for single-cell analysis, pp 85–129. https://doi.org/10.1007/978-3-662-49118-8_4

Shinde P, Kar S, Loganathan M, Chang H-Y, Tseng F-G, Nagai M, Santra TS (2020) Infrared pulse laser-activated highly efficient intracellular delivery using titanium microdish device. ACS Biomater Sci Eng 6(10):5645–5652. https://doi.org/10.1021/acsbiomaterials.0c00785

Shinde A, Shinde P, Kar S, Illath K, Dey S, Mahapatra NR, Nagai M, Santra TS (2023) Metallic micro-ring device for highly efficient large cargo delivery in mammalian cells using infrared light pulses. Lab Chip 23(9):2175–2192. https://doi.org/10.1039/D2LC00899H

Sivaraman A, Banga AK (2017) Novel in situ forming hydrogel microneedles for transdermal drug delivery. Drug Deliv Transl Res 7(1):16–26. https://doi.org/10.1007/s13346-016-0328-5

Sizikov AA, Nikitin PI, Nikitin MP (2021) Magnetofection in vivo by nanomagnetic carriers systemically administered into the bloodstream. Pharmaceutics 13(11):1927. https://doi.org/10.3390/pharmaceutics13111927

Smith HW (1919) On dichloroethylsulfide (mustard gas) IV. The mechanism of absorption by the skin. J Pharmacol Exp Ther 13:1–30

Steptoe PC, Edwards RG (1978) Birth after the reimplantation of a human embryo. Lancet 312(8085):366. https://doi.org/10.1016/S0140-6736(78)92957-4

Subhra T, Wang P-C, Gang F (2013) Electroporation based drug delivery and its applications. In: Advances in micro/nano electromechanical systems and fabrication technologies. InTech. https://doi.org/10.5772/55369

Takeuchi K, Kim B (2018) Functionalized microneedles for continuous glucose monitoring. Nano Converg 5(1):28. https://doi.org/10.1186/s40580-018-0161-2

Tseng F-G, Santra TS (eds) (2016) Essentials of single-cell analysis. Springer Berlin Heidelberg. https://doi.org/10.1007/978-3-662-49118-8

Thomas CE, Ehrhardt A, Kay MA (2003) Progress and problems with the use of viral vectors for gene therapy. Nat Rev Genet 4(5):346–358. https://doi.org/10.1038/nrg1066

Truong HH, de Sonneville J, Ghotra VPS, Xiong J, Price L, Hogendoorn PCW, Spaink HH, van de Water B, Danen EHJ (2012) Automated microinjection of cell-polymer suspensions in 3D ECM scaffolds for high-throughput quantitative cancer invasion screens. Biomaterials 33(1):181–188. https://doi.org/10.1016/j.biomaterials.2011.09.049

Tsulaia TV, Prokopishyn NL, Yao A, Victor Carsrud ND, Clara Carou M, Brown DB, Davis BR, Yannariello-Brown J (2003) Glass needle-mediated microinjection of macromolecules and transgenes into primary human mesenchymal stem cells. J Biomed Sci 10(3):328–336. https://doi.org/10.1007/BF02256452

van der Maaden K, Jiskoot W, Bouwstra J (2012) Microneedle technologies for (trans)dermal drug and vaccine delivery. J Control Release 161(2):645–655. https://doi.org/10.1016/j.jconrel.2012.01.042

van der Maaden K, Trietsch SJ, Kraan H, Varypataki EM, Romeijn S, Zwier R, van der Linden HJ, Kersten G, Hankemeier T, Jiskoot W, Bouwstra J (2014) Novel hollow microneedle technology for depth-controlled microinjection-mediated dermal vaccination: a study with polio vaccine in rats. Pharm Res. https://doi.org/10.1007/s11095-013-1288-9

Viigipuu K, Kallio P (2004) Microinjection of living adherent cells by using a semi-automatic microinjection system. Altern Lab Anim 32(4):417–423. https://doi.org/10.1177/026119290403200414

Waghule T, Singhvi G, Dubey SK, Pandey MM, Gupta G, Singh M, Dua K (2019) Microneedles: a smart approach and increasing potential for transdermal drug delivery system. Biomed Pharmacother 109:1249–1258. https://doi.org/10.1016/j.biopha.2018.10.078

Wang W, Liu X, Gelinas D, Ciruna B, Sun Y (2007) A fully automated robotic system for microinjection of zebrafish embryos. PLoS ONE 2(9):e862. https://doi.org/10.1371/journal.pone.0000862

Wang P-C, Paik S-J, Chen S, Rajaraman S, Kim S-H, Allen MG (2013) Fabrication and characterization of polymer hollow microneedle array using UV lithography into micromolds. J Microelectromech Syst 22(5):1041–1053. https://doi.org/10.1109/JMEMS.2013.2262587

Wilmut I, Schnieke AE, McWhir J, Kind AJ, Campbell KHS (1997) Viable offspring derived from fetal and adult mammalian cells. Nature 385(6619):810–813. https://doi.org/10.1038/385810a0

Wong FK, Haffner C, Huttner WB, Taverna E (2014) Microinjection of membrane-impermeable molecules into single neural stem cells in brain tissue. Nat Protoc 9(5):1170–1182. https://doi.org/10.1038/nprot.2014.074

Yang NS, Burkholder J, Roberts B, Martinell B, McCabe D (1990) In vivo and in vitro gene transfer to mammalian somatic cells by particle bombardment. Proc Natl Acad Sci 87(24):9568–9572. https://doi.org/10.1073/pnas.87.24.9568

Yoon Y-K, Park J-H, Allen MG (2006) Multidirectional UV lithography for complex 3-D MEMS structures. J Microelectromech Syst 15(5):1121–1130. https://doi.org/10.1109/JMEMS.2006.879669

Zhang X, Godbey W (2006) Viral vectors for gene delivery in tissue engineering. Adv Drug Deliv Rev 58(4):515–534. https://doi.org/10.1016/j.addr.2006.03.006

Zhang Y, LeBlanc AC (2002) Microinjections to study the specific role of proapoptotic proteins in neurons. In: Apoptosis techniques and protocols. Humana Press, pp 083–106. https://doi.org/10.1385/1-59259-188-4:083

Zhang Y, Ballas CB, Rao MP (2012) Towards ultrahigh throughput microinjection: MEMS-based massively-parallelized mechanoporation. In: 2012 Annual international conference of the IEEE engineering in medicine and biology society, pp 594–597. https://doi.org/10.1109/EMBC.2012.6346001

Zheng Z, Chen L, Yang T, Yu H, Wang H, Qin J (2018) Multiple pregnancies achieved with IVF/ICSI and risk of specific congenital malformations: a meta-analysis of cohort studies. Reprod Biomed Online 36(4):472–482. https://doi.org/10.1016/j.rbmo.2018.01.009

Zhi D, Yang T, Zhang T, Yang M, Zhang S, Donnelly RF (2021) Microneedles for gene and drug delivery in skin cancer therapy. J Control Release 335:158–177. https://doi.org/10.1016/j.jconrel.2021.05.009

Zhu T, Zhang W, Jiang P, Zhou S, Wang C, Qiu L, Shi H, Cui P, Wang J (2022) Progress in intradermal and transdermal gene therapy with microneedles. Pharm Res 39(10):2475–2486. https://doi.org/10.1007/s11095-022-03376-x

Magnetic Nanoparticles for Advanced Drug Delivery

James F. Leary

Abstract Magnetic nanoparticles are an attractive alternative for advanced drug delivery. Magnetic fields can be used to concentrate the nanoparticles at a bodily site. An oscillating magnetic field can be used to excite nanoparticles bound to diseased cells for controlled drug release and single-cell hyperthermia at the site of the diseased cells. Superparamagnetic ferric oxide nanoparticles are of very low nanotoxicity and are easily biodegraded in-vivo. They are also excellent MRI contrast and x-ray contrast imaging agents that can be used to locate tumors by non-invasive imaging. Hence, they are inherently theranostic, providing both diagnostic and therapeutic components. In addition, magnetic fields can be used for magnetoporation of drugs, including pulsed and controlled release of drugs, for transfection of genes into selected single cells and for single-cell lysis. These more sophisticated multi-layered systems can be used for multi-step targeting of drugs. Advanced multi-layered nanomedical systems can be easily constructed and purified with simple magnetic fields rather than ultracentrifugation. Hence these magnetic properties of these nanoparticles are of great practical utility during the entire manufacturing and purification processes.

1 Introduction—Theranostic, "Smart" Nanoparticles

Use of nanoparticles of various formulations for advanced and targeted drug delivery to single cells is the new frontier of drug delivery. Optimal design of these targeted nanodelivery of drugs is the topic of discussion in a recent major book on the subject by this author (Leary 2022). If designed appropriately those nanoparticles can be theranostic, meaning that they can be used for both diagnostics and therapeutics, or "theranostics". This represents an important advance for medicine and will have a major impact on healthcare (Leary 2010). As with every major advance there is a period of over-promising and under-delivering. While this has also been true

J. F. Leary (✉)
Purdue University, West Lafayette, USA
e-mail: jfleary@purdue.edu

T. S. Santra and A. U. S. Shinde (eds.), *Advanced Drug Delivery*,
Studies in Mechanobiology, Tissue Engineering and Biomaterials 26,
https://doi.org/10.1007/978-981-99-6564-9_8

of nanomedicine, this is not merely "nano-hype" (Leary 2013). It is real and will become increasingly common as old drugs are re-packaged in nanodelivery systems that improve the performance and extend the existing patents of pharmaceutical firms.

Depending on their composition they can be used with a variety of non-invasive imaging modalities for very sensitive diagnostics. Most of the same principles for design of magnetic nanoparticles for diagnostics and therapeutics follow similar designs to nanoparticles of other compositions with the added feature of being able to interact with the magnetic nanoparticle using external magnetic fields.

Once targeted to diseased cells, they can be used to deliver drugs only to the diseased cells without harming nearby normal cells, so-called "smart nanoparticles" as shown in Fig. 1.

Due to their composition magnetic nanoparticles also excellent x-ray and MRI contrast agents for non-invasive imaging, as discussed later in this chapter.

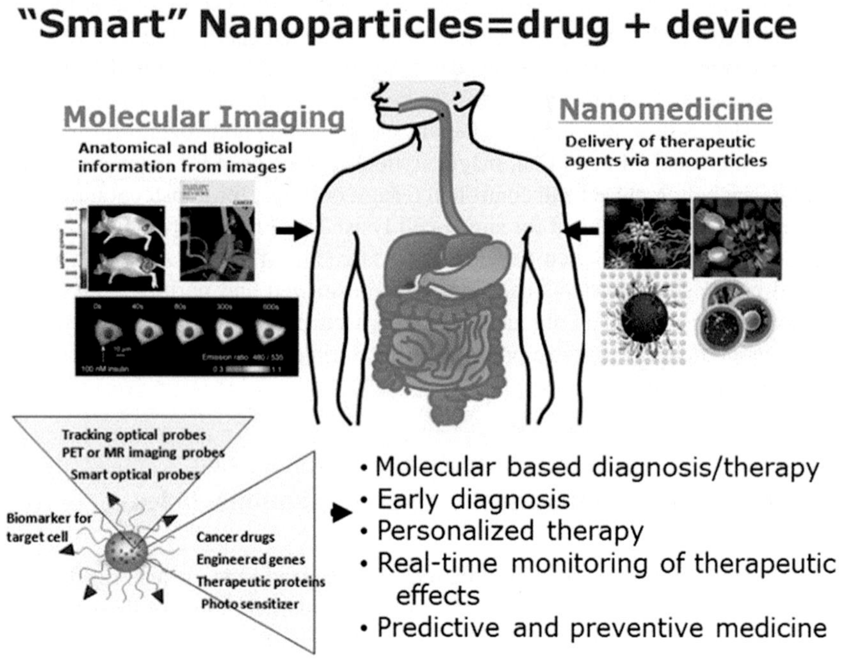

Fig. 1 "Smart" nanoparticles capable of both diagnostics and therapeutics will allow a new stage of "theranostics" in modern medicine. *Source* Adapted from Leary teaching

2 "Biomimicry" Inspired Designs of Nanodelivery Systems

"Biomimicry" is using the designs provided by Nature (Benyus 1997). Since these designs have already demonstrated at least partial success in specific situations, it is often convenient to use them as a starting point. Just because they occur in Nature does not mean that they are necessarily the optimal designs. It is the job of scientists to understand how these Nature inspired designs work and then try to improve upon them.

The pharmaceutical industry makes extensive use of biomimicry. They study naturally occurring compounds that have specific therapeutic effects. Then they make alterations in these naturally occurring molecules to enhance their performance and also to make them easier to manufacture.

In my own work I have worked extensively with ferric oxide superparamagnetic nanoparticles for drug delivery. My use of biomimicry for magnetic nanoparticles involved studying how viruses can target and enter single cells. My synthetic nanoparticles then mimicked the viruses without the problematic extra and unwanted features of viruses.

3 Magnetic Nanoparticles Are Subject to Endocytosis in Terms of Non-specific Uptake

Magnetic nanoparticles are an especially attractive approach to advanced drug delivery because they can be controlled in-vivo using ex-vivo modulating magnetic fields which can cause cells to become transiently permeable to drugs (Leary 2022). This is in addition to the natural endocytic (Bergtrom 2020) processes of phagocytosis (Aderem and Underhill 1999), for larger nanoparticles, and pinocytosis (Stillwell 2016), for smaller nanoparticles as shown in Fig. 2. Phagocytosis and pinocytosis appear to be independent of nanoparticles composition and depend on only size and shape. This means that they will be taken up non-specifically by single cells. To prevent this non-specific uptake, one can manipulate the electrical charge in terms of its zeta potential as described in Sect. 9.

4 Magnetic Nanoparticles Can Be Made in a Variety of Sizes and Shapes

Magnetic nanoparticles are an especially attractive approach to advanced drug delivery because they can be controlled in-vivo using ex-vivo modulating magnetic fields which can cause cells to become transiently permeable to drugs (Leary 2022).

Magnetic nanoparticles can be made in a wide variety of shapes and sizes with specific advantages using these sizes and shapes. Spherical (Haglund et al. 2009),

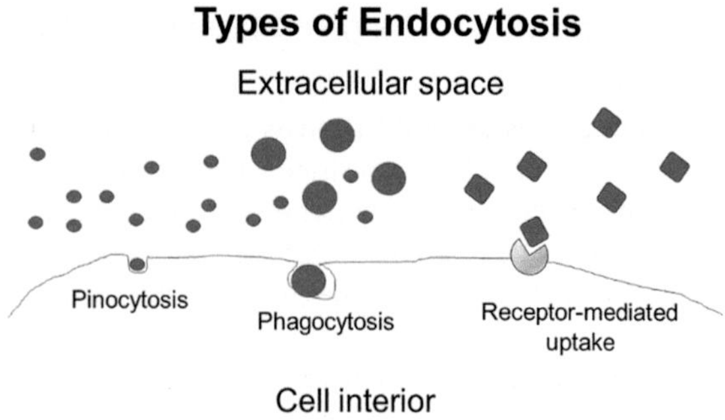

Fig. 2 All nanoparticles, including magnetic ones, will be taken up non-specifically by endocytosis unless care is taken to make sure that they are at least moderately negatively charged to be electrostatically repelled by the negatively charged cell surface. *Source* Leary teaching

cubic (Key et al. 2016), and nanoflake shaped (Cervadoro et al. 2014) nanoparticles can be synthesized. The nanoflake versions were specifically designed to enhance the T2 contrast ratio for higher performance MRI. Nanoparticles can also be synthesized using top-down fabrication (Aryal 2019).

When a magnetic nanoparticle becomes sufficiently large, it will lose its super-paramagnetic properties. One way to get around this problem is to make a "raisins in a bun" design whereby smaller superparamagnetic nanoparticles are brought together within the larger nanoparticle structure (Key et al. 2016). This is important to increase the response of superparamagnetic nanoparticles to external magnetic fields. The magnetic manipulation of magnetic nanoparticles in-vivo depends both on the strength of the external magnetic field and the magnetic power of the nanoparticles.

5 Magnetic Field Effects on Magnetic Nanoparticles

To prevent spontaneous aggregations in-vivo that can cause serious embolisms, it is important to make magnetic nanoparticles that are super-paramagnetic (only magnetized when in a magnetic field) rather than permanently magnetic. When the magnetic field is taken away, superparamagnetic nanoparticles will readily become monodispersed.

Rod-shaped magnetic nanoparticles can also be spun inside living cells by external magnetic fields, providing another way to destroy diseased cells. Magnetic fields, either constant or oscillating, can be used to concentrate the magnetic nanoparticles to cultured cells (Du et al. 2018) a specific region or organ. If the magnetic fields are oscillating, the oscillation of the nanoparticles near cells can produce hypothermia of single diseased cells as shown in Fig. 3.

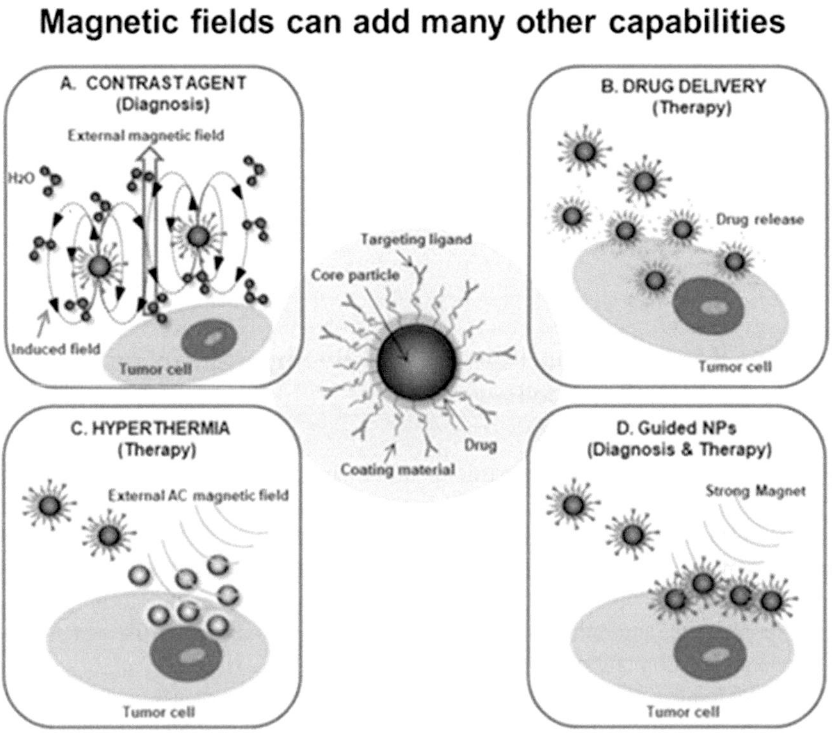

Fig. 3 Magnetic nanoparticles are useful as **a** X-ray and MRI contrast agents, **b** drug delivery using targeting molecules and membrane permeating peptides, **c** single cell hyperthermia using oscillating magnetic nanoparticles and produce heat, and **d** drug delivery concentrators by using magnetic fields to first guide the magnetic nanoparticles to the general area of the diseased cells and then stronger magnetic fields to drive the magnetic nanoparticles through the cell membrane. *Source* Adapted from Leary teaching

6 Importance of Using "Green Chemistry" to Manufacture Magnetic Nanoparticles

Magnetic nanoparticles also greatly reduce the complexity of the manufacturing process because they can be easily purified after each of a multistep layering process by use of a simple magnet rather than requiring ultracentrifugation. Unfortunately many scientists choose non-magnetic nanoparticles because the surface chemistries for attaching biomolecules can be much simpler (e.g. attaching biomolecules to gold nanoparticles via cysteine residues). They are sacrificing all of the advantages of magnetic nanoparticles in order to simplify the biofunctionalization. The surface chemistry of magnetic nanoparticles is only slightly more difficult and has been described in detail by Murcia and Naumann (2005) and Kumar (2005).

It is important to use so-called "green chemistry" (Duan et al. 2015; Gurunathan et al. 2015; Malik et al. 2014) throughout the entire process of attaching molecules to the magnetic nanoparticles. Green chemistry uses gentle aqueous chemistry rather than harsh organic solvents. This is important for at least two reasons. First, it is kinder to the environment in not producing chemically hazardous waste. Second, it prevents these toxic chemicals being carried with the manufactured nanoparticles into in-vivo situations. It is much easier to avoid use of these solvents than it is to try to take them back out. Human cells can be adversely affected at the level of a few parts per million of these toxic chemicals and organic solvents. Instead, aqueous chemical reactions are used.

7 Why These Nanoparticles Should Be Superparamagnetic for In-Vivo Drug Delivery

For in-vivo use, magnetic nanoparticles in humans should be superparamagnetic, meaning that they are only aggregated in the presence of a magnetic field (Barnes et al. 2007, Eustaquio and Leary 2012). When the magnetic field is turned off the superparamagnetic nanoparticles will disperse. That is not the case for paramagnetic nanoparticles which become permanently magnetized leading to agglomeration. Spontaneous agglomeration of nanoparticles in-vivo can lead to life-threatening embolisms. Agglomeration is driven by zeta potential described in Sect. 8.

In order to have enough superparamagnetic material in the magnetic nanoparticle to be controlled in-vivo by external magnetic fields, it is necessary to use a few tricks in assembly. When superparamagnetic nanoparticles get too big, they lose their superparamagnetic properties. To get around this problem one can imbed multiple smaller superparamagnetic particles within a larger nanoparticle. This approach we called "raisins-in-a-bun" construction (Key et al. 2012) as shown in Fig. 4.

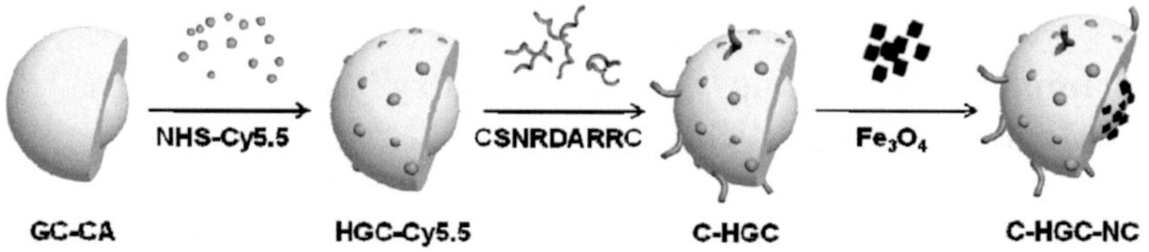

Fig. 4 Schematic of construction of larger and more complex HGC nanoparticles that maintain superparamagnetic properties due to a "raisin-in-a-bun" assembly of smaller superparamagnetic nanoparticles. *Source* Key et al. (2012)

8 Roadmap for Design of These Nanoparticles for Targeted Drug Delivery

All nanodelivery system designs for drug delivery, including magnetic nanoparticles, follow a similar roadmap of self-assembly (Leary 2022) as shown in Fig. 5.

This roadmap is a useful and methodical guide for the design of targeted nanodelivery systems. The important take-home message is that good nanodelivery systems are multicomponent systems that need to be constructed in an ordered assembly. The possibility of using magnetoporation to deliver drugs into targeted cells does not significantly change the design of the nanoparticles, except that there is an additional mechanism (magnetoporation) for driving cell entry. This magnetoporation force may, or may not, obviate the need for receptor mediated uptake or use of cell membrane permeating peptides. Indeed, magnetoporation my act synergistically with these two factors to improve the uptake of drugs into selected cells.

(1) Magnetic nanoparticles for magnetoporation are inherently part autonomous and part non-autonomous. This is because they must have some targeting molecules to get close enough to the cells for magnetoporation. But they are also non-autonomous because they can be guided and controlled by an external magnetic field. The core material must be magnetic and have superparamagnetic properties.
(2) The obvious external modulator for magnetic nanoparticles is an external magnetic field. It can be either constant or oscillating. The advantage of an oscillating external magnetic field is that it can vibrate the nanoparticles to

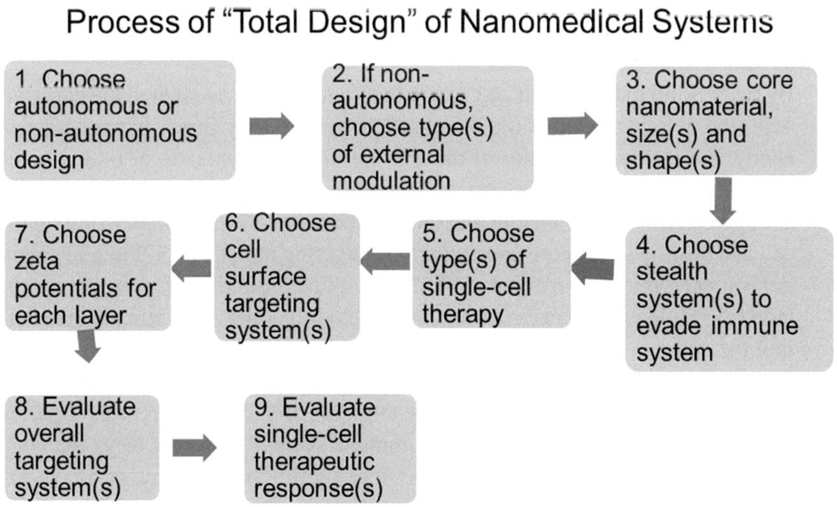

Fig. 5 Each of these design process steps can require multiple decisions and have multiple outcomes. Steps can be skipped if irrelevant to the design, and one process can sometimes cover multiple steps. *Source* Adapted from Leary (2022)

produce heat as single cell hyperthermia. Cells normally at 37C will start to die above about 40–42C. If the nanoparticles are not spherical, an oscillating magnetic field can cause the nanoparticles to rotate which can kill cells.

(3) A good core material for magnetic nanoparticles is ferric oxide. It is inexpensive, and well tolerated by the body. In the end, after degradation, it can simply be taken up by the cells as free ferric oxide. There are of course other materials for constructing magnetic nanoparticles.

(4) An essential part of any nanoparticles introduced into the body is the so-called "stealth layer". Without such a layer, the immune system will rapidly eliminate the magnetic nanoparticles. Additionally, materials such as ferric oxide are very hydrophobic. To get them to disperse, rather than aggregate, in-vivo requires a hydrophilic layer. There are many different materials that can function as stealth layers. The most common are appropriate PEG (Polyethylene Glycol) or chitosan, the material that makes the hard surface of crustaceans. The basic idea is to coat the magnetic nanoparticles with a layer that reduces or eliminates deposition of blood proteins onto the nanoparticles. This is because the immune system uses a process called "opsonification" to attach blood proteins to the nanoparticle so that the immune system can recognize the nanoparticle as "foreign" material that should be eliminated from the body.

(5) There are many type(s) of potential therapy for the diseased cells. They can be killed by a toxin carried by the nanoparticle, but mistargeting will cause some number of healthy cells to be eliminated. A better way is for the nanoparticle to carry apoptosis-inducing molecules that are in the inner core of the nanoparticle system and will only be activated if the nanoparticle makes it inside a cell.

(6) To see that the nanoparticle goes to a diseased, rather than normal, cell, specific targeting molecules on the cell surface of the diseased cells should be identified. These targeting molecules (antibodies, peptides, aptamers, or ligands) need to be attached to the nanoparticle.

(7) By the time all of these molecules have been added layer by layer to the nanoparticle, the zeta potential of the nanoparticle system may have changed significantly. If so, other molecules of the appropriate charge need to be added to the nanoparticle to get its zeta potential to approximately -15 mV. This insures enough electrostatic repulsion that the nanoparticles will not simply stick to the cell and destroy the effectiveness of all the targeting molecules. If the zeta potential is much greater than about -60 mV, it will be difficult or impossible for the targeting molecules or external magnetic field to get the magnetic nanoparticle into the cells.

(8) Once all of these steps have been taken to design a nanomedical system, it is time to evaluate the effectiveness of the targeting and to determine what fraction of the nanoparticles have evaded the immune system and other ways the body eliminates foreign materials. This may mean going back to any, or all, of the design steps in order to improve the overall design in terms of targeting. Good targeting can lead to good diagnostic through non-invasive imaging either by CAT or MRI.

(9) Lastly, if the purpose of designing the system is therapeutic (perhaps in addition to diagnostic), then the effectiveness of eliminating the diseased cells must be evaluated. Again, this may involve going back to particular, or all, of the design stems to improve therapeutic outcomes.

9 The Importance of Multistep Targeting of Magnetic Nanoparticles for Drug Delivery

While magnetoporation is the specific focus of this chapter, it is important to discuss all of the many steps necessary to bring magnetic nanoparticles to the diseased cells in vivo. First, appropriate biocoatings must be used to help the magnetic nanoparticles evade the opsonification elimination of these nanoparticles by the immune system. Without this important first step there will be little or no magnetoporation. Then a complex multi-step targeting process using sequential targeting molecules to bring the magnetic nanoparticles first to the organ or interest, then to the diseased cells in that organ and then finally to bring the nanoparticles inside those diseased cells (Haglund et al. 2008; 2009, Seale et al. 2007a, b). It is easy to see the parallels between a multilayered construction whose order of assembly produces a multi-step targeting process as shown in Fig. 6.

Attaching these targeting molecules to the magnetic nanoparticles requires biofunctionalization of the surface of the magnetic nanoparticles. This biofunctionalization process is discussed in an entire book on the subject (Kumar 2005) as well as in journal articles (Gupta and Gupta 2005; Murcia and Naumann 2005) and as a chapter in an entire book on nanodelivery systems by this author (Leary 2022). While biofunctionalization of magnetic nanoparticles is more complex than attachment of biomolecules to gold nanoparticles through cysteine residues, the rewards of magnetic nanoparticles are worth the effort both in manufacturing and also in

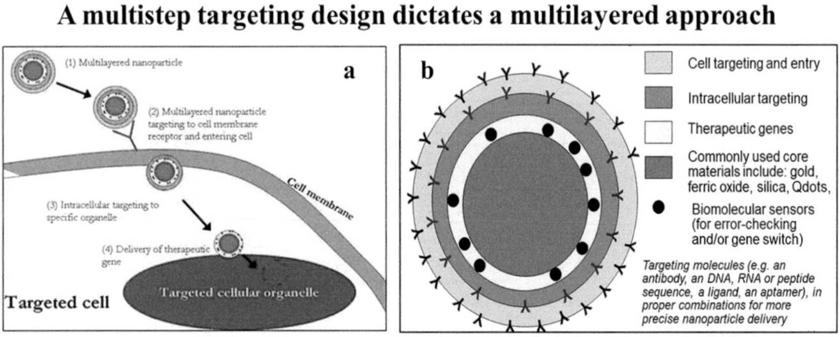

Fig. 6 **a** Nanoparticles must be guided to the cells of interest in-vivo by a multi-step targeting process. **b** This suggests that the best way to accomplish this might be through a multilayered nanodelivery system whereby each layer performs its targeting function by an outside-in layered structure. *Source* Adapted from Haglund et al. (2009) and, Seale et al. (2007ab)

the ability to modulate the functions of these nanoparticles deeply in tissue. Gold nanoparticles have interesting properties in terms of resonances, but only if light can be shown directly on the gold nanoparticles at close range which is frequently difficult or impossible.

There are many ways nanoparticles, magnetic or not, can be driven across the cell membrane in diseased cells. In the absence of magnetic fields, common strategies employ cell membrane permeating peptides (Bechara and Sagan 2013). Magnetoporation can only occur when the magnetic nanoparticles are first brought to the immediate vicinity of the targeted cells. This requires a multistep targeting strategy which then dictates a multilayered approach to the problem.

10 Electrostatics of Magnetic Nanoparticles Will Dominate Nanoparticle-Cell Interactions

It is important to recognize that magnetoporation will NOT happen unless the electrical charge properties of the nanoparticles are in a permissible range. Magnetic fields alone will NOT allow delivery of magnetic nanoparticles into cells. Before applying magnetic fields, the electrostatics must be in a range to permit magnetoporation. Electrostatics in solution (usually close to physiological saline for in-vivo medical use) are governed by solution electrostatics known as zeta potential as shown in Fig. 7.

Charged nanoparticles in a solution will attract counterions (electrically opposite charge). Virtually all cell surfaces are electrostatically negative charge (negative zeta potential). These nanoparticles in solution will then interact with cell surfaces that are usually negatively charged due to cell surface sialic acid and other molecules. If the

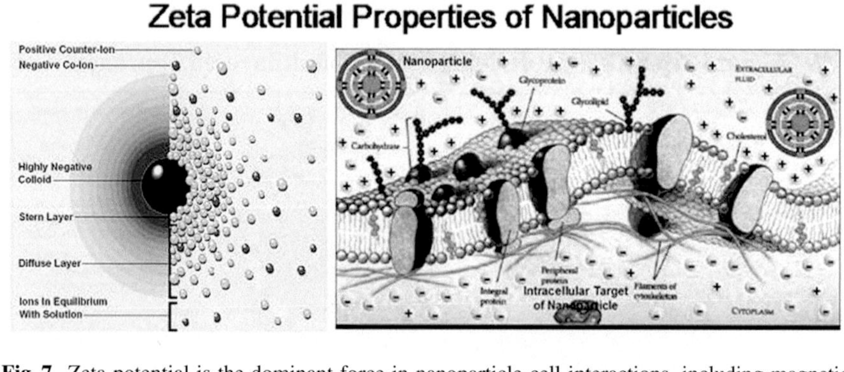

Fig. 7 Zeta potential is the dominant force in nanoparticle-cell interactions, including magnetic nanoparticles. The zeta potential is the net charge at a distance of a moving nanoparticle, including all layers, of a nanoparticle and its opposite charge "counterions". That nanoparticle interacts with a cell which also has a zeta potential depending on its surface charges and general composition. *Source* Adapted from Leary teaching

magnetic nanoparticles have a net positive charge, they will attach non-specifically to all cells, negating any attempts, including application of magnetic fields, to target diseased cells and allow those nanoparticles to penetrate the cell membrane. For these reasons the magnetic nanoparticles through their surface biocoatings must have a net negative zeta potential (usually about −15 to −20 mV). If these conditions are satisfied, a magnetic field can then be applied to drive these nanoparticles across the electronic charge barrier. This is in addition to any cell permeating means such as receptor-mediated uptake or use of cell membrane-penetrating peptides on the surface of the magnetic nanoparticles. These interactions are described in considerable detail in a chapter of a book recently written by this author (Leary 2022). If the zeta potential of the magnetic nanoparticle is not right, then there will be no magnetoporation!

11 Low Nanotoxicity of These Magnetic Nanoparticles

For magnetic nanoparticles be used therapeutically in-vivo, the toxicity (i.e., "nanotoxicity") must be very low. Nanotoxicity can be due to many different causes including size and shape of nanoparticles in addition to their nanomaterial content (Reineke 2012; Wani et al. 2011) One advantage of iron oxide superparamagnetic magnetic nanoparticles is their very low toxicity (Eustaquio and Leary 2012). Nanotoxicity of these magnetic nanoparticle systems included single-cell tests for necrosis, apoptosis, cell cycle disturbances, and changes in cell differentiation patterns. It is important to test for all these cases rather than simple tests for necrosis. Nanotoxicity-induced apoptosis is particularly overlooked. If a cell is tipped onto a pathway for eventual apoptosis ("programmed cell death"), it might take time for that to be recognized. Subtle changes in cell cycle and differentiation are also difficult, but important, to detect. All of these aspects of nanotoxicity of magnetic nanoparticles are extensively discussed in a book chapter on this subject (Eustaquio and Leary 2012). Importantly, magnetic nanoparticles are very low in nanotoxicity making them ideal candidates for nanodelivery systems.

We spent much time and applied multiple single-cell tests for nanotoxicity: necrosis based on cell membrane permeability using dye exclusion assays with trypan blue or propidium iodide (Kamiloglu et al. 2020), apoptosis (Darzynkiewicz et al. 2008) by TUNEL assay by annexin V links to phosphatidylserine (Engeland et al. 1998), generation of reactive oxygen species (ROS) (Haglund et al. 2008), disturbances in cell cycle (Darzynkiewicz et al. 1996) by measuring cyclins, and changes in cell differentiation patterns. We also looked at the potential masking effects of biocoatings. We discovered than quantum dot nanoparticles when stripped of their biocoatings and exposed to light became cytotoxic by ROS assays (Haglund et al. 2008). It frequently took several tests to discover the toxic effects of nanomaterials. No one test proved sufficient for the task. An entire book is devoted to the topic of nanotoxicity (Reineke 2012).

12 Biocoatings Can Mask Nanotoxicity of Nanoparticles

Ferromagnetic nanoparticles are generally of low toxicity. But biocoatings can mask toxicity assays as we and others have previously described (Eustaquio and Leary 2012; Ryman-Rasmussen et al. 2007) for many other types of nanoparticles (e.g. quantum dots). A detailed and extensive discussion of nanotoxicity and the many types of single cell assays is detailed for magnetic nanoparticles in multiple publications by this author (Leary 2022; Eustaquio and Leary 2012; Haglund et al. 2008) and others (Reineke 2012; Wani et al. 2011). These biocoatings can include molecules that permit relatively insoluble magnetic nanoparticles to exist as single dispersed nanoparticles in aqueous solutions. Targeting molecules (e.g. antibodies, peptides, ligands) and cell entry molecules (e.g. cell penetrating peptides) all can mask nanotoxicity. The problem is that once these biocoatings fall off during nanoparticle-cell interactions, the nanotoxicity of the underlying nanomaterials becomes important. This was particularly true for quantum dots (Haglund et al. 2008). Examples of some types of biocoatings are shown in Fig. 8.

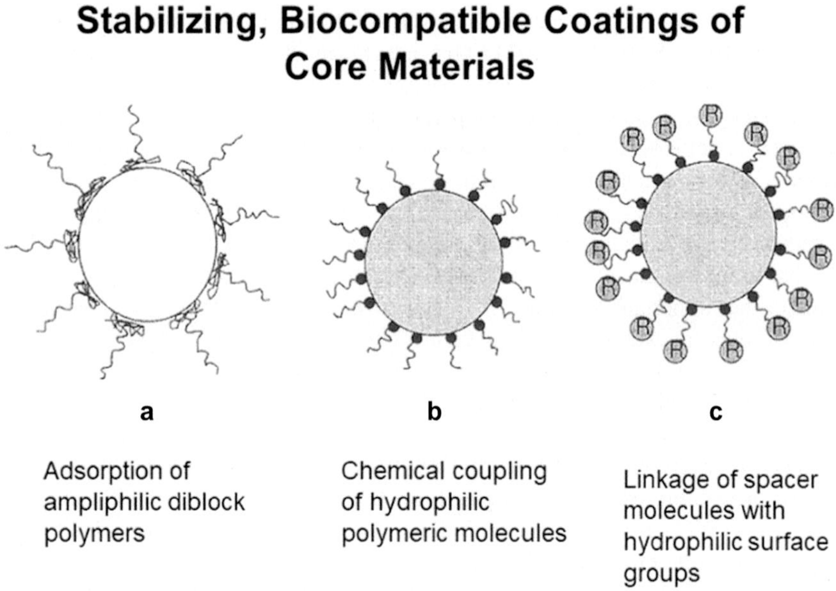

Fig. 8 Biocoatings to magnetic nanoparticles are of at least three general types: **a** simple adsorption of ampliphilic polymers through charge interactions with the nanoparticle surface, **b** chemical coupling of hydrophilic molecules necessary for hydrophobic magnetic nanoparticles to disperse into aqueous solutions, and **c** linkage of spacer molecules with hydrophilic surface groups added to hydrophobic magnetic nanoparticles (Kumar 2005)

13 Tethered Gene Expression on Magnetic Nanoparticles

An exciting alternative approach is to not deliver a drug but rather a gene manufacturing template to the cell. Transcription of therapeutic gene sequences occurs under the control of an upstream molecular biosensor in a feedback control guided process as is conceptually shown in Fig. 9 and described in some of our earlier work (Prow et al. 2005, 2006a, b). The advantage of this approach is that the cell always receives the correct therapeutic dosage regardless of how many nanoparticles reach their target. The implementation of this process has been completed using fluorescent reporter gene constructs manufactured in situ within living cells using upstream molecular biosensors and responding to feedback control mechanisms for biosensing of reactive oxygen species (ROS) molecules within the cells. We have previously demonstrated that genes tethered to ferric oxide nanoparticles can transcribe copies of genes inside living cells. For visualization purposes we transcribed eGFP and DsRed reporter genes tethered to ferric oxide nanoparticles which were used to transfect cells. Results were visualized using confocal microscopy (Fig. 9).

14 Use of Magnetic Fields to Interact with Magnetic Nanoparticles

Magnetic fields can be used to externally control in-vivo magnetic nanoparticles (Key et al. 2016; Du et al. 2018; Leary 2022). These magnetic nanoparticles can be guided to the site of interest whereupon drugs can be remotely released (Thomas et al. 2010). If modulated with an oscillating magnetic field, the nanoparticles can also generate heat for single-cell hyperthermia as shown in Fig. 10.

15 What Is "Magnetoporation" or "Magnetofection"?

"Magnetoporation" refers to the use of magnetic fields to open pores in the cell membrane in order to introduce molecules inside a living cell. Weak magnetic fields (e.g. 40–75 mT) have been shown to either to enhance cell membrane poration ("magnetoporation") or to ablate cultured human tumor cells ("magnetolysis") (Liu et al. 2012). Whether these techniques can be accomplished in-vivo is a more challenging question due to the much larger separations between external magnetic fields and the magnetic nanoparticles in-vivo.

"Magnetofection" is a term used mostly by molecular biologists who think of the process as another form of transfection. A recent interesting paper by Sizikov et al. (2021) demonstrates the potential utility of using magnetic nanoparticle complexes to transfect molecules into cells in-vivo. Unfortunately, the nanoparticles were not constructed to avoid non-specific capture in the lungs and other organs. Still the

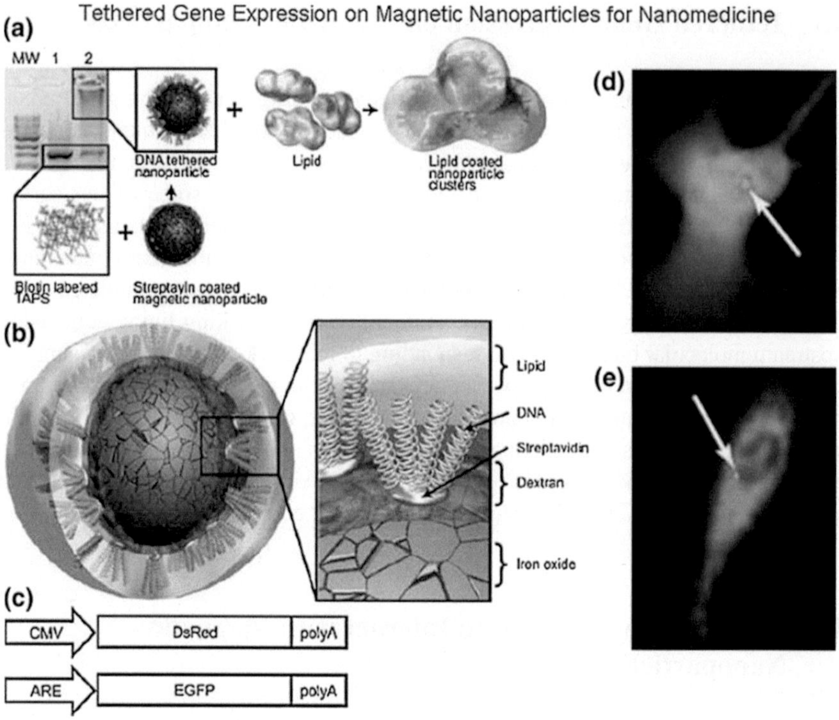

Fig. 9 Construction and anatomy of magnetic nanoparticles. **a** Conjugation of biotin-labeled transcriptionally active PCR products (TAP) DNA to streptavidin-coated magnetic nanoparticles (MNP). A 0.8% agarose gel stained with ethidium bromide was used to visualize DNA and DNA tethered nanoparticles. The leftmost lane are molecular weight markers, from 1 to 10 kb (MW). Lane 1 contains only biotin-tagged TAP DNA (Black rectangular outline). Lane 2 is a solution containing DNA from Lane 1 combined with streptavidin-coated magnetic nanoparticles. The black rectangular outline in Lane 2 highlights TAP tethered magnetic nanoparticles. **b** Schematic of the construction of the MNP. The layered anatomy of a lipid The layered anatomy of a lipid coated DNA tethered nanoparticle. **c** Schematics of the two DNA constructs used to assess transfection and ARE activity. Lipid-coated nanocrystal transfected human retinal epithelium cells. Cells were cultured with lipid-coated nanocrystals tethered to either EGFP (green in **d**) or DsRed (red in **e**) for 48 h or 10 days, respectively. Confocal microscopy was used to visualize nanocrystals and tethered fluorescent gene expression. The nanocrystals are marked by white arrows. Adapted from our previously published work (Haglund et al. 2009)

paper demonstrates the promise of the overall technique and will be discussed in more detail in Sect. 17.

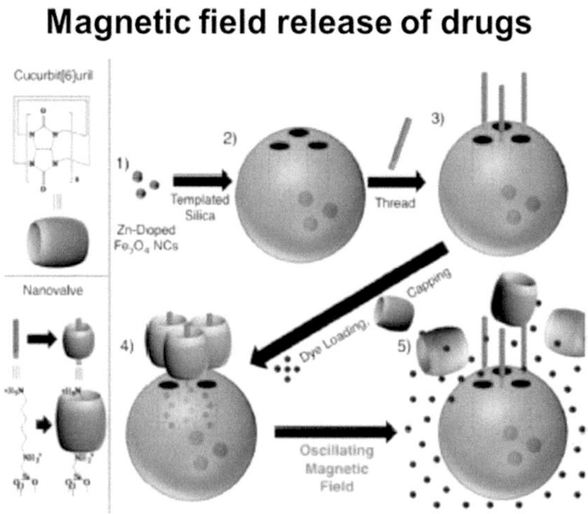

Fig. 10 Magnetic nanoparticles can be used to either release a drug at the site of a single diseased cell or and/or perform single cell hyperthermia (Thomas et al. 2010)

16 Magnetoporation of Drugs into Single Cells

Magnetic fields can be used to transfect genes and other molecules into cells (Ito and Kamihira 2011) as shown in Fig. 11.

The molecular mechanisms of magnetic transfection are not well understood. Some scientists believe that the magnetic field induces an electric field which then

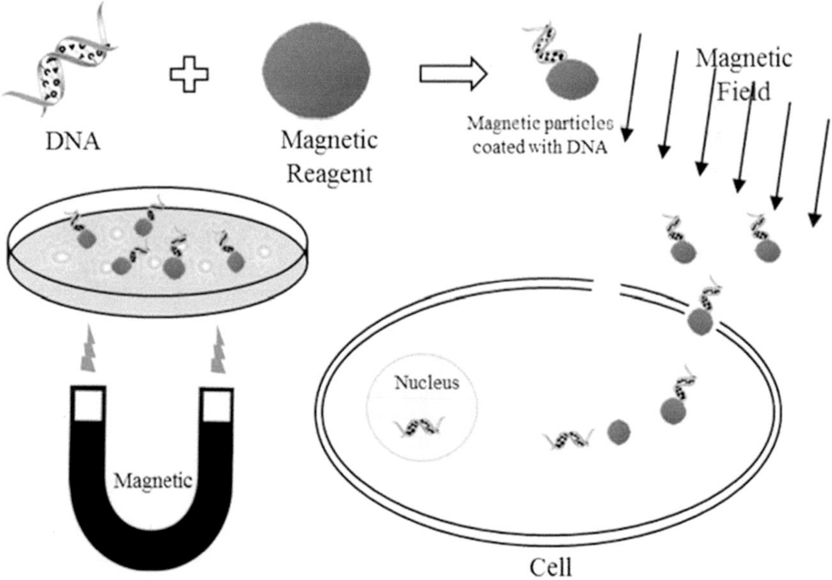

Fig. 11 Magnetic field transfection of DNA into cells using magnetic nanoparticles

becomes a process of electroporation. the principle of magnetoporation is similar to electroporation. They deem that the magnetic field induces an electric field which changes the transmembrane potential of cell membrane. When the transmembrane potential reaches to a certain threshold, the cell membrane is perforated.

In any case, regardless of the actual mechanism involves the mixing of nucleic acids with a magnetofection reagent to form a biomolecule-magnetic reagent complex. Then the entire complex is delivered into cell in the force of magnetic field. Under the effect of magnetic field, it is believed by some that the natural endocytosis and pinocytosis of cell membranes are also enhanced.

17 Magnetofection of Molecules to Cells In-Vivo

A major challenge of magnetofection of drugs or DNA into cells in-vivo (Sizikov et al. 2021) using magnetic nanoparticles with various cationic polymer coatings (Fig. 12) is producing enough magnetic force at the depth of an in-vivo single cell containing targeted magnetic nanoparticles. That is a function of both the necessary size of the superparamagnetic nanoparticles and the strength and depth of the magnetic field. But not all in-vivo magnetofection need be through deep tissues. If the targeted cells are either in a body cavity or approachable through minimally invasive surgery, then it would appear possible to accomplish this within specific diseased organs of interest even deep inside the human body.

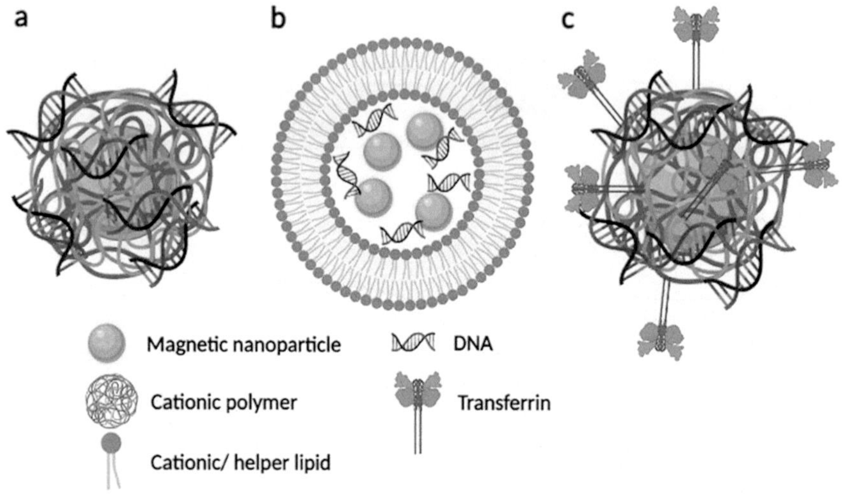

Fig. 12 Schematic examples of different magnetic nanocarriers for magnetofection in vivo: **a** magnetic polyplex, **b** magnetic liposome, **c** magnetic polyplex modified with transferrin (Sizikov et al. 2021)

While Sizikov et al. (2021) demonstrate magnetic localization of these magnetic nanoparticles to a region of the chest of animals as directed by a magnetic field, there is more complexity involved. Just getting magnetic nanoparticles to a region of the body in-vivo does not mean that the nanoparticles can be directed to specific diseased cells. Indeed, the use of cationic coated nanoparticles will insure that the nanoparticles in this general region will attach non-specifically to all cells in the region due to the inherently negatively charged cell surfaces of virtually all cells. To target to specific diseased cells in-vivo requires using antibodies, peptides, or aptamers that will first bring these magnetic nanoparticles close to the cell surface of only the diseased cells. The fundamental interaction between nanoparticles and diseased cells is governed by the zeta potential—the electrical interaction of charged nanoparticles and cells. Successful magnetoporation can only happen if the magnetic force can overcome possible electrical repulsions of the zeta potential. An extensive review of the importance of zeta potential in targeting magnetic nanoparticles to specific diseased cells is discussed in detail in a new book on the design of nanoparticles for targeted drug delivery (Leary 2022).

18 Magnetic Nanoparticles Are Useful for Simultaneous Non-invasive Imaging

Non-invasive imaging represents a huge advance for medicine. Before the invention of these non-invasive imaging modalities, it was difficult or impossible for a physician to know what was going on inside a patient without what used to be euphemistically called "exploratory surgery". Any surgery has a risk of infection, so it should only be performed for good reason. These non-invasive imaging types (CAT, MRI, PET), required significant advances in imaging and vast increases in computational capabilities to produce accurate 3D images.

Magnetic nanoparticles have valuable properties for non-invasive imaging. First, since they are electron-dense materials, they scatter x-rays and can therefore serve as CAT scan imaging contrast agents to improve the diagnostic sensitivity on this form of non-invasive imaging as shown in Fig. 13.

CAT scans with x-rays are especially sensitive for imaging bone, but less so for soft tissue an muscle because these body parts contain much lower concentrations of electron-dense materials. But by adding electron-dense contrast agents, improvements can be made for these other body parts. With much more sensitive x-ray detector technology, patients are exposed to much less radiation.

Magnetic resonance imaging (MRI) a particularly important form of non-invasive imaging because it mostly just interacts with protons in the body. As such it is inherently safer for the patient than x-rays. The major source of protons in the body that are easy to excite is water which is present in most body parts. Essentially an MRI works by using radiofrequency pulses and intense magnetic fields to flip the polarity of protons and then measuring the "relaxation time", which is the time that it

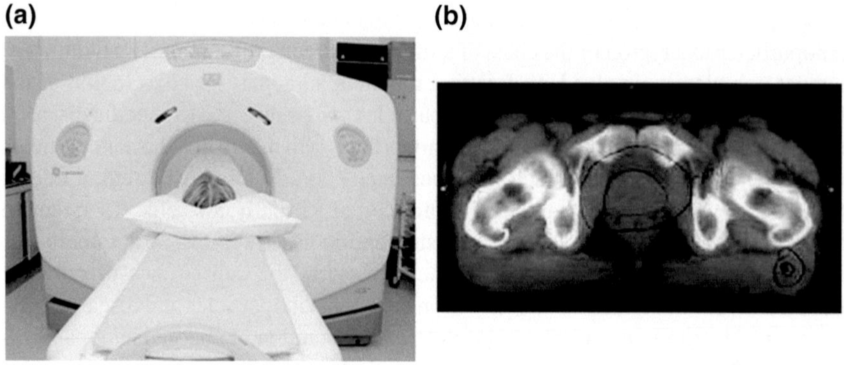

Fig. 13 **a** CAT scanner whereby the patient is transported into the circular core and given x-rays at multiple angles so that a 3D image can be generated by a computer, **b** CAT image whereby the strongest signals are in regions with intrinsic high electron density or higher electron density as produced by high electron density contrast agents. *Source* Adapted from photos on Internet

takes to revert to its initial polarity. The relaxation time is a measure of the viscosity of the tissue.

Before ferrimagnetic nanoparticles were used, the major uses of contrast agents were so-called "T1 contrast agents". But the initial use of just gadolinium led to serious toxicity for the patients. After some time, so-called "gadolinium chelates" were synthesized to reduce this toxicity. While toxicity of gadolinium has been greatly reduced by constructing gadolinium chelates to provide T1 contrast agents, it is perhaps better to use ferric oxide compounds which are not toxic and are consumed by cells at the elemental level. The T2 contrast agents provided by ferric oxide nanoparticles are negative contrast agent meaning that the image gets darker rather than brighter like T1 contrast agents. While it is a little harder to interpret, it can be equally sensitive.

Magnetic nanoparticles act as contrast agents to greatly improve the sensitivity and resolution of the images. There are two major types of contrast agents, so called "T1" and "T2". Ferrimagnetic versions of these magnetic nanoparticles are also T2 contrast agents as shown (Xu et al. 2011) as shown in Fig. 14.

19 Conclusions and Summary

Magnetic nanoparticles can be useful for nanodelivery systems of drugs. These nanoparticles should be superparamagnetic if used in-vivo since they will easily disperse in the absence of a magnetic field. Otherwise, paramagnetic nanoparticles can be permanently magnetized which can lead to large aggregations of nanoparticles that can cause embolisms in-vivo. Superparamagnetic nanoparticles can be easily concentrated by placing a magnet near the body part desired.

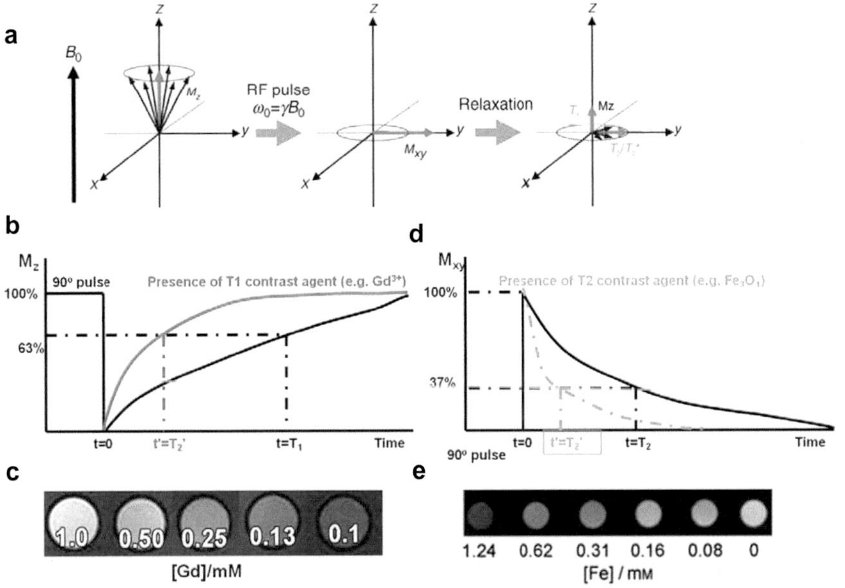

Fig. 14 **a** Protons in a magnetic field can rotate under magnetic fields, and once rotated these protons will relax back into their original orientations according to their microenvironment. **b** The MRI signal will increase in the presence of a T1 contrast agent, whereas it will decrease in the presence of a T2 contrast agent. **c** the resulting image will be either brighter (T1) or darker (T2) in the presence of these magnetic nanoparticle contrast agents. Adapted from Xu et al. (2011)

There is a design process for constructing magnetic nanoparticles for drug delivery. A series of decisions need to be made at every stage of construction. In general this means that the design process starts with the end stage and then builds backwards, in inverse order, such that the final outer layers of the magnetic nanoparticle are the first stage where the nanoparticle starts its interactions with the human body.

But magnetoporation in-vivo is a much more complex process, particularly in-vivo. Before magnetic fields can be applied for magnetoporation in-vivo, the magnetic nanoparticles must first evade the immune system. Otherwise, they will be eliminated from the body before ever reaching the targeted diseased cells. Then the magnetic nanoparticles must be brought very close to the diseased cells of interest before magnetoporation can take place. Lastly, the magnetic nanoparticles must be big enough to not require extreme external magnetic fields to provide a sufficient magnetoporation in-vivo to permit single cell entry.

As described earlier in this chapter, magnetoporation can perhaps best be used in conjunction with other cell entry techniques such as receptor mediated uptake and cell membrane permeating peptides, taking into account the zeta potential of the magnetic nanoparticle-cell interactions. It appears to act synergistically with natural mechanisms such as endocytosis. It is possible that magnetoporation is another form of electroporation whereby the magnetic field induces electric fields.

Magnetoporation and magnetofection need to be considered in the context of a complex multistep interaction with cells. If these other steps are not taken first to get these magnetic nanoparticles to the diseased cells of interest, then successful magnetoporation or magnetofection will not occur. That is why this chapter describes these techniques in a larger context.

References

Aderem A, Underhill DM (1999) Mechanisms of phagocytosis in macrophages. Annu Rev Immunol 17:593–623

Aryal S, Park H, Leary JF, Key J (2019) Top-down fabrication-based nano/microparticles for molecular imaging and drug delivery. Int J Nanomed 14:6631–6644. https://doi.org/10.2147/IJN.S212037

Barnes AL, Wassel RA, Mondalek F, Chen K, Dormer KJ, Kopke RD (2007) Magnetic characterization of superparamagnetic nanoparticles pulled through model membranes. Biomagn Res Technol 5:1–10

Bechara C, Sagan S (2013) Cell-penetrating peptides: 20 years later, where do we stand? FEBS Lett 587:1693–1702

Benyus JM (1997) Biomimicry: innovation inspired by nature. HarperCollins Publishers Inc., New York

Bergtrom C (2020) Endocytosis and exocytosis chapter. In: Libretexts biology

Cervadoro A, Cho M, Key J, Cooper C, Stigliano C, Aryal S, Brazdeikis A, Leary JF, Decuzzi P (2014) Synthesis of multifunctional magnetic nanoflakes for magnetic resonance imaging, hyperthermia, and targeting. ACS Appl Mater Interfaces 6:12939–12946

Darzynkiewicz Z, Gong J, Juan G, Ardelt B, Traganos F (1996) Cytometry of cyclin proteins. Cytometry 25:1–13

Darzynkiewicz Z, Galkowski D, Zhao H (2008) Analysis of apoptosis by cytometry using TUNEL assay. Methods 44:250–254

Du X, Wang J, Zhou Q, Zhang L, Wang S, Zhang Z, Yao C (2018) Advanced physical techniques for gene delivery based on membrane perforation. Drug Deliv 25:1516–1525

Duan H, Wang D, Li Y (2015) Green chemistry for nanoparticle synthesis. Chem Soc Rev 44:5778–5792

Eustaquio T, Leary JF (2012) Single-cell nanotoxicity assays of superparamagnetic iron oxide nanoparticles. Methods Mol Biol 926:69–85

Gupta AK, Gupta M (2005) Synthesis and surface engineering of iron oxide nanoparticles for biomedical applications. Biomaterials 26:3995–4021

Gurunathan S, Kim E, Han JW, Park JH, Kim JH (2015) Green chemistry approach for synthesis of effective anticancer palladium nanoparticles. Molecules 20:22476–22498

Haglund E, Seale-Goldsmith MM, Leary JF (2009) Design of multifunctional nanomedical systems. Ann Biomed Eng 37:2048–2063

Haglund EM, Seale-Goldsmith M-M, Dhawan D, Stewart J, Ramos-Vara J, Cooper CL, Reece LM, Husk T, Bergstrom D, Knapp D, Leary JF (2008) Peptide targeting of quantum dots to human breast cancer cells. SPIE

Ito A, Kamihira M (2011) Tissue engineering using magnetite nanoparticles. Prog Mol Biol Transl Sci 104:355–395

Kamiloglu S, Sari G, Ozdal T, Capanoglu E (2020) Guidelines for cell viability assays. Food Front 1:332–349

Key J, Dhawan D, Knapp DK, Kim K, Kwon IC, Choi K, Leary JF (2012) Multimodal in-vivo MRI and NIRF imaging of bladder tumor using peptide conjugated glycol chitosan nanoparticles. In: Imaging, manipulation, and analysis of biomolecules, cells, and tissues X.

Key J, Dhawan D, Cooper CL, Knapp DW, Kim K, Kwon IC, Choi K, Park K, Decuzzi P, Leary JF (2016) Multicomponent, peptide-targeted glycol chitosan nanoparticles containing ferrimagnetic iron oxide nanocubes for bladder cancer multimodal imaging. Int J Nanomed 11:4141–4155

Kumar CSSR (2005) Biofunctionalization of nanomaterials. In: Kumar CSSR (ed) Nanotechnologies for the life sciences, 1st edn. Wiley-VCH Verlag GmbH & Co. KGaA, Weinheim

Leary JF (2010) Nanotechnology: what is it and why is small so big? Can J Ophthalmol 45:449–456

Leary JF (2013) Nanomedicine—reality will trump hype! Nanomed Biother Discov 4:e125

Leary JF (2022) Fundamentals of nanomedicine. England, Cambridge University Press, Cambridge

Liu D, Wang L, Wang Z, Cuschieri A (2012) Magnetoporation and magnetolysis of cancer cells via carbon nanotubes induced by rotating magnetic fields. Nano Lett 12:5117–5121

Malik P, Shankar R, Malik V, Sharma N, Mukherjee TK (2014) Green chemistry based benign routes for nanoparticle synthesis. J Nanoparticles 2014:1–14

Murcia MJ, Naumann CA (2005) Biofunctionalization of fluorescent nanoparticles. In: Kumar CSSR (ed) Biofunctionalization of nanomaterials. Wiley-VCH Verlag GmbH & Co, Weinheim

Prow TW, Rose WA, Wang N, Reece LM, Lvov Y, Leary JF (2005) Biosensor-controlled gene therapy/drug delivery with nanoparticles for nanomedicine. In: SPIE advanced biomedical and clinical diagnostic systems III. SPIE

Prow T, Grebe R, Merges C, Smith JN, Mcleod DS, Leary JF, Lutty GA (2006a) Nanoparticle tethered antioxidant response element as a biosensor for oxygen induced toxicity in retinal endothelial cells. Mol Vis 12:616–625

Prow T, Smith JN, Grebe R, Salazar JH, Wang N, Kotov N, Lutty GA, Leary JF (2006b) Construction, gene delivery, and expression of DNA tethered nanoparticles. Mol Vis 12:606–615

Reineke J (2012) Nanotoxicity: methods and protocols. Humana Press, Springer

Ryman-Rasmussen JP, Riviere JE, Monteiro-Riviere NA (2007) Surface coatings determine cytotoxicity and irritation potential of quantum dot nanoparticles in epidermal keratinocytes. J Invest Dermatol 127:143–153

Seale M-M, Haglund E, Cooper CL, Reece LM, Leary JF (2007a) Design of programmable multilayered nanoparticles with in situ manufacture of therapeutic genes for nanomedicine. In: Advanced biomedical and clinical diagnostic systems V. SPIE

Seale M-M, Zemlyanov D, Cooper CL, Haglund E, Prow TW, Reece LM, Leary JF (2007b) Multifunctional nanoparticles for drug/gene delivery in nanomedicine. In: SPIE-nanoscale imaging, spectroscopy, sensing, and actuation for biomedical applications IV

Sizikov AA, Nikitin PI, Nikitin MP (2021) Magnetofection in vivo by nanomagnetic carriers systemically administered into the bloodstream. Pharmaceutics 13

Stillwell W (2016) Membrane transport. In: An introduction to biological membranes

Thomas CR, Ferris DP, Lee J-H, Choi E, Cho MH, Kim ES, Stoddart FJ, Shin J-S, Cheon J, Zink JI (2010) Noninvasive remote-controlled release of drug molecules in vitro using magnetic actuation of mechanized nanoparticles. JACS Commun 132:10623–10625

Van Engeland M, Nieland LJW, Ramaekers FCS, Schutte B, Reutelingsperger CPM (1998) Annexin V-affinity assay: a review on an apoptosis detection system based on phosphatidylserine exposure. Cytometry 31:1–9

Wani MY, Hashim MA, Nabi F, Malik MA (2011) Nanotoxicity: dimensional and morphological concerns. Adv Phys Chem 2011:1–15

Xu C, Mu L, Roes I, Miranda-Nieves D, Nahrendorf M, Ankrum JA, Zhao W, Karp JM (2011) Nanoparticle-based monitoring of cell therapy. Nanotechnology 22:494001

Microfluidic Diagnostics and Drug-Delivery Platforms for the Early Diagnosis and Treatment of Bacterial Diseases

Didem Rodoplu Solovchuk and Chia-Hsien Hsu

Abstract This chapter presents an overview of the strategy, design and utility of microfluidic technology in drug delivery, along with a comprehensive review of diagnostic testing for bacterial diseases. Microfluidics refers to the engineering of manipulating small volumes of fluids (10^{-9}–10^{-18} L), offering precise and programable control in microchannels. In drug delivery, microfluidics are advantageous for drug preparation and formulation, drug release optimization, drug testing, and controlled drug delivery. This chapter presents current state-of-art and future perspectives of advanced diagnostics incorporating viable bacteria testing, high-throughput testing, point-of-care testing (POCT) for environmental and clinical testing of bacteria, as well as rapid antibiotic susceptibility testing. Finally, we emphasize the potential advantages of single-bacterium microfluidics for testing phenotypic drug testing, and wearable/implantable ex-vivo/in-vivo devices for drug delivery applications for bacterial diseases. In conclusion, microfluidics has the potential to revolutionize the diagnosis and treatment of bacterial diseases by providing rapid and more precise diagnostic tools, together with enabling targeted and controlled drug delivery to infected tissues.

D. R. Solovchuk · C.-H. Hsu (✉)
Institute of Biomedical Engineering and Nanomedicine, National Health Research Institutes, Zhunan, Miaoli 35053, Taiwan
e-mail: chsu@nhri.edu.tw

D. R. Solovchuk
e-mail: didemrodoplu@gmail.com

C.-H. Hsu
Institute of Nanoengineering and Microsystems, National Tsing Hua University, Hsinchu 30013, Taiwan

Tissue Engineering and Regenerative Medicine, National Chung Hsing University, Taichung 40227, Taiwan

T. S. Santra and A. U. S. Shinde (eds.), *Advanced Drug Delivery*,
Studies in Mechanobiology, Tissue Engineering and Biomaterials 26,
https://doi.org/10.1007/978-981-99-6564-9_9

1 Introduction to Bacterial Diseases

Bacteria are prokaryotic single-celled organisms, and reproduce by dividing in half through a process called binary fission. They are abundant in the environment. Many bacterial species are beneficial to humans and the environment, playing a role in nutrient cycling, soil formation, and food production. Although most bacteria are harmless to the host, some pathogenic bacteria can cause disease by producing toxins that damage cells and tissues or by triggering an immune response that damages cells and tissues. When the immune system is unable to control pathogenic bacteria, it can lead to severe illnesses such as bacteremia, sepsis, multiple-organ failure, and potentially fatal outcomes (Sweeney et al. 2019; Rodoplu et al. 2021). Pathogenic bacteria may enter the body through various means including inhalation, ingestion, injection or direct contact with an open wound. Bacterial infections can affect the skin, ear (otitis media), brain and spinal cord (meningitis), upper respiratory tract, lungs (pulmonary tuberculosis and pneumonia), stomach (food poisoning and gastritis), sinus tissue (sinusitis), eye, urinary tract (urinary tract infection (UTI)), and can be sexually transmitted (Syphilis and Gonorrhea). Shown in Table 1, some of the bacterial diseases that are common throughout the world are listed according to the mortality rates, symptoms, form of spread, and treatment methods (Gorzynski et al. 2022; Gambushe et al. 2022; Boulnois 1992; Organization 2021, 2008; Soda et al. 2015; VanDemark 2013; Sack and Siddique 1998; Srisa-Art et al. 2018).

The development of **advanced testing methods** is important to aid environmental monitoring and clinical testing, particularly for the rapid detection of bacteria. **Environmental safety** monitoring may limit consumption or human contact in contaminated water (Tok et al. 2019; Rodoplu et al. 2022a), air (Jing et al. 2013) or food (Ekici and Dümen 2019; Rodoplu Solovchuk et al. 2023), and this is especially important in preventing the spread of infectious diseases where there are limited resources for hygiene, safety monitoring, diagnosis and treatment (Soda et al. 2015; Sack and Siddique 1998). To prevent an outbreak of water-borne diseases, the detection of live pathogens in water sources plays a vital role, as live and culturable microorganisms pose a threat to human health. Thus, culture-based methods are carried out on a regular basis to identify possible fecal-indicator microbial contaminants in water. Likewise, microbial food safety monitoring necessitates culture-based bacteria detection to ensure microbial safety in food processing and food products (Spatola Rossi et al. 2013). For microbial safety monitoring, conventional culture plating techniques necessitates days to conclude (Fig. 1). On the contrary, microfluidic technology offers promising features such as high sensitivity and selectivity, minimum supplement consumption, cost-reduction and point-of-care testing (POCT) to aid microbial safety in remote areas (Rodoplu et al. 2022a; Liu et al. 2019; Hassan et al. 2020). Concentrating large volumes of water sample into a small microfluidic volume allows "detection and enrichment" of wide-range concentrations of bacteria, and provides sensitive microbial water testing with limit-of-detection (LOD) of 2 CFU/100 ml *E. coli* that can aid water safety monitoring in remote areas (Rodoplu et al. 2022a).

Table 1 Some of the bacterial diseases

Bacterial disease	Symptoms	Spread through	Treatment	Mortality rate	Countries affected	Refs.
Legionellosis	Fever, cough, shortness of breath, muscle aches, headache, confusion	Airborne transmission through droplets	Antibiotics, rest, and supportive care	9–25%	Worldwide	Gorzynski et al. (2022), Soda et al. (2015)
Bacterial meningitis	Sudden fever, headache, stiff neck, sensitivity to light, confusion, vomiting, and difficulty concentrating	Close contact with an infected person's respiratory or throat secretions	Antibiotics, corticosteroids, and supportive care such as pain relief and management of fluid and electrolyte balance	5–40% of children and 20–50% of adults	Worldwide	VanDemark (2013)
Streptococcus pneumonia	Cough, fever, shortness of breath, chest pain	Airborne transmission through respiratory droplets or contact with contaminated surfaces	Antibiotics, rest, and fluids	5–7% higher in vulnerable populations	Worldwide, with higher incidence in DC	Boulnois (1992)
Tuberculosis	Coughing, coughing up blood, chest pain, fatigue, fever, chills, night sweats, loss of appetite	Airborne transmission through respiratory droplets	Combination of antibiotics over several months; drug-resistant strains require longer treatment and more expensive drugs	14%	Worldwide, with higher incidence in India, China, and South Africa	Organization (2021)

(continued)

Table 1 (continued)

Bacterial disease	Symptoms	Spread through	Treatment	Mortality rate	Countries affected	Refs.
Cholera	Watery diarrhea, vomiting, muscle cramps, dehydration	Contaminated water or food	Rehydration therapy, antibiotics in severe cases	1% with treatment 50% without treatment	Endemic in many developing countries, with periodic outbreaks in other regions	Sack and Siddique (1998)
E. coli infection	Diarrhea (often bloody), abdominal pain, nausea, vomiting, fever	Contaminated food or water, contact with infected animals	Supportive care, rehydration therapy, and sometimes antibiotics	<1% overall, higher in vulnerable populations	Worldwide, with higher incidence in DC	Gambushe et al. (2022)
Salmonellosis	Nausea, vomiting, diarrhea, fever, abdominal cramps	Contaminated food or water, or contact with infected animals or their feces	Rest, increased fluid intake, and sometimes antibiotics	<1%	Worldwide, with higher incidence in DC	Srisa-Art et al. (2018)
MRSA	Skin infections (boils, abscesses), pneumonia, bloodstream infections	Direct contact with infected wounds or contaminated objects	Antibiotics, wound care, and in severe cases, intravenous antibiotics and hospitalization	Varies depending on the severity of infection and patient population	Worldwide, with higher incidence in healthcare settings	
Measles	High fever, cough, runny nose, red, watery eyes, and a rash	Airborne transmission through respiratory droplets	Supportive care, rest, and symptom management	0.1–0.2%	Worldwide, with higher incidence in DC	

(continued)

Table 1 (continued)

Bacterial disease	Symptoms	Spread through	Treatment	Mortality rate	Countries affected	Refs.
Typhoid	High fever, stomach pain, weakness, headache, loss of appetite	Contaminated food or water	Antibiotics, rest, and increased fluid intake	1% with treatment, 12–30% without treatment, 10 times fatal for children under age of 4	Endemic in many DC	Organization (2008)
Dysentery	Abdominal pain, diarrhea, fever, and blood in the stool	Contaminated food or water, or contact with infected feces	Antibiotics, rehydration therapy, and rest	Varies depending on the severity of infection and patient population	Worldwide	

Abbreviations: *Escherichia coli* (*E. coli*); Methicillin-resistant *Staphylococcus aureus* (MRSA); developing countries (DC); Reference paper (Refs.)

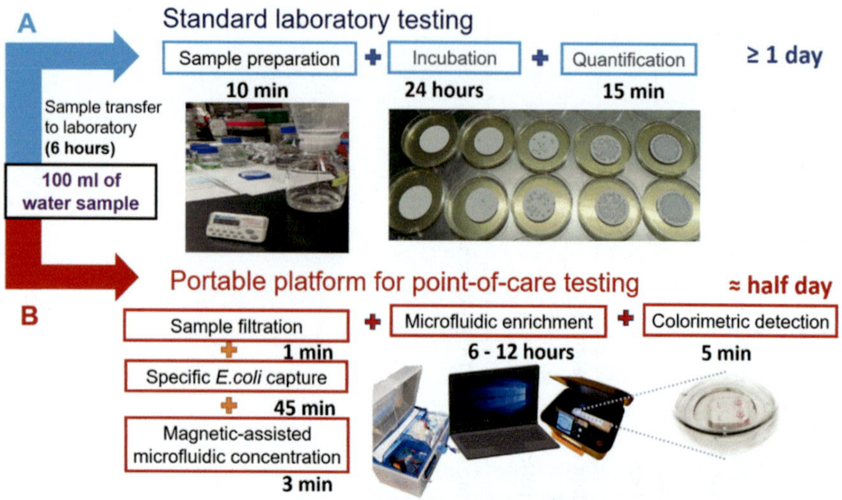

Fig. 1 Viable *E. coli* detection in water **a** membrane filtration method showing the time for colony counting on the agar plates, **b** microfluidic POCT method showing the rapid steps of microfluidic-enrichment and colorimetric testing of bacteria (Figure 1B adapted from Rodoplu et al. 2022a, *Copyright © 2022 Elsevier B.V. All rights reserved*)

Despite improved preventions of bacterial diseases, due to inappropriate use of antibiotics the antibiotic resistance rates continue to rise over time, which makes the treatment of bacterial infection demanding. Depending on the strain and antibiotic susceptibility of pathogens, and the severity of the infection, the intensity and duration of the treatment of bacterial diseases may vary (Gambushe et al. 2022).

For **microbial testing** in clinics, blood culture is the gold standard that requires incubation of blood samples for 5–7 days to test microorganism growth for the diagnosis of bacteremia, which is the existence of bacteria in the bloodstream (Phu 2020). The subsequent biochemical identification tests take 3–7 days to confirm bacterial species and strains (Sweeney et al. 2019; Davey 2011). The conventional phenotypic AST, including disc diffusion tests, epsilometer test (Etest), and broth microdilution, requires 1–4 days to determine suitable antibiotic type and minimum inhibitory concentration (MIC). The MIC is the minimum drug concentration that inhibit the bulky growth of bacteria on agar plates. Nevertheless, in the presence of antibiotic-resistant pathogenic bacteria, which have high doubling times, such as *Mycobacterium tuberculosis* (18–54 h) and *Treponema pallidum* (30–33 h), tests may conclude within several weeks and cause a delay of the treatment (Organization 2021; Gill et al. 2009). Most of the bacterial infectious diseases as shown in Table 1, can be fatal when an effective treatment is delayed (Rodoplu et al. 2021a; Boulnois 1992; Phu 2020; Wolter et al. 2008). Especially, for clinical complications such as sepsis, one hour of delay in an effective treatment may increase the mortality rate by 7%. For the early diagnosis and treatment of bacterial diseases, it is crucial to develop rapid methods for detection and AST of bacteria, which can significantly decrease testing period below 6 h (Khan et al. 2019).

In addition, conventional **antibiotic susceptibility testing** (AST) methods require bulky growth of bacteria which may mask the behavior and characteristics of subpopulation including resistant, persister and phenotypic heteroresistant bacteria. Persister bacteria typically constitute 0.01–1% of an isogenic population, may continue growing when treatment ceases, therefore may cause a relapse or deterioration of the disease (Dewachter et al. 2019; Pacocha et al. 2022). The mechanisms underlying the **phenotypic heteroresistance** are complex and not fully understood, but may involve changes in gene expression, altered cell physiology, or other factors that affect the susceptibility of bacterial cells to antibiotics. In course of time, microfluidic platforms, particularly those focused on single-cell analysis, have advanced to enable fast and precise antibiotic drug testing that can be completed within 2–6 h. Besides, microfluidic-trapping of individual bacterium can provide precise control of microenvironment, therefore enlightens the behavior and characteristics of individual bacterium under dynamic conditions (such as altered nutrients, temperature, pH, chemo attractant concentration etc.) in the small microfluidic channels (Lv 2022). By determining the minimum inhibitory concentration (MIC) of the resistant subpopulation at the single-cell level, these platforms have the potential to aid enhance our comprehension of resistance mechanisms, ultimately contributing to more effective therapies (Pacocha et al. 2022; Choi et al. 2014; Borgdorff et al. 2002).

2 Microfluidics in Drug Delivery

2.1 Physical Principles of Microfluidics

Micro/nanofluidics involve the design and fabrication of devices that enable manipulating small volumes of fluids particularly 10^{-9}–10^{-18} L in channels with dimensions in the millimeter scale or less. Design of microfluidics is analogous to the design of electrical circuits. The Hagen-Poiseuille law relates constant pressure drop (dP) resulting in a constant fluid flow (Q), which is similar to Ohm's law that relates the current (I) through a wire of resistance (R) and an electrical potential drop (dV). Consequently, circuit analysis can be useful for the predictions of pressure-driven laminar flow in microchannels.

In order to allow optical imaging at single-cell resolution and reduce the background signal for image-based analysis, the height of the microfluidic channels should be decreased. On the other hand, it may lead to a reduction on the Height/width aspect ratio and an increment on the shear rate, thereby causing a physical stress on living cells. For the design of a stress-free microfluidic device for cell research, the physical principles that govern microfluidic systems that are based on fluid mechanics and the physics laws to apply at the microscale should be well-understood (Rodoplu et al. 2021a, b, 2022b; Yeh et al. 2022). In addition to physical forces, physicochemical

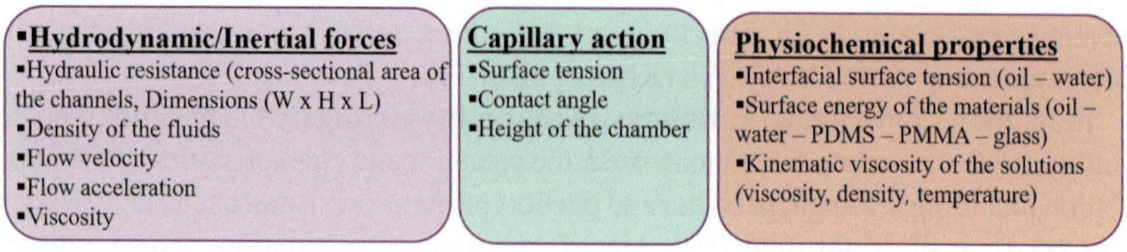

Fig. 2 Physical principles and physicochemical properties of microfluidic materials and liquids used to control the pressure drop, flow rate and shear stress in the microchannel (width (W), height (H), length (L), polymethylmethacrylate (PMMA))

properties of the microfluidic materials and liquids, and surface energy plays a role for the control of the flow rate and shear stress in the microchannel (Fig. 2).

Some of the key physical principles that are used to design and control the fluid motion, and the examples of microfluidic applications are summarized as the following.

Laminar flow is the streamlined flow of fluids through channels with low **Reynolds number** ($Re < 2300$), when inertial forces are much smaller than the viscous forces. At high Reynolds number ($Re > 4000$), inertial forces are dominant which causes chaotic mixing of fluid streamlines. Changing the physical (dimensions, shape) and chemical (material type, surface modifications, surface hydrophilicity) parameters of the channels and fluid characteristics, allows precise control over fluid movement and useful for many microfluidic applications including microfluidic flow focusing, mixing, separations, flow cytometry and flow lithography.

Capillary effect is the ability of fluids to move through narrow channels or tubes due to the adhesive and cohesive forces between the fluid and the walls of the channel. In microfluidics, capillary action can be used to move fluids through microchannels, to create microfluidic pumps, and to control the movement of fluids. Since capillary-driven microfluidics are compact and portable devices, it is of great importance for POCT diagnostics (Hassan et al. 2020).

Surface tension is the force that causes liquids to reduce the surface area and form droplets, and to adhere to solid surfaces. Controlling the surface tension is of great importance to create droplets, to manipulate the shape/size of droplets and to control the movement of fluids through microchannels (Fig. 3). For instance, (1) **T-junction** consists of two perpendicular channels, where the dispersed phase (such as water) is injected from one channel and the continuous phase (such as oil) is injected from the other channel to break-up of a continuous stream by shear from the cross flow of a second immiscible fluid stream. The droplets are formed at the intersection of the two streams due to shear stress and pressure drop. (2) **Flow-focusing junction** consists of four channels that form a cross shape, where the dispersed phase is injected from one channel and the continuous phase is injected from two lateral channels. The droplets are formed at a narrow orifice downstream of the cross junction due to the squeezing effect of the continuous phase. The flow-focusing junction has a higher degree of confinement for the dispersed phase than T junction has, which allows for smaller and more spherical droplets to be produced. Both devices can produce monodisperse droplets with different regimes depending on the flow rates, capillary effect, geometry, and interfacial tension of the fluids. (3) **Co-flow** devices allows two or more fluids are flowed through a channel using a coaxial arrangement, where the dispersed liquid phase decays into droplets via Rayleigh–Plateau instability (Nunes et al. 2013). (4) **Step-emulsification** where two immiscible fluids are introduced into a microchannel through separate inlets and flow through a series of constrictions in the microchannel, forcing continuous phase to flow around the dispersed phase, resulting in a droplet formation (Seemann et al. 2011). (5) **Hanging drop microfluidics** refers to a technique where a small droplet of cell culture medium containing cells is suspended by surface tension and gravity, upside down from the holes of the device. Enrichment of the cell suspension results with a three-dimensional structure that allows for the study of cell behavior in a more physiological-like environment. Hanging-drop microfluidics are useful for the generation and the co-culture of organoids (Rodoplu et al. 2022b), while droplet-based microfluidics are mainly used for the preparation of double **emulsions** (water-in-oil-in-water (W/O/W) or oil-in-water-in-oil (O/W/O) droplets) for single cell/bacterium research (Scheler et al. 2020), and for the preparation of nano-drug carriers for drug delivery.

Diffusion is the movement of molecules or particles from an area of high concentration to an area of low concentration. With this physical principle, microfluidic devices can be useful for precise mixing of liquids and controlled release of molecules or particles by generating a concentration gradient across a membrane.

Electrokinetics refers to the movement of fluids/particles in response to electric fields and the electric potential/current generated due to the movement of particles in

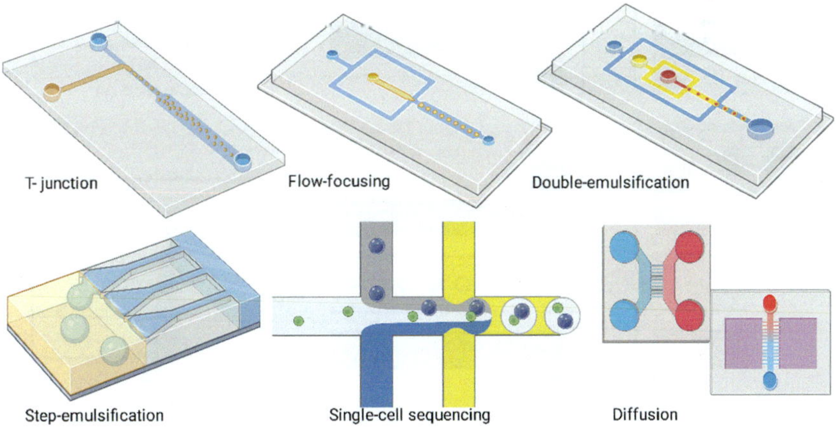

Fig. 3 Microfluidic devices for droplet generation including T-junction, flow focusing, double emulsification, step emulsification + single-cell sequencing + diffusion (Figure created with Biorender.com)

the fluids (including electrophoresis, electro-osmosis, sedimentation potential, and streaming current/potential). Microfluidic-electrokinetics can be useful to manipulate the movement of fluids, cells, bacteria and particles, and to create microfluidic pumps and mixers (Zhang et al. 2023; Sy et al. 2023). **Electrohydrodynamic (EHC) jetting** (also known as electrospraying or electrospinning) is a process that uses an electrical field to induce charge on the surface of a liquid, causing the liquid to deform and break up into droplets or jets, depending on the electrical field strength and the liquid properties (Fig. 4). Applying an electrical field that matches the natural frequency of the jet in a flow-focusing junction, can control the process of high-throughput monodisperse droplets at high speeds, and the size of the droplets can be adjusted by changing the frequency of the signal (Assche et al. 2023). EHC jetting allows precise manipulation and control of fluids at microscale which can be useful for chemical synthesis and drug delivery.

Microvortex focusing is a process that involves generating and controlling small vortices in fluid streams that can be used to mix fluids or to concentrate particles and cells. Higher flow rates can lead to larger and more turbulent vortices, while lower flow rates can result in smaller and more stable vortices. To focus particles to the trap locations in the micro-vortex manipulator device, the drag force in the z-component must be less than the net force of buoyancy and gravity. Therefore, the particle size and density can affect the interaction with the vortex and the degree of sorting that

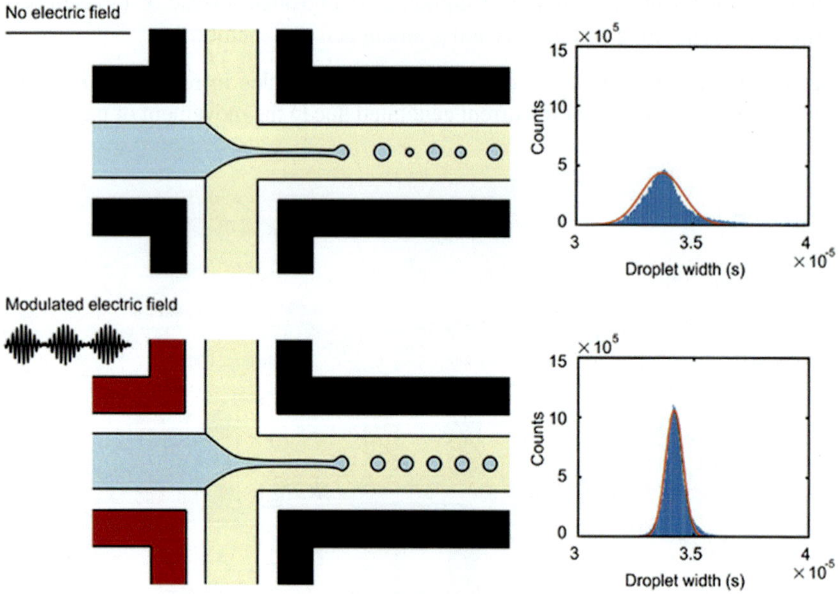

Fig. 4 Schematic of the dripping-to-jetting transition in a flow-focusing junction **a** without electrical fields, **b** under modulated electric field for the production of monodisperse droplets (Figure adapted from Assche et al. 2023, Copyright © 2023 *The Authors. Droplet published by Jilin University and John Wiley & Sons Australia, Ltd.*)

can be achieved. Particles with a lower density than the surrounding medium are lifted towards the top of the channel after being focused to the interface between two microvortices due to lateral drag forces. When the force pushing the particle upwards at the interface in the vertical direction is balanced by the downward Stokes drag and gravitational forces, the particles become trapped at that location. Similarly, particles that are denser than the surrounding medium are also focused to the bottom of the interface of the two microvortices using the same mechanism (Hsu et al. 2008).

With all these principles, microfluidic lab-on-chip (LOC) platforms can aid the bio-analytical applications including capillary electrophoresis, isoelectric focusing, immunoassays, flow cytometry, protein sample injection for mass spectrometry, PCR amplification, DNA assays, single cell/bacterium separation/isolation, and gradient formation (Sun et al. 2021; Fernandes et al. 2014; Bridle and Bridle 2014). For biomedical research, microfluidics can provide significant advantages due to nano/picoliter scale fluidic operations, as such reduce the consumption of biochemical reagents and the detection time, and allow high-throughput analysis. Moreover, the manufacturing techniques of multiplex microfluidics are relatively affordable and advantageous for mass production.

2.2 *Microfluidic Applications in Drug Delivery*

Drug delivery refers to the use of a "nano drug-carrier" such as liposomes, dendrimers, nano-emulsions, polymeric nanoparticles, carbon nanotubes, to carry a therapeutic agent and release it at a specific rate at a particular location (Jeevanandam et al. 2016). Conjugation or encapsulation of a therapeutic agent to a nano drug-carrier can improve the drug efficacy, thereby allows development of new therapies. As shown on Table 2, microfluidics can find wide range of applications in drug delivery, including the preparation and characterization of nano-drug carriers, drug testing and toxicity/viability testing, optimization of drug delivery, controllable drug delivery platforms, diagnostics, and analytical tools for the evaluation of cell function and drug discovery. A microfluidic drug delivery platform may contain (1) a microfluidic device incorporating a network of microchannels and chambers, and (2) other components such as pumps, valves, sensors, and control systems that are used to regulate the flow and delivery of drugs.

2.2.1 Microfluidic Drug-Carrier Preparation

Conventional synthesis methods incorporate mass transport of fluids, which can result with instabilities (such as turbulence) and Taylor dispersion in microchannels (Hajiani and Larachi 2012), where different components of the fluid move at different velocities, causing mixing to be inefficient. This can result in a non-uniform distribution of reactants, leading to inconsistencies in the final product. For mixing, while

Table 2 Microfluidic applications in drug delivery

Nano-drug carrier preparation	Nanomedicine characterization	Carrier-free drug delivery systems	Evaluation of cell function and drug discovery		
				System medicine	Bacteriology
• Flow focusing • Micro-vortices • Chaotic flow • Micro-droplets • Flow lithography	• Size • Morphology • Charge • Drug loading • Drug release	• Micro-needle • Micro-reservoir	• Organ-on a chip • Single-cell arrays	Blood–brain barrier, tumor, lung, liver, kidney, and heart models Heterogeneity of cell populations, metabolism, migration, growth, and differentiation of cells	Gut-on-a chip, gut microbiota, and drug testing POCT, rapid viable bacteria detection, AST of bacteria

mechanical agitation is associated with high shear forces, which can lead to the formation of non-uniform particles, diffusion can be time-consuming and inefficient. Since the inertial effect becomes negligible in microfluidics (Björnmalm et al. 2014), it has several advantages for the preparation of highly-controlled and reproducible nano-drug carriers, including precise control over reaction conditions (such as mixing, temperature, and reaction time), high throughput producing nano-drug carriers in a short time, reduced material consumption, enhanced particle properties (such as size, shape, and surface chemistry), and compatibility with a wide range of materials including polymers, lipids, and metals, making them versatile for the synthesis of different types of nano-drug carriers (Jeevanandam et al. 2016). In addition, externally excited suspended magnetic nanoparticles (MNPs) are efficient vehicles for generating liquid-phase mixing at submicron levels using low-frequency rotating magnetic fields. Hajiani et al. developed a method to reduce the Taylor dispersion in micro mixing by utilizing externally excited suspended magnetic nanoparticles (MNPs) (Hajiani and Larachi 2012). The developed technique generated liquid-phase mixing at submicron levels using low-frequency rotating magnetic fields. One limitation of the drug-carriers is that predicting the behavior of a nano drug-carrier in a complex system is difficult. A microfluidic-based multidisciplinary approach can be useful to evaluate cell and system behavior for the translation of new drug delivery vehicles from in vitro to the preclinical and clinical setting.

2.2.2 Microfluidic Drug Delivery In-Vivo

In-vivo drug delivery platforms can manipulate the flow of fluids and release of drugs in the target tissues or organs with precision and control as such reducing the risk of side effects and improving the efficacy of drug therapies. Since, there are many issues associated with microfluidic drug delivery in vivo, including the need for biocompatible materials, the potential for immune reactions or other adverse effects, and the need for precise control over drug delivery rates, microfluidics are frequently utilized for in vitro drug-delivery rather than in vivo. While in-vitro studies provide precise control of experimental conditions, in-vivo studies can provide a more comprehensive view of the biological response to drug delivery.

In-vivo bacteria research involves the study of bacteria in living organisms, typically in small animal models. These experiments can be conducted inside of a microfluidic device to allow real-time monitoring of dynamic phenotypic changes. For instance, to investigate the effects of dietary restriction on a worm mutant, Viri et al. developed an automated system allowing confinement, trapping and fast phenotyping of *Caenorhabditis elegans* (*C. elegans*) utilizing fluorescence imaging in an optofluidic device (Viri et al. 2020). This method can be useful to observe the effects of dietary restriction on a worm mutant modeling Huntington's disease. Microfluidic in-vivo drug delivery platforms shed light on bacterial pathogenesis and host response to infection, as well as the development and testing of new therapeutics.

3 Microfluidic Diagnostics and Drug Delivery for Bacterial Diseases

Microfluidic diagnostics and drug screening platforms are revolutionary tools that has a potential to replace the conventional microbiology techniques, providing the ability of monitoring bacteria at single-cell resolution in a controllable microenvironment. Current microfluidic-sensing platforms present several advantages including. (1) the devices are amenable to high-resolution imaging using benchtop microscope (Rodoplu et al. 2021a; Højris et al. 2016) as well as compact image-acquisition setups (Rodoplu et al. 2022a; Zainal Alam et al. 2018), (2) microfluidics ensure hydrodynamic trapping of individual bacterium and controllable microenvironment to envision the bacterial motility before and after a chemo-attractant, (3) microbial samples can be maintained and concentrated in small areas inside the microchannels for sensitive and viable detection (Rodoplu et al. 2021a), so that can be useful for drug screening as well. As such, lab-on-a-chip (LOC) systems represent a powerful tool for allowing rapid detection, phenotypic characterization, and wide-spectrum drug screening of viable pathogenic bacteria, so that it can be useful for clinical applications.

Microfluidic drug delivery platforms have been utilized for wide-range of bacteriology applications including antibiotic susceptibility testing, evaluation of chemotaxis and drug accumulation on the double-membrane cell envelope of gram-negative bacteria, and gut-on-a-chip (GOC) research. Microfluidic drug delivery platforms have several potential advantages: (1) precisely control the delivery of drugs to specific target sites in the body, (2) microscale dimensions of the device enable fine control over the rate, timing, and dosing of drugs, which can improve the efficacy and safety of treatments, (3) reduce the amount of drug required, minimize side effects, and improve patient compliance, (4) provide real-time monitoring of bacterial cell responses to different drugs and treatments, (5) measure the effects of drugs on bacterial cells, allowing the development of new targeted drug therapies.

3.1 Optical Sensing/Imaging Platforms

Optical methods such as surface plasmon resonance, surface enhanced Raman spectroscopy (SERS), laser induced breakdown spectroscopy (LIBS) and fluorescence microscopy are emerging technologies for rapid and sensitive testing. Scientists have been investigating new **optofluidic** sensor technologies that integrates optics and microfluidics for rapid and sensitive detection of bacteria. These sensors can utilize fluorescence, impedance, or machine learning methods to identify different types of bacteria in real-time with high accuracy. For instance, Hunter et al. developed an optofluidic **Raman** detection platform that incorporates a microfluidic-driven hollow-core photonic crystal fiber and silver nanoparticles to significantly enhance the Raman signal (Hunter et al. 2019). This fiber confines both light and cells, allowing for spectral events generated by the flowing cells to enable a unique approach to cell counting that simultaneously measures and identifies bacterial infections. A genetically optimized support vector machine learning algorithm performs the bacteria counting automatically. The microfluidic system can be reused multiple times and enables real-time detection of planktonic bacteria, with a detection limit as low as 4 CFU/mL in just 15 min. However, this system lacks viable bacteria counting. Ry et al. reported a ring-shaped interdigitated electrode (RIDE) chip and utilized a combination of dielectrophoresis (DEP) and alternating current electroosmosis (ACEO) forces to achieve the rapid collection of lactic acid bacteria *Lactobacillus plantarum* toward the detection zone of the chip, which is then detected by a CCD camera for subsequent **image acquisition** and analysis. Although this method is a rapid and affordable method of bacteria counting (Sy et al. 2023), lacks detection of target bacteria from a mixed bacteria dilution. Hojris et al. developed an optofluidic setup that consists of: (1) a fluidic-cell holding the water sample, (2) a dark field imaging setup with a light-emitting diode (LED) light source, a magnification lens, a complementary metal-oxide semiconductor (CMOS)-based camera arrangement for image acquisition and (3) a software for image analysis to distinguish target bacteria among other particles and bacteria using the morphological characteristics of different bacteria (Figure). While the limit of detection (LOD) of a **flow cytometer** method is ≥100 CFU/mL bacteria, LOD decreases to ≤10 CFU/mL bacteria when bacteria are concentrated at the microfluidic chambers using external fields (Rodoplu et al. 2021a). The transparent opto-fluidic devices, especially with a coverslip glass bottom, allows imaging with high magnification at single-bacterium resolution and facilitate rapid quantitative testing of dynamic changes (Rodoplu et al. 2021a; Højris et al. 2016). In order to detect low concentrations of viable

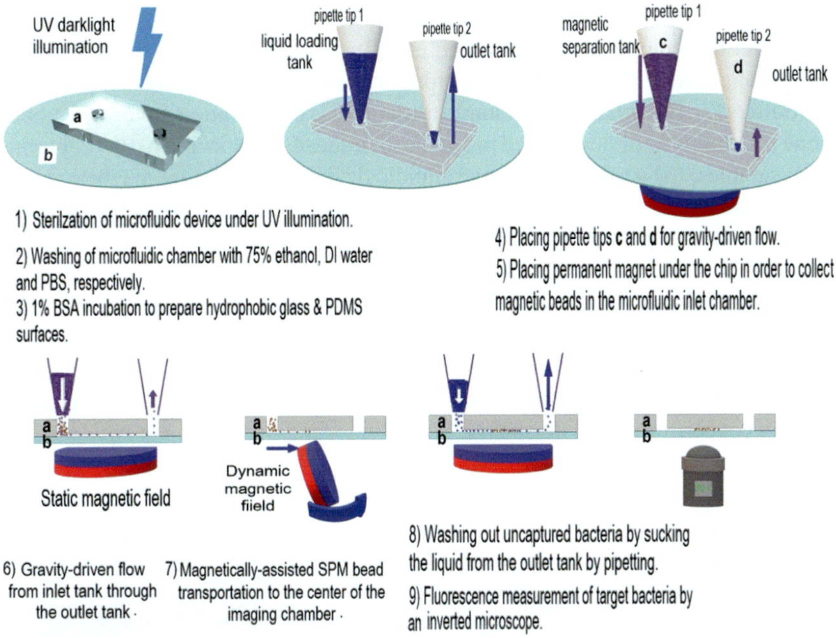

Fig. 5 Diagram of the magnet-assisted microfluidic method for bacteria testing **a** polydimethylsiloxane (PDMS) top layer with 2.3 mm thickness **b** glass coverslip with 0.17 mm thickness **c** bacteria@SPM sample loading with an initial fluid height of 25 mm **d** empty outlet tank). (Figure adapted from Rodoplu et al. (2021a) © 2021 Elsevier B.V. All rights reserved)

bacteria, Rodoplu et al. developed a **magnetic-assisted** method, which is concentrating magnetic bead-bound bacteria using a rotational motion of a magnet under the detection chamber within few seconds (Fig. 5). Due to increased signal/noise ratio, magnet-assisted method can detect 2 CFU/mL *E. coli* in1 hour (Rodoplu et al. 2021a).

3.2 *Point-of-Care Testing Platforms*

Microfluidic devices have undergone rapid development in recent years and provide a POCT solution for many biomedical and environmental applications. Microfluidic POCT devices may incorporate mobile phones, hand held readers, portable optical setups or wireless communication.

Mobile-phone based optical imaging is advantageous to replace the necessity of bulky instruments for biosensing applications, therefore could be very useful especially in low-income countries (Sack and Siddique 1998; Nunes et al. 2013). It has been gaining accelerating attention as mobile phone users are gradually increasing all around the world. Zhu et al. developed a glass capillary based POCT (Zhu et al. 2012) that revealed the potential of cell-phone based *E. coli* detection platform for screening samples in the field in resource limited environments. Rodoplu et al. developed a microfluidic-based POCT method incorporating portable instruments including (1) microfilters for 100 mL of water sample filtration, (2) the vibration motor of a mobile phone for mixing antibody-coated magnetic microbeads for specific *E. coli* capturing in the samples, (3) a micro-pupil microfluidic device for collecting magnetic beads in the testing chamber and imaging through device openings, (4) a temperature-controlled mini-incubator for enrichment of captured bacteria, (5) a mobile-phone setup and a software for the colorimetric detection of viable bacteria. All these portable and miniature instruments can be carried in a box (Fig. 6). Therefore, it can aid microbial testing in resource limited remote areas.

Rapid and sensitive detection of **food-borne pathogens** plays an important role to prevent an outbreak of foodborne diseases. *Salmonella* with 2523 serotypes receives the most concerns for food poisoning. For rapid testing of *Salmonella typhimurium*,

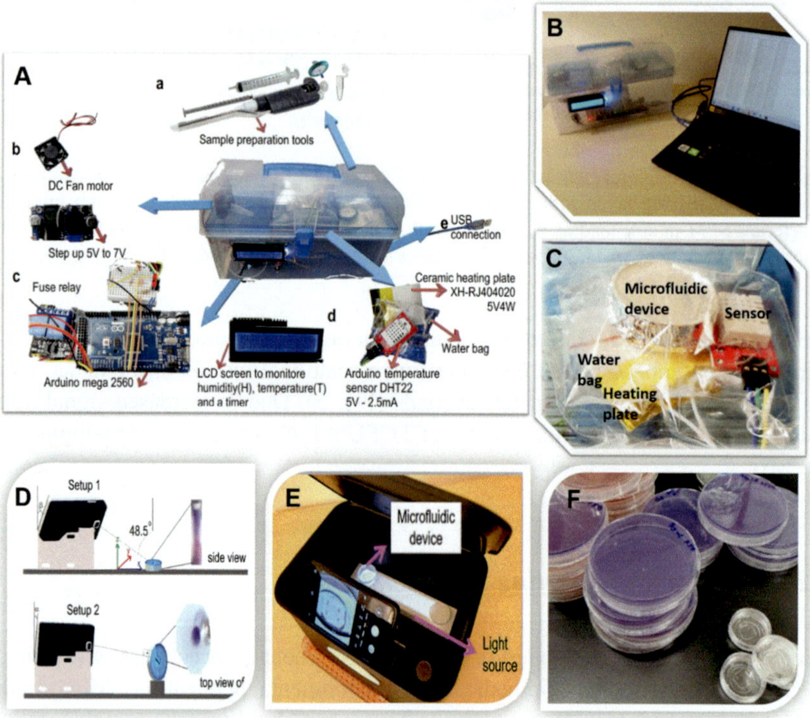

Fig. 6 Microfluidic-based POCT method for microbial water testing: **a** components of portable tool box for water testing, **b** USB powered device, **c** temperature controlled mini incubator in a plastic bag, **d** μ-pupil imaging setup, **e** black box for image acquisition, (Figure **a**, **c**, **d**, **e** modified from Rodoplu et al. 2022a *Copyright © 2022 Elsevier B.V. All rights reserved*) and **f** size comparison of ϕ: 15 cm agar plates and μ-pupil device assembled on ϕ: 35 mm culture dish

Wang et al. developed a microfluidic biosensor using a combination of immunomagnetic separation, fluorescent labeling, microfluidic chip technology, and smartphone video processing. The developed biosensor capable of detecting *Salmonella typhimurium* at concentrations as low as 58 CFU/mL within just 2 h, making it highly effective for online monitoring of foodborne pathogens (Wang et al. 2019). However, the current limitations of this biosensor are related to the video processing speed and image capturing quality. To improve the sensitivity, higher-performance of image-acquisition can be achieved with faster CPU speed and higher camera resolution to process the video, brighter fluorescence labeling to increase the signal/noise ratio, and also integration of the developed system with single-cell microfluidics for online and accurate detection of individual bacterium.

Another promising POCT devices are **wearable** or **implantable** sensors, which can detect the presence of bacteria and monitor growth-rate in real-time, for early diagnosis of bacterial infections. There are various types of wearable biosensors for bacterial diseases, including: (1) **skin patches** that are thin and flexible patches to sample interstitial fluid with microneedles, (2) **wristbands** or **smartwatches**, which are small and portable devices to detect biomarkers in sweat, blood or other bodily fluids, (3) **implantable sensors** that can be implanted under the skin, and use wireless communication to transmit information about bacterial infections to a healthcare provider.

For the fabrication of **flexible** wearable microfluidics, specially formulated variants of PDMS with optical transparency, ease of patterning, biocompatibility, favorable mechanics such as low modulus (~145 kPa) and high elasticity (up to ~200% strain at break) can be utilized (Jin et al. 2015). The soft mechanics and thin geometry of the device allow it to make non-irritating intimate contact with the skin based on van der Waals interactions which enables clinical-grade measurements of key markers of physiological health in a continuous manner. For instance, Mannor et al. developed a **wireless graphene nanosensor** that consist of a functionalized surface with a dodecapeptide graphene binding peptide, a triglycine linker and the antimicrobial peptides (AMP) odorranin-HP for bacteria recognition, electrodes and an inductive coil antenna printed on a **silk thin film** (which), for the detection of bacteria due to conductance change and wireless data transfer via resonant circuits and inductively coupled RF reader devices (Mannoor et al. 2012) (Fig. 7). The silk thin film is briefly prepared by molding fibroin solutions onto PDMS, crystallized in air, graphene monolayers were transferred onto PDMS, then integrated with gold inductive coil (Fig. 7a–c, e–f). Then, nanosensing architecture were transfered onto the surface of the tooth and muscle tissue (Fig. 7g–h). For material characterization, fluorescence and Raman spectroscopy were conducted. As shown in Fig. 7i with the absence of the fluorescence signal, complete dissolution of the silk matrix in water led to a strong attachment of the graphene–Au electrode within 15–20 min. The Raman spectrum obtained after combining graphene with tooth enamel and also after agitation in commercial mouthwash for 3 min, indicates the presence of the phosphate $\nu 1$ peak and the mechanical stability of the sensor on tooth enamel. The model bacterium *E. coli* captured on the surface by peptides self-assembled on the graphene nanotransducer (Fig. 7d, k, l). As shown in Fig. 7k–l, graphene surface functionalized with fluorescein isothiocyanate-labelled GBP–OHP, indicating a robust interaction between the immobilized peptides and the bacterium, as opposed to the bare graphene nanosensor.

Although the developed tooth-mounted sensor can detect single bacterium, it lacks quantitative enumeration of bacteria, therefore cannot be useful for growth-rate analysis and phenotypic antibiotic testing of bacteria. With future improvements of this technology, oral-cavity sensors can be useful for long-term monitoring and quantification of bacteria in saliva. The data collected by implantable biosensors can

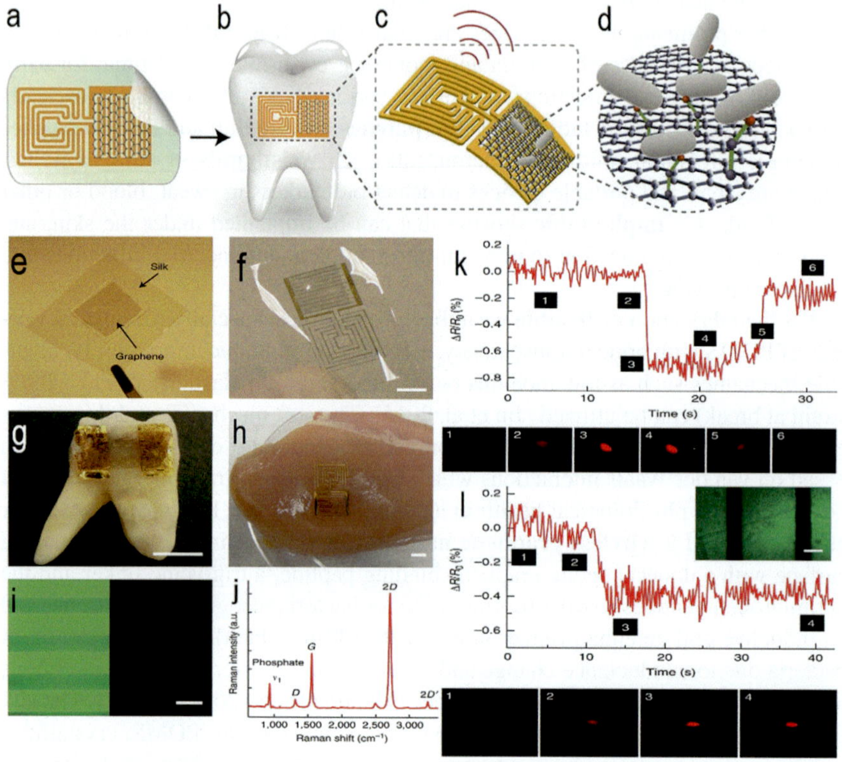

Fig. 7 Tooth-mounted ex-vivo sensor for bacteria testing in the oral cavity **a** schematic of silk thin film, **b** biotransfer onto a tooth, **c** schematic of the wireless readout, **d** pathogenic bacteria binding, **e** image of bioresorbable silk film, **f** passive wireless telemetry system, **g**, **h** nanosensor biotransferred onto a human molar and muscle tissue (Scale bars are 5 mm), **i** fluorescent signal before (left) and after (right) dissolution of silk, **j** Raman spectrum of graphene onto the tooth surface, **k**, **l** electrical resistance (upper) and fluorescence (lower) data recorded showing binding/unbinding of a single *E. coli* bacterium on a **k** bare graphene nanosensor and **l** peptide-functionalized graphene nanosensor. Inset shows fluorescent image of peptide-functionalized graphene surface (green), with the black regions representing electrodes. Scale bar is 250 μm. (Figure modified from Mannoor et al. (2012) *Copyright @ 2012, Nature Publishing Group (NPG) is a division of Macmillan Publishers Ltd. All rights reserved*)

be sent to healthcare providers and used to inform treatment decisions, such as the choice of antibiotics or the need for further testing.

3.3 High-Throughput Platforms

For certain diseases, urgent antibiotic therapy is of great importance as much as rapid isolation and identification of pathogens (Rhodes et al. 2017; Pilecky et al. 2019). High-throughput microfluidics can be useful for screening potential effects of drug candidates on bacteria and improve the treatment planning in clinics. Current state-of-art high-throughput systems includes droplet microarrays, nano-pico chamber arrays and gradient generator platforms, which are useful for running multiple tests simultaneously, either on the same sample or multiple samples, thereby allowing rapid and accurate screening of large numbers of antibiotic drug candidates. Up to date, different research groups have developed **micro/pico chamber array** systems, which utilize oil-liquid cutting for hydrodynamic trapping of single bacterium in a closed chamber, for rapid bacteria detection or AST (Avesar et al. 2017; Azizi et al. 2018). These platforms have shortened the AST time, improved the assay by providing simplicity, and decreased the volume of reagents used; however, due to small dimensions of the bacteria-testing channels sample loading could be tedious or it may require vacuum-driven flow through the microchannel. However, the antibiotic type and the antibiotic concentrations must be determined before performing the AST test, which lessens the flexibility of the whole process.

The **droplet microfluidics** generates a monodispersed layer of micro water–oil droplets or double emulsions (W/O/W) encapsulating single bacterium with growth medium and a metabolic indicator such as the resazurin dye or labeled bio receptors to quantify bacteria division. These devices allow real-time dynamic control and quantification of bacterial population in droplets (Huang et al. 2015). Since these devices demonstrate miniaturization of microbial enrichment in liquid broth medium, the time required for the testing can be decreased below several hours. An et al. developed a droplet device that could detect viable Salmonella bacterium in milk with detection limits as low as 50 CFU/mL after 5 h of incubation (An et al. 2020). One obstacle of droplet microfluidics is the difficulties in manipulating the drug concentrations in the droplets to achieve dynamic control of cellular behaviors, which can be critically important to probe bacterial dynamics in response to different environmental cues.

Microfluidic **concentration gradient generator** devices are also useful for simple, rapid, sensitive and high-throughput AST, without necessitating multiple devices (unlike picochamber and droplet arrays do) for testing different drug concentrations. The current state-of-art gradient platforms exhibited good linearity, stability, and controllability, while obtaining MIC values in one test. Sun et al. developed a platform that generates concentration of ofloxacin ranging from 0.089 to 3.2 μg/mL, and ampicillin ranging from 5.6 to 51.2 μg/mL for MIC testing of Salmonella (Sun et al. 2021). The MIC value was assessed by the fluorescence signal corresponding to the relative survival rate of the bacteria after 5 h of incubation under antibiotics. Besides, the effect of food processing on the antibiotic resistance of Salmonella was demonstrated that long-term acid-based treatment can affect the penetration of antibiotics to the bacteria cell membrane. Consequently, gradient devices possess numerous advantages for AST testing and are superior to conventional AST methods allowing monitoring dynamic antibiotic resistance changes of bacteria.

3.4 Gut-on-a Chip

Gut microbiota refers to the trillions of microorganisms that live in the gastrointestinal tract and plays a crucial role in maintaining human health by influencing digestion, nutrient absorption, production of neurotransmitters and regulation of the immune system in the body (Paone and Cani 2020). The gut-brain axis is a complex communication network that involves the immune system, the nervous system, and the endocrine system. It is linked to a range of health conditions, including inflammatory bowel disease, cancer, obesity, diabetes, autoimmune disorders such as multiple sclerosis and autism spectrum disorder, and even some neurological disorders such as Parkinson's disease and depression. The gut microbiome also strongly interacts with certain drugs, including some mental-health therapeutics, and influences their effects. The gut microbiota can produce various compounds, including neurotransmitters and hormones, that can affect the brain and the function of mental-health therapeutics (Fig. 8).

Understanding the relationship between the gut microbiota and brain disorders could lead to new treatments and interventions for these conditions. To do so, "gut-on-a-chip" or "intestine-on-a-chip" platforms consist of a microfluidic device that is designed to mimic the structure and function of the human intestine (Zhang and Qiao 2022; Nelson et al. 2021). The GOC device may contain intestinal cells, as well as gut microbiota that have been collected from human fecal samples, which can be useful to study the interactions of gut microbiota and the host, as well as to test the effects of different drugs or interventions on the gut microbiota. Yuan et al. developed a GOC device consist of upper and lower PDMS channels separated by a polytetrafluoroethylene (PTFE) membrane (as a blood-gut barrier), in order to investigate the effect of probiotics and bacterial biofilms on epithelial cell layer seeded on the upper channel (Yuan et al. 2020). Study demonstrated that *E. coli* were unable to destroy an epithelial cell layer due to probiotic B. breve protection (Fig. 9).

These platforms offer several advantages for gut microbiota research: (1) the platform uses human intestinal epithelial cells and human endothelial cells to mimic blood vessel wall, which makes it more relevant to human physiology than animal models. (2) The platform allows for precise control and manipulation of the gut environment, which allows for better reproducibility and consistency of results for better understanding complex host-gut microbiota crosstalk and inflammatory responses. (3) The platform is a cost-effective alternative to animal models, as it requires fewer resources and can produce results more quickly. (4) The platform can be used

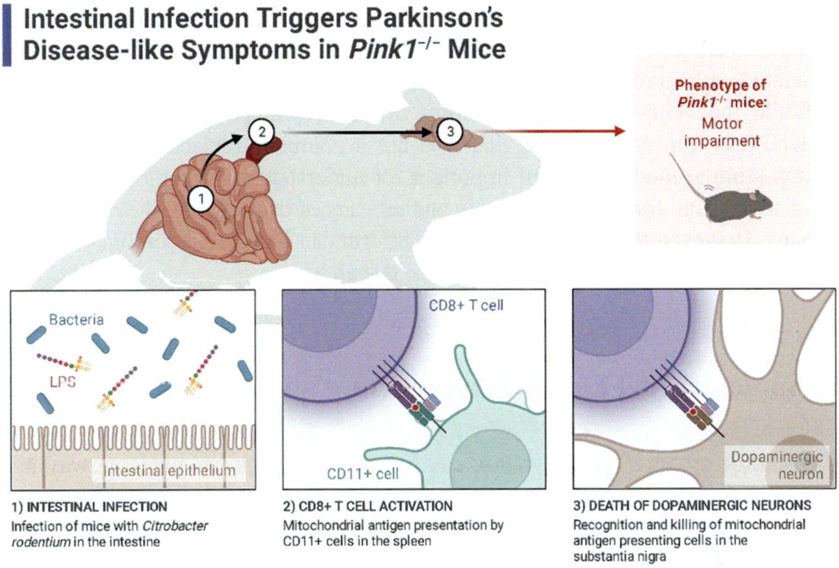

Fig. 8 Understanding gut-brain axis using animal model (Figure adapted from *intestinal infection triggers Parkinson's disease-like symptoms in Pink1 −/− Mice* (2022), by BioRender.com)

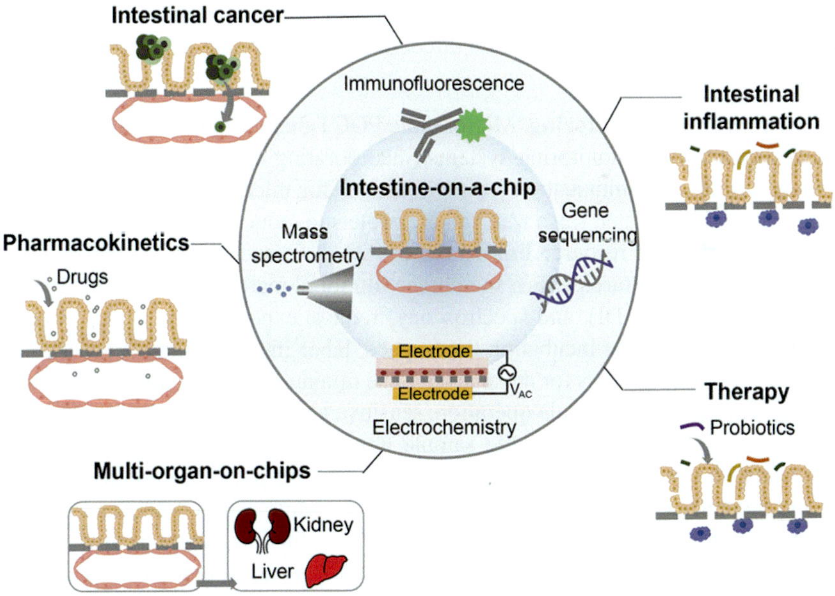

Fig. 9 Intestine-on-a-chip approaches for drug testing, intestinal disease study, and the development of intestinal disease therapy methods (Figure adapted from Zhang and Qiao (2022) © *2022 The authors, VIEW published Shanghai Fuji Technology Consulting Co., Ltd. Authorized by Professional Community of Experimental Medicine, National Association of Health Industry and Enterprise Management (PCEM) and John Wiley & Sons Australia, Ltd.*)

for high-throughput screening of drugs and interventions, which can speed up the drug discovery process and reduce the need for animal testing and ethical concerns associated with animal experimentation.

While the "Gut-on-a-Chip" platform offers several advantages over animal models, it is important to note that it is not a complete replacement for animal testing. Animal models are still important for understanding complex interactions and diseases, and for testing the safety and efficacy of drugs before they are used in humans. However, the "Gut-on-a-Chip" platform can be used in combination with animal models to complement and enhance research.

4 Summary and Future Perspectives

Microfluidic systems are a revolutionary technology which has made its way into different fields of medicine. They can assist and upgrade the fabrication of sophisticated drug carriers with specific properties, preprogrammed release profiles, and uniform sizes in the nano/micrometers scale. Microfluidic diagnostics and drug delivery platforms show great potential for improving the speed and accuracy of bacterial disease diagnosis and treatment and are likely to play an increasingly important role in healthcare in the coming years. This technology can facilitate dynamic evaluation of drug treatment and provide significant data to correlate animal studies and human clinical trials. The future directions of microfluidic platforms are listed as the following.

1. Environmental monitoring: Microfluidic POCT diagnostics has the potential to provide smart water monitoring systems (incorporating optofluidic technology for testing, wireless communication for rapid action) for microbial testing in drinking water, recreational water, fresh water and seawaters to reduce the time for testing and prevent outbreaks in resource limited remote areas. Conventional laboratory techniques (including culture plating, membrane filtration, matrix-assisted laser desorption/ionization (MALDI), and spectroscopy) require expensive and bulky systems which necessitates long incubation time, space, labor-intensive sample processing, together with trained users for the analysis. The miniaturization of the culture-based detection system offers simple operation, sensitive and accurate testing low concentrations of bacteria, increasing the sample number with high throughput testing, decreasing the turnaround time, consumable cost, and portable systems to assist environmental monitoring, along with microbial monitoring of water and food in resource limited remote areas (Fig. 10). In addition, the integration of microfluidic sensors with smart food packaging can be useful for monitoring food freshness and microbial growth in a rapid and accurate way. The vendors, distributors or food suppliers can easily receive food quality data using hand held readers for barcode reading, or wireless communication via smart phone or smart watch. This feedback mechanism can reduce the food waste, helps evaluating the contamination source,

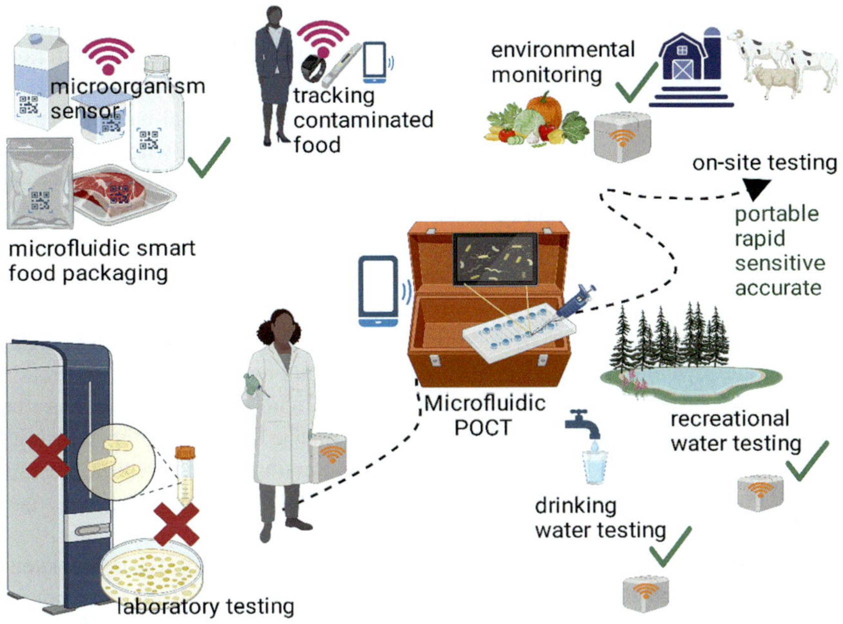

Fig. 10 Advanced food, water and environmental microbial safety monitoring (Figure created with Biorender.com)

and avoids selling contaminated/spoiled food to the customers before opening the package, thereby prevents food poisoning by taking immediate action.

2. Clinical testing: Microfluidic POCT has the potential for rapid and accurate diagnosis of diseases or conditions at the patient's bedside, in a doctor's office, or in remote locations where laboratory facilities are not available. POCT can significantly improve patient outcomes by facilitating prompt treatment and reducing the time and cost associated with traditional laboratory-based testing. POCT diagnostics offer various **bedside applications** including monitoring serum and plasma proteins, blood sugar, and reading the ECG of the patient. For instance, it can be useful to speed up the pre-analytical process by separating plasma from whole blood, volume metering, depletion of albumin, protein digestion with trypsin, and stabilization of tryptic peptides on solid-phase extraction sorbent, and to increase the performance of detection with the integration of different analytical systems such as mass spectrometry (MS). Gulquin et al. developed a microfluidic cartridge and an instrument providing fully automated fluid processing and thermal control, which conclude pre-analytical steps only within 2 h. The PepS microfluidic instrumentation has successfully demonstrated the quantification of entire pathological ranges for three protein biomarkers (alanine aminotransferase 1, C-reactive protein, and myoglobin) from blood samples (Gilquin et al. 2021). This system can be useful to streamline and simplify clinical proteomics studies for real-time monitoring health status.

Single-cell microfluidics are being developed for rapid and sensitive bacteria detection, along with testing the heterogeneity of populations, metabolism, migration, growth, and differentiation (Kaladharan et al. 2021; Deng et al. 2019; Jin et al. 2015). However, due to the small size of a bacterium, the current individual-bacterium trapping devices possess challenges for tracking individual bacterium with high resolution at large-trajectories. For example, the vast majority of rod shape bacteria has diameter of 0.5–1.5 μm, length of 1–6 μm and flagella length of 10–20 μm which allows rapid random walk (run and tumble) with speed of ~10–30 μm/s (varies depending on serotype and mutation), therefore difficult to monitoring. To tackle this issue, aside from development of **optical imaging** systems (camera, lens, CPU, software, etc.) miniaturization of microfluidic circuits plays a crucial role for single-trapping and dynamic phenotypic testing of individual bacterium. The integration of single-cell microfluidics, optical systems and **artificial intelligence** (AI) has the ability to analyze large datasets of individual bacteria in a large population with high accuracy. Therefore, the phenotypic and genotypic antibiotic susceptibility testing can be conducted within few hours which can be useful for the early diagnosis and treatment of drug-resistant infectious diseases (Fig. 11).

While the **wireless** in-vitro and in-vivo testing is still in its early stages, the potential benefits of **wearable** and implementable biosensors are significant for advanced clinical testing (Fig. 11) (Mannoor et al. 2012). In addition, small **implantable devices** with biodegradable properties can allow resorption of the implanted system and avoiding the need for surgical procedures to remove it. Therefore, it can be

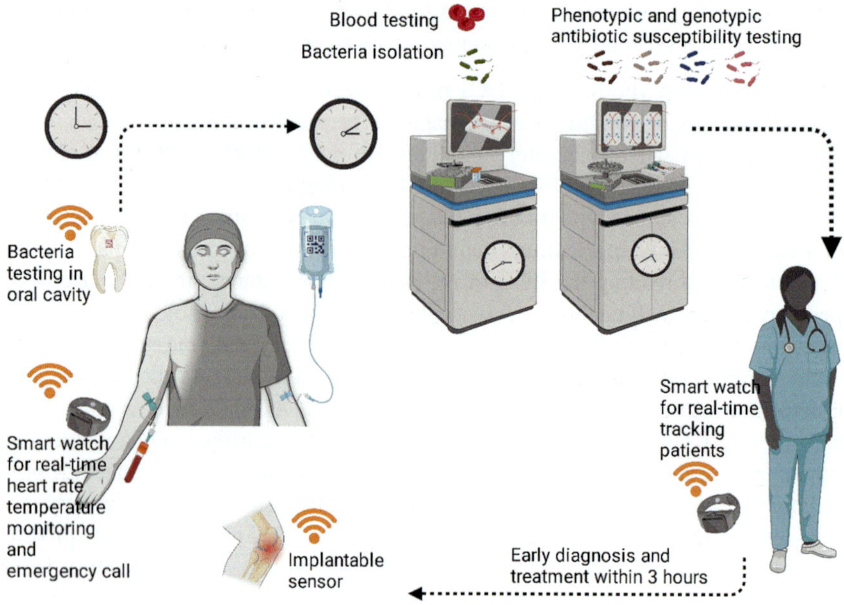

Fig. 11 Advanced microfluidic-based clinical testing (Figure created with BioRender.com)

useful for targeted and controlled antibiotic drug treatment of infected tissues. When treatment is over, the implanted system would begin to degrade in the body. On the other hand, to avoid degradation during the therapy, biodegradable materials should be covered by a protective coating layer, which will be degraded by exposure to enzymes in body fluids.

One potential application of implantable sensors is the **early diagnosis of rare diseases** such as **periprosthetic joint infection** (PJI), which is a severe complication that may occur after total joint replacement surgery. *Staphylococcus aureus* and *Staphylococcus epidermidis* are common bacteria that cause PJI primarily through its ability to adhere to implant surfaces, and produce biofilm. The diagnosis of PJI is difficult and requires supporting evidence from clinical examination, laboratory results, microbiological culture identification, histological interpretation, and intra-operative findings. While there is no gold standard method for PJI diagnosis, current PJI microbial testing methods incorporate tissue culturing, serum and synovial fluid testing rather than blood tests (Zainal Alam et al. 2018). A failure of standard cultures to identify the organism, thereby a delay to determine an appropriate drug to treat it, causes 50% of PJI cases to fail. Especially in the cardiac patients with valvular heart abnormalities, bacteria in the bloodstream may cause metastatic infections, including endocarditis, with the development of other symptoms it may progress into life-threatening conditions such as sepsis. Implantable microfluidic devices can be useful to facilitate real-time monitoring for the early diagnosis of PJI and bacteremia (Fig. 11).

3. Drug development: Microfluidic technology is promising to accelerate the drug discovery and efficiency of treatments for bacterial diseases (especially for the antibiotic drug resistant infections), as well as gut-brain axis related rare diseases. One treatment approach for drug resistant infection can be using antibiotics conjugate with aptamers, antibodies and small functional groups to specifically target the antibiotic to the pathogen, in order to improve the effectiveness of the antibiotic while minimizing the risk of effecting the body's natural microbiota. The use of targeted-antibiotic conjugates can improve the treatment, since the antibody component can help bypass bacterial mechanisms (such as efflux pumps or biofilms) that can otherwise prevent antibiotics from reaching their target (Cavaco et al. 2022). For instance, antimicrobial peptides (AMPs) are alternative therapeutics, possess numerous important properties, such as specificity, potency, low toxicity, biological diversity, and unique mechanisms of action (bacterial membrane and/or cytoplasmic), while not increasing bacterial resistance (Zhang et al. 2021; Datta and Roy 2021). In this context, microfluidic approaches can be used to create and optimize the conjugates in a controlled environment with high precision. While microfluidic droplet platforms can be useful to optimize the drug conjugate formulation, microfluidic organ-on-chip (generating a controlled physiological-like environment) and single-cell technology (dynamic monitoring individual bacteria) can facilitate drug testing, dynamic evaluation of drug accumulation on the bacterial membrane and identifying new drug targets. For instance, **lung-on-a-chip** devices can be useful to study the interaction of *Mycobacterium tuberculosis* and epithelial cells real-time, providing insights into the disease and potential therapeutic interventions.

To conclude, microfluidic technology can aid early diagnosis and treatment of bacterial diseases, as well as new drug discovery. At this time, there are no commercially available wearable or implantable microfluidic drug delivery systems for bacterial diseases. There has been ongoing research and development in this area as well.

References

An X, Zuo P, Ye B-C (2020) A single cell droplet microfluidic system for quantitative determination of food-borne pathogens. Talanta 209:120571

Van Assche D, Beneyton T, Baret J-C (2023) Rectifying jet breakup by electric forcing. Droplet e45

Avesar J et al (2017) Rapid phenotypic antimicrobial susceptibility testing using nanoliter arrays. Proc Natl Acad Sci 114(29):E5787–E5795

Azizi M et al (2018) Nanoliter-sized microchamber/microarray microfluidic platform for antibiotic susceptibility testing. Anal Chem 90(24):14137–14144

Björnmalm M, Yan Y, Caruso F (2014) Engineering and evaluating drug delivery particles in microfluidic devices. J Control Release 190:139–149

Borgdorff MW, Floyd K, Broekmans JF (2002) Interventions to reduce tuberculosis mortality and transmission in low-and middle-income countries. Bull World Health Org 80(3):217–227

Boulnois GJ (1992) Pneumococcal proteins and the pathogenesis of disease caused by Streptococcus pneumoniae. Microbiology 138(2):249–259

Bridle H (2014) Chapter ten—Miniaturized detection systems. In: Bridle H (ed) Waterborne pathogens. Academic Press, Amsterdam, pp 319–348

Cavaco M, Castanho MARB, Neves V (2022) The use of antibody-antibiotic conjugates to fight bacterial infections. Front Microbiol 13

Choi J et al (2014) A rapid antimicrobial susceptibility test based on single-cell morphological analysis. Sci Transl Med 6(267):267ra174

Datta S, Roy A (2021) Antimicrobial peptides as potential therapeutic agents: a review. Int J Pept Res Ther 27:555–577

Davey HM (2011) Life, death, and in-between: meanings and methods in microbiology. Appl Environ Microbiol 77(16):5571–5576

Deng B et al (2019) Microfluidic cell trapping for single-cell analysis. Micromachines (Basel) 10(6)

Dewachter L, Fauvart M, Michiels J (2019) Bacterial heterogeneity and antibiotic survival: understanding and combatting persistence and heteroresistance. Mol Cell 76(2):255–267

Ekici G, Dümen E (2019) Escherichia coli and food safety. In: The Universe of Escherichia coli. IntechOpen

Fernandes AC et al (2014) Lab-on-chip cytometry based on magnetoresistive sensors for bacteria detection in milk. Sensors (Basel, Switz) 14(8):15496–15524

Gambushe SM, Zishiri OT, El Zowalaty ME (2022) Review of Escherichia Coli O157: H7 prevalence, pathogenicity, heavy metal and antimicrobial resistance, African perspective. Infect Drug Resist 4645–4673

Gill WP et al (2009) A replication clock for Mycobacterium tuberculosis. Nat Med 15(2):211–214

Gilquin B et al (2021) PepS: an innovative microfluidic device for bedside whole blood processing before plasma proteomics analyses. Anal Chem 93(2):683–690

Gorzynski J et al (2022) Epidemiological analysis of Legionnaires' disease in Scotland: a genomic study. Lancet Microbe 3(11):e835–e845

Hajiani P, Larachi F (2012) Reducing Taylor dispersion in capillary laminar flows using magnetically excited nanoparticles: nanomixing mechanism for micro/nanoscale applications. Chem Eng J 203:492–498

Hassan S-U et al (2020) Capillary-driven flow microfluidics combined with smartphone detection: an emerging tool for point-of-care diagnostics. Diagnostics 10(8):509

Højris B et al (2016) A novel, optical, on-line bacteria sensor for monitoring drinking water quality. Sci Rep 6(1):23935

Hsu C-H et al (2008) Microvortex for focusing, guiding and sorting of particles. Lab Chip 8(12):2128–2134

Huang S et al (2015) Dynamic control and quantification of bacterial population dynamics in droplets. Biomaterials 61:239–245

Hunter R et al (2019) Optofluidic label-free SERS platform for rapid bacteria detection in serum. Sens Actuators B Chem 300:126907

Jeevanandam J, Chan YS, Danquah MK (2016) Nano-formulations of drugs: recent developments, impact and challenges. Biochimie 128–129:99–112

Jin D et al (2015) A microfluidic device enabling high-efficiency single cell trapping. Biomicrofluidics 9(1):014101

Jing W et al (2013) Microfluidic device for efficient airborne bacteria capture and enrichment. Anal Chem 85(10):5255–5262

Kaladharan K et al (2021) Microfluidic based physical approaches towards single-cell intracellular delivery and analysis. Micromachines (Basel) 12(6)

Khan ZA, Siddiqui MF, Park S (2019) Current and emerging methods of antibiotic susceptibility testing. Diagnostics (Basel, Switz) 9(2):49

Liu J et al (2019) Point-of-care testing based on smartphone: the current state-of-the-art (2017–2018). Biosens Bioelectron 132:17–37

Lv D et al (2022) Trapping and releasing of single microparticles and cells in a microfluidic chip. Electrophoresis

Mannoor MS et al (2012) Graphene-based wireless bacteria detection on tooth enamel. Nat Commun 3(1):763

Nelson MT et al (2021) Characterization of an engineered live bacterial therapeutic for the treatment of phenylketonuria in a human gut-on-a-chip. Nat Commun 12(1):2805

Nunes JK et al (2013) Dripping and jetting in microfluidic multiphase flows applied to particle and fiber synthesis. J Phys D Appl Phys 46(11)

Organization WH (2008) Typhoid vaccines: WHO position paper. Weekly Epidemiol Record = Relevé épidémiologique hebdomadair 83(06):49–59

Organization WH (2021) WHO operational handbook on tuberculosis: module 3: diagnosis: rapid diagnostics for tuberculosis detection

Pacocha N et al (2022) You will know by its tail: a method for quantification of heterogeneity of bacterial populations using single-cell MIC profiling. Lab Chip 22(22):4317–4326

Paone P, Cani PD (2020) Mucus barrier, mucins and gut microbiota: the expected slimy partners? Gut 69(12):2232–2243

Phu NH et al (2020) Concomitant bacteremia in adults with severe falciparum malaria. Clin Infect Dis 2020

Pilecky M et al (2019) Influence of antibiotic treatment on the detection of S. aureus in whole blood following pathogen enrichment. BMC Microbiol 19(1):180

Rhodes A et al (2017) Surviving sepsis campaign: international guidelines for management of sepsis and septic shock: 2016. Intens Care Med 43(3):304–377

Rodoplu D et al (2021a) A simple magnetic-assisted microfluidic method for rapid detection and phenotypic characterization of ultralow concentrations of bacteria. Talanta 230:122291

Rodoplu D et al (2021b) A micro-pupil device for point-of-care detection of viable Escherichia coli in tap water. SSRN Electron J

Rodoplu D et al (2022a) A micro-pupil device for point-of-care testing of viable Escherichia coli in tap water. Microchem J 178:107390

Rodoplu D, Matahum JS, Hsu C-H (2022b) A microfluidic hanging drop-based spheroid co-culture platform for probing tumor angiogenesis. Lab Chip 22(7):1275–1285

Rodoplu Solovchuk D et al (2023)A simple gradient centrifugation method for bacteria detection in skim milk. Microchem J 189:108479

Sack RB, Siddique AK (1998) Corpses and the spread of cholera. Lancet 352(9140):1570

Scheler O et al (2020) Droplet-based digital antibiotic susceptibility screen reveals single-cell clonal heteroresistance in an isogenic bacterial population. Sci Rep 10(1):3282

Seemann R et al (2011) Droplet based microfluidics. Rep Prog Phys 75(1):016601

Soda EA et al (2017) Vital signs: health care-associated Legionnaires' disease surveillance data from 20 states and a large metropolitan area—United States, 2015. Elsevier, pp 2215–2220

Spatola Rossi C et al (2023) Microfluidics for rapid detection of live pathogens. Adv Funct Mater 2212081

Srisa-Art M et al (2018) Highly sensitive detection of Salmonella typhimurium using a colorimetric paper-based analytical device coupled with immunomagnetic separation. Anal Chem 90(1):1035–1043

Sun J et al (2021) A novel concentration gradient microfluidic chip for high-throughput antibiotic susceptibility testing of bacteria. Anal Bioanal Chem 413(4):1127–1136

Sweeney TE, Liesenfeld O, May L (2019) Diagnosis of bacterial sepsis: why are tests for bacteremia not sufficient? Expert Rev Mol Diagn 19(11):959–962

Sy K et al (2023) Label-free and culture-free quantification of lactic acid bacteria using ring-shaped interdigitated electrodes. Sens Actuators B Chem 378:133160

Tok S et al (2019) Early detection of E. coli and total coliform using an automated, colorimetric and fluorometric fiber optics-based device. Lab Chip 19(17):2925–2935

VanDemark M (2013) Acute bacterial meningitis: current review and treatment update. Crit Care Nurs Clin North Am 25(3):351–361

Viri V et al (2020) An in vivo microfluidic study of bacterial transit in C. elegans nematodes. Lab Chip 20(15):2696–2708

Wang S et al (2019) A microfluidic biosensor for online and sensitive detection of Salmonella typhimurium using fluorescence labeling and smartphone video processing. Biosens Bioelectron 140:111333

Wolter A, Niessner R, Seidel M (2008) Detection of Escherichia coli O157:H7, Salmonella typhimurium, and Legionella pneumophila in water using a flow-through chemiluminescence microarray readout system. Anal Chem 80(15):5854–5863

Yeh C-F et al (2022) A portable controllable compressive stress device to monitor human breast cancer cell protrusions at single-cell resolution. Front Bioeng Biotechnol 10

Yuan L et al (2020) Visualization of bacterial colonization and cellular layers in a gut-on-a-chip system using optical coherence tomography. Microsc Microanal 26(6):1211–1219

Zainal Alam MNH et al (2018) A portable sensor for cell optical density measurement in microfluidic chips. Meas Control 51(7–8):213–222

Zhang Q-Y et al (2021) Antimicrobial peptides: mechanism of action, activity and clinical potential. Mil Med Res 8:1–25

Zhang J et al (2023) Electrokinetics in antimicrobial resistance analysis: a review. Electrophoresis 44(1–2):323–336

Zhang D, Qiao L (2022) Intestine-on-a-chip for intestinal disease study and pharmacological research. View 2022:20220037

Zhu H, Sikora U, Ozcan A (2012) Quantum dot enabled detection of Escherichia coli using a cell-phone. Analyst 137(11):2541–2544

Cytokine Response to Nanoparticles Bearing Nucleic Acid Cargo

Daniel Nisakar Meenakshi Sundaram, Mohammad Nasrullah, and Hasan Uludag

Abstract Non-viral gene delivery systems have proved to be promising for treatment of various cancers and other diseases. They are actively explored in clinical setting to deliver a range of nucleic acids to undertake diverse therapeutic actions, including the expression of therapeutic proteins using plasmid DNA, thereby restoring the defective protein, or to knockdown aberrant proteins using RNA interference (RNAi) with short interfering RNA (siRNA). The success of such delivery systems depends on various factors such as their cationic charge affecting the binding and complexation with the nucleic acids, their ability to protect against nuclease degradation, the uptake efficiency and the localized delivery inside the cell, the intracellular dissociation ability, and the properties to escape endosomal compartment. Most importantly, the ability to remain 'stealth' without causing any adverse reactions in a biological system is paramount irrespective of specific application. The cytokine response to the applied therapeutic agent is a critical component in determining the success of intervention. In this chapter, we will summarize the most recent literature on cytokine response to nanoparticulate delivery systems for nucleic acids, emphasizing the factors responsible for cytokine response. Structure–function studies will be emphasized and the implication(s) of cytokine response will be discussed in the context of specific applications. Representative medical conditions where the nanoparticles are used to control the cytokine profiles will be particularly probed in pursuit of new interventions related to inflammatory disorders. It is the goal of this

D. N. M. Sundaram and M. Nasrullah contributed equally to this chapter.

D. N. M. Sundaram · M. Nasrullah · H. Uludag (✉)
Department of Chemical and Materials Engineering, Faculty of Engineering, University of Alberta, Edmonton, AB, Canada
e-mail: huludag@ualberta.ca

M. Nasrullah · H. Uludag
Faculty of Pharmacy and Pharmaceutical Sciences, University of Alberta, Edmonton, AB, Canada

H. Uludag
Department of Biomedical Engineering, Faculty of Medicine and Dentistry, University of Alberta, Edmonton, AB, Canada

T. S. Santra and A. U. S. Shinde (eds.), *Advanced Drug Delivery*,
Studies in Mechanobiology, Tissue Engineering and Biomaterials 26,
https://doi.org/10.1007/978-981-99-6564-9_10

chapter to familiarize the reader with issues related to cytokine elicitation of non-viral nanoparticles as well as intervention strategies to address cytokine-mediated disorders.

1 Introduction

Cytokines are a class of proteins with essential functions in immunity, and other physiological processes such as cellular proliferation, differentiation, apoptosis and inflammatory reactions. They are produced and secreted by various cells in the body, including immune cells and endothelial cells in response to a variety of physiological events and pathological processes. They act as chemical messengers, which bind to specific receptors on the surface of target cells, triggering a range of downstream signaling pathways, and facilitating communications between cells of the immune system, as well as other cells in the body. Cytokines can be broadly classified into different categories, including growth factors (GFs), interleukins (ILs), interferons (IFNs), tumor necrosis factors (TNFs), and chemokines (Deckers et al. 2023).

Abnormal cytokine levels can trigger a pro-inflammatory or anti-inflammatory response, respectively, as they have been associated with a wide range of health conditions. Excessive secretion of certain cytokines has been linked to the development of autoimmune diseases such as rheumatoid arthritis, lupus, and multiple sclerosis, in which the body's immune system erroneously targets healthy cells and tissues (Moudgil and Choubey 2011). Another potentially fatal disease, sepsis, is induced by an overproduction of cytokines in response to infections. The physiology of the cancer cells can be also regulated by cytokines, depending on the type of cytokine and the stage of the disease. Certain cytokines, such as IL-15 and M-CSF may promote tumorigenesis, whereas others, such as IFN-γ, may inhibit it (Dranoff 2004). Cytokines can typically display multiple roles, such as TNF-α and IFN-γ, which can either promote tumor growth or suppress it (Wang and Lin 2008). Cytokine alterations are also the root-cause of certain chronic inflammatory disorders including ulcerative colitis, Crohn's disease, and inflammatory bowel disease. Allergens and asthma can be triggered by harmless agents such as pollen, dust, or food that alter cytokine levels. Moreover, altered cytokine levels may govern emotional and behavioral states. Depression and other mood disorders have been linked to elevated levels of pro-inflammatory cytokines (Raison et al. 2006). Therefore, it is crucial to explore how cytokines contribute to illness and develop therapies that specifically target these proteins to achieve better health outcomes for patients with altered cytokine regulations.

Nucleic acid-based therapies utilize a range of endogenous and altered nucleic acids derived from DNA and RNA to target specific genes and modify their expression or activity, which can lead to therapeutic benefits. A handful of nucleic acid therapies have gained approval and a large number of nucleic acids are now undergoing clinical testing and regulatory approval. They offer several advantages over conventional drugs (i.e., small organic molecules) and provide activities that cannot be matched by conventional drugs. They can be designed to be highly specific, avoiding off-target effects that can occur with conventional drugs. These therapies have the potential to treat diseases at the genetic level by altering the root-cause of the disease with long-lasting effects, rather than just suppressing symptoms, and allowing them for targeted and precise treatments (Kulkarni et al. 2021). Nucleic acid-based therapies derived from plasmid DNA (pDNA), messenger RNA (mRNA), small interfering RNA (siRNA), and antisense oligonucleotides (ASOs), have shown promise in treating a range of diseases, including those that are currently considered incurable. For example, FDA has recently approved an siRNA-based drug, Vutrisiran, to treat a rare genetic disease, polyneuropathy of Hereditary transthyretin-mediated amyloidosis (Mullard 2022). In 2020, the FDA granted emergency use authorization for an mRNA-based COVID-19 vaccine, which decreased the global death toll during the pandemic (Watson et al. 2022).

Nucleic acid-based therapies are well tolerated with low toxicity, which minimizes their risk of adverse reactions. With the help of recent technological advancements, nucleic acid-based therapies have become progressively feasible in terms of both production and accessibility, which has resulted an increased number of successful clinical trials and regulatory approvals. Although they are a promising avenue for future treatments, they have a few limitations, which include delivery to target cells, off-target effects, and avoiding immune system reactions (Walters et al. 2021). It is considered that the main obstacle to their success is getting the nucleic acids into the specifically targeted cells in a stable way (Uludag et al. 2019). There is continuous progress in the explorations of safe and efficient delivery means of nucleic acids, which is expected to ultimately change the medical landscape significantly and provide benefits over conventional pharmaceuticals.

Nanoparticles are being developed as a mean to deliver different nucleic acid-based drugs. This approach has shown a great potential as targeted therapy for various diseases including viral infections and cancer; however, the success of nanoparticle approach depends on the ability of nanoparticles to efficiently reach the target cells and release their payload without inducing toxicities. Widely used nucleic acid-delivering nanoparticles are branched polymers, polymeric dendrimers, ionizable lipids lipid nanoparticles (Paunovska et al. 2022), and more recently exosomes (El-Andaloussi et al. 2012). They all are non-viral in nature, but they could still interact with the immune system and subsequently trigger immune responses that can lead to adverse effects. Cytokine response is considered as one of the key immune response to nanoparticles (Zolnik et al. 2010).

When nanoparticles bearing nucleic acids enter the body, they can be recognized by the immune system as foreign particles (along the same lines as viruses) and which in turn activate immune cells like monocytes, macrophages, and dendritic cells to release defensive cytokines. Several factors, including hydrodynamic size, shape, and surface charge of nanoparticles, nucleic acid cargo, route of administration, presence of co-stimulators, and patient's genetic profile can influence the nature and the intensity of the cytokine response (Gonçalves et al. 2020; Bila et al. 2021; Gu et al. 2021). For instance, large nanoparticles with complex surface chemistry tend to induce stronger cytokine responses than those with smaller and simpler surface chemistry (Elsabahy and Wooley 2013; Weiss et al. 2021). Nanoparticle-induced cytokine responses can be detrimental and may cause life-threatening cytokine storm if triggering in an uncontrolled way. On the other hand, they can be beneficial by enhancing the immune response against tumors or pathogens, which leads to better therapeutic outcomes.

Understanding the cytokine response to nanoparticles is essential to investigate the safety and efficacy of nucleic acid-based drugs. In this report, we cover the latest developments in nanoparticle-based non-viral delivery systems, provide insights into the ideas behind their formulation, highlight technological features that facilitated their clinical translation, and bring examples of regulatory-approved drugs in clinical practice. We also discuss how these methods pave the way to innovate novel nanoparticle-delivered targeted medicines with improved formulations and minimal adverse effects.

2 Nanoparticle-Based Nucleic Acid Delivery Systems

A diverse range of biomaterials of both natural and synthetic origin has been developed to fabricate nanoparticles suitable for nucleic acid delivery (Fig. 1). Below is a short synopsis on different classes of materials employed in nanoparticle fabrication.

2.1 Peptides and Proteins

Peptides and proteins are natural or synthetic biomolecules composed of amino acids that can be engineered to form nanoparticles with nucleic acids cargo. They can facilitate nucleic acid delivery into cells through various uptake mechanisms, including cell membrane penetration and receptor-mediated internalization, and further binding to nucleic acids to protect them from degradation in the extracellular environment. Among peptides and proteins, cell-penetrating peptides (CPPs), antibodies, and viral proteins are widely used for delivering nucleic acids. The CPPs are short cationic peptides that can bind to anionic nucleic acids to facilitate their neutralization and penetrate through the cell membrane (Hoyer and Neundorf 2012). Examples of CPPs include HIV-1 transactivator of transcription (TAT) peptide, transportan, Antennapedia homeodomain protein (Antp), and penetratin. TAT peptide can enter cells because of its elevated arginine and lysine content, which helps it to bind to the anionic components of cell membranes as well as the nucleic acid. TAT peptide has been successfully applied to delivery of mRNA, siRNA and pDNA, among other nucleic acids (Piecyk et al. 2020). CPP can function on their own in delivery, as well as a part with other macromolecular carriers once they are incorporated into them (Nori et al. 2003; Langel 2021).

Antibodies can also be employed to transport nucleic acids into cells by targeting certain cell surface receptors that undergo cellular internalization. For instance, the

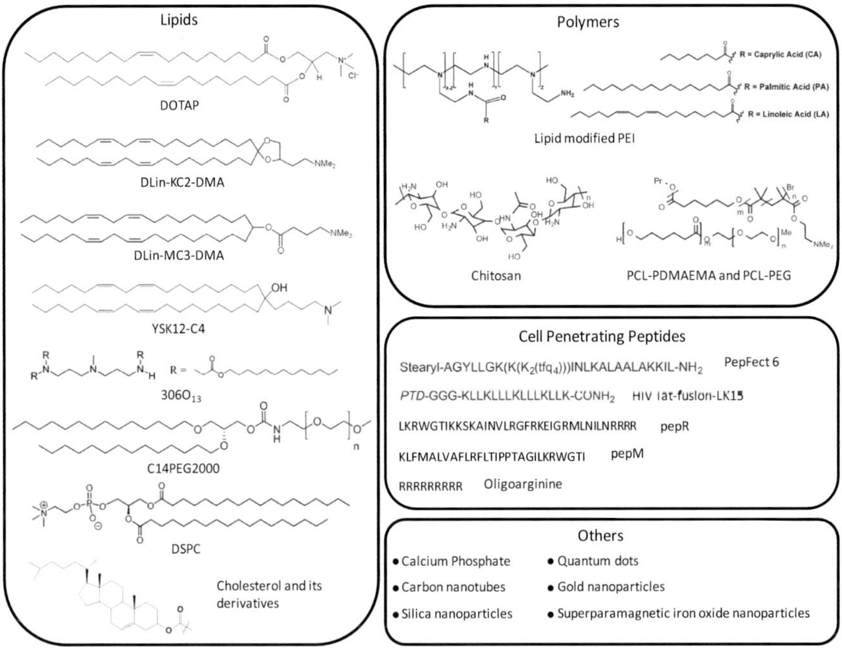

Fig. 1 Chemical structures of different classes of building blocks used for nanoparticles design in nucleic acid delivery (Ansari et al. 2017)

Her2 antibody has been fused with the siRNA for targeted delivery to Her2-positive breast cancer cells (Dou et al. 2012). Selected viral proteins can similarly bind to nucleic acids and transport them across the cell membrane. VP16 and VP22 protein display a high capacity for endosomal escape critical for successful nucleic acid delivery. Additionally, synthetic cationic polypeptides, typically composed of lysine and arginine moieties, can function along the same lines and increase nucleic acid uptake into cells (Li et al. 2022).

2.2 *Dendrimer*

Dendrimers are synthetic, highly branched, nanoscale polymeric macromolecules that have been explored for delivery of nucleic acids into cells. They have a defined three-dimensional globular shape with a core, a shell, and an outer surface that can be tailored with various functional groups including chemical groups (e.g., amino, carboxylic and hydroxyl) and targeting moieties (e.g., antibodies) to optimize delivery. They have the capacity to protect nucleic acids from nuclease degradation and increase their stability in the extracellular environment, in addition to facilitating nucleic acid transport into cells via electrostatic contact, covalent bonding, hydrogen bonding, and hydrophobic interactions. Common examples of dendrimer-based nucleic acid delivery systems include polyamidoamine (PAMAM), polypropyleneimine (PPI), polyethyleneimine (PEI), and polylysine dendrimers. The prototypical PAMAM dendrimers possess a high number of amine groups on their surface for electrostatic interactions with nucleic acids. The resulting complexes can enter cells through endocytosis and release the cargo into the cytoplasm through endosomal escape (Abedi-Gaballu et al. 2018). PPI dendrimers also function in the similar way, forming stable complexes that get transported across the cell membrane and released from endosomes. After surface modification with folate receptors, siRNA can be delivered into cancer cells that overexpress the folate receptor using dendrimers functionalized with folic acid (Xu et al. 2017).

2.3 Liposomes and Lipid Nanoparticles (LNPs)

Lipid nanoparticles (LNPs) are composed of a lipid bilayer surrounding a hydrophilic nucleic acid payload for their delivery into different cells. They have emerged as a promising delivery system for particularly mRNA-based therapeutics by protecting the payload from degradation. Their adaptability, biocompatibility, and high encapsulation efficiency have huge implications for the future of genetic medicine. When LNPs come into contact with a cell membrane, they can fuse with it and then release their cargo into the cytoplasm directly. Alternatively, they can facilitate endosomal escape with functional components (so called ionizable lipids) following the endocytosis. Antibodies and peptides that bind to receptors on the surface of desired cells can be incorporated into the LNPs to enhance the efficiency and precision of nucleic acid delivery.

The widely used LNP-based nucleic acid delivery system consists of a mixture of four lipid components: a neutral lipid (functions in nucleic acid entrapment and membrane fusion. e.g. DSPC, DPPC), cholesterol (helps in structure integrity and endosomal release), PEGylated lipid (helps to create a hydrophilic surface. e.g. DMG-PEG, DSPE-PEG-Maleimide, and DSPE-PEG), and an ionizable lipid (enables membrane fusion and endosomal escape. e.g. DLin MC3-DMA, OF-Deg-Lin, DDA, DOTAP, ALC-0315, A6, OF-02, 9A1P9, 7C1, G0-C14, L319, 306-O12B, FTT5, 4A3-SC8, and C12-200) (Verbeke et al. 2019; Wang et al. 2023). The formulation is normally prepared by microfluidic mixing method but can be also conveniently prepared in small vails for research scale. Both Moderna, and Pfizer-BioNTech COVID-19 vaccines were developed by using LNPs to encapsulate mRNA encoding the spike protein of SARS-CoV-2; Moderna used ALC-0315 lipid and Pfizer-BioNTech used SM-012 lipid in their products (Han et al. 2021). Another FDA approved drug for hereditary transthyretin-mediated amyloidosis, Patisiran, also developed by using LNP, particularly DLin MC3-DMA lipid to encapsulate siRNA that targets and silences the mutated transthyretin gene (Wood 2018). LNPs have also been used to deliver other types of nucleic acids, like ASOs.

2.4 Polymeric Systems

Polymeric systems with long chains of repeating units are another delivery platform composed of synthetic or natural biocompatible polymers. They can form complexes with nucleic acids through different mechanisms such as electrostatic interactions, covalent bonding, and hydrophobic interactions, and protect them from degradation and facilitating their delivery (Shen et al. 2017). The most widely used polymeric system for nucleic acid delivery is polyethylenimine (PEI) derived polymers, which are cationic polymers of various architectures and sizes, and have been formulated into various nanoparticles (Fig. 2). The complexes and nanoparticles formed from PEIs are able to enter cells via endocytosis and escape the endosomal via the "proton sponge" effect, in which the cationic PEI absorbs H^+ and produces osmotic swelling of the endosomes, resulting in rupture and discharge of the nucleic acids into the cytoplasm (Pei and Buyanova 2019). PEI can be conjugated with different types of lipids to improve the delivery efficiency and reducing unwanted cytotoxicity.

Poly(lactic-co-glycolic acid) (PLGA) is another example of biodegradable polymer that have been utilized to deliver nucleic acids; they are usually used as an encapsulation medium to form nanoparticles (not directly interacting with anionic nucleic acids) and can be designed to target specific tissues by modifying the surface

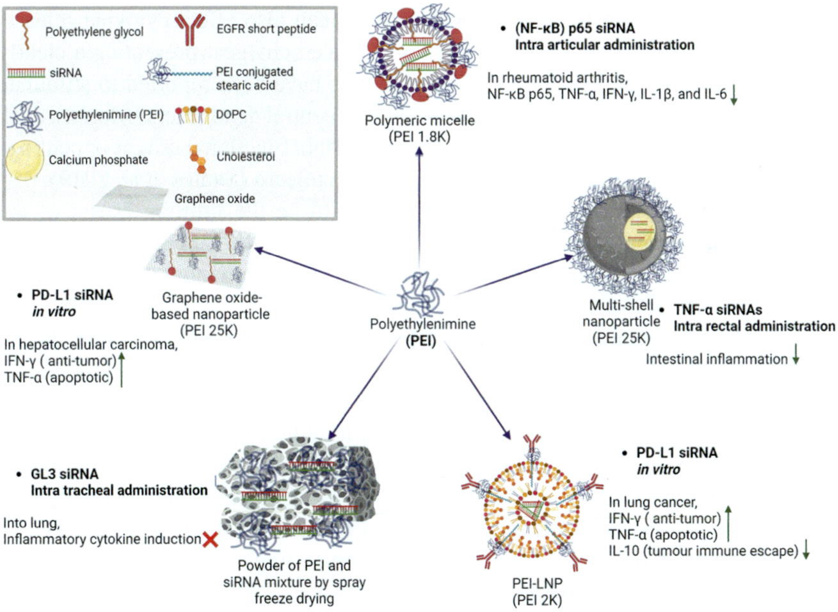

Fig. 2 Utilization of polyethylenimine (PEI) in the formulation of nucleic acid-bearing nanoparticles. PEI is employed in a wide variety of nanoparticle structures, from polymeric micelles to multi-shell nanoparticles to lipid nanoparticles (LNPs), and even solid powders, depending on their molecular weights

of PLGA nanoparticles with targeting ligands. Another polymers used for delivery is the naturally-occurring chitosan, whose endocytosed complexes can be released into the cells through endosomal escape mechanisms (Cao et al. 2019). Often times copolymers are also designed to utilize the beneficial effect of each component, such as the one with PEI and PEG copolymers used in murine lung cell lines (Beyerle et al. 2010). An advantage of the polymeric system is the ability to incorporate units that respond to 'stimuli', such as pH/temperature. Poly(N-isopropylacrylamide), or PNIPAM is a prototypical example that respond to temperature changes, by remaining soluble at low temperature but getting at body temperature to encapsulate and release the nucleic acids (Ansari et al. 2022).

2.5 Limitations of Delivery Systems

The ability of a host to mount an effective immune response and cytokine release in response of administering nucleic acids with various delivery platforms has been appreciated. Anti-inflammatory drug regimens are commonly utilized with nucleic acid therapeutics in clinical setting to minimize potential inflammatory reactions (Zhang et al. 2021b). With peptide-based carriers, the resultant size and charge of the carrier can significantly influence the cytokine release (Tsai et al. 2020). The type of dendrimer used for nucleic acid delivery can also affect cytokine release, especially with their high density of regularly spaced charges (presenting a clearly artificial surface to the host). Cationic dendrimers have been reported to stimulate cytokine release more strongly than the anionic or neutral dendrimers (Moreira et al. 2023). Surface modification of dendrimers with certain functional groups or coatings can reduce immunogenicity and decrease cytokine release (Santos et al. 2019).

The cytotoxicity and immunogenicity appear to be the limiting aspect of the dendrimer platform as well (Palmerston Mendes et al. 2017). With LNPs, special lipid compositions are being developed to minimize the immunogenicity (Tahtinen et al. 2022). The size of LNPs can also impact cytokine release; smaller LNPs have been reported to induce less cytokine release than larger LNPs, possibly due to differences in cellular uptake or intracellular processing (Hassett et al. 2021). Cationic polymeric carriers, analogues to dendrimers, can also elicit significant inflammatory cytokines and care is needed in polymer design to reduce reactive species (Beyerle et al. 2010), beyond simple PEGylation. An attempt to deliver non-silencing siRNA into BV2 microglia cells using 3 different carriers found significant stimulation of cytokines. A cationic liposome formulation (3:1 DOSPA: DOPE) stimulated TNF-α and TLR2 expression, while cationic non-liposomal amphiphile triggered TNF-α, TLR2 and IL-1b, and PAMAM dendrimer augmented TNF-α, TLR2, COX-2 and IL-1b expression. Clearly, cytokine response will be an issue with all of the non-viral delivery systems (Godinho et al. 2014).

3 Prediction of Immune Response

Elicitation of an immune response by a therapeutic agent following administration is undesirable, as it a major safety concern for patients and can negatively impact the treatment outcome. Therefore, it is imperative to estimate and/or predict the immunogenicity of the agent and follow the stringent safety guidelines structured by the regulatory bodies for the overall wellbeing of the patients. This section outlines one of the commonly used methods of immune response prediction, cytokine quantification which acts as an indicator for immunogenicity. The special focus will be on (i) non-viral systems discussed above, and (ii) clinical translation of the outcomes. We also discuss various factors that could govern this response and act as predictive factors, such as the dose, administration route and patient-to-patient variation in response to non-viral gene delivery systems.

3.1 *Investigation of Cytokine Release*

The entry of any foreign species into a host can activate the immune system through a cascade of signaling pathways, which can be detected through the changes in cytokine levels. Evaluating different cytokines and their pharmacokinetics profile provide a direct and immediate safety of the therapeutic agent. This can be accomplished through ELISA (enzyme linked immunosorbent assay), PCR (polymerase chain reaction) and cytometric bead arrays (CBA), as some of the common techniques used for the detection and quantification of cytokine levels.

i. In Vitro and In Vivo Outcomes with Various Delivery Systems

The immunogenicity of various CPPs like TP10 (Transportan 10), PF (PepFects) 3, PF4, PF6, HIV-TAT and stearyl-$(RxR)_4$ were evaluated by quantification of IL-1β, IL-18 and TNF-α levels in vitro and in vivo using ELISA. Minimal cytokine stimulation was recorded from THP-1 cell line, PBMC and mice serum with levels ranging between 5 and 50 pg/mL, showcasing the non-immunogenic nature of these peptides (Suhorutsenko et al. 2011). Similarly, another study employing HIV-TAT, Antennapedia and Transportan CPPs was carried out looking at IL-8 and IL-6 levels from various epithelial cells like A431, A549 and Caco-2. Cytokine levels were equivalent to the control treatments thereby demonstrating the safety of these CPPs and the failure to induce an immune response in vitro (Carter et al. 2013).

Among the dendrimer, the PAMAM dendrimers of G-4, G-5, and G-6 generations were tested for immunogenicity in J774A.1 mouse macrophage cells by quantification of IL-6, TNF-α and MIP-2 (macrophage inflammatory protein-2) using ELISA. Elevated levels of cytokines ranging between 100 and 450 pg/mL were reported with strong correlation towards the PAMAM generation, treatment duration and concentration, laying an important finding for therapeutic use of PAMAM (Naha et al. 2010). Likewise, G-2 and G-3 PAMAM were utilized to study the stimulation of various cytokines (IL-1β, IL-6, IL-8, IL-10, IL-12p70 and TNF) using CBA assay following exposure to human keratinocytes and fibroblast cells. A gradual and significant upregulation was recorded in IL-1β, IL-6, and IL-8 levels which were concentration dependent as in the previous report (Czarnomysy et al. 2019).

With lipid carriers, a sphingosine:cholesterol:DAPC combination having either lactosylceramide or PEG-ceramide coating were used for pDNA delivery in vivo through intravenous injection and the levels of TNF-α, IFN-γ and IL-6 were quantitated. The pDNA used in this study were specifically designed to carry minimal CpG sequences to reduce the immunogenicity, as high CpG contents were shown to be highly immunogenic (Bonnet et al. 2008; Cheng et al. 2018). Upregulation in all three cytokine levels ranging between 10^2 and 10^5 pg/mL were recorded in blood, liver, spleen, lung, and tumor (murine colon carcinoma) through ELISA. They were able to establish an important finding on the role of repeated administration of sphingosine:cholesterol:DAPC over immune response and gene expression (Betker and Anchordoquy 2022). Similarly, the effect of gadolinium chloride ($GdCl_3$) pre-treatment on the cytokine induction (TNF-α, IFN-γ and IL-12) and biodistribution of intravenously-injected pDNA-cationic liposome (DOTMA/Chol liposome) complexes were established. Cytokine levels were quantified in various organs and significant decreases of threefold to 25-fold were observed in blood, liver, spleen, and kidney with $GdCl_3$ pre-treatment, providing a significant contribution towards the development of more safer gene delivery vectors (Sakurai et al. 2002).

Among polymeric gene delivery systems, various low molecular weight (0.6, 1.2 and 2 kDa), branched PEIs were evaluated for TNF-α, IFN-γ and IL-6 secretion from PBMCs using ELISA and RT-PCR along with PMA/IO (phorbol 12-myristate 13-acetate/ionomycin) as positive controls. Cytokine levels in culture remained <2 pg/ml for all the unmodified PEI polymers irrespective of the molecular weight, while the controls reached a peak ranging between 400 and 2000 pg/ml providing a better understanding on the safety of PEI polymers and the potential to utilize them for specific therapeutic use (Meenakshi Sundaram et al. 2022). Similarly, the immunogenicity of high molecular weight (22 kDa) PEIs was examined in vivo for various cytokines such as TNF-α, IFN-γ, IL-6, IL-12/IL-23, IL-1β and IFN-β. Mice blood cytokine levels ranged between ~4 and ~10^3 pg/mL and this study helped to highlight the importance of following good manufacturing practices to ensure the safety of PEI non-viral delivery agents, as different batches of polymers showed different levels of cytokine response in both studies (Bonnet et al. 2008).

ii. Clinical Outcomes

Numerous other in vitro/in vivo studies employing cytokine quantitation as a measure of immune response can be found; however, clinical translation and relevance of these finding is important to assess. Here, we discuss clinical studies related to cytokine response and its correlation towards the safety as well as the outcomes of the treatment including viral gene carriers, given the high number of clinical trials based on viral vectors.

The normal levels of cytokines (IL-2, -7, -8, -9, -10, -12 and -16) in healthy individuals were recorded to be in the range of ~10 to ~40 pg/mL irrespective of the age groups (1–6, 7–17 and ≥18 years; n = 72). However, IL-4, IL-6, IL-13, TNF-α and IFN-γ were slightly upregulated for 7–17-year-old with levels ranging between ~10 and ~200 pg/mL, as measured using a magnetic bead-based multiplex immunoassay (Kleiner et al. 2013). Such low cytokines levels were also recorded by another study involving 112 healthy individuals, where the cytokines IL-1β, -2, -6, -10, -15, -17, 30, TNF-α and IFN-γ levels ranged between ~2 and ~30 pg/mL. However, a slight upregulation could be seen for ages ≥65 years compared to individuals <45 years old (Kim et al. 2011). It was interesting to note that some of the earlier studies have reported an age-related upregulation in IL-6 which increased 0.016 pg/mL per year as observed between 20 and 90 years old individuals (n = 59) (Hager et al. 1994). Similarly, along with IL-6, TNF-α was also seen to increase from the age group of 18–30 and 55–65 years old (n = 62) (Brüünsgaard and Pedersen 2003). Various other cytokines like IL-4, IL-5, IL-8, and IL-10 also increased (n = 17) with age pointing out the age-dependent cytokine upregulation among heathy volunteers (Cruz-Almeida et al. 2015). Moreover, the recent studies with COVID-19 elderly patients revealed age related upregulation of key cytokines like IL-6 and IL-2R during disease progression and specific inhibitors were required to decreases its levels to improve patient recovery (Jing et al. 2022).

One of the early reports of inflammatory response syndrome (a sudden increase in cytokine levels) were recorded in the 90s following an adenovirus-based gene therapy infusion into the right hepatic artery for the treatment of ornithine transcarbamylase deficiency, ultimately ending in a causality. One patient (out of 19) exhibited peak levels of IL-6 (~4500 pg/mL) at 8 h post injection and ~6000 pg/mL of IL-10 at day 2 in the serum. While the IL-10 levels dropped over time, IL-6 levels remained high till the patient's passing. Some of the symptoms included fever, back pain, frontal headache, dizziness, tachycardia, altered mental status, thrombocytopenia, hypophosphatemia, and jaundice (Raper et al. 2003; Chatenoud et al. 1989). Although one other patient also exhibited high IL-6 secretion, the levels dropped helping in complete recovery. This pilot study highlighted the importance of the immune safety of gene therapy and the need to provide an overall safety assessment before the onset of clinical trials. Since then, clinical trials have included numerous safety restrictions and guidelines. Another study with an adenovirus vector (in 2009) carrying human hepatocyte growth factor (Ad-HGF) proved successful involving 21 patients with severe coronary artery disease. 11 patients were given intracoronary administration of Ad-HGF treatment with a stent, while 10 patients were provided with stent alone, acting as control groups. IL-4 levels remained the same for both treatment and control groups while IL-10 was upregulated between 6 and 24 h reaching a peak of 210 pg/mL (vs. 150 pg/mL in control) and later decreasing to control levels. This was one of the first studies to demonstrate the clinical translation of an adenovirus vector based HGF gene therapy and the possible regulation of cytokines, as it was hypothesized that IL-8 reduction combined with IL-10 upregulation could aid in achieving a positive outcome in the treatment of coronary heart disease (Yang et al. 2009).

Correction of cystic fibrosis using a nonviral approach was carried out with lipid-DNA complexes in an aerosolized form with the hypothesis of being safer than a viral vector. The lipid used was 67A (GL-67:DOPE:DMPE–PEG_{5000}). A highly purified plasmid DNA (pDNA) carrying the cystic fibrosis transmembrane conductance regulator (CFTR) gene was used for this purpose with 4 out of 8 patients experiencing fever, muscle, and joint aches along with myalgias and arthralgia in some of them. Serum IL-6 cytokine levels were upregulated in all the patients ranging between ~10 and ~50 pg/mL although no causality was reported, while IL-1, IL-8, TNFα and IFN-γ levels were unaltered. This study emphasized the possibility of immune stimulation even with non-viral cationic liposomes used with an endotoxin-free pDNA (Ruiz et al. 2001).

A polymeric nanoparticle, CALAA-01 consisting of (i) a linear cationic cyclodextrin-based polymer, (ii) a hydrophilic stabilizer adamantane-PEG and (iii) a targeting ligand towards human transferrin receptor was used to carry an siRNA against an anti-cancer target, ribonucleotide reductase M2 subunit (RRM2). The serum cytokine studies in monkey recorded an upregulation of IL-6, IL-12 and IFN-γ levels which correlated with patient clinical trials (n = 24). Serum levels of IL-2, IL-4, IL-6, IL-10, IL-12p40, IL-12p70, TNFα and IFN-γ were evaluated at various doses and time points. The levels peaked mostly between 2 and 6 h post injection with IL-6 and IL-10 reaching ~600 pg/mL, TNFα ~ 200 pg/mL and IFN-γ ~ 50 pg/mL, but the levels dropped back to normal/baseline in 24 h. It was evident that the polymeric CALAA-01 induced an inflammatory response but was tolerated to some extent by the patients. However, dose-limiting toxic events (DLTs) were observed causing toxicity in liver and kidney as recorded by the quantitation of liver enzymes, creatinine and, blood urea nitrogen (BUN) in animal studies and this in turn help to develop mitigation methods to avoid this in patients (Zuckerman et al. 2014).

3.2 Factors Influencing Cytokine Response

Having outlined few key clinical studies on serum cytokine levels with gene delivery systems, we will below summarize some of the important factors that have been shown to influence the outcomes.

(i) Dose: Dose is an important contributing factor to cytokine response as higher doses corelate with higher cytokine levels in various cell culture, animal models, and clinical studies. Although no major cytokine upregulation was reported with CPPs, a slight upregulation could be seen in TNF-α levels between 1 and 5 mg/kg doses of TP10, stearyl-$(RxR)_4$, PF3 and PF4 in vivo (Suhorutsenko et al. 2011). Similarly, the levels of IL-1β, IL-6 and IL-8 from keratinocytes and fibroblasts increased with increase in the dose of PAMAM from 0.3 mg/mL through 3.0 mg/mL irrespective of the dendrimer's generation (G2 and G3) (Czarnomysy et al. 2019). Additionally, the treatment of macrophage cells to PAMAM G-4, G-5, and G-6 lead to a concentration dependent increase of MIP-2, TNF-α and IL-6 cytokine levels (Naha et al. 2010).

Among 7 different LMW bPEI having lipid modifications, only one polymer showed increased IL-6 secretion at higher ratio/dose ~0 pg/mL versus <5 pg/mL at lower ratio/dose, and a similar trend was observed in TNF-α secretion of ~170 versus ~100 pg/mL of low versus high polymer:pDNA used used in that study. No such concentration or ratio dependent increase in cytokine was observed with other polymers. An increase in pDNA dose (0.5–2 μg) did not alter the cytokine levels from PBMC (Meenakshi Sundaram et al. 2022). Similarly, with 10 kDa bPEI administration, very minimal change was observed as an increase in polymer ratio caused the TNF-α levels to change from ~5 to ~30 pg/mL in vivo. The TNF-α levels were unaltered with 25 kDa bPEI and linear PEI among different ratios administered and interestingly an inverse relation was observed with 70 kDa PEI as the cytokine levels decreased with increasing ratio from 7 to 15 (~50 to ~10 pg/mL). Increase in pDNA dose from 30 to 80 μg increased the serum TNF-α levels from ~5 to ~40 pg/mL (Kawakami et al. 2006). Similar observations were reported with oligodeoxynucleotides having CpG motifs as the levels of IL-6 and IFN-α increased from ~500 to ~1250 pg/mL and ~80 to ~180 pg/mL, respectively, with increase in the concentration of the oligodeoxynucleotides (50–200 nM) delivered by a 1.8 kDa PEI (Cheng et al. 2018).

Such a dose-dependent increase in IL-6, IL-10, TNF-α and IFN-γ cytokine levels could also be seen with CALAA-01 polymeric nanoparticles, as patients treated with 10 mg/m^2 displayed negligible increase compared to baseline, however with 20–30 mg/m^2 dose, there was a maximum increase of 200-fold with IL-6, 100-fold with IL-10, 15-fold with TNF-α and 20-fold with IFN-γ (Zuckerman et al. 2014). This could also be applicable to viral gene delivery systems as adenovirus mediated transduction in vivo caused a dose-dependent increase of TNF-α (Zaiss et al. 2002). In the clinical trial of adenoviral gene transfer for the treatment of ornithine transcarbamylase deficiency, a dose-dependent increase in IL-6 and IL-10 levels could be seen. IL-6 levels increased from ~10 to ~10^5 pg/mL for 0.002 to 0.6 × 10^{12} particles/kg of vector and IL-10 from ~1 to ~10^5 pg/mL (Raper et al. 2003). Altogether, use of a low dose should be a key consideration irrespective of the therapy and should be implemented even at the earliest stages of discovery and validation studies in in vitro and in vivo bioassays. High dose of the treatment (and presumably the delivery system) not only activates the immune system, but also leads to various organ-related toxicities thereby outweighing the efficacy of the therapeutic treatment. It may require additional interventions (i.e., immunosuppression) to lower the detrimental effects.

(ii) Route of Administration: Some of the earliest studies on the influence of administration route on immune response was performed by using recombinant adeno-associated viral vectors (AAV) where a comparative study between different administration routes, such as intraperitoneal (IP), intravenous (IV), subcutaneous (SC), and intramuscular (IM), disclosed some interesting findings in female C57BL/6 mice for ovalbumin delivery. The immune response following IM injection showed the lowest response as quantified by the number of ovalbumin-specific antibodies, AAV antibodies and cytotoxic T-lymphocytes (CTL) compared to the other routes (Brockstedt et al. 1999). Contrasting effects were observed in numerous other studies where IM route triggered a high immune response compared to hepatic route of administration. Although there are contrasting observations in these studies, it is evident that the route of administration affects the immune response (Sun et al. 2003). Additional studies on the difference in immune response between the oral, IV and IM route of administration have been reported with AAV and discussed elsewhere (Sun et al. 2003; Shirley et al. 2020).

Few studies could be found with non-viral gene delivery systems and the role of the route of administration on immune response, however a comparative analysis among various non-viral systems such as liposomes, chitosan nanoparticles, and PLGA for vaccination was explored. Routes such as IM, intradermal (ID), intralymphatic (ILy), and SC were compared in BALB/c mice with a model antigen (ovalbumin). All systems showed an antibody response with ILy administration. ID and IM induced a moderate response, whereas the SC route did not elicit any response. Re-stimulation of the mice splenocytes (in vitro) was performed to better understand the cellular response based on IFN-γ, IL-4 and IL-10 cytokine secretion. With liposomes, the ILy route showed an elevated 300–600 pg/mL IFN-γ secretion compared to other routes SC, ID and IM (10–50 pg/mL). The chitosan nanoparticles and PLGA did not show any major difference between the route of administrations and had ~200 and ~180 pg/mL levels of IFN-γ respectively. For IL-4, only chitosan nanoparticle showed a difference between the routes, where IM had the lowest ~20 pg/mL, SC showed ~30 pg/mL, ID showed ~50 pg/mL and i.ln ~100 pg/mL showing the highest among them. IL-10 levels were unchanged irrespective of the system and the administration route (Mohanan et al. 2010). Although this was a vaccination study, the delivery systems used here are also employed for non-viral gene delivery, so that the findings can be extrapolated to gain an insight on the immunogenicity of the non-viral delivery systems as well as their capacity to act as an adjuvant system.

The IV route is most common administration route for gene therapy in the clinical setting. Some of the clinical trials with non-viral system that utilize IV route are shown in the Table 1 along with the studies that employed the intratumor (IT) route.

Selection of administration route is dependent on the type of therapeutic intervention (antibody-based vs. viral vectors vs. non-viral vectors, etc.), in addition to the disease or cancer type being targeted. The IT route would be more beneficial for solid tumors as the treatment effect would be confined to tumor sites without affecting the other organs, thereby avoiding unwanted side effects. More utilized IV route has the advantage of targeting metastatic tumors as it can reach various organs. Nevertheless, it is still preferred to understand and establish personalized immune response screening before the onset of treatment as patient-to-patient variations can result in different treatment outcomes and immune responses.

(iii) Patient-to-Patient Variation: Varying responses among patients undergoing gene therapy trials were observed in early studies with AV-based treatment of ornithine transcarbamylase deficiency in which 2 out of 19 patients in the study developed strong immune response, of which 1 female (F) recovered in few hours but not the other male (M) patient. One of the key hypotheses for the death of the M patient is

Table 1 Summary of select non-viral gene therapy clinical trials with specific drugs and delivery system used, the cancer type, route of administration (IV—intravenous and IT—intratumor), the phase of clinical trial with the number of participants in the trial and the corresponding clinical trial number

Drug name	Disease	Route	Phase	Clinical trial No
OTX-2002 LNP + mRNA	Hepatocellular carcinoma	IV	Phase I and II N = 190	NCT05497453
mRNA-NP vaccine DOTAP liposome + mRNA	Early melanoma	IV	Phase I N = 18	NCT05264974
Rexin-G Nanoparticle + Gene	Sarcoma	IV	Phase I and II N = 36	NCT00505713
INT-1B3 LNP + microRNA (miR-193a-3p)	Advanced solid cancer	IV	Phase I N = 80	NCT04675996
DOTAP:Chol-FUS1 LNP + Fus1 gene	Lung cancer	IV	Phase I N = 32	NCT00059605
TKM-080301 PLK1 siRNA	Advanced solid tumor	IV	Phase I N = 68	NCT01262235
EphA2-targeting DOPC-encapsulated siRNA	Advanced malignant solid neoplasm	IV	Phase I N = 76	NCT01591356
NBF-006 LNP + GSTP siRNA (Glutathione S-Transferase P)	Lung cancer	IV	Phase I/Ib N = 44	NCT03819387
Mesenchymal stromal cells-derived exosomes with KRAS G12D siRNA	Pancreatic cancer	IV	Phase I N = 28	NCT03608631
CAS3/SS3 CpG-STAT3 siRNA	B-cell non-hodgkin lymphoma	IT	Phase I N = 18	NCT04995536
TriMix Dendritic cell activating protein encoding mRNA	Breast cancer	IT	Phase I N = 36	NCT03788083
mRNA-2752 LNP + mRNA (OX40L, IL-23, IL-36γ)	Solid tumors or lymphoma	IT	Phase I N = 264	NCT03739931
CYL-02 Polyethylenimine + pDNA (sst2 + DCK:UMK)	Pancreatic adenocarcinoma	IT	Phase 2 N = 68	NCT02806687

believed to be a pre-exposure to AV infection as the patient exhibited antibodies and activated T-cells to AV in subsequent analysis. The difference in gender could have been an additional factor (Raper et al. 2003). Similarly, with the lipid-DNA complex of CFTR gene therapy study involving 8 patients (4 F/4 M), only 4 exhibited immune response with elevated IL-6 levels and no gender-based differences were observed

(Ruiz et al. 2001). Moreover, in the Phase Ia/Ib studies with CALAA-01 having 24 (19 M and 5 F) participants, cancer patients experienced various dose-limiting toxic events (DLTs). 19 (M) Phase Ia patients had good tolerance to the dose escalation while 3 (F) out of 5 (F) Phase Ib patients experienced DLTs along with grade 3 toxicities. It was interesting to note that this adverse effect was speculated to the possible structural changes to CALAA-01 due to long-term storage, in particular the transferrin-targeting component. This study indicated the varying responses among patients and the need to gain a more complete understanding on the stability of the therapeutic agent before commencing clinical studies (Zuckerman et al. 2014).

Such differences among individuals could also be observed with the cytokine response studies performed in vitro with PBMC obtained from otherwise healthy individuals (i.e., blood donors); only 2 out of 9 donor PBMC exhibited very high levels with the positive treatment PMA/IO (phorbol 12-myristate 13-acetate/ionomycin) and this was also true to some extent with the LMW-bPEI treatment (Meenakshi Sundaram et al. 2022).

Substantial variation in cytokine induction was observed among 307 participants (children) towards virus and bacterial infection with differences as high as 1000-fold. Twenty-eight separate cytokines were evaluated against 15 different stimuli including various pathogens using PBMC isolated from 307 participants in the Manchester Asthma and Allergy Study children. The reason(s) for the difference could be due to (i) different receptors being recognized among the children, thereby inducing a different intracellular signaling pathway, (ii) failure to determine the specific cell type present in PBMCs before undertaking the study, (iii) use of PBMCs stored in liquid nitrogen and (iv) the difference in the time of sample collection along with genetic differences among the cells (Lin et al. 2022), which points back to the need of evaluating the patient's medical history and any underlying conditions before commencing treatment.

4 Management of Cytokine Response Against Viral and Non-viral Gene Delivery Systems

The clinical approach to circumvent immune response is mostly through the administration of immunosuppressive agents which include drugs and specific antibodies that can block the interaction between the immune cells by inhibiting the pathways involved in activation.

4.1 Antibody-Based Inhibitors

Tocilizumab (TCZ) is a humanized antibody targeting IL-6 receptor, predominantly used in treating rheumatoid arthritis (Navarro et al. 2014) and received FDA approval to address cytokine release syndrome (CRS) caused following CTL019 and KTE-C19 CAR-T therapy based on AAV with a response rate of ~70% (in 14 days) and ~50% (in 5 days) respectively (Le et al. 2018). Other studies have also shown the effectiveness of TCZ in decreasing IL-6 levels during AAV mediated gene therapy without affecting the therapeutic potential of the treatment (Kuranda et al. 2021) and has been discussed in elsewhere (Campochiaro et al. 2021; Stroud et al. 2019). Although TCZ is the leading monoclonal antibody in clinical use to address CRS, there are other immunosuppressive agents capable of blocking the immune response, such as anti-CD40, anti-CD80, anti-CD86, anti-CD40L, and CTLA4-Ig (Zhou et al. 2004).

4.2 *Kinase Inhibitors*

Ruxolitinib is a JAK/STAT kinase inhibitor which was shown to decrease various cytokines like IL-6, IL-12α, IL-15, IFN-γ and TNF-α from NK cells when exposed to helper-dependent AV-activated macrophages (Ankathatti and Hu 2015; Elli et al. 2019). Similarly, itacitinib, a JAK1 inhibitor, helped to significantly reduce IL-6, IL-12 and IFN-γ levels from CAR-T cells and is currently being investigated in Phase II studies to prevent CRS (Huarte et al. 2020). Another inhibitor ibrutinib, specific to Bruton's tyrosine kinase, was successfully used in combination with rapamycin to decrease the immune response to AAV and aided in the expansion of CD19 CAR-T cells with a strong reduction in IL-4 levels (Xiang et al. 2022; Fraietta et al. 2016). Additionally, Dasatinib a tyrosine kinase inhibitor developed specifically for BCR-ABL fusion protein was shown to exhibit immunosuppressive properties and blocked IL-2 and IFN-γ levels, demonstrating its potential to control CRS (Mestermann et al. 2019).

4.3 *Corticosteroids (CCS)*

The immunosuppressive properties of the corticosteroids are attributed to their ability to bind glucocorticoid receptors which in turn blocks various signaling pathways in immune cells thereby affecting cytokine production. Corticosteroids are considered as second-line treatment for CRS after TCZ failure as CCS are known to exert strong effects on CAR-T cells resulting in the impairment of the therapeutic effect (Brudno and Kochenderfer 2016). A recent trial (NCT05164471) with AAV gene therapy for hemophilia B aimed at achieving sustained expression of Factor IX was carried out with and without prophylactic prednisolone treatment. This study was able to achieve normal levels of Factor IX in patients and showed the need to follow an immunosuppressive regimen to maintain the therapeutic effect (Chowdary et al. 2022). Similarly, dexamethasone application was not only able to suppress the immune response but also prolong the gene expression (through AAV) in mice models along with TNF-α and IFN-β reduction (Chai et al. 2022).

4.4 *Protein-Based Inhibitors*

Among the proteins used for immune suppression, Etanercept composed of 934 amino acids (a.a) acts like a soluble TNF receptor and binds to TNF-α and TNF-β thereby inhibiting its activity. Etanercept was successfully used to address CRS in 3 patients without altering the therapeutic effect of CAR-T therapy for multiple myeloma (Zhang et al. 2021c). Similarly, the IL-1 receptor inhibitor Anakinra composed of 153 a.a was administered to treat CRS caused upon CAR-T therapy after TCZ (IL-6 inhibitor) treatment to 18 patients of which, in 17 the CRS was resolved (Wong et al. 2021). Similarly, Anakinra has been used for treating CRS while administering axicabtagene ciloleucel (CAR T-cell therapy) for treatment of B-cell lymphoma (Strati et al. 2020). More recently a chemically modified polypeptide was prepared which had phosphoserine containing peptide-conjugated AAV vector. This approach was successful in establishing stable gene expression in addition to suppressing any immune response to AAV (Yuan et al. 2022).

Other molecules such as cyclosporin and tacrolimus, which inhibit the phosphatase calcineurin, can downregulate the IL-2 cytokine levels and was shown to help prolong transgene expression from AAV (Chu and Ng 2021). Endosomal small molecular inhibitors such as chloroquine and quinacrine hydrochloride were able to decrease the IL-12 cytokine levels by 50% without altering the gene expression mediated by cationic lipids/pDNA complexes in vitro from spleen cells isolated from mice (Yew et al. 2000).

5 New Approaches to Specific Manipulation of Cytokines with Nucleic Acid Technologies

Novel approaches are being explored to formulate minimally immunogenic non-viral carriers and implementing them for therapeutic purpose in wide range of diseases (Fig. 3), as they have the capacity to modify unwanted effects into curing effects. In a variety of cellular environments, cytokine mRNA, pDNA, and siRNA have been investigated with non-viral nanoparticle systems and found promising. Table 2 shows a summary of the mRNA and pDNA-based strategies, and the siRNA-based approaches are briefly explained below.

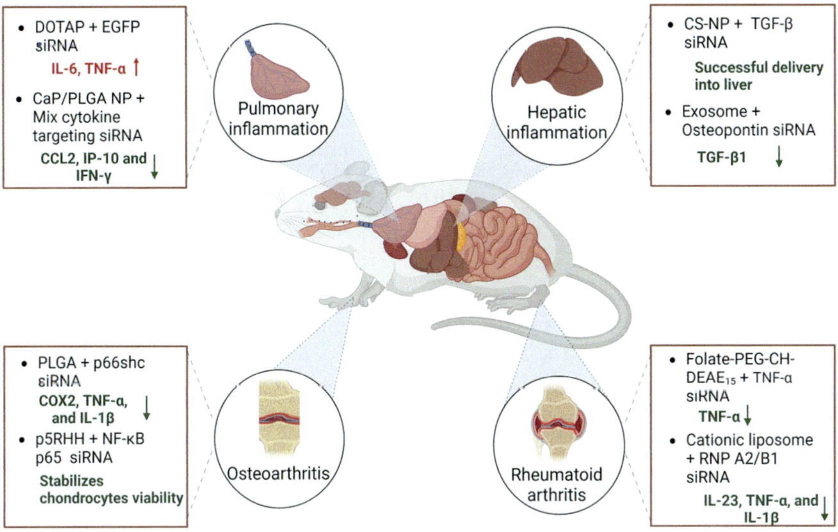

Fig. 3 Representative attempts to modulate cytokine responses using siRNA/nanoparticle formulations in different preclinical models. DOTAP: dioleoyl-3-trimethylammonium propane; EGFP: enhanced green fluorescent protein; PLGA: poly(lactic-co-glycolic acid); p66shc: a gene that regulates the level of reactive oxygen species; Cap/PLGA NP: multi-shell nanoparticle system of calcium phosphate and PLGA; p5RHH: self-assembled peptidic nanoparticle; NF-κB: nuclear Factor Kappa B; CS-NP: chitosan-based nanoparticle; TGF-β: transforming growth factor beta; TNF-α: tumor necrosis factor alpha; Folate-PEG-CH-DEAE15: conjugating folate-PEG with deacetylated chitosan (CH-DEAE15); RNP A2/B1: heterogeneous nuclear ribonucleoproteins A2/B1

5.1 *Inflammatory Disorders*

OA is a common reason for disability amongst elder people. Here, the cartilage in joints is compromised, which is mainly mediated by inflammatory cytokines. Despite its occurrence, there are limited medications available, including questionable benefits from intermittent intra-articular corticosteroid or hyaluronic acid injections and disease-modifying osteoarthritis drugs. In OA, p66shc has been shown to be a key player in the production of reactive oxygen species in mitochondria. A non-viral siRNA carrier, PLGA (50/50) nanoparticle, was developed by Shin et al. and was used to deliver p66shc siRNA in OA-induced Sprague–Dawley rats to the knee. This formulation significantly reduced inflammatory cytokines (TNF-α, IL-1β, and COX2 (Shin et al. 2020). An alternative self-assembling peptidic nanoparticle-based technology, p5RHH, was employed to carry NF-κB p65 siRNA in OA model to prevent cartilage degeneration. OA activates NF-κB signaling, which induce the release of pro-inflammatory cytokines (e.g. IL-1β) and stimulate apoptosis in articular chondrocytes. Single exposure of 500 nM p5RHH/NF-κB p65 siRNA into IL-1β treated human cartilage explants for 48 h has been found to stabilize the cell viability and significantly decrease the apoptosis (Yan et al. 2019).

Table 2 Select studies probing cytokine responses upon non-viral delivery of mRNA and pDNA. A range of non-viral delivery systems was used for nucleic acid delivery

Carrier	Cargo	Indication	Study type	Cell line/ in vivo model	Formulation activity
Cationic liposome (DOSPA/DOPE)	IL-10 mRNA	Graft versus host disease (GvHD)	In vivo	Mesenchymal stem cells (MSCs)	IV treatment of IL-10 mRNA-engineered MSCs reduced splenic CD4+ and CD8+ T cell proliferation and suppressed production of pro-inflammatory cytokines (IL1α, IL-2, IL-5, and IL-17) (Zhang et al. 2021a)
Cationic liposome (DOTMA and cholesterol)	IL-12 pDNA	Head and neck cancer	In vivo	Squamous carcinoma cells, SCC-VII, in C3H/HeJ mice	Tumor growth was significantly suppressed and anti-inflammatory cytokines IFN-γ and IL-12 were significantly increased after intralesional injection of pIL-2 compared to control plasmid (O'Malley et al. 2005)
Polymeric (Deoxycholic acid-conjugated PEI 2K)	EGFP mRNA and pDNA	Ischemic stroke	In vitro	Raw264.7 macrophage cells	The pro-inflammatory cytokine TNF-α was not induced by either EGFP-pDNA/ DA-PEI2k or EGFP-mRNA/ DA-PEI2k (Oh et al. 2020)
Chitosan nanoparticles	IL-12 pDNA	Colon carcinoma	In vitro	Murine CT-26 colon carcinoma cells	Immune system stimulation can be achieved by IL-12 for antitumor effects. The highest levels of IL-12 expression were achieved using pUMVC3-hIL12 (Hallaj-Nezhadi et al. 2011)

(continued)

Table 2 (continued)

Carrier	Cargo	Indication	Study type	Cell line/ in vivo model	Formulation activity
Cationic solid LNP	Integrin β1 pDNA	Osteoarthritis	In vitro	Rat chondrocytes	By decreasing the death of rat chondrocytes and boosting tissue healing, solid LNP/integrin 1 pDNA showed potential as a therapeutic nanomedicine (Zhao et al. 2020)
PEI "Max"/polyanion	GM-CSF pDNA	Melanoma	In vivo	Melanoma cell line, B16, in ddY mice	Intertumoral administration of GM-CSF pDNA/ PEI "Max"/ polyanion showed high expression of GM-CSF in tumors and better tumor suppression (Koyama et al. 2015)
Booster-transfection by DOPE/CHEMS lipid suspension with commercial Polyplex (Turbofect/ Lipofectamine)	GCSF-mRNA and pDNA	Differentiated neuronal and stem cells	In vitro	Human adult bone marrow mesenchymal stem cells (MSCs)	The presence of DOPE/CHEMS caused increased production of GCSF in MSCs and efficient production of dopaminergic neurons from neural stem cells (Ho et al. 2017)

(continued)

Table 2 (continued)

Carrier	Cargo	Indication	Study type	Cell line/ in vivo model	Formulation activity
Cationic polymer (comb- and sunflower-shaped pHEMA-g-pDMAEMA)	pDNA	Human T cells	In vitro	CD4+ and CD8+ primary human T cells	Nanoparticles not only alter cytokines, but cytokines also alter the efficiency of gene transfer. When IL-21 was added to T cell culture, it substantially improved the overall cell survival and transfection efficiency (Olden et al. 2018)
Cationic liposome (DOTAP and MBC)	TNF-α pDNA	Bladder cancer	In vivo	Bladder cancer cell line, MB49, in mice	The incidence of bladder tumors was substantially reduced after TNF-α delivery, which increased the number of T-cells and NK cells (Zang et al. 2004)

Cytokines are associated with the pathogenesis of RA. They promote autoimmunity, keep chronic inflammatory synovitis going, and cause the destruction of tissues next to the affected joint. For individuals who do not respond to conventional treatment, anti-TNF biotherapies, Khoury et al. have designed 3 siRNAs targeting the pro-inflammatory cytokines (IL-1, IL-6, and IL-18), and prepared nanoparticles with the RPR209120/DOPE liposome (Khoury et al. 2008) to treat RA. This system was able to silence target cytokines up to 70–75% in vitro without affecting other pro-inflammatory cytokine expressions, and impeding all pathological features of arthritis including inflammation, joint destruction and Th1 response in DBA/1 mice with collagen-induced arthritis. Another modified chitosan based nanocarrier, folate-PEG-CH-DEAE$_{15}$, was formulated by conjugating folate-PEG with deacetylated chitosan (CH-DEAE$_{15}$) to carry TNF-α siRNA into a rodent RA model. Intraperitoneal injection of folate-PEG-CH-DEAE$_{15}$/siRNA (50 μg siRNA) in collagen antibody-induced arthritis model, DBA/1 mice, significantly decreased TNF-α secretion in knee and (%) TNF-α + cells. Moreover, it also reduced the scores for cartilage destruction, bone erosion, and synovitis (Shi et al. 2018). In a distinct study, cationic liposome-based formulation was used to deliver heterogeneous nuclear RNP A2/B1 (hnRNP A2/B1)–siRNA into RA-induced DBA/1 mice. Intravenous administration of this formulation (0.5 mg/kg of siRNA) markedly reduced the secretion of pro-inflammatory cytokine (IL-23, TNF-α, and IL-1β) levels in the cells of draining lymph nodes and synovial macrophages (Herman et al. 2015).

Systemic inflammatory sepsis is a global challenge that is responsible for ~20% of all deaths worldwide (Rudd et al. 2020). A peptide-based gene carrier, prepared by conjugating cationic arginines (9R) and the TKPR (Thr-Lys-Pro-Arg) sequence from the Fc region of Immunoglobulin G (IgG), was successfully used to deliver TNF-α converting enzyme (TACE) siRNA in a murine sepsis model (C57BL/6 mice), and found to decrease pro-inflammatory cytokines (TNF-α, IL-6) with recruiting less amount of TNF-α releasing macrophage (Lee et al. 2021). For allergic rhinitis, which arises when natural allergens like pollen, ragweed, dust mites, and pet dander cause inflammation in the nose, Kim and his team developed a biocompatible, water-soluble chitosan to deliver antisense oligonucleotides (ASO) against IL5 in ovalbumin induced BALB/c mice. Upon administering 100 μg of ASO intranasally, they reduced the inflammation (IL-5 and IgE) in allergic rhinitis model (Kim and Kim 2007). For psoriasis plaques, an inflammatory conditions of the skin, TNF-α siRNA complexed with PAMAM dendrimer and DOTAP liposome improved the phenotypic and histopathological features in a psoriatic model of C57BL/6 mice, compared to the clinically-used imiquimod, with decreased amount of IL-6, TNF-α, IL-17 and IL-22 (Pandi et al. 2018).

During inflammatory diseases, macrophages are responsible for secreting a large amount of TNF-α, a major pro-inflammatory cytokine. Inhibiting TNF-α in macrophages represents a promising therapeutic approach to treat these conditions. A chitosan-based nanoparticle system was developed by Xiao et al. to transfect CD98 (over expressed in macrophages and potent inflammation inducer) siRNA into macrophages. Chitosan, coupled with urocanic acid, was complexed with CD98 siRNA and delivered to murine macrophage cell line, Raw 264.7 in vitro. This complex did not show any apparent cytotoxicity and downregulated TNF-α significantly (Xiao et al. 2016). The same group reported another chitosan-based nanoparticle system (HTPP) next year by conjugating glycidyltrimethylammonium chloride/amine group with chitosan. This HTPP was complexed with TNF-α siRNA in the presence of tripolyphosphate and transfected into Raw 264.7 cells. This complex (100 nM siRNA) significantly (~90%) reduced TNF-α expression, demonstrating their utility in macrophages (Xiao et al. 2017).

Wu et al. reported the synthesis of a different carrier, PEG-PAsp(DETA)-Lys-CA2, where PEG-block-poly(L-aspartic acid) polymer was grafted with diethylenetriamine, lysine, and cholic acid to deliver siRNA for controlling cytokine release. They delivered Notch1 siRNA into RAW264.7 and were able to significantly impede LPS-activated pro-inflammatory IL-6 expression, and increase anti-inflammatory IL-10 expression (Wu et al. 2018). In another study, Ying et al. reported successful cytosolic siRNA delivery into macrophages to knockdown cytokine expressions in animal model. They developed nanoparticle-stabilized nanocapsule (NPSC) technology, which utilizes nanocapsule stabilization within 150 nm via interaction between the hydrophobic/anionic component and the cationic gold nanoparticle shell, to bind siRNA electrostatically. Upon IV injection of low dose (0.56 mg/kg) of TNF-α siRNA and NPSC complex, they showed 60% downregulation of TNF-α in LPS-challenged (5 mg/kg) BALB/c mouse (Jiang et al. 2018).

5.2 *Neurological Disorders*

Several neurological diseases are characterized by inflammatory response mediated by TNF-α due to neuronal apoptosis. siRNA approach can alleviate the inflation by specific suppression of TNF-α expression. Sang et al. employed a short peptide (RVG-9dR) derived from the rabies virus glycoprotein to deliver TNF-α siRNA. They were able to effectively inhibit LPS-induced TNF-α production in primary macrophages and microglia cells in vitro. In addition, the LPS-induced levels of TNF-α in blood and brain were lowered after IV injection of this complex in mice, resulting in a marked suppression of neuroinflammation in vivo (Kim et al. 2010). To address excitotoxic cell death, where neurons gradually stop working and die, and neurodegenerative diseases like Alzheimer's disease and multiple sclerosis could emerge, c-Jun N-terminal kinase pathway was targeted to improve pathological changes. Cationic liposomes associated to transferrin (Tf-lipoplexes) was utilized to deliver c-Jun siRNA into C57/BL6 mouse hippocampus. Tf-lipoplexes was able to deliver c-Jun siRNA successfully and decreased seizure activity effectively. There was also reduction in inflammatory cytokine (IL-1β, IL-6 and TNF-α) levels, which confirmed the nanoparticle safety (Cardoso et al. 2010).

Huntington disease is another type of neurological disorder caused by mutation in the Huntingtin (HTT) gene, and a non-viral delivery of HTT siRNA has shown potential therapeutic results. A hybrid lipid-coated nanoparticle was prepared by fusion of liposome (prepared with phospholipids DPPC and DOPC and CBD by film hydration method) and nanoparticle (prepared by ionotropic gelation of chitosan with siRNA) to transfer E30 siRNA (targeting exon 30 of HTT gene) in bone marrow mesenchymal stem cells, which express human HTT gene. Significant reduction of HTT expression and IL-6 secretion were achieved with this formulation (Fihurka et al. 2022).

5.3 *Gastrointestinal Disorders*

IBD is a chronic relapsing gastrointestinal disorder that can lead to colorectal cancer if inflammation is not effectively suppressed. Crohn's disease and ulcerative colitis are common IBDs. To address the ulcerative colitis, a non-viral nanoparticle-based carrier, scCD98-PEG-UAC, was developed by Xiao et al. to deliver therapeutic siRNA in mouse model. scCD98-PEG-UAC is a nanoparticle in hydrogel (chitosan/alginate) system formulated by self-assembly procedure using PEI (2 kDa), PEG, and urocanic acid modified chitosan (UAC), which also contains single-chain CD98 antibody in the surface and loaded with CD98 siRNA. Oral administration of this formulation (1 mg/kg) in chronic colitis-induced RAG1–/– mice downregulated CD98 expression significantly (~65%) in the colonic lumen and reduced chronic colitis by lowering colonic myeloperoxidase level. Subsequently, it also significantly inhibited the expression of pro-inflammatory cytokines, TNF-α (59.9%), IL-6 (80.4%), and IL-12 (31.8%). This nanoparticle in hydrogel formulation demonstrated the promise of orally administered nucleic acid-based therapy for altering cytokine responses in colitis (Xiao et al. 2014).

In a related approach, Peer et al. formulated liposome-based targeted (β7 integrin) stabilized nanoparticles (tsNPs) to entrap siRNAs and deliver them into gut leukocytes. They entrapped cyclin D1 siRNA into tsNPs that was systemically administered into colitis induced (by dextran sodium sulfate) C57BL/6 mice (2.5 mg/kg). Silencing of cyclin D1 downregulated T helper cell 1 cytokine (IFN-γ, IL-2, TNF-α) expressions, suppressed leukocyte proliferation, and ultimately impeded induced colitis in mice (Peer et al. 2008). Another multi-shell nanoparticle system, (PLGA/CaP), consisting of a calcium phosphate core coated with siRNA directed against pro-inflammatory mediators (TNF-α, KC or IP-10), encapsulated into PLGA and further coated with a final outer layer of PEI, was developed and tested in colitis-induced BALB/c mice. This approach decreased the expression of target genes in colonic biopsies significantly, confirming its potential to treat intestinal inflammation (Frede et al. 2016).

5.4 Pulmonary Disorders

Wu et al. conducted a safety and efficacy investigation of PLGA nanoparticles loaded with lipoplexes (LPXs) and lipid–polymer hybrid nanoparticles (LPNs) in murine pulmonary disease model. Two types of nanoparticles were prepared: LPX by assembling enhanced green fluorescent protein (EGFP) siRNA with DOTAP, and LPN by co-entrapping with siRNA/DOTAP in PLGA nanoparticles. Both preparations were nebulized with a hand-hold aerosolizer at 5 μg/dose of siRNA into EGFP transgenic mice. They found elevated IL-6 and TNF-α levels by LPX administration, indicating cationic lipid DOTAP induced acute inflammation. Interestingly, LPN did not stimulate IL-6 and TNF-α levels, suggesting that PLGA in LPNs could decrease inflammatory effects. LPN also resulted in longer EGFP silencing than LPX, indicating their feasibility for use in lung inflammatory diseases (Wu et al. 2021).

A different new system, dry siRNA/PEI powder, was developed to deliver GL3 siRNA into lungs of metastatic mice. The spherical powder with highly porous structure enhanced the aerosol deposition into lungs. Intratracheal administration of 3 μg of siRNA provided excellent target gene knockdown in lungs. No markable stimulation of inflammatory cytokine (IL-6, IL-10, IFN-γ, and TNF-α) were found, making them safe for certain cytokine-induced toxicity (Okuda et al. 2018). The PLGA/Ca preparation developed by Frede et al., discussed above, was also studied in a pulmonary inflammation model; CCL2, IP-10 and IFN-γ siRNAs was delivered with this system into SPC-HA transgenic mice and indicated reduced inflammation upon treatment with siRNA/PLGA/CaP (Frede et al. 2017).

5.5 Fibrotic Disorders

Idiopathic pulmonary fibrosis is a serious, long-term lung disease that affects the tissue around the alveoli. IL-11 is a profibrotic cytokine, which is essential for the differentiation of fibroblasts but also implicated in the development of idiopathic pulmonary fibrosis. A new type of nanoparticle was prepared by Bai et al. based on self-assembly of PLGA-PEG and the cationic G0-C14 (reacting 1,2-epoxytetradecane with ethylenediamine core–PAMAM dendrimer), named PPGC nanoparticles. In a bleomycin-induced idiopathic pulmonary fibrosis model in C57BL6 mice, a single intratracheal injection of IL-11 siRNA/PPGC nanoparticles at 0.75 or 1.5 mg/kg of siRNA via inhalation route led to significantly improved pulmonary function and reduced fibrosis development without causing toxicities (Bai et al. 2022). To deliver siRNAs, targeting androgen receptor gene, in a pulmonary fibrosis model, Yoon et al. developed a second non-viral nanoparticle system called SAMiRNA (self-assembled micelle inhibitory RNA), which is composed of individually biconjugated siRNAs with a hydrophilic polymer and lipid on their ends. Bleomycin challenged fibrotic C57BL/6 mice were treated with SAMiRNA (3 mg/kg) intravenously. Three doses of treatment reduced the collagen content in lungs drastically. NIH3T3 fibroblast cells were used to test the immunogenicity of SAMiRNA. No stimulation of cytokines (TNF-, MCP-1, IFN-, IL-12, and IL-6) was observed (Yoon et al. 2016).

Another potent pro-fibrotic cytokine, TGF-β, is responsible for fibrosis in inflamed liver. Inhibiting the expression of this cytokine would, therefore, result in a reduced activated phenotype and, perhaps, a reversal of hepatic fibrosis. Chitosan-based nanoparticle (CS-NP) was fabricated to deliver TGF-β siRNA into activated hepatic stellate cells (main orchestrators of the hepatic fibrotic cascade) by adding TGF-β siRNA with chitosan nanoparticle, and then modifying them with platelet-derived growth factor receptor-beta-binding peptides. Intravenous administration of CS-NP (0.8 mg per mouse) into CCl_4-induced fibrotic Swiss albino mice caused significant distribution in the liver, but not the healthy mice (Azzam et al. 2020). In the hepatic fibrosis model, another nucleic acid delivery system based on exosomes also showed a significant potential. MSCs from adipose tissue were extracted to prepare exosomes, which were then tailored to deliver osteopontin (an oxidant stress-sensitive cytokine) siRNA into primary hepatic stellate cells. Exosome-mediated siRNA delivery system effectively improved liver function compared to naked siRNA group, significantly reduced a common cytokine that causes fibrosis (TGF-1) by lowering HMGB1 (Tang et al. 2022).

5.6 Ocular and Kidney Disorders

Nanoparticles-based nucleic acid delivery also showed potential for eye and kidney inflammatory disease. Autoimmune uveitis is a complicated eye disease, often related to the immunological responses to retinal proteins. The TLR signalling plays an important role in this disorder's progression. Chen and colleagues developed a chitosan-based nanocarrier, pSP-CS, to deliver TLR3-siRNA in the subretinal region of experimental autoimmune uveitis (EAU) induced B10RIII mice. Treatment with pSP-CS/TLR3-siRNA complex resulted in very mild EAU compared to the control group, and a well-preserved retinal structure. Also, this treatment greatly reduced the amount of cytokines (IL-7 and IFN-γ) that eye infiltrated cells released (Chen et al. 2013).

Cytokines play crucial role in the pathogenesis of lupus nephritis, a kidney disease, that causes widespread mortality and mortality, from its initiation to its advanced stages. Proliferation of a glomerular mesangial cell line, MMC, is a sign of multiple kidney diseases. HMGB1 is a major cytokine in the pathogenesis of lupus nephritis, and it causes mesangial cells to proliferate and is linked to kidney damage. A method was developed to transfer HMGB1 siRNA and dihydroartemisinin to MMC cells that had been stimulated by LPS. DOTAP, DHA, and DSPE-PEG-DOTAP were mixed together with TAT peptide to make PEGylated TAT peptide-cationic liposome (TAT-CL), which was used to deliver HMGB1 siRNA in MMC cells. This formula tion greatly attenuated LPS-induced MMC cell proliferation and reduced HMGB1 translocation from the nucleus to the cytoplasm, demonstrating their therapeutic potential for lupus nephritis (Diao et al. 2019).

5.7 Cancer Therapy

Manipulating cytokine responses could be also exploited for cancer treatment. A lipid nanoparticle system was prepared by grafting PEI, PD-L1 siRNA and EGFR short peptide vaccine adjuvant onto LNPs (DOPC, PEI-SA and Cholesterol). The delivery system efficiently delivered the PD-L1 siRNA into T-cell pretreated-EGFR-positive lung cancer model, A549, and significantly increased IFN-γ and TNF-α levels to induce anti-tumor immunity and cell death. Interestingly, this system also reduced IL-10 level, which is implicated in tumor immune escape (Yang et al. 2022). Along similar lines, where activating cytokine response may be critical to secure an anti-tumor response, treatment of bone marrow-derived dendritic cells with PLGA nanoparticles bearing IL-10 siRNA elevated the ratio of the Th1 cytokine IL-12 to the Th2 cytokine IL-10, which prepared the cells for efficient paclitaxel chemotherapy (Heo et al. 2015).

Using a cationic liposome, activating cytokines also helped to suppress a different highly tumorigenic and invasive breast cancer tumor model. Cationic liposome, LPP-P4-Ep, was formulated by encapsulating PD-L1 siRNA and CD47 siRNA into LPP (a mixture of DC-Chol, DOPE and MAL-PEG-DOPE), and then conjugating this LPP with HS-EpCAM (epithelial cell adhesion molecule). Then, LPP-P4-Ep (0.15–1.2 mg/kg of siRNA) was injected subcutaneously in local 4 T1 cells xenograft Balb/c mice. Six doses of injections significantly increased the secretion of IFN-γ and IL-6 in both blood and tumor to enhance anti-tumor immunity. To inhibit tumor growth, IFN-γ collaborates with other cytokines available in the tumor microenvironment. By promoting the recruitment of effector T cells in the tumor microenvironment, IL-6 keeps anti-tumor immunity functioning effectively in both the lymph nodes, where lymphocyte priming occurs, and in tumor nests. In addition to cytokine stimulation, this treatment also reduced tumor volume substantially without affecting the body weight, confirming the feasibility and efficacy of this non-viral method (Lian et al. 2019).

6 Conclusions and Perspectives

The concerns related to cytokine response, or cytokine release syndrome, have been well appreciated in early gene therapy work where viral vectors have been used to implement nucleic acid based therapeutic options. This issue has been also critical when it comes to non-viral systems; although it has been possible to fine-tune this response and minimize it at times (Cron et al 2023), it is still a concern with some non-viral delivery systems. Non-viral delivery systems has lagged viral systems for clinical entry, and it is not clear if immunological reactions are the primary reason for this delay (more likely, this was driven by relatively lower efficacy obtained with non-viral systems). However, it is well appreciated that some delivery systems are associated with heightened sensitivity to cytokine reactions. Relatively minor measures can be taken to adjust the dose, administration route and especially pre-screen patients to cytokine sensitivity, that will ultimately help with the host response. This will not only be beneficial for the overall safety and outcome of the clinical trials but also save lives of the patients. It may be possible to intervene with the cytokine responses very precisely, by relying on transient suppression of gene expression by using RNAi or more permanent measures such as CRISPR/Cas9 system (Ottaviano et al 2022). Transient expression may be sufficient for controlling initial host response, which has a tendency to subside as long as it is not life-threatening. A permanent stoppage of any cytokine response may give additional complications on the long term in managing host response to foreign invaders and/or other normal physiological events. RNAi can also make it possible to intervene with a spectrum of cytokines in a convenient way, by using a mixture of specific siRNAs. Considering the summary of attempts undertaken in several clinical indications provided in this short review, it is clear that there is a wide range of inflammatory and cytokine-associated diseases that will directly benefit from the RNAi approach. Where localized application is feasible (e.g., colitis with defined anatomical involvement), it may be safer to utilize nanoparticulate systems to locally deliver the cytokine interfering agents. Where systemic intervention might be required (e.g., sepsis), it is preferred to utilize 'stealth' nanoparticulate systems that do not exasperate the disease any further. While PEG incorporation into nanoparticles (or nanoparticle building blocks) has been the 'go-to' approach for this end, this modification typically reduced the efficacy of nanoparticulate system and needs to be implemented in a way that the dose administered are not compromised due to lower efficacy. Coating the surface of nanoparticles or including additives that change the zeta-potential (from highly cationic surface to neutral surface, in our hands) could be a simpler measure to reduce inflammatory cytokine response (Meenakshi Sundaram et al. 2022).

The use of realistic models, such as human PBMCs, are important in assessing the cytokine response, possibly more so than the animal models. We found the cells lines such as Jurkat T-cells to be more uniformly responsive as well as much more sensitive than the primary PBMCs, which may lead to an erroneous assessment of the cytokine response if used routinely. Individual variations in cytokine response is obviously better modeled by the use of PBMCs in culture (Fig. 4). It will be important to tackle this with a representative sampling of PBMCs to estimate the true potential of delivery systems to induce inflammatory cytokines. This may on the long run save a lot of effort by focusing on minimally reactive delivery systems, where the clinical safety is utmost important.

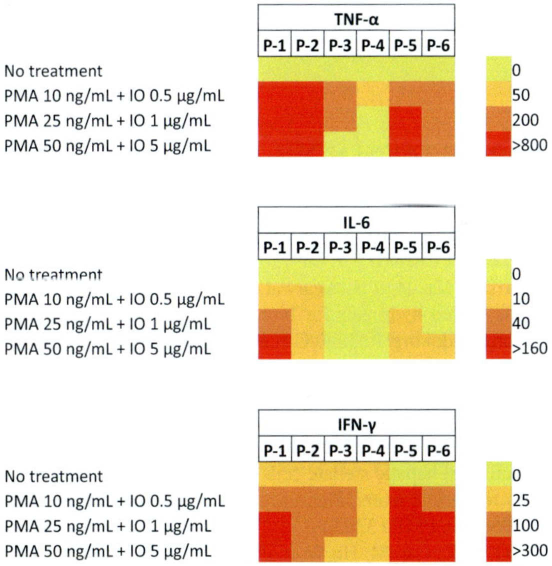

Fig. 4 Secretion of inflammatory cytokines (TNF-α, IL-6 and IFN-γ) from PBMCs obtained from 6 separate, healthy donors in culture. The cells were treated with PMA/IO combinations at indicated concentrations and the cytokine secretion was determined 3 days afterwards (by specific ELISAs). Untreated cells served as the reference. Note that some PBMCs were unresponsive to the stimulation (especially in IL-6 and TNF-α secretion) while other PBMCs were hyper-sensitive to the stimulation. Significant donor-to-donor variation was evident among the PBMCs when it comes to cytokine secretion upon the stimulation (Meenakshi Sundaram et al. 2022)

While a great deal of efforts have been placed on minimizing cytokine response, it may be useful in certain indications to utilize nanoparticles that can elicit a specific cytokine response. As eluted in experimental cancer applications above, induction of cytokine response can aid in therapeutic effects for certain kinds of cancers, as in the case of sensitizing malignant cells to chemotherapy. In that way, better understanding of structural features of the nanoparticles that can elicit a strong and/or selective cytokine response will help to design novel systems for therapy. Induction of a certain cytokine cocktail has the potential to alter the physiology of local immune cells so that the effects might be long lasting and more potent upon such an 'engineered' cytokine response. For example, with lipid-substituted PEI polymers, we have observed that cytotoxicity, hemocompatibility (i.e., RBC lysis potency) and inflammatory cytokine (TNF-α, IL-6 and IFN-γ) secretion was differentially regulated with the nature of the lipid substituted on the PEI (Meenakshi Sundaram et al. 2022), that it may be possible to fine-tune these properties independent of each other. While these cytokines are lumped as inflammatory cytokines, we also saw differential up-regulation of the cytokines in response to non-viral carrier exposure, in that one or more of this class of cytokines could be selectively stimulated, so that it may be possible to fine tune the response among a physiologically similar group of cytokines as a function of carrier features.

References

Abedi-Gaballu F, Dehghan G, Ghaffari M, Yekta R, Abbaspour-Ravasjani S, Baradaran B et al (2018) PAMAM dendrimers as efficient drug and gene delivery nanosystems for cancer therapy. Appl Mater Today 12:177–190. https://doi.org/10.1016/J.APMT.2018.05.002. PMID 30511014

Ansari A, Santerre JP, Uludağ H (2017) Biomaterials for polynucleotide delivery to anchorage-independent cells. J Mater Chem B 5:7238–7261

Ansari MJ, Rajendran RR, Mohanto S, Agarwal U, Panda K, Dhotre K et al (2022) Poly(N-isopropylacrylamide)-based hydrogels for biomedical applications: a review of the state-of-the-art. Gels 8(7). https://doi.org/10.3390/GELS8070454. PMID 35877539

Ankathatti M, Hu J (2015) Transient blocking of NK cell function with small molecule inhibitors for helper dependant adenoviral vector-mediated gene delivery. Cell Biosci 5:29. https://doi.org/10.1186/s13578-015-0023-0. PMID 26085921

Azzam M, El Safy S, Abdelgelil SA, Weiskirchen R, Asimakopoulou A, de Lorenzi F et al (2020) Targeting activated hepatic stellate cells using collagen-binding chitosan nanoparticles for siRNA delivery to fibrotic livers. Pharmaceutics 12(6):1–16. https://doi.org/10.3390/PHARMACEUTICS12060590. PMID 32630415

Bai X, Zhao G, Chen Q, Li Z, Gao M, Ho W et al (2022) Inhaled siRNA nanoparticles targeting IL11 inhibit lung fibrosis and improve pulmonary function post-bleomycin challenge. Sci Adv 8(25):eabn7162. https://doi.org/10.1126/SCIADV.ABN7162. PMID 35731866

Betker JL, Anchordoquy TJ (2022) The effect of repeat administration of lipoplexes on gene delivery, biodistribution, and cytokine response in immunocompetent tumor-bearing mice. J Pharm Sci 111(7):1926–1936. https://doi.org/10.1016/j.xphs.2021.12.017. PMID 34929156, PMCID PMC 9206043

Beyerle A, Merkel O, Stoeger T, Kissel T (2010) Pegylation affects cytotoxicity and cell-compatibility of poly(ethylene imine) for lung application: structure-function relationships. Toxicol Appl Pharmacol 242(2):146–154. https://doi.org/10.1016/J.TAAP.2009.10.001. PMID 19822165

Bila D, Radwan Y, Dobrovolskaia MA, Panigaj M, Afonin KA (2021) The recognition of and reactions to nucleic acid nanoparticles by human immune cells. Molecules 26(14). https://doi.org/10.3390/MOLECULES26144231. PMID 34299506

Bonnet ME, Erbacher P, Bolcato-Bellemin AL (2008) Systemic delivery of DNA or siRNA mediated by linear polyethylenimine (L-PEI) does not induce an inflammatory response. Pharm Res 25(12):2972–2982. https://doi.org/10.1007/s11095-008-9693-1. PMID 18709489

Brockstedt DG, Podsakoff GM, Fong L, Kurtzman G, Mueller-Ruchholtz W, Engleman EG (1999) Induction of immunity to antigens expressed by recombinant adeno-associated virus depends on the route of administration. Clin Immunol 92(1):67–75. https://doi.org/10.1006/clim.1999.4724. PMID 10413654

Brudno JN, Kochenderfer JN (2016) Toxicities of chimeric antigen receptor T cells: recognition and management. Blood 127(26):3321–3330. https://doi.org/10.1182/blood-2016-04-703751. PMID 27207799, PMCID PMC 4929924

Brüünsgaard H, Pedersen BK (2003) Age-related inflammatory cytokines and disease. Immunol Allergy Clin North Am 23(1):15–39. https://doi.org/10.1016/s0889-8561(02)00056-5. PMID 12645876

Campochiaro C, Farina N, Tomelleri A, Ferrara R, Lazzari C, De Luca G et al (2021) Tocilizumab for the treatment of immune-related adverse events: a systematic literature review and a multicentre case series. Eur J Intern Med 93:87–94. https://doi.org/10.1016/j.ejim.2021.07.016. PMID 34391591

Cao Y, Tan YF, Wong YS, Liew MWJ, Venkatraman S (2019) Recent advances in chitosan-based carriers for gene delivery. Mar Drugs 17(6). https://doi.org/10.3390/MD17060381. PMID 31242678

Cardoso ALC, Costa P, de Almeida LP, Simões S, Plesnila N, Culmsee C et al (2010) Tf-lipoplex-mediated c-Jun silencing improves neuronal survival following excitotoxic damage in vivo. J Control Release 142(3):392–403. https://doi.org/10.1016/J.JCONREL.2009.11.004. PMID 19913061

Carter E, Lau CY, Tosh D, Ward SG, Mrsny RJ (2013) Cell penetrating peptides fail to induce an innate immune response in epithelial cells in vitro: implications for continued therapeutic use. Eur J Pharm Biopharm 85(1):12–19. https://doi.org/10.1016/j.ejpb.2013.03.024. PMID 23958314

Chai Z, Zhang X, Dobbins AL, Samulski RJ, Merricks EP, Nichols TC et al (2022) Dexamethasone transiently enhances transgene expression in the liver when administered at late-phase post long-term adeno-associated virus transduction. Hum Gene Ther 33(3–4):119–130. https://doi.org/10.1089/hum.2021.083. PMID 34617445, PMCID PMC 8885437

Chatenoud L, Ferran C, Reuter A, Legendre C, Gevaert Y, Kreis H et al (1989) Systemic reaction to the anti-T-cell monoclonal antibody OKT3 in relation to serum levels of tumor necrosis factor and interferon-gamma [corrected]. N Engl J Med 320(21):1420–1421. https://doi.org/10.1056/NEJM198905253202117. Erratum in: N Engl J Med 321(1):63. https://doi.org/10.1056/NEJM198907063210127. PMID 2785642

Chen S, Yan H, Sun B, Zuo A, Liang D (2013) Subretinal transfection of chitosan-loaded TLR3-siRNA for the treatment of experimental autoimmune uveitis. Eur J Pharm Biopharm 85(3 Pt A):726–735. https://doi.org/10.1016/j.ejpb.2013.09.005. PMID 24055691

Cheng T, Miao J, Kai D, Zhang H (2018) Polyethylenimine-mediated CpG oligodeoxynucleotide delivery stimulates bifurcated cytokine induction. ACS Biomater Sci Eng 4(3):1013–1018. https://doi.org/10.1021/acsbiomaterials.8b00049. PMID 33418784

Chowdary P, Shapiro S, Makris M, Evans G, Boyce S, Talks K et al (2022) Phase 1–2 trial of AAVS3 gene therapy in patients with hemophilia B. N Engl J Med 387(3):237–247. https://doi.org/10.1056/NEJMoa2119913. PMID 35857660

Chu WS, Ng J (2021) Immunomodulation in administration of rAAV: preclinical and clinical adjuvant pharmacotherapies. Front Immunol 1(12):658038. https://doi.org/10.3389/fimmu.2021.658038. PMID 33868303, PMCID PMC 8049138

Cron RQ, Goyal G, Chatham WW (2023) Cytokine storm syndrome. Annu Rev Med 74:321–337. https://doi.org/10.1146/annurev-med-042921-112837. PMID 36228171

Cruz-Almeida Y, Aguirre M, Sorenson HL, Tighe P, Wallet SM, Riley JL 3rd (2015) Age differences in cytokine expression under conditions of health using experimental pain models. Exp Gerontol 72:150–156. https://doi.org/10.1016/j.exger.2015.09.017. PMID 26456458, PMCID PMC 4664177

Czarnomysy R, Bielawska A, Bielawski K (2019) Effect of 2nd and 3rd generation PAMAM dendrimers on proliferation, differentiation, and pro-inflammatory cytokines in human keratinocytes and fibroblasts. Int J Nanomed 3(14):7123–7139. https://doi.org/10.2147/IJN.S211682. PMID 31564869, PMCID PMC 6731979

Deckers J, Anbergen T, Hokke AM, de Dreu A, Schrijver DP, de Bruin K et al (2023) Engineering cytokine therapeutics. Nat Rev Bioeng 2023:1–18. https://doi.org/10.1038/s44222-023-00030-y

Diao L, Tao J, Wang Y, Hu Y, He W (2019) Co-delivery of dihydroartemisinin and HMGB1 siRNA by TAT-modified cationic liposomes through the TLR4 signaling pathway for treatment of lupus nephritis. Int J Nanomed 14:8627–8645. https://doi.org/10.2147/IJN.S220754. PMID 31806961

Dou S, Yao YD, Yang XZ, Sun TM, Mao CQ, Song EW et al (2012) Anti-Her2 single-chain antibody mediated DNMTs-siRNA delivery for targeted breast cancer therapy. J Control Release 161(3):875–883. https://doi.org/10.1016/J.JCONREL.2012.05.015. PMID 22762887

Dranoff G (2004) Cytokines in cancer pathogenesis and cancer therapy. Nat Rev Cancer 4(1):11–22. https://doi.org/10.1038/nrc1252. PMID 14708024

El-Andaloussi S, Lee Y, Lakhal-Littleton S, Li J, Seow Y, Gardiner C et al (2012) Exosome-mediated delivery of siRNA in vitro and in vivo. Nat Protoc 7(12):2112–2126. https://doi.org/10.1038/nprot.2012.131. PMID 23154783

Elli EM, Baratè C, Mendicino F, Palandri F, Palumbo GA (2019) Mechanisms underlying the anti-inflammatory and immunosuppressive activity of Ruxolitinib. Front Oncol 7(9):1186. https://doi.org/10.3389/fonc.2019.01186. PMID 31788449, PMCID PMC 6854013

Elsabahy M, Wooley KL (2013) Cytokines as biomarkers of nanoparticle immunotoxicity. Chem Soc Rev 42(12):5552–5576. https://doi.org/10.1039/C3CS60064E. PMID 23549679

Fihurka O, Sava V, Sanchez-Ramos J (2022) Dual-function hybrid nanoparticles with gene silencing and anti-inflammatory effects. Nanomedicine (Lond) 17(9):577–590. https://doi.org/10.2217/NNM-2021-0458. PMID 35373577

Fraietta JA, Beckwith KA, Patel PR, Ruella M, Zheng Z, Barrett DM et al (2016) Ibrutinib enhances chimeric antigen receptor T-cell engraftment and efficacy in leukemia. Blood 127(9):1117–1127. https://doi.org/10.1182/blood-2015-11-679134. PMID 26813675, PMCID PMC 4778162

Frede A, Neuhaus B, Klopfleisch R, Walker C, Buer J, Müller W et al (2016) Colonic gene silencing using siRNA-loaded calcium phosphate/PLGA nanoparticles ameliorates intestinal inflammation in vivo. J Control Release 222:86–96. https://doi.org/10.1016/J.JCONREL.2015.12.021. PMID 26699423

Frede A, Neuhaus B, Knuschke T, Wadwa M, Kollenda S, Klopfleisch R et al (2017) Local delivery of siRNA-loaded calcium phosphate nanoparticles abates pulmonary inflammation. Nanomedicine 13(8):2395–2403. https://doi.org/10.1016/J.NANO.2017.08.001. PMID 28800875

Godinho BMDC, McCarthy DJ, Torres-Fuentes C, Beltrán CJ, McCarthy J, Quinlan A et al (2014) Differential nanotoxicological and neuroinflammatory liabilities of non-viral vectors for RNA interference in the central nervous system. Biomaterials 35(1):489–499. https://doi.org/10.1016/J.BIOMATERIALS.2013.09.068. PMID 24138827

Gonçalves A, Machado R, Gomes AC, Costa Ad (2020) Nanotechnology solutions for controlled cytokine delivery: an applied perspective. Appl Sci 10(20):7098. https://doi.org/10.3390/APP10207098

Gu P, Zhang Y, Cai G, Liu Z, Hu Y, Liu J et al (2021) Administration routes of polyethylenimine-coated PLGA nanoparticles encapsulating Angelica sinensis polysaccharide vaccine delivery system affect immune responses. Mol Pharm 18(6):2274–2284. https://doi.org/10.1021/acs.molpharmaceut.1c00090. PMID 33926191

Hager K, Machein U, Krieger S, Platt D, Seefried G, Bauer J (1994) Interleukin-6 and selected plasma proteins in healthy persons of different ages. Neurobiol Aging 15(6):771–772. https://doi.org/10.1016/0197-4580(94)90066-3. PMID 7534383

Hallaj-Nezhadi S, Valizadeh H, Dastmalchi S, Baradaran B, Jalali MB, Dobakhti F et al (2011) Preparation of chitosan-plasmid DNA nanoparticles encoding interleukin-12 and their expression in CT-26 colon carcinoma cells. J Pharm Pharm Sci 14(2):181–195. https://doi.org/10.18433/J3TP4T. PMID 21733408

Han X, Zhang H, Butowska K, Swingle KL, Alameh MG, Weissman D et al (2021) An ionizable lipid toolbox for RNA delivery. Nat Commun 12(1):7233. https://doi.org/10.1038/s41467-021-27493-0. PMID 34903741

Hassett KJ, Higgins J, Woods A, Levy B, Xia Y, Hsiao CJ et al (2021) Impact of lipid nanoparticle size on mRNA vaccine immunogenicity. J Control Release 335:237–246. https://doi.org/10.1016/J.JCONREL.2021.05.021. PMID 34019945

Heo MB, Kim SY, Yun WS, Lim YT (2015) Sequential delivery of an anticancer drug and combined immunomodulatory nanoparticles for efficient chemoimmunotherapy. Int J Nanomed 10:5981–5992. https://doi.org/10.2147/IJN.S90104. PMID 26451105

Herman S, Fischer A, Presumey J, Hoffmann M, Koenders MI, Escriou V et al (2015) Inhibition of inflammation and bone erosion by RNA interference-mediated silencing of heterogeneous nuclear RNP A2/B1 in two experimental models of rheumatoid arthritis. Arthritis Rheumatol 67(9):2536–2546. https://doi.org/10.1002/ART.39223. PMID 26017221

Ho YK, Zhou LH, Tam KC, Too HP (2017) Enhanced non-viral gene delivery by coordinated endosomal release and inhibition of β-tubulin deactylase. Nucleic Acids Res 45(6):e38. https://doi.org/10.1093/NAR/GKW1143. PMID 27899629

Hoyer J, Neundorf I (2012) Knockdown of a G protein-coupled receptor through efficient peptide-mediated siRNA delivery. J Control Release 161(3):826–834. https://doi.org/10.1016/J.JCONREL.2012.05.017. PMID 22626516

Huarte E, O'Connor RS, Peel MT, Nunez-Cruz S, Leferovich J, Juvekar A et al (2020) Itacitinib (INCB039110), a JAK1 inhibitor, reduces cytokines associated with cytokine release syndrome induced by CAR T-cell therapy. Clin Cancer Res 26(23):6299–6309. https://doi.org/10.1158/1078-0432.CCR-20-1739. PMID 32998963, PMCID PMC 7895329

Jiang Y, Hardie J, Liu Y, Ray M, Luo X, Das R et al (2018) Nanocapsule-mediated cytosolic siRNA delivery for anti-inflammatory treatment. J Control Release 283:235–240. https://doi.org/10.1016/J.JCONREL.2018.06.001. PMID 29883695

Jing X, Xu M, Song D, Yue T, Wang Y, Zhang P et al (2022) Association between inflammatory cytokines and anti-SARS-CoV-2 antibodies in hospitalized patients with COVID-19. Immun Ageing 19(1):12. https://doi.org/10.1186/s12979-022-00271-2. PMID 35248063, PMCID PMC 8897556

Kawakami S, Ito Y, Charoensit P, Yamashita F, Hashida M (2006) Evaluation of proinflammatory cytokine production induced by linear and branched polyethylenimine/plasmid DNA complexes in mice. J Pharmacol Exp Ther 317(3):1382–1390. https://doi.org/10.1124/jpet.105.100669. PMID 16522808

Khoury M et al (2008) Combined anti-inflammatory therapy using a novel siRNA formulation successfully prevents and cures mice from arthritis. Mol Ther 16:S386. https://doi.org/10.1016/s1525-0016(16)40431-4

Kim ST, Kim CK (2007) Water-soluble chitosan-based antisense oligodeoxynucleotide of interleukin-5 for treatment of allergic rhinitis. Biomaterials 28(22):3360–3368. https://doi.org/10.1016/J.BIOMATERIALS.2007.03.029. PMID 17459468

Kim SS, Ye C, Kumar P, Chiu I, Subramanya S, Wu H et al (2010) Targeted delivery of siRNA to macrophages for anti-inflammatory treatment. Mol Ther 18(5):993–1001. https://doi.org/10.1038/MT.2010.27. PMID 20216529

Kim HO, Kim HS, Youn JC, Shin EC, Park S (2011) Serum cytokine profiles in healthy young and elderly population assessed using multiplexed bead-based immunoassays. J Transl Med 20(9):113. https://doi.org/10.1186/1479-5876-9-113. PMID 21774806, PMCID PMC 3146842

Kleiner G, Marcuzzi A, Zanin V, Monasta L, Zauli G (2013) Cytokine levels in the serum of healthy subjects. Mediators Inflamm 2013:434010. https://doi.org/10.1155/2013/434010. PMID 23533306, PMCID PMC 3606775

Koyama Y, Sugiura K, Yoshihara C, Inaba T, Ito T (2015) Highly effective non-viral antitumor gene therapy system comprised of biocompatible small plasmid complex particles consisting of pDNA, anionic polysaccharide, and fully deprotected linear polyethylenimine. Pharmaceutics 7(3):152–164. https://doi.org/10.3390/pharmaceutics7030152. PMID 26213961

Kulkarni JA, Witzigmann D, Thomson SB, Chen S, Leavitt BR, Cullis PR et al (2021) The current landscape of nucleic acid therapeutics. Nat Nanotechnol 16(6):630–643. https://doi.org/10.1038/s41565-021-00898-0. PMID 34059811

Kuranda K, Weisshaar C, Chen Y, Smith C, Beck H, Kahle K et al (2021) Tocilizumab as a novel immunomodulatory regimen for hemophilia gene therapy. Blood 138(Suppl 1):1853. https://doi.org/10.1182/blood-2021-147511

Langel Ü (2021) Cell-penetrating peptides and transportan. Pharmaceutics 13(7):987. https://doi.org/10.3390/PHARMACEUTICS13070987. PMID 34210007

Le RQ, Li L, Yuan W, Shord SS, Nie L, Habtemariam BA et al (2018) FDA approval summary: tocilizumab for treatment of chimeric antigen receptor T cell-induced severe or life-threatening cytokine release syndrome. Oncologist 23(8):943–947. https://doi.org/10.1634/theoncologist.2018-0028. PMID 29622697, PMCID PMC 6156173

Lee J, Son W, Hong J, Song Y, Yang CS, Kim YH (2021) Down-regulation of TNF-α via macrophage-targeted RNAi system for the treatment of acute inflammatory sepsis. J Control Release 336:344–353. https://doi.org/10.1016/J.JCONREL.2021.06.022. PMID 34147573

Li P, Holliger P, Tagami S (2022) Hydrophobic-cationic peptides modulate RNA polymerase ribozyme activity by accretion. Nat Commun 13(1):3050. https://doi.org/10.1038/s41467-022-30590-3. PMID 35665749

Lian S, Xie R, Ye Y, Xie X, Li S, Lu Y et al (2019) Simultaneous blocking of CD47 and PD-L1 increases innate and adaptive cancer immune responses and cytokine release. EBioMedicine 42:281–295. https://doi.org/10.1016/J.EBIOM.2019.03.018. PMID 30878596

Lin L, Curtin JA, Regis E, Hirsman A, Howard R, Tutino M et al (2022) A systems immunology approach to investigate cytokine responses to viruses and bacteria and their association with disease. Sci Rep 12(1):13463. https://doi.org/10.1038/s41598-022-16509-4. PMID 35931775, PMCID PMC 9356009

Meenakshi Sundaram DN, Plianwong S, Kc R, Ostergaard H, Uludağ H (2022) In vitro cytotoxicity and cytokine production by lipid-substituted low molecular weight branched PEIs used for gene delivery. Acta Biomater 148:279–297. https://doi.org/10.1016/j.actbio.2022.06.030. PMID 35738388

Mestermann K, Giavridis T, Weber J, Rydzek J, Frenz S, Nerreter T et al (2019) The tyrosine kinase inhibitor dasatinib acts as a pharmacologic on/off switch for CAR T cells. Sci Transl Med 11(499):eaau5907. https://doi.org/10.1126/scitranslmed.aau5907. PMID 31270272, PMCID PMC7523030

Mohanan D, Slütter B, Henriksen-Lacey M, Jiskoot W, Bouwstra JA, Perrie Y et al (2010) Administration routes affect the quality of immune responses: a cross-sectional evaluation of particulate antigen-delivery systems. J Control Release 147(3):342–349. https://doi.org/10.1016/j.jconrel.2010.08.012. PMID 20727926

Moreira DA, Santos SD, Leiro V, Pêgo AP (2023) Dendrimers and derivatives as multifunctional nanotherapeutics for Alzheimer's disease. Pharmaceutics 15(4):1054. https://doi.org/10.3390/PHARMACEUTICS15041054

Moudgil KD, Choubey D (2011) Cytokines in autoimmunity: role in induction, regulation, and treatment. J Interferon Cytokine Res 31(10):695–703. https://doi.org/10.1089/JIR.2011.0065. PMID 21942420

Mullard A (2022) FDA approves fifth RNAi drug—Alnylam's next-gen hATTR treatment. Nat Rev Drug Discov 21(8):548–549. https://doi.org/10.1038/D41573-022-00118-X. PMID 35794463

Naha PC, Davoren M, Lyng FM, Byrne HJ (2010) Reactive oxygen species (ROS) induced cytokine production and cytotoxicity of PAMAM dendrimers in J774A.1 cells. Toxicol Appl Pharmacol 246(1–2):91–99. https://doi.org/10.1016/j.taap.2010.04.014. PMID 20420846

Nasrullah M, Meenakshi Sundaram DN, Claerhout J, Ha K, Demirkaya E, Uludag H (2023) Nanoparticles and cytokine response. Front Bioeng Biotechnol 11. https://doi.org/10.3389/fbioe.2023.1243651

Navarro G, Taroumian S, Barroso N, Duan L, Furst D (2014) Tocilizumab in rheumatoid arthritis: a meta-analysis of efficacy and selected clinical conundrums. Semin Arthritis Rheum 43(4):458–469. https://doi.org/10.1016/j.semarthrit.2013.08.001. PMID 24262929

Nori A, Jensen KD, Tijerina M, Kopecková P, Kopecek J (2003) Tat-conjugated synthetic macromolecules facilitate cytoplasmic drug delivery to human ovarian carcinoma cells. Bioconjug Chem 14(1):44–50. https://doi.org/10.1021/bc0255900. PMID 12526691

O'Malley BW, Li D, McQuone SJ, Ralston R (2005) Combination nonviral interleukin-2 gene immunotherapy for head and neck cancer: from bench top to bedside. Laryngoscope 115(3):391–404. https://doi.org/10.1097/00005537-200503000-00002. PMID 15744147

Oh J, Kim SM, Lee EH, Kim M, Lee Y, Ko SH et al (2020) Messenger RNA/polymeric carrier nanoparticles for delivery of heme oxygenase-1 gene in the post-ischemic brain. Biomater Sci 8(11):3063–3071. https://doi.org/10.1039/d0bm00076k. PMID 32348398

Okuda T, Morishita M, Mizutani K, Shibayama A, Okazaki M, Okamoto H (2018) Development of spray-freeze-dried siRNA/PEI powder for inhalation with high aerosol performance and strong pulmonary gene silencing activity. J Control Release 279:99–113. https://doi.org/10.1016/J.JCONREL.2018.04.003. PMID 29627404

Olden BR, Cheng Y, Yu JL, Pun SH (2018) Cationic polymers for non-viral gene delivery to human T cells. J Control Release 282:140–147. https://doi.org/10.1016/J.JCONREL.2018.02.043. PMID 29518467

Ottaviano G, Georgiadis C, Gkazi SA, Syed F, Zhan H, Etuk A et al (2022) Phase 1 clinical trial of CRISPR-engineered CAR19 universal T cells for treatment of children with refractory B cell leukemia. Sci Transl Med 14(668):eabq3010. https://doi.org/10.1126/scitranslmed.abq3010. PMID 36288281

Palmerston Mendes LP, Pan J, Torchilin VP (2017) Dendrimers as nanocarriers for nucleic acid and drug delivery in cancer therapy. Molecules 22(9). https://doi.org/10.3390/MOLECULES22091401. PMID 28832535

Pandi P, Jain A, Kommineni N, Ionov M, Bryszewska M, Khan W (2018) Dendrimer as a new potential carrier for topical delivery of siRNA: a comparative study of dendriplex vs. lipoplex for delivery of TNF-α siRNA. Int J Pharm 550(1–2):240–250. https://doi.org/10.1016/J.IJPHARM.2018.08.024. PMID 30165098

Paunovska K, Loughrey D, Dahlman JE (2022) Drug delivery systems for RNA therapeutics. Nat Rev Genet 23(5):265–280. https://doi.org/10.1038/s41576-021-00439-4. PMID 34983972

Peer D, Park EJ, Morishita Y, Carman CV, Shimaoka M (2008) Systemic leukocyte-directed siRNA delivery revealing cyclin D1 as an anti-inflammatory target. Science 319(5863):627–630. https://doi.org/10.1126/SCIENCE.1149859. PMID 18239128

Pei D, Buyanova M (2019) Overcoming endosomal entrapment in drug delivery. Bioconjug Chem 30(2):273–283. https://doi.org/10.1021/ACS.BIOCONJCHEM.8B00778. PMID 30525488

Piecyk K, Pietrow P, Arnold T, Worch R, Korneeva NL, Jankowska-Anyszka M (2020) Effect of HIV-1 TAT peptide fusion on 5' mRNA cap analogs cell membrane permeability and translation inhibition. Bioconjug Chem 31(4):1156–1166. https://doi.org/10.1021/acs.bioconjchem.0c00080. PMID 32227927

Raison CL, Capuron L, Miller AH (2006) Cytokines sing the blues: inflammation and the pathogenesis of depression. Trends Immunol 27(1):24–31. https://doi.org/10.1016/j.it.2005.11.006. PMID 16316783

Raper SE, Chirmule N, Lee FS, Wivel NA, Bagg A, Gao GP et al (2003) Fatal systemic inflammatory response syndrome in a ornithine transcarbamylase deficient patient following adenoviral gene transfer. Mol Genet Metab 80(1–2):148–158. https://doi.org/10.1016/j.ymgme.2003.08.016. PMID 14567964

Rudd KE, Johnson SC, Agesa KM, Shackelford KA, Tsoi D, Kievlan DR et al (2020) Global, regional, and national sepsis incidence and mortality, 1990–2017: analysis for the global burden of disease study. Lancet 395(10219):200–211. https://doi.org/10.1016/S0140-6736(19)32989-7. PMID 31954465

Ruiz FE, Clancy JP, Perricone MA, Bebok Z, Hong JS, Cheng SH et al (2001) A clinical inflammatory syndrome attributable to aerosolized lipid-DNA administration in cystic fibrosis. Hum Gene Ther 12(7):751–761. https://doi.org/10.1089/104303401750148667. PMID 11339892

Sakurai F, Terada T, Yasuda K, Yamashita F, Takakura Y, Hashida M (2002) The role of tissue macrophages in the induction of proinflammatory cytokine production following intravenous injection of lipoplexes. Gene Ther 9(16):1120–1126. https://doi.org/10.1038/sj.gt.3301784. PMID 12140741

Santos A, Veiga F, Figueiras A (2019) Dendrimers as pharmaceutical excipients: synthesis, properties, toxicity and biomedical applications. Materials (Basel) 13(1):65. https://doi.org/10.3390/MA13010065. PMID 31877717

Shen ZL, Xia YQ, Yang QS, Tian WD, Chen K, Ma YQ, Lin Z et al (2017) Polymer-nucleic acid interactions. Top Curr Chem (Cham) 375(2):44. https://doi.org/10.1007/S41061-017-0131-X. PMID 28357711

Shi Q, Rondon-Cavanzo EP, Dalla Picola IP, Tiera MJ, Zhang X, Dai K et al (2018) In vivo therapeutic efficacy of TNFα silencing by folate-PEG-chitosan-DEAE/siRNA nanoparticles in arthritic mice. Int J Nanomed 13:387–402. https://doi.org/10.2147/IJN.S146942. PMID 29391796

Shin HJ, Park H, Shin N, Shin J, Gwon DH, Kwon HH et al (2020) p66shc siRNA nanoparticles ameliorate chondrocytic mitochondrial dysfunction in osteoarthritis. Int J Nanomed 15:2379–2390. https://doi.org/10.2147/IJN.S234198. PMID 32308389

Shirley JL, de Jong YP, Terhorst C, Herzog RW (2020) Immune responses to viral gene therapy vectors. Mol Ther 28(3):709–722. https://doi.org/10.1016/j.ymthe.2020.01.001. PMID 31968213, PMCID PMC 7054714

Strati P, Ahmed S, Kebriaei P, Nastoupil LJ, Claussen CM, Watson G et al (2020) Clinical efficacy of anakinra to mitigate CAR T-cell therapy-associated toxicity in large B-cell lymphoma. Blood Adv 4(13):3123–3127. https://doi.org/10.1182/bloodadvances.2020002328. PMID 32645136, PMCID PMC 7362369

Stroud CR, Hegde A, Cherry C, Naqash AR, Sharma N, Addepalli S et al (2019) Tocilizumab for the management of immune mediated adverse events secondary to PD-1 blockade. J Oncol Pharm Pract 25(3):551–557. https://doi.org/10.1177/1078155217745144. PMID 29207939

Suhorutsenko J, Oskolkov N, Arukuusk P, Kurrikoff K, Eriste E, Copolovici DM et al (2011) Cell-penetrating peptides, PepFects, show no evidence of toxicity and immunogenicity in vitro and in vivo. Bioconjug Chem 22(11):2255–2262. https://doi.org/10.1021/bc200293d. PMID 21978260

Sun JY, Anand-Jawa V, Chatterjee S, Wong KK (2003) Immune responses to adeno-associated virus and its recombinant vectors. Gene Ther 10(11):964–976. https://doi.org/10.1038/sj.gt.3302039. PMID 12756417

Tahtinen S, Tong AJ, Himmels P, Oh J, Paler-Martinez A, Kim L et al (2022) IL-1 and IL-1RA are key regulators of the inflammatory response to RNA vaccines. Nat Immunol 23(4):532–542. https://doi.org/10.1038/s41590-022-01160-y. PMID 35332327

Tang M, Guo C, Sun M, Zhou H, Peng X, Dai J et al (2022) Effective delivery of osteopontin small interference RNA using exosomes suppresses liver fibrosis via TGF-β1 signaling. Front Pharmacol 13:882243. https://doi.org/10.3389/FPHAR.2022.882243. PMID 36120332

Tsai SJ, Black SK, Jewell CM (2020) Leveraging the modularity of biomaterial carriers to tune immune responses. Adv Funct Mater 30(48). https://doi.org/10.1002/ADFM.202004119. PMID 33692662

Uludag H, Ubeda A, Ansari A (2019) At the intersection of biomaterials and gene therapy: progress in non-viral delivery of nucleic acids. Front Bioeng Biotechnol 7(JUN):131. https://doi.org/10.3389/FBIOE.2019.00131. PMID 31214586

Verbeke R, Lentacker I, De Smedt SC, Dewitte H (2019) Three decades of messenger RNA vaccine development. Nano Today 28:100766. https://doi.org/10.1016/J.NANTOD.2019.100766

Walters AA, Dhadwar B, Al-Jamal KT (2021) Modulating expression of inhibitory and stimulatory immune 'checkpoints' using nanoparticulate-assisted nucleic acid delivery. EBioMedicine 73:103624. https://doi.org/10.1016/j.ebiom.2021.103624. PMID 34688033

Wang X, Lin Y (2008) Tumor necrosis factor and cancer, buddies or foes? Acta Pharmacol Sin 29(11):1275–1288. https://doi.org/10.1111/J.1745-7254.2008.00889.X. PMID 18954521

Wang X, Liu S, Sun Y, Yu X, Lee SM, Cheng Q et al (2023) Preparation of selective organ-targeting (SORT) lipid nanoparticles (LNPs) using multiple technical methods for tissue-specific mRNA delivery. Nat Protoc 18(1):265–291. https://doi.org/10.1038/s41596-022-00755-x. PMID 36316378

Watson OJ, Barnsley G, Toor J, Hogan AB, Winskill P, Ghani AC (2022) Global impact of the first year of COVID-19 vaccination: a mathematical modelling study. Lancet Infect Dis 22(9):1293–1302. https://doi.org/10.1016/S1473-3099(22)00320-6. PMID 35753318

Weiss M, Fan J, Claudel M, Sonntag T, Didier P, Ronzani C et al (2021) Density of surface charge is a more predictive factor of the toxicity of cationic carbon nanoparticles than zeta potential. J Nanobiotechnol 19(1):5. https://doi.org/10.1186/s12951-020-00747-7. PMID 33407567

Wong SW, Richard S, Lin Y, Madduri D, Jackson CC, Zudaire E et al (2021) Anakinra targeting cytokine release syndrome associated with chimeric antigen receptor T-cell therapies. Blood 138(Suppl 1):2812. https://doi.org/10.1182/blood-2021-150592

Wood H (2018) FDA approves patisiran to treat hereditary transthyretin amyloidosis. Nat Rev Neurol 14(10):570. https://doi.org/10.1038/s41582-018-0065-0. PMID 30158559

Wu T, Tan M, Gong H, Wang Y, Shuai X (2018) Co-delivery of andrographolide and Notch1-targeted siRNA to macrophages with polymer-based nanocarrier for enhanced anti-inflammation. Chin J Polym Sci 36(12):1312–1320. https://doi.org/10.1007/s10118-018-2158-z

Wu L, Wu LP, Wu J, Sun J, He Z, Rodríguez-Rodríguez C et al (2021) Poly(lactide-co-glycolide) nanoparticles mediate sustained gene silencing and improved biocompatibility of siRNA delivery systems in mouse lungs after pulmonary administration. ACS Appl Mater Interfaces 13(3):3722–3737. https://doi.org/10.1021/ACSAMI.0C21259. PMID 24503126

Xiang Z, Kuranda K, Quinn W, Chekaoui A, Ambrose R, Hasanpourghai M et al (2022) The effect of rapamycin and ibrutinib on antibody responses to adeno-associated virus vector-mediated gene transfer. Hum Gene Ther 33(11–12):614–624. https://doi.org/10.1089/hum.2021.258. PMID 35229644

Xiao B, Laroui H, Viennois E, Ayyadurai S, Charania MA, Zhang Y et al (2014) Nanoparticles with surface antibody against CD98 and carrying CD98 small interfering RNA reduce colitis in mice. Gastroenterology 146(5):1289–300.e1. https://doi.org/10.1053/J.GASTRO.2014.01.056. PMID 24503126

Xiao B, Ma P, Viennois E, Merlin D (2016) Urocanic acid-modified chitosan nanoparticles can confer anti-inflammatory effect by delivering CD98 siRNA to macrophages. Colloids Surf B Biointerfaces 143:186–193. https://doi.org/10.1016/J.COLSURFB.2016.03.035. PMID 27011348

Xiao B, Ma P, Ma L, Chen Q, Si X, Walter L et al (2017) Effects of tripolyphosphate on cellular uptake and RNA interference efficiency of chitosan-based nanoparticles in Raw 264.7 macrophages. J Colloid Interface Sci 490:520–528. https://doi.org/10.1016/J.JCIS.2016.11.088. PMID 27918990

Xu L, Yeudall WA, Yang H (2017) Folic acid-decorated polyamidoamine dendrimer exhibits high tumor uptake and sustained highly localized retention in solid tumors: its utility for local siRNA delivery. Acta Biomater 57:251–261. https://doi.org/10.1016/J.ACTBIO.2017.04.023. PMID 28438704

Yan H, Duan X, Pan H, Akk A, Sandell LJ, Wickline SA et al (2019) Development of a peptide-siRNA nanocomplex targeting NF-κB for efficient cartilage delivery. Sci Rep 9(1):442. https://doi.org/10.1038/S41598-018-37018-3. PMID30679644

Yang ZJ, Xu SL, Chen B, Zhang SL, Zhang YL, Wei W et al (2009) Hepatocyte growth factor plays a critical role in the regulation of cytokine production and induction of endothelial progenitor cell mobilization: a pilot gene therapy study in patients with coronary heart disease. Clin Exp Pharmacol Physiol 36(8):790–796. https://doi.org/10.1111/j.1440-1681.2009.05151.x. PMID 19215239

Yang G, Zhou D, Dai Y, Li Y, Wu J, Liu Q et al (2022) Construction of PEI-EGFR-PD-L1-siRNA dual functional nano-vaccine and therapeutic efficacy evaluation for lung cancer. Thorac Cancer 13(21):2941–2950. https://doi.org/10.1111/1759-7714.14618. PMID 36117149

Yew NS, Zhao H, Wu IH, Song A, Tousignant JD, Przybylska M et al (2000) Reduced inflammatory response to plasmid DNA vectors by elimination and inhibition of immunostimulatory CpG motifs. Mol Ther 1(3):255–262. https://doi.org/10.1006/mthe.2000.0036. PMID 10933941

Yoon PO, Park JW, Lee CM, Kim SH, Kim HN, Ko Y et al (2016) Self-assembled micelle interfering RNA for effective and safe targeting of dysregulated genes in pulmonary fibrosis. J Biol Chem 291(12):6433–6446. https://doi.org/10.1074/jbc.M115.693671. PMID 26817844

Yuan Z, Li B, Gu W, Luozhong S, Li R, Jiang S (2022) Mitigating the immunogenicity of AAV-mediated gene therapy with an immunosuppressive phosphoserine-containing zwitterionic peptide. J Am Chem Soc 144(44):20507–20513. https://doi.org/10.1021/jacs.2c09484. PMID 36301739

Zaiss AK, Liu Q, Bowen GP, Wong NC, Bartlett JS, Muruve DA (2002) Differential activation of innate immune responses by adenovirus and adeno-associated virus vectors. J Virol 76(9):4580–4590. https://doi.org/10.1128/jvi.76.9.4580-4590.2002. PMID 11932423, PMCID PMC 155101

Zang Z, Mahendran R, Wu Q, Yong T, Esuvaranathan K (2004) Non-viral tumor necrosis factor-alpha gene transfer decreases the incidence of orthotopic bladder tumors. Int J Mol Med 14(4):713–717. https://doi.org/10.3892/IJMM.14.4.713/HTML. PMID 15375606

Zhang C, Delawary M, Huang P, Korchak JA, Suda K, Zubair AC (2021a) IL-10 mRNA engineered MSCs demonstrate enhanced anti-inflammation in an acute GvHD model. Cells 10(11). https://doi.org/10.3390/CELLS10113101. PMID 34831324

Zhang MM, Bahal R, Rasmussen TP, Manautou JE, Zhong XB (2021b) The growth of siRNA-based therapeutics: updated clinical studies. Biochem Pharmacol 189:114432. https://doi.org/10.1016/J.BCP.2021.114432. PMID 33513339

Zhang L, Wang S, Xu J, Zhang R, Zhu H, Wu Y et al (2021c) Etanercept as a new therapeutic option for cytokine release syndrome following chimeric antigen receptor T cell therapy. Exp Hematol Oncol 10(1):16. https://doi.org/10.1186/s40164-021-00209-2. PMID 33608054, PMCID PMC 7893957

Zhao Y, Chen H, Wang L, Guo Z, Liu S, Luo S (2020) Cationic solid lipid nanoparticles loaded by integrin β1 plasmid DNA attenuates IL-1β-induced apoptosis of chondrocyte. Aging 12(22):22527–22537. https://doi.org/10.18632/AGING.103656. PMID 33289706

Zhou HS, Liu DP, Liang CC (2004) Challenges and strategies: the immune responses in gene therapy. Med Res Rev 24(6):748–761. https://doi.org/10.1002/med.20009. PMID 15250039

Zolnik BS, González-Fernández A, Sadrieh N, Dobrovolskaia MA (2010) Nanoparticles and the immune system. Endocrinology 151(2):458–465. https://doi.org/10.1210/EN.2009-1082. PMID 20016026

Zuckerman JE, Gritli I, Tolcher A, Heidel JD, Lim D, Morgan R et al (2014) Correlating animal and human phase Ia/Ib clinical data with CALAA-01, a targeted, polymer-based nanoparticle containing siRNA. Proc Natl Acad Sci USA 111(31):11449–11454. https://doi.org/10.1073/pnas.1411393111. PMID 25049380, PMCID PMC 4128111

Controlled Drug Delivery System

Akhilraj Pillai, Dhanashree Bhande, and Vinal Pardhi

Abstract This comprehensive chapter delves into the world of controlled drug delivery systems, exploring their fundamental aspects, advancements, and future possibilities. The chapter begins by emphasizing the importance of controlled drug delivery in optimizing therapeutic outcomes and enhancing patient compliance. Various drug delivery systems are discussed, including oral, injectable, transdermal, and implantable systems, each offering unique characteristics and applications. The working mechanisms of these systems are thoroughly examined, including diffusion-controlled, dissolution-controlled, osmotic pressure-controlled, and targeted drug delivery mechanisms. The advantages of controlled drug delivery systems are highlighted, such as prolonged drug release, improved therapeutic efficacy, reduced dosing frequency, minimized side effects, and increased patient convenience. However, challenges and limitations also exist, requiring attention and consideration. Formulation complexities, manufacturing intricacies, regulatory considerations, and patient variability are hurdles in developing and implementing controlled drug delivery systems. Recent advancements in the field are explored, including the emergence of intelligent drug delivery systems that respond to specific cues, nanotechnology-based platforms enabling precise drug targeting, innovative combination therapy approaches, and bio-responsive drug delivery systems that adapt to physiological changes. These advancements pave the way for promising future applications, such as personalized medicine, targeted drug delivery, remote-controlled systems, and the integration of bioprinting and 3D printing technologies. In conclusion, this chapter provides a comprehensive understanding of controlled drug delivery systems, showcasing their potential to revolutionize drug therapy and improve patient outcomes. Researchers and practitioners can unlock new avenues in pharmaceutical

A. Pillai
Department of Biological Sciences, St. John's University, New York, USA

D. Bhande
Resolved Diagnostic Pvt Ltd., Mumbai, India

V. Pardhi (✉)
Scientity Services, Thane, India
e-mail: vinal@scientityservices.com

T. S. Santra and A. U. S. Shinde (eds.), *Advanced Drug Delivery*,
Studies in Mechanobiology, Tissue Engineering and Biomaterials 26,
https://doi.org/10.1007/978-981-99-6564-9_11

science and create innovative solutions for controlled drug delivery by addressing the challenges, harnessing recent advancements, and exploring prospects.

1 Introduction

Controlled drug delivery systems have emerged as transformative approaches in medicine, revolutionizing how therapeutic agents are delivered within the body. These systems enable precise and targeted delivery, offering significant advantages over conventional drug dosage forms. They enhance therapeutic efficacy, minimize side effects, and improve patient compliance by optimizing drug release rates, achieving sustained drug levels, and facilitating targeted drug delivery.

One key advantage of controlled drug delivery systems is their ability to enhance therapeutic efficacy. By precisely controlling the delivery of drugs, these systems ensure optimal rates and concentrations, maximizing their effectiveness in treating diseases or conditions. They can maintain sustained drug levels within the body, leading to a consistent therapeutic effect over an extended period. Moreover, they can minimize side effects by preventing drug concentration spikes and reducing the likelihood of adverse reactions. This is particularly beneficial for drugs with narrow therapeutic windows or high toxicity.

Additionally, controlled drug delivery systems improve patient compliance and convenience. Traditional drug dosage forms often require frequent dosing, which can be inconvenient and lead to non-adherence. Controlled delivery systems can be designed to release drugs over an extended duration, reducing the frequency of administration and simplifying the treatment regimen. This enhances patient convenience and improves treatment outcomes by ensuring consistent drug intake. Furthermore, these systems enable targeted drug delivery, releasing drugs specifically at the site of action. This minimizes drug exposure to healthy tissues and organs, increasing the drug's concentration at the desired location and improving therapeutic effects while reducing the potential for systemic side effects.

Controlled drug delivery systems have revolutionized medicine by enabling precise and targeted drug delivery. They enhance therapeutic efficacy, minimize side effects, and improve patient compliance. These systems offer significant advantages over conventional drug dosage forms by optimizing drug release rates, achieving sustained drug levels, and facilitating targeted drug delivery. They represent a vital area of research and development in modern healthcare, aiming to improve drug delivery further and advance medical treatments.

1.1 Background

Advancements in pharmaceutical sciences and technology have driven the development of drug delivery systems. Torchilin (2006) emphasizes the pivotal role of these systems in effectively delivering drugs to their intended targets. Over the years, extensive research has focused on designing carriers and strategies that improve drug bioavailability, overcome physiological barriers, and enhance therapeutic outcomes.

Advances in pharmaceutical sciences, biomaterials, and nanotechnology have significantly contributed to developing controlled drug delivery systems. These multidisciplinary efforts have allowed researchers to design carriers with precise control over drug release profiles, spatial targeting, and temporal delivery. Incorporating biomaterials with desirable properties, such as biocompatibility and biodegradability, has facilitated translation of these systems from the laboratory to clinical applications.

Controlled drug delivery systems have applications in various therapeutic areas, including cancer, chronic diseases, pain management, and infectious diseases. By providing sustained and targeted drug delivery, these systems improve therapeutic outcomes, reduce side effects, and enhance patient compliance. The versatility of controlled drug delivery systems also allows for personalized medicine, tailoring treatments to individual patient needs and optimizing therapeutic efficacy.

The advancements in pharmaceutical sciences, biomaterials, and nanotechnology present exciting opportunities for further innovation and refinement of controlled drug delivery systems. Researchers continue to explore new materials, technologies, and approaches to enhance drug delivery's precision, efficiency, and safety. Future developments may involve incorporating stimuli responsive materials that can respond to specific cues within the body to trigger drug release. Additionally, integrating imaging modalities or sensors within these systems could enable real-time drug release and patient response monitoring.

Moreover, the emerging field of nanomedicine holds promise for developing more sophisticated controlled drug delivery systems. Nanoscale carriers, such as nanoparticles or liposomes, offer unique advantages in their ability to encapsulate and deliver drugs with high precision. Further advancements in nanotechnology may lead to the design of intelligent nanocarriers that can actively target specific cells or tissues, increasing drug delivery efficiency and reducing off-target effects.

As researchers delve deeper into the intricacies of controlled drug delivery systems, there is a growing realization that these systems have the potential to revolutionize the field of medicine even further. By unraveling the complexities of drug release, targeting, and personalized medicine, scientists can unlock new avenues for improved therapies and patient care. The ongoing exploration and discussion surrounding controlled drug delivery systems are critical to maximizing their potential and driving the future of healthcare innovation.

1.2 Significance of Controlled Drug Delivery Systems

Controlled drug delivery systems offer distinct advantages compared to conventional dosage forms, making them indispensable in modern healthcare. These systems enable targeted drug delivery, ensuring therapeutic agents reach specific tissues or cells precisely (Reis et al. 2006) highlights the potential of controlled drug delivery systems in improving the efficacy of cancer therapies by selectively targeting tumor tissues and reducing systemic toxicity. Modulating drug release rates also allows for sustained drug levels, particularly crucial for chronic diseases such as diabetes and cardiovascular disorders (Allen and Cullis 2004) underscores the importance of controlled drug delivery systems in industrial chemistry and biotechnology, as they enhance drug performance, increase patient compliance, and improve therapeutic outcomes.

This section aims to provide a comprehensive overview of controlled drug delivery systems, encompassing their design principles and applications in modern healthcare. We delve into controlled drug delivery systems, including sustained-release, stimuli-responsive, and targeted delivery systems. Furthermore, we explore the factors that influence the design and development of these systems, such as drug properties, release kinetics, and biocompatibility. By discussing the challenges and prospects associated with controlled drug delivery systems, we aim to provide valuable insights to researchers and healthcare professionals. Our discussion serves as a resource to foster innovation, facilitate advancements, and contribute to the continued development of drug delivery strategies.

2 Fundamentals of Controlled Drug Delivery

2.1 Principles of Drug Delivery

The field of drug delivery encompasses a range of principles and techniques that aim to optimize the delivery of therapeutic agents to their intended targets. Torchilin provides a comprehensive overview of drug delivery systems, highlighting the importance of understanding the principles underlying drug release, distribution, and pharmacokinetics (Torchilin et al. 2007. These principles include routes of administration, such as oral, transdermal, parenteral, and inhalation routes, which have been developed to accommodate different therapeutic needs (Davis et al. 2008). Various ways of drug administration are shown in Fig. 1 (Adepu and Ramakrishna 2021). The preferred route of drug administration depends on three main factors: The part of the body being treated, the way the drug works within the body, and the solubility and permeability of the drug. Drug formulation, release kinetics, and bioavailability are crucial considerations in designing drug delivery systems. These aspects determine the drug's release rate, duration, and concentration at the target site (Torchilin 2006).

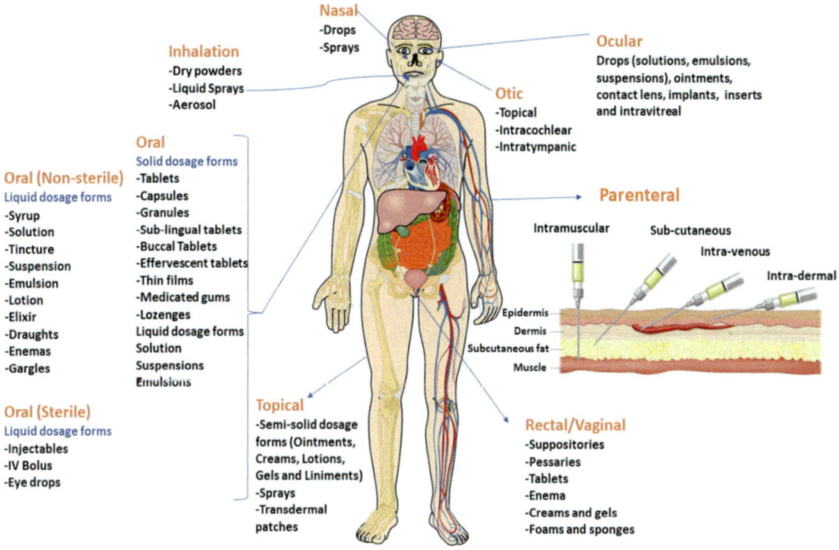

Fig. 1 A schematic representation of various routes of drug administration (©2021 by the authors. Licensee MDPI, Basel, Switzerland)

2.2 *Factors Influencing Drug Delivery*

Multiple factors can influence the effectiveness of drug delivery systems. Mitragotri et al. discusses the challenges of administering biopharmaceuticals, including barriers to drug transport, immune responses, and stability issues (Allen and Cullis 2013). Physicochemical properties of drugs, such as solubility, molecular weight, and stability, play a crucial role in their delivery and release (Chakravarthi 2012). Furthermore, physiological factors, such as pH, temperature, and enzymatic activity, can affect drug behavior and release kinetics (Brigger et al. 2002). The formulation, including the choice of carriers, polymers, and excipients, also influences drug delivery and its therapeutic outcomes (Peer et al. 2007). The selection of appropriate drug delivery systems must consider these factors to ensure optimal drug performance (Reis et al. 2006).

2.3 *Challenges in Conventional Drug Delivery*

Conventional drug delivery methods often face challenges that limit their efficacy and therapeutic potential. Prausnitz et al. (2004) discuss the limitations of traditional delivery approaches, including low bioavailability, poor patient compliance, and systemic side effects (Desai et al. 1996). Difficulties in achieving targeted drug delivery and sustaining therapeutic drug levels over time further restrict the effectiveness of conventional systems (Soppimath et al. 2001; Langer 1998). These challenges have spurred the development of controlled drug delivery systems that offer solutions to overcome these limitations (Jong and Borm 2008; Oerlemans et al. 2010). Controlled drug delivery systems aim to improve drug efficacy by delivering drugs in a controlled manner, optimizing drug release profiles, and minimizing systemic exposure to reduce side effects (Allen and Cullis 2004).

Controlled drug delivery systems exhibit advantages that are correlated in the above figure. The formulated drug candidate shows better bioavailability and thus required in minimum doses quantity which in-turn reduces side effects thereby increasing the efficacy of the formulation. The drug stability can also be correlated with better bioavailability as well as efficacy.

2.4 Advantages of Controlled Drug Delivery Systems

Controlled drug delivery systems offer numerous advantages over conventional approaches shows in Fig. 2 (Saikia et al. 2015). Langer (1998) highlights their potential for improving drug efficacy, reducing side effects, and enhancing patient compliance (Saikia et al. 2015; Oerlemans et al. 2010). These systems precisely control drug release rates, achieving sustained and controlled drug concentrations at the target site (Anderson and Shive 1997). Additionally, they enable targeted delivery to specific tissues or cells, enhancing therapeutic outcomes and minimizing off-target effects (Reis et al. 2006). The use of nanotechnology in drug delivery, as discussed by Farokhzad and Langer, has further expanded the possibilities of controlled drug release and targeting (Chakravarthi 2012; Panyam and Labhasetwar 2003). By leveraging these advantages, controlled drug delivery systems have the potential to revolutionize the field of medicine and improve patient care (Jong and Borm 2008; Langer 1990).

3 Sustained-Release Systems

Sustained-release systems are a vital component of controlled drug delivery, as they enable the prolonged and controlled release of drugs over an extended period. These systems offer significant advantages over conventional immediate-release formulations, making them highly desirable in various therapeutic applications.

One key advantage of sustained-release systems is their ability to enhance therapeutic efficacy. By releasing drugs gradually and maintaining a consistent drug concentration within the body, these systems can optimize the drug's effectiveness in treating diseases or conditions. This sustained release allows for a continuous drug supply at the desired site of action, ensuring a prolonged therapeutic effect. In

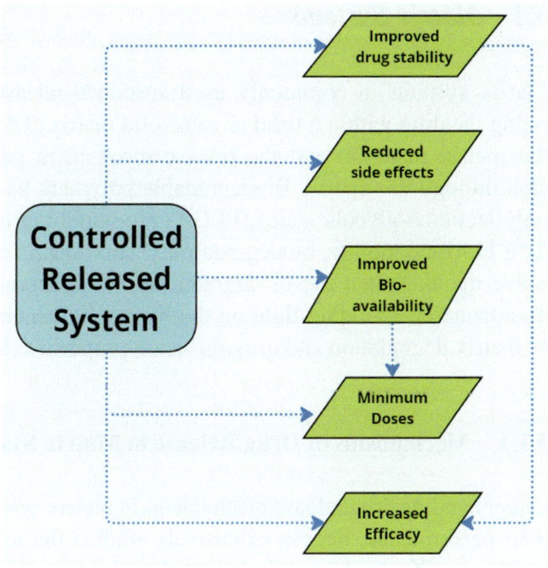

Fig. 2 Interdependent advantages of controlled drug delivery system

contrast, immediate-release formulations may result in fluctuating drug levels, which can lead to suboptimal therapeutic outcomes.

Reduced dosing frequency is another benefit of sustained-release systems. Patients must take medication multiple times daily with conventional drug dosage forms to maintain therapeutic levels. This can be inconvenient and contribute to non-adherence to the prescribed treatment regimen. On the other hand, sustained-release systems can be designed to release drugs gradually over an extended period, thereby reducing dosing frequency. This improves patient convenience and promotes better compliance, as patients are more likely to adhere to a simplified dosing schedule.

Minimizing side effects is another crucial aspect of sustained-release systems. Immediate-release formulations can sometimes lead to drug concentration spikes followed by rapid clearance, which may increase the risk of adverse reactions. In contrast, sustained-release systems provide a controlled release profile that avoids concentration peaks and maintains drug levels within the therapeutic range. This controlled delivery helps minimize side effects associated with high drug concentrations or rapid fluctuations in drug levels.

Enhanced patient compliance is a significant advantage of sustained-release systems. Simplifying the dosing regimen by reducing the frequency of administration can improve patient adherence to the prescribed treatment. Patients are more likely to comply with a medication schedule requiring fewer daily administrations. Moreover, sustained-release systems can provide a consistent drug supply, reducing the risk of missed or forgotten doses. Improved patient compliance ultimately leads to better treatment outcomes and overall patient health.

3.1 Matrix Systems

Matrix systems, a commonly used sustained-release approach, involve incorporating the drug within a solid or semisolid matrix. The drug is dispersed throughout the matrix material, and the release mechanism primarily relies on drug diffusion through the matrix. Biodegradable polymers like poly(lactic acid) (PLA) and poly(lactic-co-glycolic acid) (PLGA) are widely utilized in matrix systems due to their biocompatibility, biodegradability, and tunable release profiles. Anderson and Shive demonstrated the biodegradation and biocompatibility of PLA and PLGA microspheres, shedding light on the factors influencing drug release kinetics, such as matrix degradation and drug diffusion properties (Siepmann and Peppas 2001).

3.1.1 Mechanisms of Drug Release in Matrix Systems

Understanding drug release mechanisms in matrix systems is essential for optimizing their performance. Peppas extensively studied the release of drugs from polymer-based matrices and contributed to the characterization of drug release profiles, including Fickian and non-Fickian behavior (Ritger and Peppas 1987). Ritger and Peppas developed a simple equation to describe drug release from swellable devices, which aids in the understanding of Fickian and anomalous release phenomena (Korsmeyer et al. 1983). These studies provided valuable insights into the diffusion and release kinetics within matrix systems, facilitating the formulation design and optimization process.

3.1.2 Formulation Strategies and Examples

Formulation strategies play a critical role in achieving desired drug release profiles and maximizing the therapeutic potential of sustained-release systems. Siepmann and Peppas focused on modeling drug release from hydroxypropyl methylcellulose (HPMC)-based matrix systems, providing valuable tools for predicting drug release profiles and optimizing formulation parameters (Singh and Lillard 2009). Gupta et al. developed Eudragit-based matrix tablets for the sustained delivery of nifedipine, demonstrating the versatility and effectiveness of specific polymer matrices in oral drug delivery (Kim et al. 2011). These studies highlight the importance of considering the selection of matrix materials, their physicochemical properties, and the incorporation of drug release modifiers to achieve the desired release kinetics.

3.2 *Reservoir Systems*

Reservoir systems are sustained-release systems that involve encapsulating the drug within a reservoir or core surrounded by a rate-controlling membrane (Siepmann and Peppas 2001). The membrane plays a crucial role in regulating the release of the drug, ensuring controlled and predictable drug delivery. Reservoir systems offer several advantages, including achieving specific release profiles, maintaining drug stability, and providing flexibility in drug loading and release rates.

3.2.1 Design Principles of Reservoir Systems

Designing effective reservoir systems requires careful consideration of various principles to achieve the desired drug release characteristics. Some fundamental design principles include:

(a) Membrane Selection: The choice of membrane material is critical in determining the drug release rate and permeability. Different polymers, such as ethyl cellulose, polyethylene, and polyvinyl chloride, can create membranes with varying permeability properties (Singh and Lillard 2009). The selection of the membrane material should consider factors such as drug solubility, desired release kinetics, and drug stability within the reservoir.

(b) Membrane Thickness: The membrane's thickness can influence the drug's diffusion and release rate. Thinner membranes generally result in faster drug release, while thicker membranes can provide a slower and more controlled release. Optimizing the membrane thickness ensures that the desired release profile is achieved.

(c) Drug loading and Reservoir Capacity: The drug loading capacity of the reservoir and size directly affect the amount of drug that can be delivered and the duration of release. Determining the appropriate drug loading level and reservoir size is essential to achieve the desired therapeutic effect and course of action.

(d) Membrane Permeability Modification: The membrane's permeability can be modified by incorporating additives or changing the composition of the membrane. Modifying membrane permeability allows fine-tuning the drug release rate and achieving the desired release kinetics (Siepmann and Peppas 2001).

3.2.2 Examples and Applications

Reservoir systems find applications in various therapeutic areas and drug delivery approaches. Some notable examples and applications include:

(a) Transdermal Drug Delivery: Reservoir systems are widely used in transdermal drug delivery to provide controlled and sustained release of drugs through the skin. For example, reservoir patches containing drugs such as nicotine, estrogen, or pain medications are used for long-term delivery, providing consistent therapeutic levels over an extended period (Xue et al. 2011).

(b) Implantable Devices: Reservoir systems are utilized in implantable devices, such as drug-eluting stents or implants for localized drug delivery. These devices release drugs directly at the target site, ensuring optimal therapeutic concentrations and reducing systemic side effects. Examples include drug-eluting cardiovascular stents that prevent restenosis and implantable contraceptives that provide long-term contraception (Zhuang et al. 2013).

(c) Ophthalmic Drug Delivery: Reservoir systems are employed in ophthalmic drug delivery to achieve prolonged drug release in the eye. This allows for less frequent administration and improved patient compliance. Reservoir-based ocular inserts or contact lenses loaded with drugs are used for conditions such as glaucoma or dry eye syndrome (Aman et al. 2015).

(d) Controlled Release Injections: Reservoir systems are utilized in controlled-release injectables, where the drug is encapsulated within the reservoir and released over an extended period. This approach eliminates the need for frequent injections and provides sustained therapeutic effects. Examples include long-acting depot formulations for contraception or treating chronic conditions like schizophrenia (Lai et al. 2015).

In summary, reservoir systems provide controlled and predictable drug release by encapsulating the drug within a reservoir surrounded by a rate-controlling membrane. Design principles such as membrane selection, thickness optimization, drug loading, and membrane permeability modification are essential in achieving the desired release profiles. Reservoir systems find applications in various drug delivery approaches, including transdermal delivery, implantable devices, ophthalmic delivery, and controlled-release injections.

4 Stimuli-Responsive Systems

4.1 pH-Responsive Systems

pH-responsive drug delivery systems have gained significant attention due to their ability to release drugs in response to changes in pH within specific target environments. Torchilin (2011) discussed the development of multifunctional, stimuli-sensitive nanoparticulate systems for drug delivery, including pH-responsive nanoparticles. Jain et al. (2008) explored the use of pH-sensitive liposomes as a novel carrier for drug delivery. Furthermore, Kim et al. (2016) focused on pH-sensitive polymeric nanoparticles designed for tumor-targeting drug delivery.

Designing pH-responsive systems involves considering nanoparticle composition, surface charge, and pH-dependent drug release mechanisms. Li and Huang (2010) discussed nanoparticles that can evade the reticuloendothelial system, emphasizing the supported bilayer's role. Choi et al. (2010) developed self-assembled hyaluronic acid nanoparticles for active tumor targeting, utilizing pH responsiveness in their design. Hu et al. (2014) also explored enzyme-responsive nanomaterials for controlled drug delivery, highlighting pH-responsive properties.

4.2 *Temperature-Responsive Systems*

Temperature-responsive systems utilize thermosensitive polymers that undergo reversible phase transitions in response to changes in temperature, enabling triggered drug release. Peppas et al. (2000) provided an overview of hydrogels in pharmaceutical formulations, including thermosensitive polymers. Schild (1992) extensively discussed poly(N-isopropyl acrylamide) (PNIPAAm), a commonly used thermosensitive polymer, in terms of its experimental and theoretical aspects. Furthermore, Zhang et al. (2016) developed temperature-responsive polymer-grafted mesoporous silica nanoparticles for remotely triggered drug release.

Applications of temperature-responsive systems include controlled drug delivery and enhanced drug targeting. Chen et al. (2018) reviewed synthetic and biopolymeric pH-responsive nanoparticles for drug delivery applications. Fan et al. (2012) developed an intelligent pH-responsive nano-carrier for the controlled release of 5-fluorouracil. Moreover, Liu et al. (2019) investigated temperature-triggered gelation of poly(N-isopropyl acrylamide)-based microgels to achieve controlled drug delivery.

4.3 *Enzyme-Responsive Systems*

Enzyme-responsive drug delivery systems offer the potential for targeted and triggered drug release in response to specific enzymes in pathological conditions. Khullar et al. (2012) extensively reviewed enzyme-responsive drug delivery systems in cancer therapy, focusing on various approaches and mechanisms. Kopeček (2013) discussed innovative and genetically engineered biomaterials and drug delivery systems, including enzyme-responsive plans.

Understanding enzyme-triggered drug release mechanisms is crucial for designing effective enzyme-responsive systems. Shi et al. (2014) reported complete regression of xenograft tumors upon targeted delivery of paclitaxel via Π-Π stacking stabilized polymeric micelles. Chen et al. (2018) provided an overview of recent advances in stimuli-responsive release function drug delivery systems for tumor treatment. Lee et al. (2005) developed a super pH-sensitive multifunctional polymeric micelle for enzyme-responsive drug delivery.

5 Targeted Delivery Systems

Targeted delivery systems are crucial for improving the efficacy and specificity of drug delivery by selectively delivering therapeutic agents to the desired site of action. This section explores active targeting strategies, passive targeting strategies, and combination targeting approaches.

5.1 Active Targeting Strategies

Active targeting strategies involve ligand-based targeting approaches and targeted nanoparticles and liposomes to deliver specific drugs to the desired site.

Ligand-based targeting approaches have emerged as a promising avenue for targeted drug delivery (Torchilin 2006; Reis et al. 2006). These approaches utilize ligands on the surface of nanoparticles to facilitate specific interactions with receptors overexpressed on target cells. Incorporating ligands onto liposomes has also been explored to enable active targeting and improve drug delivery (Reis et al. 2006).

Targeted nanoparticles and liposomes have gained significant attention as carriers for active drug targeting (Allen and Cullis 2004; Langer 1998). These systems utilize surface modifications with targeting ligands to enhance nanoparticle accumulation at the desired site, enabling improved therapeutic outcomes (Langer 1998).

5.2 Passive Targeting Strategies

Passive targeting strategies rely on the enhanced permeability and retention (EPR) effect and system for enhancing passive targeting efficiency.

The EPR effect is pivotal in passive targeting by utilizing the abnormal vasculature and impaired lymphatic drainage in tumors (Hu et al. 2014). This effect leads to improved tumor uptake, reduced systemic toxicity, and distinct tumor imaging.

Strategies for enhancing passive targeting efficiency have been explored to maximize the therapeutic potential of this approach (Yang et al. 2017). These strategies involve optimizing nanoparticle properties such as size, surface charge, and shape to improve tumor penetration and accumulation (Jain et al. 2021). Leveraging interactions with the tumor microenvironment and extracellular matrix components has also been proposed to enhance passive targeting efficiency (Peer et al. 2007).

5.3 *Combination Targeting Approaches*

Combination targeting approaches aim to synergistically utilize active and passive targeting strategies to enhance drug delivery efficacy.

Dual-responsive systems offer the potential for targeted and stimuli-responsive drug release. These systems demonstrate responsiveness to multiple stimuli, such as pH and reducing conditions, enabling efficient drug delivery to tumor cells (Sun et al. 2016). Enzyme-responsive nanomaterials have also been investigated for controlled drug delivery, and their combination with other targeting strategies shows promise in enhancing therapeutic outcomes (Sun et al. 2016).

Multi-stage delivery systems involve the sequential release of therapeutic agents at different stages of the disease. Temperature-responsive and pH-responsive nanoparticles have been explored for multi-stage delivery, enabling precise control over drug release and sequential delivery to different body regions disease (Peppas et al. 2000; Yang et al. 2017).

By combining active and passive targeting strategies, these approaches hold great promise for achieving improved drug delivery efficiency and therapeutic outcomes.

6 Advances in Controlled Drug Delivery

Controlled drug delivery has witnessed remarkable advancements by integrating various technologies and materials. This section explores the application of nanotechnology, biomaterials, and emerging technologies in controlled drug delivery, highlighting key references.

6.1 *Nanotechnology in Drug Delivery*

Nanoparticles and nanostructured carriers have revolutionized drug delivery by enabling precise targeting and enhanced therapeutic efficacy (Blanco et al. 2015). Nanoparticles offer unique properties such as small size, large surface area, and tunable surface chemistry, allowing improved drug solubility, prolonged circulation time, and targeted delivery to specific tissues or cells. Blanco et al. (2015) discuss the principles of nanoparticle design for overcoming biological barriers to drug delivery, emphasizing the importance of nanoparticle size, surface modification, and encapsulation techniques in optimizing drug delivery efficiency. Shi and Votruba (2013) provide insights into using nanoparticles for cancer treatment, highlighting their potential to improve drug delivery to tumor sites through passive and active targeting mechanisms (Shi et al. 2013).

6.2 Biomaterials and Drug Delivery

Biomaterials are critical in controlled drug delivery systems, offering tailored properties and functionalities to optimize drug release. Biodegradable polymers and hydrogels are widely used biomaterials due to their biocompatibility and ability to control drug release kinetics. Langer (2000) shares valuable experiences utilizing biomaterials for drug delivery and tissue engineering, providing critical insights into their applications and the design considerations for achieving controlled release. Chen and Hoffman (1995) discuss temperature-responsive graft copolymers that exhibit phase transitions over a wide range of pH, offering the potential for pH-responsive drug release. Their work highlights the importance of biomaterial design in achieving precise control over drug release based on environmental stimuli (Chen and Hoffman 1995).

6.3 Emerging Technologies and Future Trends

Emerging technologies hold immense promise in advancing controlled drug delivery systems. One such technology is 3D printing, also known as additive manufacturing, which has garnered attention for its ability to fabricate drug delivery systems with customized geometries and drug release profiles. In a study by Goyanes et al. (2014), the use of 3D printing in fabricating modified-release tablets was explored, showcasing the potential for personalized drug delivery solutions. The researchers demonstrated the capability of 3D printing to create complex drug delivery systems with controlled release characteristics. This technology opens up new avenues for tailoring drug formulations to specific patient needs, optimizing therapeutic outcomes, and improving patient compliance (Goyanes et al. 2014).

Microfabrication and surface engineering, discussed by Mitragotri and Lahann (2009), offer physical approaches to biomaterial design that enable precise control over drug release kinetics. These techniques involve fabricating microscale structures and modifying material surfaces to achieve controlled drug release (Mitragotri and Lahann 2009). Researchers can tailor drug release rates and patterns by manipulating factors such as material composition, topography, and surface chemistry. This level of control allows for customized drug delivery systems that optimize therapeutic efficacy and minimize side effects. The research highlights the potential of these emerging techniques in advancing controlled drug delivery systems and expanding the design possibilities for drug formulations.

In addition to 3D printing and microfabrication, nanotechnology has played a transformative role in controlled drug delivery. Nanoparticles, at the forefront of nanotechnology-based drug delivery systems, can deliver multiple therapeutic agents and achieve multimodal synergistic cancer therapy. Gao and Chan review the use of nanotechnology in combination therapy, emphasizing the potential of nanoparticles to overcome multidrug resistance and enhance treatment outcomes. Synergistic effects can be achieved by encapsulating and delivering different therapeutic agents within a single nanoparticle, leading to improved efficacy in cancer treatment. The ability of nanoparticles to precisely target tumor tissues further enhances their therapeutic potential (Gao et al. 2014).

While these advancements in nanotechnology, biomaterials, and emerging technologies hold great promise for controlled drug delivery systems, there are also regulatory considerations to address. Ventola (2014a) discusses the progress in 3D printing and its regulatory aspects, highlighting the challenges and opportunities in manufacturing and regulating 3D-printed drug delivery systems. Ensuring these systems' safety, efficacy, and quality is essential for their successful translation from the laboratory to clinical use. Regulatory bodies such as the FDA and EMA play a critical role in establishing guidelines and evaluating the safety and performance of these innovative drug delivery systems (Ventola 2014a).

In summary, emerging technologies such as 3D printing, microfabrication, surface engineering, and nanotechnology are transforming the landscape of controlled drug delivery. These technologies offer opportunities for personalized drug delivery, precise control over release kinetics, and synergistic therapy in cancer treatment. However, regulatory considerations are crucial to ensure these technologies' safe and effective use. The advancements in nanotechnology, biomaterials, and emerging technologies hold great promise in advancing the field of controlled drug delivery, enabling targeted and efficient therapeutic interventions for improved patient outcomes.

7 Regulatory Considerations and Clinical Applications

Regulatory considerations and clinical applications are crucial in developing and translating controlled drug delivery systems. This section, referencing relevant studies, explores these systems' safety and toxicity assessments, regulatory guidelines, and clinical applications.

7.1 *Safety and Toxicity Assessments*

Safety and toxicity assessments are paramount in evaluating the potential risks associated with controlled drug delivery systems. Kim et al. (2018) employed a high-throughput screening approach to comprehensively assess the nanotoxicity of complex carbon nanomaterials, shedding light on their potential adverse effects and guiding safety assessments. Nel et al. (2006) provided valuable insights into the complex physicochemical interactions at the nano-bio interface, aiding in understanding nanomaterial toxicity and facilitating the development of robust safety evaluation strategies.

7.2 Regulatory Guidelines for Controlled Drug Delivery Systems

Regulatory guidelines are pivotal in ensuring controlled drug delivery systems' quality, safety, and efficacy. The U.S. Food and Drug Administration (FDA) provides product-specific guidance documents for generic drug development, encompassing considerations specific to controlled drug delivery systems (U.S. Food and Drug Administration 2017). Additionally, the European Medicines Agency (EMA) has published guidelines on water quality for pharmaceutical use, offering regulatory guidance to ensure the safety and quality of drug delivery systems (European Medicines Agency 2015).

7.3 Clinical Applications and Case Studies

Clinical applications and case studies showcase the real-world impact of controlled drug delivery systems. Farra et al. (2012) conducted groundbreaking first-in-human testing of a wirelessly controlled drug delivery microchip, demonstrating the potential of advanced technologies in personalized medicine and targeted drug delivery. Pritchard et al. (2018) investigated the clinical application of nanoparticle-based drug delivery systems in peritoneal mesothelioma, highlighting the selective tumor localization of nanoparticles, disruption of autophagosomal trafficking, and prolonged drug delivery for improved patient outcomes.

These studies and regulatory guidelines provide valuable insights into controlled drug delivery systems' safety, regulatory considerations, and clinical applications. Addressing safety concerns and complying with regulatory guidelines can translate these systems into effective therapies that improve patient outcomes.

8 Challenges and Future Directions

Controlled drug delivery systems face several challenges and offer exciting prospects for future development. This section discusses the challenges associated with stability and long-term release, scale-up and manufacturing, personalized medicine and precision drug delivery, and integrating theranostic approaches, referencing relevant studies.

8.1 *Stability and Long-Term Release*

Achieving stability and long-term release of drugs within controlled delivery systems is crucial for their efficacy. A comprehensive review by Pohlmann et al. (2018) highlights the challenges and state-of-the-art techniques in achieving stable drug-loaded polymeric nanoparticles. The study emphasizes the importance of understanding the physicochemical properties of nanoparticles and the influence of formulation factors on stability. Additionally, Mo et al. (2014) demonstrated enhanced anticancer efficacy by utilizing ATP-mediated liposomal drug delivery, which improves stability and enhances therapeutic outcomes by targeted and controlled drug release.

8.2 *Scale-Up and Manufacturing Challenges*

The scale-up and manufacturing of controlled drug delivery systems pose significant challenges. Farah et al. (2016) provided a comprehensive review of the physical and mechanical properties of poly(lactic acid) (PLA), an extensively used biodegradable polymer, discussing its challenges and applications in controlled drug delivery systems. The study highlights the need to consider the scalability of manufacturing processes and the importance of maintaining the desired drug release characteristics during scale-up. Ventola (2014b) discussed the current and projected uses of 3D printing in the medical field, shedding light on its potential to overcome scale-up and manufacturing challenges. The study emphasizes the customization and flexibility offered by 3D printing technology for fabricating complex drug delivery systems.

8.3 *Personalized Medicine and Precision Drug Delivery*

Personalized medicine and precision drug delivery aim to tailor treatments based on individual patient characteristics. Agarwal et al. (2013) developed macrophage-like nanoparticles capable of absorbing endotoxins and proinflammatory cytokines for sepsis management. The study highlights the potential of personalized approaches in disease treatment by selectively targeting and removing specific harmful substances in the body. Lim et al. (2018) highlighted the precision delivery of nanomaterials for inflammation and infection control. The research showcases the importance of targeted therapies in achieving precise drug delivery to specific disease sites, thereby enhancing treatment efficacy while minimizing side effects.

8.4 *Integration of Theranostic Approaches*

Integrating theranostic approaches, combining therapy and diagnostics, holds great promise in controlled drug delivery. Jokerst et al. (2017) discussed the PEGylation of nanoparticles for imaging and treatment, enabling simultaneous drug delivery and imaging capabilities. The study emphasizes the potential of theranostic approaches in providing real-time monitoring of drug delivery and therapeutic response, enhancing treatment outcomes. Chen et al. (2019) explored the application of theranostic nanomedicine for cancer imaging and therapy. The research highlights the multi-functionality of theranostic systems, combining diagnostic imaging and targeted drug delivery to achieve personalized and precise cancer treatment.

9 Conclusion

In conclusion, the chapter on controlled drug delivery systems has provided a comprehensive understanding of this field's principles, strategies, and applications. By summarizing the key points discussed throughout the chapter and exploring future perspectives and areas of research, we can appreciate the significance of controlled drug delivery systems and their potential for advancing healthcare.

9.1 *Summary of Key Points*

Throughout this chapter, we have delved into various aspects of controlled drug delivery systems. We began by establishing the fundamentals, including the principles of drug delivery and the factors that influence drug release. We explored the challenges of conventional drug delivery methods and highlighted the advantages of controlled drug delivery systems.

9.1.1 Sustained-Release Systems

Sustained-release systems, including matrix and reservoir systems, were thoroughly examined. Matrix systems involve the dispersion of drugs within a polymeric matrix, enabling controlled release over time. We discussed drug release mechanisms in matrix systems, such as diffusion, erosion, and combination. Formulation strategies, such as varying polymer types and drug loading methods, were explored, along with examples of sustained-release formulations and their applications. Reservoir systems, conversely, involve encapsulating drugs within a reservoir or compartment that controls drug release through diffusion or osmosis. We discussed the design principles of reservoir systems, including the selection of membrane materials and the influence of reservoir geometry on drug release kinetics. Specific examples and applications, such as transdermal patches and implantable devices, were presented to highlight the versatility and effectiveness of sustained-release systems.

9.1.2 Stimuli-Responsive Systems

The chapter also focused on stimuli-responsive systems, which can respond to specific triggers such as pH, temperature, or enzymes. pH-responsive drug delivery systems are designed to release drugs in response to changes in the pH environment of the target site. We explained the mechanisms of pH-responsive drug release, such as pH-dependent swelling or dissolution of polymeric carriers, and discussed formulation strategies and examples of pH-responsive systems. Temperature-responsive systems, utilizing thermosensitive polymers, undergo reversible phase transitions in response to temperature changes. We explored the use of thermosensitive polymers in temperature-responsive systems, highlighting their applications and future directions. Furthermore, we examined enzyme-responsive systems, which exploit the presence of specific enzymes in disease sites for triggered drug release. We discussed the mechanisms of enzyme-triggered drug release, including enzyme-cleavable linkers or coatings, and presented case studies and promising developments in this field.

9.1.3 Targeted Delivery Systems

Targeted delivery systems, both active and passive, were another important area of discussion. Functional targeting approaches utilize specific ligands, such as antibodies or peptides, to actively guide drug carriers to desired sites. We explored ligand-based targeting approaches and discussed targeted nanoparticles and liposomes that selectively bind to target cells' receptors. Passive targeting strategies exploit the enhanced permeability and retention (EPR) effect, which takes advantage of leaky blood vessels and poor lymphatic drainage in tumor tissues. We discussed the principles of passive targeting and highlighted plans for enhancing passive targeting efficiency, such as modifying carrier size and surface properties. Additionally, we explored combination targeting approaches, including dual-responsive and multistage delivery systems, which offer enhanced targeting capabilities by integrating multiple targeting mechanisms.

9.1.4 Advances in Controlled Drug Delivery

Advances in controlled drug delivery were a significant focus of this chapter. We discussed the application of nanotechnology in drug delivery, specifically nanoparticles and nanostructured carriers. Nanoparticles, such as liposomes, polymeric, and solid lipid nanoparticles, have unique properties that enable efficient drug encapsulation and controlled release. We highlighted their potential applications, including targeted delivery, imaging, and combination therapy. Additionally, we explored nanostructured carriers, such as dendrimers and mesoporous materials, for controlled drug delivery and their potential impact on enhancing drug stability and release kinetics. Biomaterials and their role in drug delivery were also examined, focusing on biodegradable polymers, hydrogels, and innovative materials for controlled release. We discussed their biocompatibility, biodegradability, and tunable properties that make them suitable for controlled drug delivery applications.

9.1.5 Regulatory Considerations and Clinical Applications

Regulatory considerations and clinical applications were crucial topics covered in this chapter. We emphasized the importance of safety and toxicity assessments in evaluating the potential risks associated with controlled drug delivery systems. Understanding the behavior of carriers and their degradation products within the body is essential for ensuring patient safety. Furthermore, we discussed regulatory guidelines for these systems, focusing on the guidelines provided by regulatory authorities such as the U.S. Food and Drug Administration (FDA) and the European Medicines Agency (EMA). Compliance with these guidelines is essential for successfully translating controlled drug delivery systems from the laboratory to clinical use. Additionally, we presented clinical applications and case studies demonstrating the real-world impact and efficacy of controlled drug delivery systems in various medical conditions, such as cancer, infectious diseases, and chronic disorders.

9.2 Challenges and Future Directions

While controlled drug delivery systems have shown tremendous promise, several challenges and future directions warrant further exploration. Stability and long-term release of drugs within these systems are crucial considerations. Addressing formulation design, particle stability, and release kinetics will be necessary for optimal stability and sustained drug release. Scale-up and manufacturing challenges, including process optimization and quality control measures, must be overcome to ensure efficient and cost-effective production of these systems.

Personalized medicine and precision drug delivery offer exciting avenues for tailoring treatments to individual patient characteristics. Further research can explore innovative approaches such as advanced targeting strategies, personalized dosing regimens, and integration with diagnostic technologies to enhance therapeutic outcomes. Integrating theranostic processes, combining therapy and diagnostics, holds great potential. Investigating novel biomaterials, nanotechnology, and advanced manufacturing techniques can open new horizons for controlled drug delivery systems.

In conclusion, controlled drug delivery systems have the potential to revolutionize healthcare by improving drug efficacy, minimizing side effects, and enabling targeted therapies. This chapter's comprehensive exploration of key points has highlighted the versatility and advantages of controlled drug delivery systems. By addressing challenges and focusing on future directions, we can unlock their full potential and pave the way for advancements in drug delivery, benefiting patients worldwide.

References

Adepu S, Ramakrishna S (2021) Controlled drug delivery systems: current status and future directions. MDPI 26(19):1–45

Agarwal P et al (2013) Macrophage-like nanoparticles concurrently absorbing endotoxins and proinflammatory cytokines for sepsis management. Proc Natl Acad Sci 110(26):10799–10804

Allen TM, Cullis PR (2004) Drug delivery systems: entering the mainstream. Science 303(5665):1818–1822

Allen TM, Cullis PR (2013) Liposomal drug delivery systems: from concept to clinical applications. Adv Drug Deliv Rev 65(1):36–48

Aman MW et al (2015) Chronopharmaceutics: a promising approach for drug targeting. Pharm Dev Technol 20(5):517–528

Anderson JM, Shive MS (1997) Biodegradation and biocompatibility of PLA and PLGA microspheres. Adv Drug Deliv Rev 28(1):5–24

Blanco E, Shen H, Ferrari M (2015) Principles of nanoparticle design for overcoming biological barriers to drug delivery. Nat Biotechnol 33(9):941–951

Brigger I, Dubernet C, Couvreur P (2002) Nanoparticles in cancer therapy and diagnosis. Adv Drug Deliv Rev 54(5):631–651

Chakravarthi SS (2012) Controlled drug delivery systems: a comprehensive review. Crit Rev Food Sci Nutr 52(7):670–679

Chen G, Hoffman AS (1995) Graft copolymers that exhibit temperature-induced phase transitions over a wide range of pH. Nature 373(6515):49–52

Chen X et al (2018) Recent advances in stimuli-responsive release function drug delivery systems for tumor treatment. Molecules 23(7):1–24

Chen H et al (2019) Theranostic nanomedicine for cancer imaging and therapy. J Mater Chem B 7(23):3496–3519

Choi KY, Chung H, Min KH, Yoon HY, Kim K, Park JH, Kwon IC, Jeong SY (2010) Self-assembled hyaluronic acid nanoparticles for active tumor targeting. Biomaterials 31(1):106–114

Davis ME, Chen Z, Shin DM (2008) Nanoparticle therapeutics: an emerging treatment modality for cancer. Nat Rev Drug Discov 7(9):771–782

De Jong WH, Borm PJ (2008) Drug delivery and nanoparticles: applications and hazards. Int J Nanomed 3(2):133–149

Desai MP, Labhasetwar V, Amidon GL, Levy RJ (1996) Gastrointestinal uptake of biodegradable microparticles: effect of particle size. Pharm Res 13(12):1838–1845

European Medicines Agency (2015) Guideline on the quality of water for pharmaceutical use

Fan Q et al (2012) A smart pH-responsive nano-carrier as a drug delivery system for the controlled release of 5-fluorouracil. RSC Adv 2(13):5618–5625

Farah S, Anderson DG, Langer R (2016) Physical and mechanical properties of PLA, and their functions in widespread applications—a comprehensive review. Adv Drug Deliv Rev 107:367–392

Farra R, Sheppard Jr NF, McCabe L, Neer RM, Anderson JM, Santini Jr JT, Cima MJ, Langer R (2012) First-in-human testing of a wirelessly controlled drug delivery microchip. Sci Transl Med 4(122):122ra21

Gao FO et al (2014) Nanotechnology for multimodal synergistic cancer therapy. Chem Rev 114(21):10970–11020

Goyanes A, Buanz AB, Hatton GB, Gaisford S, Basit AW (2014) 3D printing of modified-release aminosalicylate (4-ASA and 5-ASA) tablets. Eur J Pharm Sci 58:36–42
Hu Q, Katti PS, Gu Z (2014) Enzyme-responsive nanomaterials for controlled drug delivery. Nanoscale 6(21):12273–12286
Jain SK et al (2008) pH-sensitive liposomes: a novel carrier for drug delivery. Crit Rev Ther Drug Carrier Syst 25(6):551–573
Jain KOP et al (2021) An enzyme-responsive sequential drug delivery system for targeted therapy in breast cancer. ACS Biomater Sci Eng 7(6):3117–3132
Jokerst JV, Lobovkina T, Zare RN, Gambhir SS (2017) Nanoparticle PEGylation for imaging and therapy. Nanomedicine 12(7):1717–1740
Khullar R et al (2012) Enzyme-responsive drug delivery systems in cancer therapy. J Pharm Pharmacol 64(2):162–181
Kim S et al (2011) Self-assembled hyaluronic acid nanoparticles for active tumor targeting. Biomaterials 32(27):7181–7187
Kim YJ et al (2016) pH-sensitive polymeric nanoparticles for tumor-targeting drug delivery. Biomater Sci 4(6):890–899
Kim J et al (2018) Nanotoxicity evaluations of hard carbon nanomaterials by a high-throughput screening approach. Nano Converg 5(1):1–12
Kopeček J (2013) Smart and genetically engineered biomaterials and drug delivery systems. Eur J Pharm Sci 50(1):8–18
Korsmeyer RW, Gurny R, Doelker E, Buri P, Peppas NA (1983) Mechanisms of solute release from porous hydrophilic polymers. Int J Pharm 15(1):25–35
Lai WF et al (2015) Controlled-release hydrogel of platelet-rich plasma in suppressing biliary fibrosis after hepatic injury. J Control Release 206:37–47
Langer R (1990) New methods of drug delivery. Science 249(4976):1527–1533
Langer R (1998) Drug delivery and targeting. Nature 392(6679):5–10
Langer R (2000) Biomaterials in drug delivery and tissue engineering: one laboratory's experience. Acc Chem Res 33(2):94–101
Lee ES, Na K, Bae YH (2005) Super pH-sensitive multifunctional polymeric micelle. Nano Lett 5(2):325–329
Li SD, Huang L (2010) Nanoparticles evading the reticuloendothelial system: role of the supported bilayer. Biochimica et Biophysica Acta (BBA) Biomembranes 1788(10):2259–2266
Lim J et al (2018) Precision delivery of nanomaterials for inflammation and infection control. Nanomedicine 14(3):913–924
Liu LX et al (2019) Temperature-triggered gelation of poly(N-isopropylacrylamide)-based microgels for controlled drug delivery. Int J Biol Macromol 124:763–772
Mitragotri S, Lahann J (2009) Physical approaches to biomaterial design. Nat Mater 8(1):15–23
Mo R, Jiang T, Gu Z (2014) It enhanced anticancer efficacy by ATP-mediated liposomal drug delivery. Angew Chem Int Ed 53(19):5815–5820
Nel AE, Mädler L, Velegol D, Xia T, Hoek EM, Somasundaran P, Klaessig F, Castranova V, Thompson M (2006) Understanding biophysicochemical interactions at the nano-bio interface. Nat Mater 8(7):543–557
Oerlemans C, Bult W, Bos M, Storm G, Nijsen JF, Hennink WE (2010) Polymeric micelles in anticancer therapy: targeting, imaging and triggered release. Pharm Res 27(12):2569–2589
Panyam J, Labhasetwar V (2003) Biodegradable nanoparticles for drug and gene delivery to cells and tissue. Adv Drug Deliv Rev 55(3):329–347
Peer D, Karp JM, Hong S, Farokhzad OC, Margalit R, Langer R (2007) Nanocarriers as an emerging platform for cancer therapy. Nat Nanotechnol 2(12):751–760
Peppas NA, Bures P, Leobandung WS, Ichikawa H (2000) Hydrogels in pharmaceutical formulations. Eur J Pharm Biopharm 50(1):27–46
Pohlmann AR et al (2018) Stability study of drug-loaded polymeric nanoparticles: a comprehensive review of the state of the art. Mater Sci Eng 92:983–993

Prausnitz MR (2004) Microneedles for transdermal drug delivery. Adv Drug Delivery Rev 56(5):581–587. https://doi.org/10.1016/j.addr.2003.10.023
Pritchard EM et al (2018) Nanoparticle tumor localization, disruption of autophagosomal trafficking, and prolonged drug delivery improve survival in peritoneal mesothelioma. Biomaterials 159:93–104
Reis CP, Neufeld RJ, Ribeiro AJ, Veiga F (2006) Nanoencapsulation I. Methods for preparation of drug-loaded polymeric nanoparticles. Nanomedicine 2(1):8–21
Ritger PL, Peppas NA (1987) A simple equation for description of solute release II. Fickian and anomalous release from swellable devices. J Control Release 5(1):37–42
Saikia C, Gogoi P, Maji TK (2015) Chitosan: a promising biopolymer in drug delivery applications. J Mol Genet 2–10
Schild HG (1992) Poly(N-isopropylacrylamide): experiment, theory and application. Prog Polym Sci 17(2):163–249
Shi Y, Van Der Meel R, Theek B, Oude Blenke E, Pieters EH, Fens MH, Ehling J, Schiffelers RM, Storm G, Van Nostrum CF, Lammers T (2014) Complete regression of xenograft tumors upon targeted delivery of paclitaxel via Π-Π stacking stabilized polymeric micelles. ACS Nano 8(4):37
Shi J, Votruba AR (2013) Nanoparticles for drug delivery in cancer treatment. Comprehensive Reviews in Food Science and Food Safety, 12(3):204–218
Siepmann J, Peppas NA (2001) Modeling of drug release from delivery systems based on hydroxypropyl methylcellulose (HPMC). Adv Drug Deliv Rev 48(2–3):139–157
Singh R, Lillard JW Jr (2009) Nanoparticle-based targeted drug delivery. Exp Mol Pathol 86(3):215–223
Soppimath KS, Aminabhavi TM, Kulkarni AR, Rudzinski WE (2001) Biodegradable polymeric nanoparticles as drug delivery devices. J Control Release 70(1–2):1–20
Sun H, Guo B, Cheng R, Meng F, Liu H, Zhong Z (2016) Biodegradable micelles with sheddable poly(ethylene glycol) shells for triggered intracellular release of doxorubicin. Biomaterials 102:295–305
Torchilin VP (2006) Multifunctional nanocarriers. Adv Drug Deliv Rev 58(14):1532–1555
Torchilin VP (2011) Multifunctional, stimuli-sensitive nanoparticulate systems for drug delivery. Nat Rev Drug Discov 10(10):787–808
U.S. Food and Drug Administration (2017) Product-specific guidance documents for generic drug development. Center for Drug Evaluation and Research
Ventola CL (2014a) Progress in 3D printing: FDA regulation of 3D-printed medical products. Pharm Ther 39(5):346–352
Ventola CL (2014b) Medical applications for 3D printing: current and projected uses. Pharm Ther 39(10):704–711
Xue H et al (2011) Development of a thermosensitive hydrogel for doxorubicin delivery. Int J Pharm 409(1–2):280–287
Yang C et al (2017) Synthetic and biopolymeric pH-responsive nanoparticles for drug delivery. Bioconjug Chem 28(4):907–917
Zhang C et al (2016) Temperature-responsive polymer-grafted mesoporous silica nanoparticles for remotely triggered drug release. ACS Appl Mater Interfaces 8(1):179–187
Zhuang B et al (2013) Fabrication and characterization of sodium alginate/gelatin hydrogel scaffolds with sustained VEGF-release. J Biomed Mater Res A 101(3):852–859

Biomaterials in Drug Delivery Systems

Harshala S. Naik, Parvindar M. Sah, and Rajesh W. Raut

Abstract Drug delivery research is advancing day by day due to the advancement of science and technology, which helps cure many diseases. Traditionally, drugs are administered to the body either orally or through injections. It is imperative to take several factors into account when delivering a specific drug with pharmacologically active ingredients. These factors include the drug's cellular efficiency, pharmacokinetics, therapeutic efficiency, cellular uptake, and its effect on metabolism, exercise, and cellular toxicity. The use of biomaterials can enhance both injectable and oral drug delivery, as well as ocular, pulmonary, nasal, and transdermal drug delivery. Metals, glass, ceramics, and oscular are examples of biomaterials used in drug delivery. These materials can also be found in orthopedic devices, contact lenses, heart valves, and pacemakers. Biomaterials are categorized as either naturally driven or synthetically driven. It is most common to use synthetically driven polymers for drug delivery systems. As the polymeric drug is inoculated inside the body, its morphology changes from a sol to a gel. It is imperative to note that these gel formulations react differently to temperature, ultraviolet irradiation, ions, and pH changes. This book chapter presents a comprehensive analysis of polymeric systems used in drug delivery, their types, formulations, advancements, and drawbacks.

1 Introduction

The development of new drugs holds great promise for improving human health. However, the conventional methods of drug delivery often come with unwanted side effects. Patients who lack sufficient knowledge about their medication may inadvertently consume higher doses, leading to temporary relief but also causing the accumulation of drugs beyond the body's normal tolerance level (Adeosun et al. 2020). Additionally, some drugs require repeated dosing due to their low concentration

H. S. Naik · P. M. Sah · R. W. Raut (✉)
Department of Botany, The Institute of Science, Dr. Homi Bhabha State University, 15 Madam Cama Road, Mumbai, Maharashtra 400032, India
e-mail: rajesh@iscm.ac.in

T. S. Santra and A. U. S. Shinde (eds.), *Advanced Drug Delivery*,
Studies in Mechanobiology, Tissue Engineering and Biomaterials 26,
https://doi.org/10.1007/978-981-99-6564-9_12

reaching the target organ. Moreover, during drug consumption and action, metabolic reactions in the body can result in drug modification or decreased concentration, leading to toxic effects or delayed recovery (Han et al. 2022).

To overcome these challenges, drug carriers offer a compelling solution by providing precise control over drug delivery. These carriers are made from materials that allow the release of medications without modification or degradation. They can be used in various ways, such as implantable devices, direct injections, or a combination of both. It is crucial to select materials that can release drugs over extended therapeutic windows, ranging from days to years, while maintaining their effectiveness (Fenton et al. 2018). By directly implanting or injecting these carriers into affected tissues or cells, the efficiency of drug delivery is significantly enhanced. Furthermore, incorporating affinity ligands onto the surfaces of biomaterials promotes tissue retention and utilization (Fenton et al. 2018).

In drug delivery, two main types of biomaterials are used: natural and synthetic (Adeosun et al. 2020). Traditional biomaterials derived from natural sources, including silk (Numata and Kaplan 2010), collagen (Sahithi et al. 2013), gelatin (Santoro et al. 2014), and alginate (Lee and Mooney 2012) have been utilized in medical applications since the 1950s. While these materials have significantly improved patients quality of life, advances in molecular cell biology and polymer chemistry have led to the development of biomaterials with improved properties (Fenton et al. 2018; Sood et al. 2021; Choukaife et al. 2022). These biomaterials are constantly evolving to meet the demands of emerging areas such as drug delivery, scaffolding, tissue engineering and bioprinting. Even biologically inert materials used in clinical settings have been found to elicit harmful cellular and immune responses (Salimi et al. 2022; Jumelle et al. 2020; Srinivasarao et al. 2019), highlighting the importance of biomaterials that interact minimally with surrounding tissues and cells to promote tissue healing and regeneration (Sood et al. 2021). Furthermore, for advanced applications, biologically inert and passive biomaterials can be enriched with drugs (Ma et al. 2012), growth factors, or vectors for gene delivery (Saeed et al. 2020), enhancing their therapeutic efficacy (Zhang et al. 2013).

The use of biomaterials in drug delivery presents an exciting opportunity to overcome the limitations of conventional drug administration methods (Jumelle et al. 2020). Through the development of efficient drug carriers and the advancement of both natural and synthetic biomaterials, it becomes possible to achieve controlled and targeted drug delivery without compromising patient safety or therapeutic efficacy. The continuous evolution of biomaterials holds immense potential in revolutionizing the field of drug delivery and improving patient outcomes in various medical applications (Saini 2015).

In this chapter, we will explore the various biomaterials utilized in drug delivery and their significant contributions to enhancing the effectiveness of therapeutic treatments.

2 Mechanism of Biomaterial in Drug Delivery System

Biomaterials play a crucial role in drug delivery systems by providing a platform for controlled and targeted release of therapeutic agents. The mechanism of biomaterials in drug delivery involves their ability to encapsulate drugs, protect them from degradation, control their release rate, and target specific tissues or cells. Here is a general overview of the mechanism.

Encapsulation: Biomaterials can encapsulate drugs within their structure, either through physical entrapment or chemical conjugation. This encapsulation helps protect the drug from degradation and enhances its stability during storage and transportation. Encapsulated biomaterials used in drug delivery include liposomes (Xu et al. 2007), polymeric nanoparticles, micelles, dendrimers, nanogels, and mesoporous silica nanoparticles.

Hydroxyapatite (HA) coated liposomes (HACL) have been developed to encapsulate the hydrophobic drug indomethacin (IMC). The HA coating on the liposomes reduces the release rate of IMC compared to uncoated liposomes. The drug release profiles show that approximately 70% of the drug is released from uncoated liposomes in about 5 h, while HA-coated liposomes extend the release time to over 20 h. Furthermore, the release behaviour is influenced by the pH of the surrounding environment (Xu et al. 2007).

liposomes-hydrogel linked with natural biopolymer (Grijalvo et al. 2016) and used in drug delivery, including chitosan (Lee et al. 2012; Berger et al. 2004; Shukla et al. 2013), gelatin (Ciobanu et al. 2014; Song and Leeuwenburgh 2014; Phromsopha and Baimark 2014), mupirocin (Hurler et al. 2012) hyaluronic acid (Sannino et al. 2004; Leach and Schmidt 2005), pullulan (Akiyoshi et al. 1998), polysaccharide. Figure 1 indicated structure of drug loaded chitosan liposome under rest condition and shear condition. The shear-thinning characteristic of these gels allows for easy injection at a desired site, while their gel-like behaviour when at rest ensures localization of the material at the intended location. Furthermore, the interior of the liposomes can encapsulate drugs, enabling sustained release at the localized site over an extended period (Lee et al. 2012).

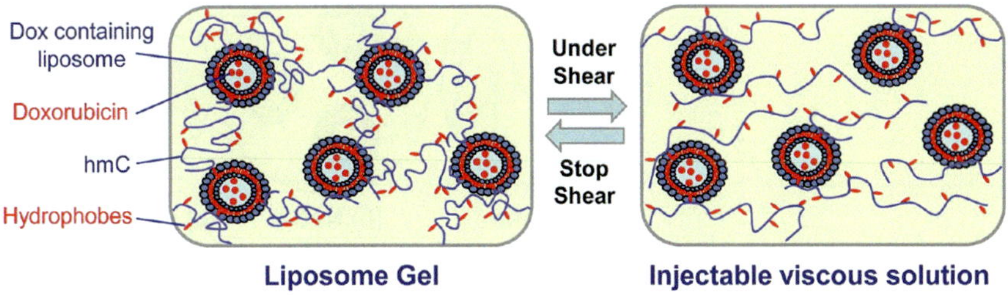

Fig. 1 Structure of an injectable chitosan liposome gel left side at rest and right side under shear. Reprinted with permission from Lee et al. (2012). Copyright 2012, American Chemical Society

Figure 2 shows NaoA liposome vesicle encapsulated with gelatin gel. The pH-sensitive vesicles made of sodium oleate (NaOA) within gelatin gels. At pH 8.3, the gel contains vesicles, but when exposed to a pH 10 buffer, the vesicles transform into micelles, resulting in a visible change in turbidity. This pH-responsive system can be used for controlled release, as the transition from vesicles to micelles allows hydrophilic solutes to be released from the gel into the surrounding environment (Dowling et al. 2009).

Supramolecular hydrogels by combining poly (ethylene glycol) (PEG) containing polymers and α-cyclodextrin (α-CDs), which can incorporate magnetic nanoparticles (MNPs) and gold nanoparticles (AuNPs). These hybrid hydrogels exhibit reversible sol–gel transitions and have potential applications in biomaterials and drug delivery due to the gradual release of the embedded nanoparticles (Niu et al. 2017).

Targeted delivery: Biomaterials can be modified to achieve targeted drug delivery to specific tissues, cells, or organs. This is often achieved by incorporating targeting ligands, such as antibodies, peptides, or aptamers, onto the surface of the biomaterial. These ligands can recognize and bind to specific receptors or markers on the target cells, facilitating the selective uptake of the drug-loaded biomaterial (Zhang et al. 2013).

For effective target drug release in the body there are four way such as stimuli responsive carrier (Agrawal et al. 2012), passive targeting (Calzoni et al. 2019), Nanomaterial-induced endothelial leakiness (Han et al. 2022) and active targeting with specific antibodies or other ligands (Shevtsov et al. 2018).

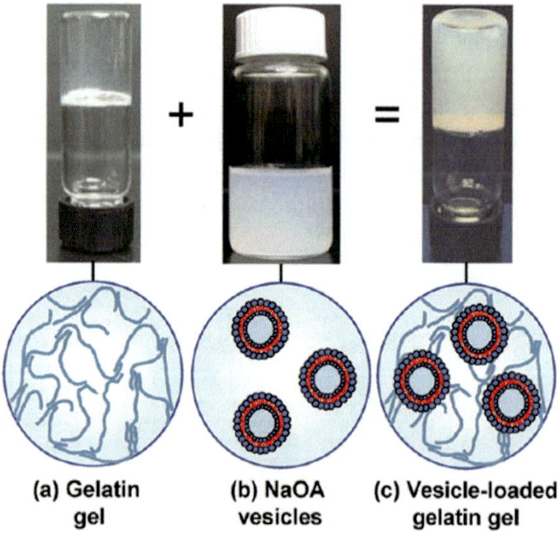

Fig. 2 NaoA liposome vesicle encapsulated with gelatin gel. Reprinted with permission from Dowling et al. (2009). Copyright 2009, American Chemical Society

2.1 Stimuli-Responsive

In drug delivery systems, two types of stimulus-responsive mechanisms can be utilized: endogenous and exogenous. These mechanisms allow for controlled and targeted drug release based on specific triggers present either within the body (endogenous) or introduced from external sources (exogenous). Let us explore these mechanisms in more detail.

Endogenous stimulus-responsive mechanism: Endogenous stimuli are naturally occurring triggers present within the body. These mechanisms take advantage of physiological conditions or pathological changes to trigger drug release. Examples of endogenous stimuli include.

pH-responsive systems: pH changes can occur in various physiological environments, such as tumor tissues, inflamed areas, or intracellular compartments. pH-responsive biomaterials can be designed to undergo conformational changes or dissolution in response to these pH variations, leading to drug release at the target site (Zhuo et al. 2020).

The pH-responsive materials discussed in the various report have shown promising properties for various applications. Carboxylic acid and amine-modified cellulose nanocrystals exhibited pH-dependent behaviour, offering potential for drug delivery and nanocomposite films (Way et al. 2012). The developed aldehyde hyaluronic acid-cisplatin nanoparticles demonstrated sustained release and targeted delivery (Cheng et al. 2019), while pH-sensitive polymeric vesicles showed enhanced uptake and endosome escape (Zhu et al. 2012). Additionally, the pH-responsive amphiphilic polycations displayed cationic surface switching, suggesting their suitability for targeted drug delivery (Gu et al. 2008). These all findings contribute to the development of innovative pH-responsive systems with potential applications in various fields.

Cellulose nanocrystals (CNCs) modified with carboxylic acid (CNC–CO_2H) or amine (CNC–NH_2) groups exhibit pH-responsive behaviour. The amine-modified CNCs form aqueous dispersions at low pH due to electrostatic repulsion, while transitioning to hydrogels at higher pH driven by hydrogen bonding. On the other hand, carboxylic acid-modified CNCs disperse at high pH and gel in an acidic environment. As well as incorporating pH-responsive CNCs into poly(vinyl acetate), these nanocomposites can be mechanically adaptive (Way et al. 2012).

Aldehyde hyaluronic acid-cisplatin (A-HA-CDDP) complex nanoparticles were successfully prepared, showing sustained release and pH sensitivity. They demonstrated effective tumour cell inhibition, good biocompatibility, and potential for targeted cisplatin delivery. Further in vivo experiments will be conducted to evaluate their circulation, targeting, and tumour inhibition capabilities (Cheng et al. 2019).

The pH-sensitive polymeric vesicles composed of cholate grafted poly (l-lysine) (PLL-CA) and poly (ethylene glycol)–doxorubicin conjugate (PEG-DOX) was developed. The vesicles exhibited tunable permeability based on the PLL-CA/PEG-DOX ratio and pH conditions. At lower pH levels, the vesicles destabilized, enabling enhanced uptake by cancer cells in the acidic tumor environment and efficient escape from endosomes (Zhu et al. 2012).

The pH-responsive amphiphilic polycations were synthesized by PEGylation using a benzoic imine linker. These "stealth" polycations formed polymeric micelles with a pH-switchable cationic surface. The micelles exhibited reduced cytotoxicity and haemolysis at physiological pH, but in mild acidic conditions, they exposed their cationic surface, retaining membrane-disrupting properties, making them a potential system for targeted drug delivery to tumors (Gu et al. 2008).

Enzyme-responsive systems: Enzymes are highly specific biomolecules present in the body. Certain diseases or conditions are associated with altered enzyme activity. Enzyme-responsive drug delivery systems employ biomaterials that can be cleaved or degraded by specific enzymes, resulting in drug release at sites where these enzymes are overexpressed or highly active (Hu et al. 2014). Enzyme-responsive drug release from nanoparticles allows for controlled release of drugs in the body, particularly in tumor microenvironments where specific enzymes trigger controlled changes in nanoparticle structure (Hu et al. 2014).

The Fig. 3 shows a glucose-mediated release strategy using an acid-degradable polymeric network, composed of insulin-loaded dextran nanoparticles and glucose-specific enzymes, and enables the controlled release of insulin in response to hyperglycaemia. In vivo studies on diabetic mice showed that a single subcutaneous injection of this nano-network provided improved glucose control for up to 10 days, stabilizing blood glucose levels in the normoglycemic state (Gu et al. 2013).

Redox-responsive systems: Redox potential, characterized by oxidative and reductive conditions, can vary in different tissues or disease states. Redox-responsive biomaterials can be designed to undergo changes in their structure or properties in response to redox imbalances, triggering drug release in specific cellular or tissue environments (Guo et al. 2018).

Redox-responsive delivery systems in nanotechnology have shown promise in biomedicine. They remain stable in circulation but rapidly release their cargo in response to high intracellular levels of glutathione, targeting tumor tissue or specific cells. These systems improve drug concentrations in targeted cells, enhance therapeutic efficiency, and reduce side effects or toxicity of primary drugs. Examples include polymeric nanoparticles with disulphide bonds (Wang et al. 2011), liposomes with redox-responsive components (Ong et al. 2008), polymer-drug conjugates with redox-sensitive linkages (Su et al. 2015), and functionalized inorganic nanoparticles (Guo et al. 2018). These systems provide targeted and controlled drug release, enhancing therapeutic efficacy and minimizing off-target effects (Guo et al. 2018).

(1) Exogenous stimulus-responsive mechanism: Exogenous stimuli are external triggers applied to the body to activate drug release. These stimuli are intentionally introduced from outside sources and can be controlled by the healthcare provider or patient. Examples of exogenous stimuli include:

    1. Light-responsive systems: Light-responsive biomaterials utilize light energy, such as ultraviolet, visible, or near-infrared light, to trigger drug release. Photosensitive molecules or nanoparticles incorporated into the

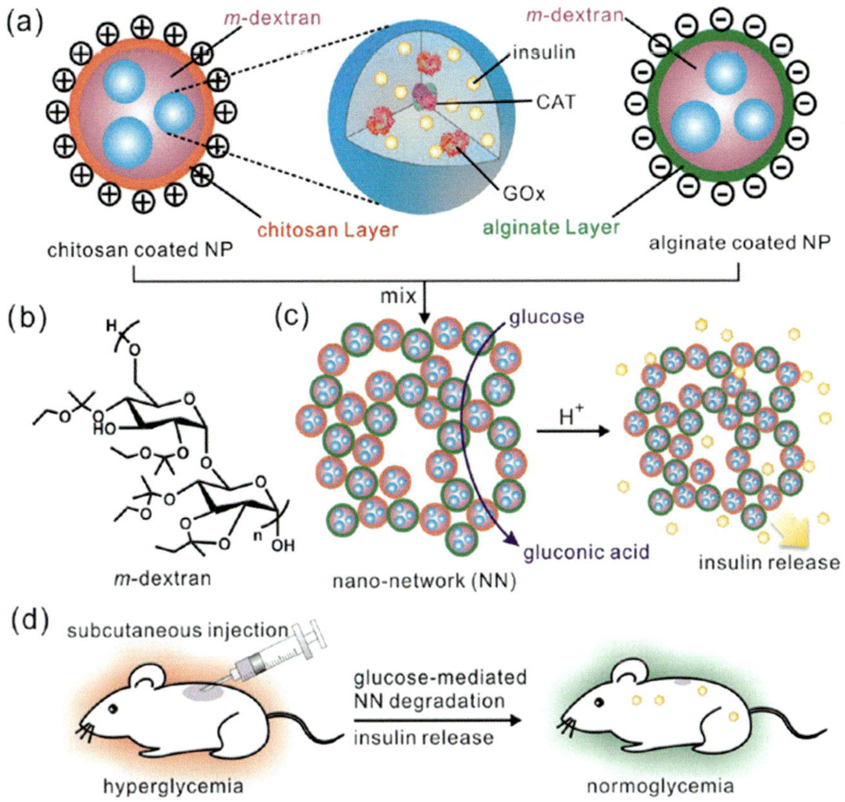

Fig. 3 Injectable glucose-responsive nano-network for insulin delivery. **a** Ketal modified dextran-based nanoparticle encapsulated with insulin, Gox and catalase. **b** Decoration of nanoparticles with chitosan and alginate, respectively. **c** Formation of Nano-network (NN) via electrostatic interaction and dissociation via the catalytic conversion of glucose into gluconic acid. **d** Schematic of glucose-responsive insulin delivery for type 1 diabetes treatment using the STZ-induced diabetic mice model. Gu et al. (2013) Reproduced with permission from, Copyright 2013 American Chemical Society

biomaterial can absorb light and induce changes in their structure or properties, leading to drug release (Luo et al. 2019; Li et al. 2020a; Wang et al. 2021).

2. Temperature-responsive systems: Temperature changes can be induced externally using external heat sources or locally applied energy. Temperature-responsive biomaterials undergo phase transitions or structural changes at specific temperature thresholds, resulting in controlled drug release (Gu et al. 2013; Karimi et al. 2016; Debabrata and Chakrabarti 2018).
3. Magnetic-responsive systems: Magnetic fields can be applied externally to trigger drug release in magnetic-responsive systems. Magnetic nanoparticles incorporated into the biomaterial can be manipulated or triggered to release

drugs through the application of an external magnetic field (Nappini et al. 2016; Sharifianjazi et al. 2021; Qian et al. 2020).

2.2 *Nanomaterial-Induced Endothelial Leakiness*

Nanomaterial-induced endothelial leakiness refers to the disruption of the endothelial barrier, which is composed of endothelial cells lining the inner surface of blood vessels. Certain nanomaterials, when exposed to the endothelial cells, can lead to increased permeability or leakiness of the endothelial barrier (Setyawati et al. 2017).

Nanomaterials are particles or structures with dimensions typically in the range of 1–100 nm. They possess unique physical and chemical properties that make them useful in various applications, including medicine, electronics, and energy. However, some nanomaterials have been found to have adverse effects on biological systems, including the endothelial barrier (Setyawati et al. 2017).

When nanomaterials interact with endothelial cells, they can trigger a series of cellular responses that disrupt the integrity of the endothelial barrier. This disruption can result in increased permeability, allowing molecules and particles to pass through the endothelial layer more easily. As a consequence, substances that are normally restricted from crossing the endothelium, such as plasma proteins and immune cells, can leak into the surrounding tissues (Setyawati et al. 2017).

The mechanisms underlying nanomaterial-induced endothelial leakiness can vary depending on the specific properties of the nanomaterial. For instance, certain nanomaterials may induce oxidative stress in endothelial cells, leading to the activation of inflammatory pathways and the disruption of tight junction proteins that maintain the integrity of the endothelial barrier. Other nanomaterials may directly interact with the endothelial cell membrane, causing structural changes and impairing barrier function (Setyawati et al. 2017).

Endothelial leakiness induced by nanomaterials can have implications for various physiological processes and disease conditions. It can contribute to tissue inflammation, edema formation, impaired organ function, and compromised immune responses. Additionally, the increased permeability of the endothelial barrier may facilitate the transport of nanomaterials themselves into the underlying tissues, potentially leading to further tissue damage or adverse effects (Setyawati et al. 2017).

2.3 *Active Targeting with Specific Antibodies or Other Ligands*

Active targeting involves using specific ligands, such as antibodies, to bind to receptors or molecules on target cells. Ligands are attached to nanoparticles or drug delivery systems, which are then administered to the body. The ligands recognize and bind to the target, allowing for targeted drug delivery or imaging. This approach enhances selectivity and improves therapeutic outcomes (Attia et al. 2019).

The drug targeting potential of antibody-conjugated modified immunodendrimers was investigated for ovarian cancer. Bio distribution studies revealed efficient targeting and increased accumulation of these immunodendrimers in ovarian cancer cells expressing mesothelin protein. These findings suggest that immunodendrimers have the capability to deliver a greater amount of bioactive molecules, leading to improved therapeutic outcomes for ovarian cancer (Jain et al. 2015).

2.4 Passive Targeting

The enhanced permeability and retention (EPR) effect seems to be another term for such a passive drug targeting mechanism. The carriers having specific chemical and physical properties, which include dimensions, structure, surface characteristics, stability, and surface energy, affect the speed and rate at which drug delivering to the targeted sites. Simple nanostructures can optimize penetration of drugs into the hypo vascular cancer cell core, continuing to increase the therapeutic effects of active ingredients against numerous multi-drug resistance cancerous cells whereas reducing cytotoxic effects (Attia et al. 2019).

Aniruddha et al. found that nanoparticle with size 20 nm formed by formation of antimicrobial compounds and Propylene glycol with acetylation of carboxymethyl cellulose used to have a 5–20-fold increment in cancer cell targeting delivery system, which already had strong absorption rate in normal hepatocytes. Whenever it happens to come to surface properties, extremely either negative or positive nanoparticles are much more likely to be taken up by macrophages than neutral ones. For e.g. poly-amino esters (PBDEs) which are positively charged are employed as therapeutic drugs method for connective tissue by effectively utilizing electrostatic attraction among both positively charged nanocarriers and oppositely charged extracellular constituents of connective tissue such as Extracellular matrix (ECM) (Setyawati et al. 2017).

In general, rigid, and spherical carriers were easier to internalize than soft and cylindrical carriers. However, due to the off-target absorption of certain drugs, such as those found in the spleen and liver such strategies remain far from ideal. Besides that, the incorporation forms might very well differ because of the involvement of various material combinations or any other absorption methodologies. Moreover, good potential EPR impacts in animal melanoma cells did not reconstruct in people with cancer, limiting their clinical application (Attia et al. 2019).

3 Classification of Biomaterial Used in Drug Delivery

The selection of Biomaterial for drug delivery is crucial part because every cell or tissue or organ have of different physiochemical and biological behavior. It acts differently on different targeted site including hard tissue (like teeth and bone) and soft tissue such as cartilage, skin, ligament, nerve, muscle, and tendon). Typical composites, metal alloys, as well as ceramic materials of limited features are now being altered by much more advanced materials, and polymeric biomaterials are substituting by stable prosthetics due to fears the about relatively low biocompatibility and the need for modification treatments. The polymeric biomaterial is produced from natural or synthetic source which having multiple application in drug delivery, gene therapy, nanotechnology as well as tissue engineering and tissue regeneration field (Jiann Chong et al. 2022).

3.1 Naturally Extracted

Natural biomaterials are divided into subcategories based on their chemical makeup that included polysaccharide-based materials like alginate cellulose, chitosan, and gum-based biomaterial as well as protein-based materials (Adeosun et al. 2020). Figure 4 shows classification of naturally extracted biomaterial used in drug delivery.

3.1.1 Protein Based Biomaterial

In nature multiple protein including structural and non- structural protein act as essential role in living organism which are recently used for production of biomaterial for drug delivery. The protein-based biomaterial used in drug delivery because of structural chemistry and easily adapted for cell-to-cell communication. The protein-based biomaterial such as silk, collagen, fibrin, elastin, gelatin, and keratin are used in drug delivery and tissue engineering (Kim et al. 2014).

Collagen

Collagen is basic fibrous structural and abundant protein in animal kingdom. It involves 30% part of mammalian protein and important part of extracellular matrix which having alpha triple helix structure with common repeating sequence. The fibrous nature of collagen helps in the supporting of both hard tissue as well as soft tissue like cartilages, tendon, blood vessels, and ligaments (Sahithi et al. 2013).

Collagen sponge are usually synthesis from insoluble collagen of animal origin such as collagen of pig, cows which are good candidate for wound dressing. The frame work of sponge is made up acid and freeze thawing alkali which having 5% dry

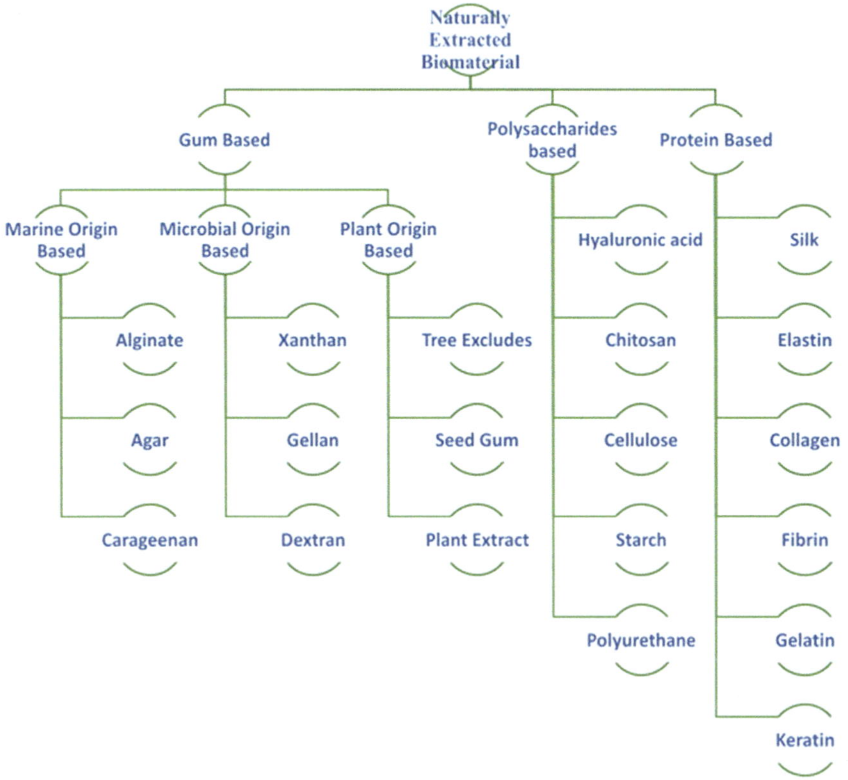

Fig. 4 Classification of naturally extracted biomaterial used in drug delivery system

matter. The temperature of freezing is crucial part for synthesis of collagen sponges. The porosity is depended on freezing temperature, the lower freezing temperature leads to small pore channel where some drug molecules are not passed and unable to interact with cell surface. Whereas high temperature leads to large pore size of sponges where chance to be more accumulation of drug as well as more chance of infection because of large open pore structure is favor to more infection (Chen et al. 2008).

Collagen based sponges are effective against burn infection. The one more used of collagen is preparation of Sponges which are ideal for wound dressing. These sponges are place easily on surface and maintain moist environment without bacterial infection (Takemoto et al. 2008). The collagen sponge with functionalized molecules like glutaraldehyde are used for extra support it enhances mechanical strength of collagen sponges (Peng et al. 2017). Sponge applications also included the healing of burns, diabetic foot ulcers, as well as donor sites because collagen stimulates the penetration of immune system cells like macrophages. Exogenous growth factors, including fibroblast growth factor (FGF) and platelet-derived growth factor (PDGF) that promote the capillary generation and epidermal wound repair, respectively, may also be injected into scaffolds to improve healing of wounds. Collagen scaffolds were also used to deliver retinoic acid intravaginal to repair cervical dysplasia without causing systemic side effects (Surwit et al. 1982).

Beside sponge's collagen also used in preparation of hydrogel. Basically, hydrogel is 3D network made up of polymers that hold sufficient number of fluids (Surwit et al. 1982).

Collagen film is producing from mixing of collagen with dried bovine telopeptide. Their primary uses are as therapeutic drug delivery systems that simultaneously serve as barriers to safeguard wounds or ulcers. Covalent or hydrogen bonding, simple entrapment, or other drug delivery mechanisms can all be used to load drugs onto films, which can then be sterilized without changing their mechanical properties. The use of collagen films for wound dressing, supporting weakened tissues, and directing tissue regeneration is well documented (Oechsle et al. 2016; Bahrami et al. 2019).

Gelatin/Elastin

Gelatin is a polymer that is well-characterized, biocompatible, and biodegradable. This is produced by denatured proteins and puncturing natural collagen, which is typically of porcine or bovine origin. It is commonly utilized as a less costly collagen alternative in foodstuff and cosmetics manufacturing, and it has numerous pharmaceutical purposes. Collagen is made of type I fibrillar collagen that comprises water, mineral salts, and up to 92% pure protein. In therapeutic applications, it has a few benefits over its own parent collagen, including reduced immunogenicity, enhanced dispersion, ease of transformation from solutions to gel (Foox and Zilberman 2015).

Mammalian gelatin is highly recommended for in vivo applications due to its abundance of cell-binding domains, making it an excellent substrate for mobilization and attachment of adherent cells (Young et al. 2005). A smaller proportion of phenylalanine as well as the lack of tryptophan and tyrosine can generate ring structures and radical that could trigger an immune response, making gelatin less antigenic than collagen (Sahoo et al. 2015).

Gelatin microparticles have also been used in the delivery drug which is essential to decreasing sensitivities and adverse reactions from oncology treatments. A review suggests that incorporating an epidermal growth factor receptor (EGFR) into the gelatin backbone scaffold offers advantages over the gelatin amino acid backbone (Magadala and Amiji 2008). This introduction of the growth factor has the potential to improve the safety and effectiveness of hepatocellular cancer treatment. Gelatin is a versatile biomaterial that can be utilized for various tissue engineering and delivery applications. Moreover, it is easily accessible and biocompatible (Young et al. 2005).

The pH-responsive properties of gelatin gels by incorporating pH-sensitive nanosized vesicles, specifically sodium oleate (NaOA). The vesicles transition from vesicles to micelles when exposed to a pH higher than approximately 10. By combining NaOA and gelatin, a vesicle-loaded gel is formed, and the vesicles can be transformed into micelles by introducing a pH 10 buffer solution, enabling controlled release of solutes from the gel based on pH (Dowling et al. 2009).

Silk

Fibroin solutions effortlessly give rise to silk films through the application of spin-coating or layer-by-layer methods. These films have emerged as promising scaffolds within the realm of tissue engineering, finding application in a plethora of endeavors. Notably, the chemical bonding of RGD domains to silk films has proven instrumental in fostering bone formation upon the introduction of osteoblasts to the scaffolds. Interestingly, both silk and collagen films have exhibited comparable prowess in facilitating cell binding, encouraging cellular differentiation, and preserving physiological morphology when subjected to human cell culture (Numata and Kaplan 2010).

An artificial skin material based on chitosan/silk composite film was demonstrated to promote soft tissue wound healing in murine models by seeding it with human adipose stem cells (Pankaew et al. 2015).

Recombinant spider silk coatings have been employed as "bio shields" to mitigate the foreign body response that arises from the utilization of silicone in cosmetic and medical interventions (Salehi et al. 2020). The development of a protective capsule around the implanted material can result in the shrinkage of the implant, which can lead to various issues such as tissue distortion, infection, pain, and unfavorable cosmetic outcomes. These complications occur in approximately 10–26% of individuals who receive silicone implants, particularly when the biocompatibility of the implanted material diminishes over time (Zeplin et al. 2014). The application of a micro-thin layer of recombinant *Araneus diadematus* spider silk protein eADF4 (C16) on silicone implants has shown a significant enhancement in their biocompatibility during the crucial early months after implantation. The absence of immune reactions triggered by the spider silk protein eADF4 (C16) makes it an excellent choice for enhancing biocompatibility, making it an ideal candidate for this purpose (Zeplin et al. 2014).

Fibrin

In addition to being biocompatible and biodegradable, fibrin polymer is a natural material with many uses. A proposed method involves harnessing the biochemical reaction between fibrinogen and thrombin to produce the fibrin polymer. This fibrin polymer can then be utilized as a delivery system, where drugs are either encapsulated within it or coated onto its surface. These systems can be designed for parenteral injection, with the drugs entrapped within them, or in the form of sheets for surgical implantation (Noori et al. 2017).

The distinctive characteristics of the fibrin microsphere render it a practical option for drug delivery. The fibrin microsphere undergoes biodegradation through a natural and physiological process, which makes it well-suited as a delivery device that can be either surgically implanted or injected (Noori et al. 2017). The release of proteins from fibrin microspheres is primarily influenced by factors such as the molecular size of the macromolecules, the preparation medium used for the fibrin beads, and the resulting size distribution. Furthermore, charge interactions can also affect the release of macromolecules. Consequently, achieving the desired release rate in drug delivery systems can only be accomplished by carefully controlling these factors during the engineering process (Gila-Vilchez et al. 2022).

Keratin

Keratin biomaterials have garnered significant attention in the field of biomaterials research due to their unique properties and potential applications in various biomedical fields. Biocompatibility: Keratin biomaterials are biocompatible, meaning they are unlikely to cause any toxic or immunogenic responses in the body (Feroz et al. 2020).

Examples of keratin biomaterials include:

1. Keratin hydrogels: These are three-dimensional networks of keratin proteins that can be used as scaffolds for tissue engineering and wound healing applications. They provide a supportive environment for cell growth and regeneration (Konop et al. 2021).
2. Keratin nanoparticles: These are tiny particles made from keratin proteins that can encapsulate drugs and facilitate their controlled release. They are used in drug delivery systems to improve the efficiency and targeted delivery of therapeutic agents (Ferroni and Varchi 2021).
3. Keratin-based films: Films made from keratin proteins can be used as coatings for medical devices or as wound dressings. They provide a protective barrier, promote healing, and can be tailored to have antimicrobial properties (Lee et al. 2015).
4. Keratin-based scaffolds: These three-dimensional structures made from keratin proteins mimic the extracellular matrix and provide a framework for tissue regeneration. They can be used in tissue engineering to support the growth and organization of cells into functional tissues (Guzman and Rabbany 2016).
5. Keratin-based fibres: Keratin fibres can be spun into various forms such as threads or meshes, which can be used as scaffolds for tissue engineering, wound healing, or as sutures in surgical procedures. They offer mechanical strength and biocompatibility (Zhang et al. 2020).

These are just a few examples of the wide range of keratin biomaterials that have been developed and explored for various biomedical applications.

3.1.2 Polysaccharide Based Biomaterial

A polysaccharide-based biomaterial is a type of biomaterial that is made up of polysaccharides, which are long chains of carbohydrates. Polysaccharides are a natural component of many organisms, including plants, animals, and bacteria, and they have several desirable properties for use in biomaterials. Polysaccharide-based biomaterials have several potential applications in the field of biomedicine, including drug delivery, tissue engineering, and wound healing. They are biocompatible, meaning that they can be safely used within the body without causing an immune response or toxicity. They are also biodegradable, which means that they can be broken down and eliminated from the body over time (Barclay et al. 2019).

Examples of polysaccharide-based biomaterials include chitosan, hyaluronic acid, alginate, and cellulose. These materials can be processed into a variety of forms, including gels, films, and fibres, and can be modified to enhance their properties for specific applications. Overall, polysaccharide-based biomaterials offer a promising avenue for the development of new and innovative biomedical technologies that can improve human health and well-being (Dattilo et al. 2023).

Hyaluronic Acid

Glucuronic acid and N-acetyl glucosamine form dual saccharide units that alternately connect to form hyaluronic acid, a linear macromolecule mucopolysaccharide. It is highly viscoelastic, biodegradable, and has good biocompatibility. tumor cells but expressed at low levels on the surface of epithelial, hematopoietic, and neural cells. Additionally, hydroxyl, carboxyl, and N-acetyl groups in hyaluronic acid are good candidates for chemical modification. Therefore, the use of hyaluronic acid and its derivatives as drug carriers helps to improve drug targeting, and prolonged release (Huang and Huang 2018a).

Pulmonary elastin is protected against inflammation by hyaluronic acid, which is naturally present in the lungs and has a mucoadhesive function. The study found that hyaluronic acid may be utilized to deliver medications to the nose and lungs, and that hyaluronic acid-based particles might extend the average retention period of medications in the lung's primary absorption site. For instance, in the research encapsulating recombinant human insulin with hyaluronic acid, the dry powder was rendered acceptable for inhalation by spray drying, and the body's insulin level and the dog's insulin-induced glucose level were monitored (How et al. 2020).

Pro drugs created by covalently attaching small molecule anticancer medications to hyaluronic acid are known as hyaluronic acid-drug conjugates. In the blood, these covalent bonds are difficult to dissolve, but once they reach the target, they do hydrolysis or enzymolysis, allowing the drug to be released. Hyaluronic acid-drug conjugates can increase a drug's solubility, alter its half-life and distribution in vivo, increase tumor tissue accumulation by boosting the osmotic retention effect, and more effectively exercise their effects (Huang and Huang 2018b).

Chitosan (CS)

Chitosan (CS) is a remarkable nanomaterial derived from a biological and positively charged polysaccharide. It is primarily composed of (1,4)-linked D-glucosamine

(GlcN) and N-acetyl-D-glucosamine (GlcNAc), obtained from the abundant chitin found in the exoskeletons of arthropods, insects, crab shells, and fungi. At physiological or neutral pH, CS exhibits cationic properties due to the presence of GlcN.

CS holds great promise in the field of medicine due to its numerous beneficial attributes such as biocompatibility, bioactivity, biodegradability, and mucoadhesion. When CS is broken down by internal enzymes like lysozymes and chitosanases, it yields oligosaccharides and monosaccharides that can be easily absorbed by the body. However, despite these advantageous qualities, CS faces limitations in its application in clinical settings. It is hindered by weak mechanical characteristics, low solubility, and distinctive physicochemical and biological properties (Sahithi et al. 2013).

To overcome these challenges, various approaches have been developed to modify CS and address its shortcomings. These methods aim to enhance its mechanical strength, solubility, and other desirable characteristics, thus expanding its potential applications in clinical settings (Chen et al. 2008).

A variety of CS derivatives have been created with increased solubility due to their high affinity with functional proteins and capacity for self-assembly. As a result, CS has been widely used in a variety of biomedical and pharmaceutical activities, including the transport of drugs, genes, and vaccines, tissue engineering, wound healing, and the production of cosmetics (Wang et al. 2022). The Table 1 shows used of chitosan in drug delivery.

The synthesis of CCHNs involves creating complexes of chitosan and CDs using ethylenediaminetetraacetic acid (EDTA) monomers in an aqueous solution. The miniature CDs are immobilized within the chitosan frameworks through selective crosslinking of chitosan chains in the nanoparticles shown in Fig. 5. This results in CCHNs with enhanced colloidal stability, high doxorubicin (DOX) loading capacity, strong and persistent UV to NIR fluorescence, efficient NIR photothermal conversion, and intelligent drug delivery responsive to both NIR light and pH changes (Wang et al. 2017).

In order to bind Ca^{2+} and trigger the deposition of apatite, carboxymethyl chitosan (CMCS), a water-soluble derivative of chitosan, has greater biodegradability and bioactivity. Additionally, CMCS nanofibers avoid the acidic salt removal as compared to electro-spun chitosan because they use water as the solvent. The polyethylene oxide (PEO) to successfully create homogenous CMCS nanofibers. A mouse bone marrow stromal cell (mBMSC) differentiation was assessed on the nanofiber scaffolds. Cell tests showed that the ALP activity was boosted by CMCS-HA composite nanofibers (Zhao et al. 2017).

Cellulose

Cellulose, being the most abundant biopolymer found in nature, has found extensive applications in the production of various essential products in our modern lives. It is composed of long chains of anhydro-D-glucopyranose units (AGU) that are covalently linked together through β1,4-glycosidic bonds between the equatorial

Table 1 Chitosan biomaterial used in drug delivery

Sr. No	Types of chitosan biomaterials	Appearance of biomaterial	Preparation technique	Application	Refs.
1	Chitosan-carbon dot hybrid nanogels	Nanogel	Covalent cross-linking	Photothermal chemo therapy	Wang et al. (2017)
2	Biomimetic mineralization of carboxymethyl chitosan nanofibers	Nanofibers	Electrospinning process	Osteogenic drug delivery	Zhao et al. (2017)
3	Chitosan-dextran	Magnetic nanoparticle	Self-modification	Drug delivery for glioblastoma cancer	Shevtsov et al. (2018)
4	Chitosan microsphere	Microsphere	Hydrolysis	Brain cancer drug delivery	Lara-Velazquez et al. (2020)
5	Chitosan-collagen hybrid material	Sponge	Freeze-drying	Antibacterial drug delivery and wound dressing	Zhang et al. (2021a)
6	Chitosan gelatin sheet	Sheet	Hydrolysis	Wound dressing	Ciobanu et al. (2014)
7	Chitosan nanoparticle	Drug delivery vehicle	Ionotropic crosslinking of linker molecules	Drug delivery and tracker	Saeed et al. (2020)

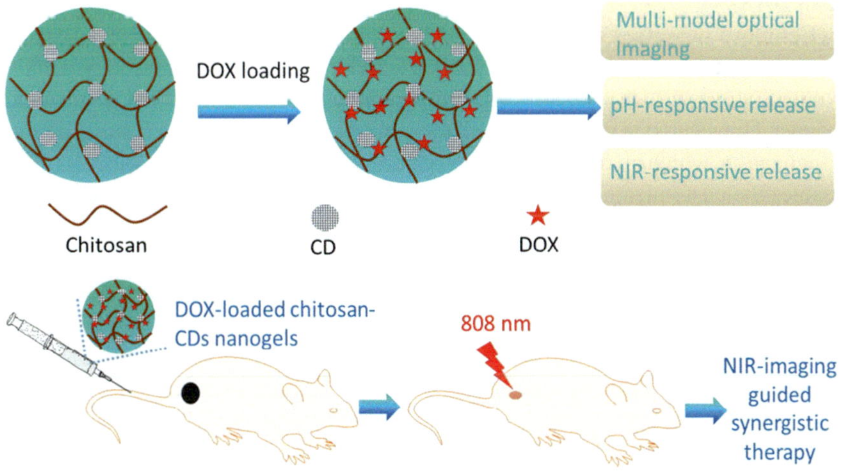

Fig. 5 Chitosan loaded with DOX. Reprinted with permission from Wang et al. (2017). Copyright 2017, American Chemical Society

group of the C1 carbon atom and the C4 carbon atom. These linear chains can extend to form molecules containing approximately 1000–1500 β-glucose units (Sun et al. 2018). Cellulose and its derivatives have found extensive application in various drug delivery systems including hydroxypropyl methyl cellulose (HPMC), hydroxypropyl cellulose (HPC), sodium carboxymethyl cellulose (CMC), ethyl cellulose (EC), and methylcellulose (MC) (Sun et al. 2018).

Cellulose can undergo different treatments to yield modified forms such as hydroxypropyl cellulose (HPC) and hydroxypropyl methylcellulose (HPMC), which are commonly employed in drug delivery applications. Hydroxypropyl cellulose (HPC) is produced by reacting propylene oxide with alkali cellulose on the anhydrous glucose chain. HPC exhibits solubility in various organic solvents and even cold water, making it highly suitable for drug delivery purposes due to its broad solubility range (Phromsopha and Baimark 2014).

Since the 1960s, hydroxypropyl methylcellulose (HPMC) has been utilized in the formulation of sustained-release oral drug tablets. HPMC is a water-soluble cellulose ether characterized by its hydrophilic polymeric structure. When a drug is encapsulated within HPMC, water from the surrounding environment penetrates the matrix and hydrates the polymer chains. As a result, the HPMC matrix gradually disintegrates, leading to the release of the active ingredients (Sun et al. 2018).

The release of the drug is generally governed by two primary mechanisms: (1) diffusion through the gelled layer of HPMC, and (2) erosion of the HPMC matrix. In cases where the drug exhibits high water solubility, the overall drug release primarily occurs through the diffusional mechanism, where the drug molecules diffuse through the gel-like HPMC layer. It is widely acknowledged that the relative contribution of these mechanisms determines the rate of drug release from the HPMC matrix (Sun et al. 2018).

Cellulose possesses certain inherent limitations, such as inadequate resistance to creasing, undesired solubility in solvents, and a lack of thermoplasticity. Unlike synthetic polymers, the natural cellulose structure cannot be modified in the same manner. However, one viable approach to overcome these challenges is surface functionalization, which involves introducing external groups to cellulose molecules. This enables the customization of cellulose surface chemistry, promoting self-assembly, controlled dispersion within a wide range of polymer matrices, and enhancing both inter-particle and particle–matrix bonding strength while preserving its desirable intrinsic properties. Functionalization opens up opportunities for cellulose to be converted into various derivatives, expanding its potential applications (Troy et al. 2021; Way et al. 2012).

Starch

Starch, in addition to cellulose, is a renewable biopolymer that is abundantly available and has been extensively studied for its potential in developing novel drug dosage forms. The appealing characteristics of starch, such as its affordability, widespread availability, versatility, biocompatibility, biodegradability, and low immunogenicity, make it an appealing choice for drug delivery applications. The chemical composition of native starches is typically determined by the ratio of amylose to amylopectin, as well as intermediate materials, which can vary depending on factors such as the botanical source, inherent variations, and environmental conditions (Lemos et al. 2021).

Starch can be extracted from various plant-based sources and undergoes processes like hydrolysis, esterification, cross-linking, and oxidation to make it porous. Coating tablets, capsules, and facilitating controlled drug delivery in solid dosage forms are the primary applications of commercial starch products. Examples of such products include Lycoat RS 720®, which is a hydroxyl propylated pea starch, and Amprac 01®, an acetylated corn starch. Another starch derivative known as Contramid®, obtained through dual chemical modifications, serves as an excipient in solid dosage forms. It possesses properties well-suited for oral controlled-release formulations. These starch-based products provide numerous benefits, including safety, effective disintegration, and protection for delicate active ingredients (Lemos et al. 2021).

Polyurethane (PURs)

It is a versatile group of polymers renowned for their ability to finely tune their physicochemical properties through careful control of their chemical composition. Segmented PURs have been meticulously engineered to exhibit well-defined degradation and mechanical characteristics, while also showcasing remarkable biocompatibility. As a result, they emerge as an appealing choice for the development of cutting-edge medical devices. Tissue engineering, and drug delivery systems. Several types of PURs, such as PEURs, poly (ester urethanes), PCURs, PSURs, surface-modified PURs, and composite PURs, have been created for various biomedical applications. Ongoing research efforts are focused on developing PURs with specific biodegradability and biocompatibility properties for drug delivery and tissue regeneration purposes (Sannino et al. 2004).

3.1.3 Gum Based Biomaterial

There are three kinds of gum material biomaterial used in drug delivery. These gum are extracted from algae, microbes, and plants (Amiri et al. 2021).

Marine Gum-Based Biomaterial

Alginate

Alginate is a biopolymer extracted from brown seaweeds, commonly used in various applications due to its biocompatibility and low toxicity. Brown algae Alginate is a polysaccharide such as *Laminaria hyperborea*, *Laminaria digitata*, *Laminaria japonica*, *Ascophyllum nodosum*, and *Macrocystis pyrifera* extracted from brown seaweeds, including. This polysaccharide has a sequence of α-L-guluronate (G) and β-D-mannuronate (M) monomers linked by (1Ñ4) bonds, and its M/G ratios vary depending on the seaweed source. Alginate is widely used as an excipient in drug delivery systems and stabilizer in pharmaceutical formulations. Alginate hydrogels are used in wound healing, tissue regeneration, and cell encapsulation. Recent studies have shown that encapsulating drugs in alginate hydrogels improve the bioavailability and efficacy of healing. Alginate-based capsules are suitable for encapsulating different types of cells, making them useful in cryotherapy treatments and cellular microcultures. Alginate hydrogels are increasingly popular in tissue engineering and regenerative medicine applications due to their advantages as appropriate materials (Leach and Schmidt 2005).

Agar

Agar, a biomaterial derived from red algae, is extracted to cater to a range of biomedical applications, such as drug delivery. Its versatility allows for easy customization to meet specific drug delivery needs. Numerous agar-based biomaterials have been created to address various biomedical requirements. Presented below are a few noteworthy examples. Agar hydrogels: Agar hydrogels can be used as a drug delivery system for various types of drugs, including proteins, peptides, and small molecules. They can also be used as a scaffold for tissue engineering, wound healing, and other regenerative medicine applications (Gila-Vilchez et al. 2022).

1. Agar microspheres: Agar microspheres can be used as drug carriers for the sustained release of drugs. They can also be used for targeted drug delivery to specific tissues and organs (Gila-Vilchez et al. 2022).
2. Agarose beads: Agarose beads can be used for chromatography and other separation techniques in the laboratory. They are also used in various medical applications, such as drug delivery, tissue engineering, and cell culture (Salati et al. 2020).
3. Agar films: Agar films can be used for wound healing, drug delivery, and other biomedical applications. They are flexible and biocompatible, making them an attractive option for medical devices (Onofre-Cordeiro et al. 2018).

Carrageenan

Carrageenan is a family of linear sulfated polysaccharides that are derived from red seaweeds. The sulfated groups present in the carrageenan structure give it its characteristic properties, such as its ability to form gels and stabilize emulsions (Qureshi et al. 2019). The Table 2 shows carrageenan-based biomaterial used in drug delivery.

Table 2 Carrageenan biomaterial functionalized with drug molecules for various diseases

Sr. No	Component of carrageenan	Cross linker	Drug	Application	Refs.
1	κ-Carrageenan and sodium carboxymethyl cellulose	$AlCl_3$	Ibuprofen	Controlled drug delivery	Lohani et al. (2016)
2	κ-Carrageenan and sunflower oil	–	Curcumin	Drug delivery against selected lung cancer cells (A549)	Sathuvan et al. (2017)
3	κ-Carrageenan and Ag nanoparticles	KCl	–	Controlled drug delivery	Azizi et al. (2017)
4	κ-Carrageenan and NaCMC	Genipin	–	Gastro-intestinal tract drug release	Hezaveh and Muhamad (2012)
5	Gelatin and i-carrageenan	Glutaraldehyde, ethanol and HCl	Ciprofloxacin	Delivery or delivery of anticancer drugs	Padhi et al. (2016)
6	Carboxymethylated κ carrageenan	Polyglutaraldehyde	Insulin	Oral insulin delivery	Leong et al. (2011)
7	κ-Carrageenan and sulfuric acid	–	Lappaconitine	Delivery of hydrophobic drugs	Sun et al. (2016)

Microbes Gum-Based Biomaterial

Dextran

Dextran is a natural polysaccharide that can be used as a biomaterial in various applications due to its unique physicochemical properties. Dextran-based biomaterials have been extensively studied for their biocompatibility, biodegradability, and ability to enhance wound healing and tissue regeneration (Bos et al. 2005).

Dextran-based biomaterials can be used as drug delivery systems, scaffolds for tissue engineering, and wound dressings. Dextran can be cross-linked to form hydrogels that can be used as drug-delivery vehicles. These hydrogels can be designed to release drugs in a controlled manner over an extended period. Dextran-based hydrogels can also be used as scaffolds for tissue engineering. The hydrogels provide a three-dimensional structure that mimics the extracellular matrix and can support the growth and differentiation of cells. The hydrogel can also provide a barrier against infection and can absorb excess exudate from the wound. Overall, dextran-based biomaterials have great potential in biomedical applications due to their biocompatibility, biodegradability, and versatility (Bos et al. 2005; Xiao et al. 2009; Cabral et al. 2015).

Xanthan (XG)

Xanthan gum (XG) has garnered considerable attention as a biomaterial utilized in drug delivery systems due to its exceptional properties and versatility. Derived from microbial fermentation, XG is a polysaccharide that possesses numerous advantages for drug delivery applications. One of its notable features is the capacity to form hydrogels. Through the crosslinking of XG molecules, hydrogels can be engineered with customizable properties like gel strength, porosity, and swelling behavior. These hydrogels possess a three-dimensional network structure that can encapsulate drugs and enable controlled release. The gel formation of XG can be triggered by various means, such as temperature, pH, and the presence of specific ions, thereby allowing tailored drug release profiles (Singhvi et al. 2019).

XG hydrogels also possess excellent water absorption capacity, which is crucial for drug delivery systems. They can absorb and retain a significant amount of water, ensuring stability and hydration of the formulation. This property is particularly advantageous in wound healing applications, where maintaining a moist environment is crucial for proper healing. Furthermore, XG exhibits mucoadhesive properties, meaning it can adhere to mucosal surfaces. This property is highly beneficial for drug delivery systems targeting mucosal routes, such as oral, nasal, ocular, and vaginal delivery. By enhancing the residence time and contact with the mucosal membrane, XG-based formulations can improve drug absorption and bioavailability (Benny et al. 2014).

The biocompatibility of XG is another important aspect for its application as a drug delivery biomaterial. Numerous studies have demonstrated the non-toxic nature of XG, both in vitro and in vivo. It has shown minimal cytotoxicity and immunogenicity, making it suitable for biomedical applications. Additionally, the biodegradability of XG ensures its gradual breakdown and elimination from the body, minimizing potential long-term complications (Singhvi et al. 2019; Benny et al. 2014).

Gellan

Gellan gum finds application as a biomaterial in the field of tissue engineering, particularly in the creation of scaffolds. These scaffolds, composed of gellan gum, possess a porous and three-dimensional structure that closely resembles the extracellular matrix of tissues. They serve as a supportive framework for cells, promoting attachment, proliferation, and differentiation, thereby facilitating the regeneration of damaged or diseased tissues. Gellan gum-based scaffolds have been successfully utilized in diverse tissue engineering areas such as bone regeneration, cartilage repair, and skin tissue engineering. The porosity and interconnected pore structure of these scaffolds enable the diffusion of nutrients and oxygen, as well as the removal of waste products, crucial for sustaining cell growth and tissue regeneration. Additionally, gellan gum exhibits biocompatibility and biodegradability, making it suitable for in vivo applications. Over time, the scaffold gradually degrades, allowing the newly formed tissue to take its place (Milivojevic et al. 2019).

In addition to its structural properties, gellan gum can be modified to incorporate bioactive molecules, growth factors, or drugs into the scaffold. This allows for controlled release of therapeutic agents, promoting tissue regeneration and healing. The versatility and tunability of gellan gum as a biomaterial make it a promising choice for tissue engineering applications and the development of advanced regenerative medicine strategies (Das and Giri 2020).

Plant Gum and Mucilage-Based Biomaterial

Mucilaginous plants have been utilized in traditional medicine for over 4000 years, and in modern times, Mucilages have found applications in pharmaceutical formulations as binding agents and drug excipients. Compared to synthetic materials, mucilage has superior bonding properties. It is commonly used in tablet production as emulsifying and suspending agents, bio adhesive agents, and gelling and thickening agents. Mucilages have also been used in medicinal formulations to treat inflammatory processes of the gastrointestinal tract by forming a protective layer over mucous membranes, protecting them from nerve stimulation (Amiri et al. 2021).

In a study conducted in 2020 by Haile et al., the potential medicinal properties of the bark mucilage of *Grewia ferruginea* Hochst. Ex A. Rich. (Malvaceae) were investigated. The aim of the study was to explore the feasibility of using this mucilage as a medicinal excipient. The mucilage was extracted from the inner stem bark of the plant using water, precipitated with ethanol, dried, and then powdered. The powdered mucilage underwent various physical and chemical tests, including moisture absorption, microbial load, acute oral toxicity, pH, viscosity, ash content, solubility, and swelling index (Amiri et al. 2021).

The results of the study showed that the final yield percentage of dried and powdered Grewia ferruginea mucilage was 11.96% (w/w), and it exhibited good powder flow properties. The mucilage displayed false current behavior, and its moisture absorption, solubility, and swelling improved with temperature. The pH of the mucilage was almost neutral, and it had a low microbial load, making it a safe component in medicinal formulations. Acute oral toxicity testing indicated that it was safe up to 2000 mg/kg. Based on these findings, it can be concluded that the bark mucilage of *Grewia ferruginea* can be used as a potential adjunct in medicinal formulations (Amiri et al. 2021).

In another study, mucilage extracted from quince seeds was used to create a porous physical structure for medicinal purposes. The process involved extraction, molding, and freezing of quince seed mucilage to create porous scaffolds with an interconnected structure. Human-adipose-derived mesenchymal stem cells were then seeded on these scaffolds, and cell viability was evaluated using the 3-(4,5-dimethylthiazol-2-yl)-2,5-diphenyl-2H-tetrazolium bromide (MTT) assay. The scaffolds did not have a cytotoxic effect on the cells, as observed in the MTT assay. Adhesion and migration of human-adipose-derived mesenchymal stem cells on the scaffolds derived from interconnected seed mucilage were assessed using histological staining and scanning electron microscopy analysis. The results indicated that the transplanted seed-mucilage-derived scaffolds have the potential to replace common polysaccharides in regenerative medical applications (Amiri et al. 2021).

Plant Extract

The clinical impact of nanocarriers, such as microemulsions and nanoemulsions, is significant as they improve drug efficacy, safety, and patient compliance. Several successful examples highlight their effectiveness. For instance, nano emulsions have been used to enhance solubility and oral absorption of herbal extracts like *Vitex agnus-castus* and *Silybum marianum*. Nanoencapsulation of *Phyllanthus amarus* extract has demonstrated improved hepatoprotective activity. In addition, an injectable nanoparticulate system has been formulated to achieve sustained release of components from *Ginkgo biloba* extract. Furthermore, *Panax notoginseng* Saponins have been successfully loaded into core–shell hybrid liposomal vesicles, leading to enhanced bioavailability and protective effects (Rahman et al. 2020).

3.2 Synthetically Driven

Synthetically driven biomaterials play a crucial role in drug delivery systems, providing controlled release and targeted delivery of therapeutic agents. These biomaterials offer advantages such as tunable properties, biocompatibility, and stability, making them ideal candidates for drug delivery applications. One example of synthetically driven biomaterials for drug delivery is polymeric nanoparticles. These nanoparticles can be engineered with precise control over their size, shape, surface properties, and drug loading capacity. They can encapsulate a wide range of drugs, including small molecules, proteins, and nucleic acids, protecting them from degradation and enhancing their therapeutic efficacy. Polymeric nanoparticles have been extensively studied for various applications, such as cancer treatment, gene therapy, and vaccination (Fenton et al. 2018).

4 Preparation of Polymer

Polymeric drug delivery systems have been extensively researched, and in situ, gel systems have gained considerable attention in recent years (Guo et al. 2018). The advantages of in situ forming polymeric delivery systems include ease of administration, reduced frequency of administration, improved patient compliance, and comfort (Wang et al. 2011). It has been reported that several patents have been filed for the use of in situ gel-forming systems in a variety of biomedical applications, such as drug delivery. Sol–gel transitions occur when smart polymeric systems are administered, which make them potential drug delivery systems (Ong et al. 2008). It is possible for gel formation to be triggered in situ by changes in pH, temperature, solvent exchange, or by a combination of these stimuli. The mechanisms of gel formation from the sol forms, evaluation, and characterization of in situ polymeric formulations, and the different types of smart polymers are discussed in this review Since the early 1970s, extensive research has focused on the utilization of biodegradable polymers in clinical applications. Both natural and synthetic polymers have been explored for the development of in situ forming drug delivery systems. The classification of in situ polymeric systems varies depending on the route of administration, and this will be further detailed in the following section (Chhetri et al. 2021).

4.1 *In Situ Forming Polymeric Systems for Oral Administration*

The development of in situ gel systems for drug delivery has gained significant interest in recent years due to their potential advantages such as ease of administration and improved patient compliance. In the field of oral drug delivery, natural polymers such as pectin, xyloglucan, and gellan gum have been extensively investigated as in situ forming polymeric systems. Pectin, for instance, is polysaccharide composed mainly of α-(1–4)-D-galacturonic acid residues and it's in situ gelation can be induced by divalent ions such as calcium ions present in the stomach. A study reported sustained release of paracetamol using an orally administered in situ gelling pectin formulation (Kedir et al. 2022).

Xyloglucan, another natural polymer, is derived from tamarind seeds and has a (1–4)-β-D-glucan backbone chain with (1–6)-α-D xylose branches partially substituted by (1–2)-β-D-galactoxylose. The gelation of xyloglucan is thermally reversible and can be induced by β-galactosidase (Zhang et al. 2021b).

Gellan gum is a deacetylated exocellular polysaccharide secreted by *Pseudomonas elodea* with a tetra saccharide repeating unit. Its gelation can be induced by temperature or cations and involves the formation of double helical junction zones followed by aggregation of the double-helical segments to form a three-dimensional network. A study reported increased bioavailability and sustained drug release of theophylline using an in-situ gelling gellan formulation for oral delivery (Das and Giri 2020; Milivojevic et al. 2019).

4.2 *Ocular Delivery*

In recent years, there has been significant interest in the use of in situ gels for ocular drug delivery, mainly due to the limitations of conventional delivery systems that often result in low bioavailability and therapeutic response. Natural polymers like gellan gum, alginic acid, and xyloglucan are commonly employed in the formulation of in situ gel-based ocular delivery systems. Gellan gum has undergone extensive research for its pharmaceutical applications in ophthalmic drug delivery. When applied, an aqueous solution of gellan gum undergoes a transition into a gel state triggered by the temperature and ionic conditions (specifically, the presence of calcium ions) in the tear fluid. This gel formation prolongs the precorneal contact time compared to conventional eye drops, thereby facilitating sustained drug release. Another natural polymer, alginic acid, is a linear block copolymer comprising β-D-mannuronic acid and α-L-glucuronic acid residues. It can form robust gels in dilute aqueous solutions when di- and trivalent metal ions are added. Due to its biodegradability and non-toxic nature, alginic acid has been selected as a vehicle for ophthalmic formulations. (Vasile et al. 2020).

The mucoadhesive properties of alginic acid contribute to its ability to remain in precorneal residence and to increase drug penetration into the eye (Agrawal et al. 2012). Xyloglucan, a natural polymer with a mitotic response, has been used for the formulation of in situ gels for ocular delivery (Lee and Mooney 2012). The in-situ precipitation of polymeric drugs using a PH-induced system is also used for the in situ delivery of ophthalmic drugs, such as Carbopol system-hydroxypropyl methylcellulose system and poly (methacrylic acid)-poly (ethylene glycol). These systems are designed to stay in solution form at acidic pH but form a low viscosity gel at alkaline pH, providing sustained release of drugs (Li et al. 2018). A study on the ophthalmic delivery system for indomethacin for the treatment of uveitis showed sustained release of the drug for a period of 8 h in vitro, making it an excellent candidate for ocular delivery (Gupta et al. 2022). Overall, in situ, gels-based ocular delivery has shown promise in overcoming the limitations of conventional delivery systems and improving drug bioavailability and therapeutic response in ophthalmic applications.

4.3 Rectal and Vaginal Delivery

Miyazaki et al. explored the application of xyloglucan-based thermo-reversible gels for delivering indomethacin rectally to rabbits. The findings of the study revealed that the polymeric system, when administered in situ, offered an extended period of drug retention and a wider absorption peak in comparison to commercially available suppositories. Additionally, the in-situ gel formulation exhibited a notable decrease in drug C max, indicating a reduced likelihood of adverse effects on the nervous system (Agrawal et al. 2012; Miyazaki et al. 1998).

Biomaterials play a vital role in rectal and vaginal drug delivery. Hydrogels are commonly used due to their ability to form a gel-like consistency and sustain drug release. Mucoadhesive polymers enhance drug absorption by adhering to mucosal surfaces. Liposomes offer stability, penetration, and controlled release of drugs. Solid suppositories provide controlled drug release and are commonly used for local and systemic delivery. Polymeric nanoparticles protect drugs, improve stability, and allow targeted delivery. Films and inserts offer comfort and gradual drug release. The choice of biomaterial depends on the drug, release profile, patient comfort, and regulations. Research in this field continues to improve therapeutic outcomes and patient compliance (Valle et al. 2017).

To improve therapeutic efficacy and patient compliance, a mucoadhesive, thermosensitive, and prolonged-release vaginal gel incorporating a clotrimazole-β-cyclodextrin complex was formulated for the treatment of vaginitis (Misbah et al. 2021) (Gupta et al. 2022). Mucoadhesive rectal drug delivery has been successfully employed to enhance drug absorption and provide localized therapeutic effects in various medical conditions, including cancer therapy, anti-hypertension, anti-infection/anti-inflammation, and peptide drug delivery. These benefits are achieved through prolonged retention time and close contact between the drug delivery systems and the rectal mucosa. However, despite the potential of mucoadhesive formulations in treating colon-related diseases, limited research exists on their application in ulcerative colitis (UC) treatment, with only two studies identified. Therefore, the SSZ/Cat-CS hydrogel discussed in this study represents the first rectal mucoadhesive formulation specifically designed for UC treatment. Additionally, it is noteworthy as the first catechol-induced mucoadhesive drug delivery system developed for rectal drug administration (Sreekanth et al. 2021).

4.4 Nasal Drug

Biomaterials commonly used for nasal drug delivery include chitosan, hydrogels, liposomes, nanoparticles, cyclodextrin, and in situ gelling systems. Chitosan offers mucoadhesive properties, while hydrogels provide sustained release and controlled drug delivery. Liposomes protect drugs and enhance penetration, while nanoparticles improve drug absorption and targeting. Cyclodextrin enhance solubility and stability of drugs. In situ gelling systems undergo gelation upon administration, improving drug retention and absorption. The choice of biomaterial depends on factors such as the specific drug, desired release profile, and safety considerations. Ongoing research aims to optimize nasal drug delivery for improved therapeutic outcomes (Moinuddin et al. 2019).

These studies highlight the potential of in situ gel systems and thermosensitive hydrogels as drug delivery systems for the nasal administration of drugs. These systems offer advantages such as improved bioavailability, prolonged drug release, and reduced dosing frequency compared to traditional nasal drug delivery systems. The nasal route of drug delivery has gained increased attention due to its non-invasive nature, rapid onset of action, and avoidance of first-pass metabolism, making it an ideal route for drug absorption. Further research is needed to evaluate the safety and efficacy of these systems in clinical trials (Wu et al. 2007).

5 Application of Biomaterial in Drug Delivery

5.1 *Cancer*

Scientific research reveals that cancer is a leading cause of mortality in developed nations. Despite advancements, traditional cancer treatments like surgery, radiotherapy, and chemotherapy have notable drawbacks in terms of effectiveness and harmful side effects. Fortunately, Nanomedicine offers a hopeful solution for combating cancer by specifically delivering drugs to cancer cells, minimizing toxicity, and enhancing therapeutic outcomes. Promising results have been achieved by scientists who have utilized polymeric nanoparticles (NPs) to successfully transport chemotherapy drugs like doxorubicin, paclitaxel, and camptothecin to various cancer types (Karandish et al. 2016).

Nanoparticles are utilized in drug delivery through two approaches: passive targeting and active targeting, where the latter relies on the former. Passive targeting takes advantage of the enhanced permeation and retention (EPR) effect commonly observed in human cancers. In this method, nanoparticles exploit the tumour characteristics to accumulate at the site. Active targeting involves functionalizing nanoparticles with specific antibodies, proteins, or peptides that bind to the tumor cell surface, which is typically overexpressed in a uniform manner. Immunotherapy, on the other hand, is a promising treatment strategy that harnesses molecules to enhance the immune system's ability to detect and eliminate cancer cells. Employing polymeric nanoparticles in immunotherapy is particularly intriguing, as Toll-like receptor (TLR) agonists can stimulate dendritic cells, resulting in their increased activation and expansion when compared to free-form agonists in animal models. Consequently, Nanomedicine emerges as a highly promising avenue for cancer treatment (Cheng et al. 2021).

5.2 *Neurodegenerative Disorder*

Neurological disorders refer to a range of conditions that impact the brain and central nervous system. These disorders encompass neurodegenerative, neuroinflammatory, and neoplastic diseases. As people age, the prevalence of neurological disorders is on the rise, making them increasingly expensive and life-threatening medical conditions globally (Feigin et al. 2019). Presently, available treatments for neurodegenerative diseases like Alzheimer's and Parkinson's merely address the symptoms and cannot provide a cure, primarily due to the limitations imposed by the blood–brain barrier (BBB). The BBB restricts the passage of therapeutic substances into the brain, hindering effective treatment. Consequently, researchers have been concentrating on the development of nanoparticle systems that facilitate the transport of drugs across the BBB and into the central nervous system (Calzoni et al. 2019).

Nanoparticles made from functionalized PGA, PLA, and PLGA have been extensively studied for their ability to traverse the BBB and release drugs for the treatment of neural disorders. These nanoparticles can be modified during production to interact with the components of the BBB. For instance, surface-coated PLGA-NPs with polylobate 80 and poloxamer 188 have demonstrated an enhanced capability to penetrate the central nervous system (Lara-Velazquez et al. 2020).

In recent years, researchers have also explored the use of functionalized nanoparticles for diagnosing and treating neurodegenerative diseases like Alzheimer's disease (AD) and Parkinson's disease (PD). In a study conducted by Zhang et al., a nanoparticle drug delivery system based on PEGylated PLA-NPs was developed for treating AD. The system exhibited a strong affinity for the beta 1–42 peptide associated with AD. Similarly, functionalized PLGA-NPs loaded with curcumin have been investigated to disrupt A beta aggregates, a hallmark of AD. These studies indicate that functionalized nanoparticles hold significant potential as targeted drug delivery systems for the diagnosis and treatment of neurodegenerative diseases (Cheng et al. 2022).

5.3 *Buccal Disease*

Recent research has established various strategies for delivering antibiotics to specific areas within the body, such as implant coatings made from biocompatible materials like chitosan, gelatin, silica, polylactic acid and sol–gel. These materials can serve as absorbable drug carriers, degrading the coating and eliminating infections, while also promoting Osseointegration. There is no resistance to any strain of bacteria when antimicrobial coatings are made using these biomaterials. Drug delivery systems using natural polymeric forms, including gelatin, chitosan, alginate, calcium phosphate, and hyaluronic acids, have also been developed to enhance drug release control, individualized targeting, prolonged contact time, and reduced dosage frequency (Thouvenin et al. 2022).

Chitosan is a natural polymer that has shown promise as a drug carrier due to its degradability, biocompatibility, and nontoxicity. Its mucoadhesive properties make it useful in various routes of administration, including buccal, ocular, pulmonary, and nasal routes. However, chitosan's insolubility in physiological environments has made it challenging to use for drug delivery, but various modifications such as thiolation, carboxymethylation, quaternization, and acetylation have been used to enhance its solubility. Various types of drug carriers based on chitosan have been developed to deliver antibiotics, growth factors, chemotherapy drugs, vaccines, and anti-inflammatory agents to specific cells. These chitosan-based drug delivery systems come in forms such as films, fibers, sponges, gels, hydrogels, and nano/microparticles. In the field of dentistry, chitosan-based drug carriers have been utilized for the treatment of tooth caries, endodontics, peri-implantitis, periodontitis, and local anaesthesia (Thouvenin et al. 2022).

Chitosan-based microspheres have demonstrated effectiveness against saliva digestive enzymes and have the advantage of providing controlled and sustained release in the subgingival area. Additionally, chitosan nanoparticles can be utilized to attach DNA/siRNA through adsorption, ionic gelation, and simple complexation techniques. However, chitosan-based drug nano/microparticles do have certain limitations, such as initial drug rupture and stability efficacy. These challenges can be addressed by coating the particles with other anionic biomaterials. Polyelectrolyte complexation, including chitosan-alginate microspheres, has been employed to enhance controlled drug release in acidic environments. Overall, chitosan-based drug delivery systems have demonstrated promising results in delivering drugs to the oral cavity for the treatment of dental diseases (Thouvenin et al. 2022).

5.4 *Gastrointestinal Disease*

The gastrointestinal tract (GI) is responsible for converting food into energy and eliminating waste from the body. The GI tract comprises several long components with various functions, making it prone to many disorders and assessments for GI diseases. Recent bioengineering projects focus on targeted GI drug delivery and bio-mimicked reconstruction of GI organs to improve their efficiency and biocompatibility. Oral drug delivery is the preferred route, but the poor bioavailability of drugs is a challenge. Nano/microparticulate systems can help overcome this challenge by interacting with the mucosal surface. Bioengineered tissue for the GI tract with emphasis on organ-specific cell types, scaffold components, and biocompatibility has been investigated. In addition, studies have focused on using biomaterials such as super-elastic nitinol and acrylic resin to treat dysphagia, a common symptom in the pharynx and oral cavity disorders. Furthermore, acellular dermal matrix (AlloDerm) and sternocleidomastoid muscle flap have been used to replace pharyngeal defects in patients with squamous cell carcinoma. Hybrid collagen-cell penetrating peptides (COL/CPP) have been explored as a carrier for cancer drugs, and recent studies have shown promising results for the targeted delivery of cancer drugs to hypopharyngeal carcinoma (Choukaife et al. 2022).

6 Evolution of Biomaterial

Over the years, biomaterials have played an important role in drug delivery systems. Biomaterials are materials that interact with biological systems and can be used to deliver drugs to specific sites in the body. The use of biomaterials has revolutionized drug delivery, leading to improved efficacy and reduced toxicity. One of the earliest biomaterials used in drug delivery was gelatin. Gelatin was used to encapsulate drugs and protect them from the harsh environment of the digestive system. This allowed for the oral delivery of drugs that would otherwise be destroyed in the stomach. As research progressed, other biomaterials were discovered that could be used for drug delivery. Polymers such as poly (lactic-co-glycolic acid) (PLGA) and polyethene glycol (PEG) were found to be biocompatible and could be used to deliver drugs over a sustained period of time (Vallet-Regí 2022).

Recent advances in biomaterials have led to the development of new drug delivery systems that are more targeted and efficient. For example, nanoparticles made from biomaterials such as lipids or polymers can be used to encapsulate drugs and deliver them to specific cells or tissues in the body. Another important development has been the use of biomaterials to create scaffolds for tissue engineering. These scaffolds can be used to deliver drugs to specific tissues while also promoting tissue regeneration. Overall, the evolution of biomaterials in drug delivery has led to significant improvements in drug efficacy and safety. With continued research, it is likely that biomaterials will continue to play an important role in the development of new drug delivery systems. Furthermore, advancements in nanotechnology have led to the development of more sophisticated drug delivery systems. Nanoparticles, liposomes, dendrimers, and carbon nanotubes are some examples of nanomaterials that have been used in drug delivery. These materials have unique physical and chemical properties that allow for targeted drug delivery and controlled release. For instance, nanomaterials can be engineered to bind specifically to certain cells or tissues, enabling drug delivery to those specific locations. The use of biomaterials has also led to the development of implantable drug-delivery devices. These devices can be implanted into the body and release drugs over an extended period of time, avoiding the need for frequent dosing (Zhang et al. 2013).

This is particularly useful in cases where long-term treatment is required, such as in chronic diseases. Moreover, biomaterials have been developed to respond to various physiological stimuli, such as changes in pH or temperature. These materials can be used to create smart drug delivery systems that release drugs in response to specific stimuli. This approach can improve drug delivery efficiency and reduce unwanted side effects. In conclusion, the evolution of biomaterials in drug delivery has significantly advanced the field of medicine. By improving drug efficacy and safety, biomaterials have enabled the development of new treatment options for a wide range of diseases. With continued research, it is likely that biomaterials will continue to play a vital role in the development of innovative drug delivery systems. Another important aspect of biomaterials in drug delivery is their ability to modulate the immune response. This is particularly relevant in immunotherapy, where the goal is to enhance or suppress the immune system to treat diseases such as cancer. Biomaterials can be designed to interact with immune cells, allowing for more targeted and effective immunotherapy (Fenton et al. 2018; Zhang et al. 2013).

Moreover, biomaterials can be used to deliver nucleic acids such as DNA and RNA, which have significant potential for the treatment of genetic disorders. Delivery of nucleic acids is challenging due to their large size and susceptibility to degradation. However, biomaterials such as lipids and polymers can be used to protect and deliver these molecules to specific cells or tissues. The use of biomaterials in drug delivery is also driving the development of personalized medicine. By tailoring drug delivery systems to the specific needs of individual patients, personalized medicine aims to improve treatment outcomes and reduce adverse reactions. Biomaterials can be designed to interact with specific biomarkers, enabling targeted drug delivery to specific patient populations. Overall, biomaterials are a critical component of drug delivery systems and are driving the development of new treatment options in medicine. The continued evolution of biomaterials is expected to lead to even more sophisticated drug delivery systems with enhanced efficacy, safety, and targeted delivery capabilities (Vallet-Regí 2022).

7 Advantages and Disadvantages

These polymers can be used as biomaterials for a variety of applications, including delivering cells, drugs, and genes. Biocompatible, readily available, and easy to process, natural polymers offer several advantages. Additionally, they closely resemble the extracellular matrix found in tissues, making them more effective at mimicry. There are, however, some disadvantages associated with natural polymers, including limited supply, high cost, batch-to-batch variation, and cross-contamination potential (Valle et al. 2017; Belda Marín et al. 2020; Vidya and Rajagopal 2021; Su and Wang 2015; Diller and Tabor 2022). Table 3 provides a detailed overview of the properties, advantages, and disadvantages of biomaterials.

8 Conclusion

In summary, drug delivery systems have been innovated to enhance drug effectiveness and safety. Biomaterials play a crucial role in drug delivery across various medical applications, such as cancer treatment, neurodegenerative disorders, buccal diseases, gastrointestinal conditions, and more. Nonetheless, there are persisting challenges in biomaterial development, primarily centered around issues of biocompatibility and biodegradability. The evolution of biomaterials has seen significant progress, driven by technological advancements and extensive research, resulting in the creation of novel and enhanced materials for drug delivery. In general, biomaterial-controlled release offers several advantages, including targeted drug delivery and improved patient compliance. However, it also presents drawbacks, such as potential toxicity, stability concerns, and complexity in preparation. Therefore, researchers remain committed to exploring and refining biomaterials for drug delivery to ultimately enhance patient outcomes.

Table 3 The properties, advantages, and disadvantages of natural biomaterials

Sr. No	Biomaterials	Properties	Advantages	Disadvantages	Citations
1	Chitosan	It is used for various purposes including skin, cartilage, bone and vascular grafts, substrates, and mammal cell culture	The materials are biodegradable, biocompatible, non-antigenic, non-toxic, bio-functional, and bio adhesive	Limitations in drug loading capacity, exhibit burst release effects, show variable release profiles, be sensitive to environmental conditions	Shukla et al. (2013)
2	Fibrin	In addition to its important role in haemostasis, fibrin acts as a scaffold for tissue repair and provides important cues for directing cell phenotype following injury	High biocompatibility, Improved cellular interaction	Maintaining structural integrity is challenging	Belda Marín et al. (2020)
3	Silk fibroin	Collagen-derived, insoluble in water	The material is biocompatible, slowly degradable, and has excellent mechanical properties	The production of spider silk is very low	Vidya and Rajagopal (2021)
4	Gelatin	Extracellular matrix component (ECM)	Low immunogenicity, biodegradable and biocompatible in a physiological environment	The material has low immunogenicity, biodegradability, and biocompatibility in physiological condition	Santoro et al. (2014)
5	Collagen	In the natural ECM, this component plays a role. An important role in wound healing Biologically compatible, able to recognize cell	Collagen biomaterial offers exceptional biocompatibility, versatility, and regenerative potential	Properties of poor mechanical strength	Wang et al. (2022)

(continued)

Table 3 (continued)

Sr. No	Biomaterials	Properties	Advantages	Disadvantages	Citations
6	Hyaluronic acid	The product is derived from seaweed	Hyaluronic acid biomaterial enables efficient drug delivery due to its high-water retention capacity and biodegradability	Properties that are poor mechanically	How et al. (2020)
7	Alginate	Structurally like natural glycosaminoglycan	Recognition of cells with good accuracy	Properties that are poor mechanically	Tønnesen and Karlsen (2002)
8	Bulk Biodegradable polymers like (poly lactic acid, polyglycolic acid, poly propylene fumarate)	Variations in polymer segments can alter mechanical and degradation properties	The biocompatibility, solubility, and biodegradation rates are excellent	From products, create an acidic environment. There is a possibility of inflammation. Compression strength is poor due to poor cell adhesion	Liechty et al. (2010)
9	Poly (ethylene glycol)	A gel that can be injected, Degradation and mechanical damage	Poly (ethylene glycol) biomaterial exhibits excellent biocompatibility, tunable physical properties, and resistance to protein adsorption, making it a versatile choice for various biomedical applications	Its potential for limited drug loading capacity, which may restrict the delivery of high drug doses	Wu et al. (2007)
10	Surface biodegradable polymers (polyorthoester, polyanhydrides, polyphoshazen)	Surface erosion properties make it ideal for advanced drug delivery	Biocompatibility is excellent. Maintaining mechanical integrity is possible	New bone tissue cannot replace them completely	Ha and Gardella (2005)

(continued)

Table 3 (continued)

Sr. No	Biomaterials	Properties	Advantages	Disadvantages	Citations
11	Calcium phosphates (hydroxyapatite, Tricalcium phosphate)	The minerals phase of bone is naturally found as a component of the minerals phase of bone and has a composition like the minerals phase of bone	It has excellent biocompatibility, good osteoconductivity, and adequate mechanical strength	Lacks mechanical properties, slow degradability, brittleness, and non-resorb ability	Eliaz and Metoki (2017)
12	Polymer-ceramics	Bioengineering with natural and synthetic polymers	Mechanical degradation, tailoring ability, and biological properties	The best qualities of each component are compromised	Li et al. (2020b)

Acknowledgements H.S.N expressed gratitude to the CSIR-UGC (Council of Scientific and Industrial Research-University Grants Commission) for providing a junior researcher scholarship (JUNE18-359427) that supported their Ph.D. work.

Author Contribution H.S.N the first author of the book chapter and was responsible for writing the manuscript. P.M.S, the second author, contributed to the book chapter by drawing tables and obtaining permission to use photographs from respected journals. R.W.R reviewed the manuscript, provided valuable feedback, and made necessary revisions to enhance the overall quality and coherence of the content.

References

Adeosun SO, Ilomuanya MO, Gbenebor OP, Dada MO, Odili CC (2020) Biomaterials for drug delivery: sources, classification, synthesis, processing, and applications. In: Advanced functional materials, pp 1–25

Agrawal AK, Das M, Jain S (2012) In situ gel systems as 'smart' carriers for sustained ocular drug delivery. Expert Opin Drug Deliv 9(4):383–402. https://doi.org/10.1517/17425247.2012.665367

Akiyoshi K et al (1998) Self-assembled hydrogel nanoparticle of cholesterol-bearing pullulan as a carrier of protein drugs: complexation and stabilization of insulin. J Control Release 54(3):313–320. https://doi.org/10.1016/S0168-3659(98)00017-0

Amiri MS, Mohammadzadeh V, Yazdi MET, Barani M, Rahdar A, Kyzas GZ (2021) Plant-based gums and mucilages applications in pharmacology and nanomedicine: a review. Molecules 26(6):1–23. https://doi.org/10.3390/molecules26061770

Attia MF, Anton N, Wallyn J, Omran Z, Vandamme TF (2019) An overview of active and passive targeting strategies to improve the nanocarriers efficiency to tumour sites. J Pharm Pharmacol 71(8):1185–1198. Wiley. https://doi.org/10.1111/jphp.13098.

Azizi S, Mohamad R, Abdul Rahim R, Mohammadinejad R, Bin Ariff A (2017) Hydrogel beads bio-nanocomposite based on Kappa-Carrageenan and green synthesized silver nanoparticles for

biomedical applications. Int J Biol Macromol 104:423–431. https://doi.org/10.1016/j.ijbiomac.2017.06.010
Bahrami N, Nouri Khorasani S, Mahdavi H, Ghiaci M, Mokhtari R (2019) Low-pressure plasma surface modification of polyurethane films with chitosan and collagen biomolecules. J Appl Polym Sci 136(21). https://doi.org/10.1002/APP.47567
Barclay TG, Day CM, Petrovsky N, Garg S (2019) Review of polysaccharide particle-based functional drug delivery. Carbohydr Polym 221:94–112. NIH Public Access. https://doi.org/10.1016/j.carbpol.2019.05.067
Belda Marín C, Fitzpatrick V, Kaplan DL, Landoulsi J, Guénin E, Egles C (2020) Silk polymers and nanoparticles: a powerful combination for the design of versatile biomaterials. Front Chem 8:1–22. https://doi.org/10.3389/fchem.2020.604398
Benny IS, Gunasekar V, Ponnusami V (2014) Review on application of Xanthan gum in drug delivery. Int J Pharmtech Res 6(4):1322–1326
Berger J, Reist M, Mayer JM, Felt O, Peppas NA, Gurny R (2004) Structure and interactions in covalently and ionically crosslinked chitosan hydrogels for biomedical applications. Eur J Pharmaceut Biopharmaceut 57(1):19–34. Elsevier. https://doi.org/10.1016/S0939-6411(03)00161-9
Bos GW et al (2005) Tissue reactions of in situ formed dextran hydrogels crosslinked by stereocomplex formation after subcutaneous implantation in rats. Biomaterials 26(18):3901–3909. https://doi.org/10.1016/j.biomaterials.2004.10.008
Cabral JD et al (2015) Characterization of the in vivo host response to a bi-labeled chitosan-dextran based hydrogel for postsurgical adhesion prevention. J. Biomed. Mater. Res. Part A 103(8):2611–2620. https://doi.org/10.1002/jbm.a.35395
Calzoni E, Cesaretti A, Polchi A, Di Michele A, Tancini B, Emiliani C (2019) Biocompatible polymer nanoparticles for drug delivery applications in cancer and neurodegenerative disorder therapies. J Funct Biomater 10(1):1–15. MDPI AG. https://doi.org/10.3390/jfb10010004
Chen G, Kawazoe N, Tateishi T (2008) Collagen-based scaffolds for tissue engineering. In: Natural-based polymers for biomedical applications. Woodhead Publishing, pp 396–415
Cheng Z, Li M, Dey R, Chen Y (2021) Nanomaterials for cancer therapy: current progress and perspectives. J Hematol Oncol 14(1):1–27. https://doi.org/10.1186/s13045-021-01096-0
Cheng C, Meng Y, Zhang Z, Chen J, Zhang Q (2019) Imine bond- and coordinate bond-linked pH-sensitive cisplatin complex nanoparticles for active targeting to tumor cells. J Nanosci Nanotechnol 19(6):3277–3287. https://doi.org/10.1166/jnn.2019.16314
Cheng G et al (2022) Anti-Parkinsonian therapy: strategies for crossing the blood–brain barrier and nano-biological effects of nanomaterials. 14(1). Springer Singapore
Chhetri P, Chakraborty P, Das D, Afnan T (2021) In-situ forming polymeric drug delivery systems for ophthalmic use: an overview. J Drug Deliv Ther 11(3-S):98–103. https://doi.org/10.22270/jddt.v11i3-s.4874
Choukaife H, Seyam S, Alallam B, Doolaanea AA, Alfatama M (2022) Current advances in chitosan nanoparticles based oral drug delivery for colorectal cancer treatment. Int J Nanomed 17:3933–3966. https://doi.org/10.2147/IJN.S375229
Ciobanu BC, Cadinoiu AN, Popa M, Desbrières J, Peptu CA (2014) Modulated release from liposomes entrapped in chitosan/gelatin hydrogels. Mater Sci Eng C 43:383–391. https://doi.org/10.1016/j.msec.2014.07.036
Das M, Giri TK (2020) Hydrogels based on gellan gum in cell delivery and drug delivery. J Drug Deliv Sci Technol 56:101586. https://doi.org/10.1016/j.jddst.2020.101586
Dattilo M, Patitucci F, Prete S, Parisi OI, Puoci F (2023) Polysaccharide-based hydrogels and their application as drug delivery systems in cancer treatment: a review. J Funct Biomater 14(2):55. Multidisciplinary Digital Publishing Institute. https://doi.org/10.3390/jfb14020055
de Guzman RC, Rabbany SY (2016) PEG-immobilized keratin for protein drug sequestration and pH-mediated delivery. J Drug Deliv 2016:1–9. https://doi.org/10.1155/2016/7843951

Debabrata GD, Chakrabarti G (2018) Thermoresponsive drug delivery systems, characterization and application. In: Applications of targeted nano drugs and delivery systems: nanoscience and nanotechnology in drug delivery. Elsevier, pp 133–155

Del Valle LJ, Díaz A, Puiggalí J (2017) Hydrogels for biomedical applications: cellulose, chitosan, and protein/peptide derivatives. Gels 3(3):1–28. https://doi.org/10.3390/gels3030027

Diller RB, Tabor AJ (2022) The role of the extracellular matrix (ECM) in wound healing: a review. Biomimetics 7(3):14–16. https://doi.org/10.3390/biomimetics7030087

Dowling MB, Lee JH, Raghavan SR (2009) pH-responsive jello: gelatin gels containing fatty acid vesicles. Langmuir 25(15):8519–8525. https://doi.org/10.1021/la804159g

Eliaz N, Metoki N (2017) Calcium phosphate bioceramics: a review of their history, structure, properties, coating technologies and biomedical applications. Materials (Basel) 10(4). https://doi.org/10.3390/ma10040334

Feigin VL et al (2019) Global, regional, and national burden of neurological disorders, 1990–2016: a systematic analysis for the Global Burden of Disease Study 2016. Lancet Neurol 18(5):459–480. https://doi.org/10.1016/S1474-4422(18)30499-X

Fenton OS, Olafson KN, Pillai PS, Mitchell MJ, Langer R (2018) Advances in biomaterials for drug delivery. Adv Mater 30(29). Wiley-VCH Verlag. https://doi.org/10.1002/adma.201705328

Feroz S, Muhammad N, Ranayake J, Dias G (2020) Keratin-based materials for biomedical applications. Bioactive Mater 5(3):496–509. Elsevier. https://doi.org/10.1016/j.bioactmat.2020.04.007

Ferroni C, Varchi G (2021) Keratin-based nanoparticles as drug delivery carriers. Appl Sci (Switz) 11(20):9417. Multidisciplinary Digital Publishing Institute. https://doi.org/10.3390/app11209417

Foox M, Zilberman M (2015) Drug delivery from gelatin-based systems. Expert Opin Drug Deliv 12(9):1547–1563. https://doi.org/10.1517/17425247.2015.1037272

Gila-Vilchez C et al (2022) Biocompatible short-peptides fibrin co-assembled hydrogels. ACS Appl Polym Mater. https://doi.org/10.1021/ACSAPM.2C02164/SUPPL_FILE/AP2C02164_SI_002.MP4

Grijalvo S, Mayr J, Eritja R, Díaz DD (2016) Biodegradable liposome-encapsulated hydrogels for biomedical applications: a marriage of convenience. Biomater Sci 4(4):555–574. https://doi.org/10.1039/c5bm00481k

Gu J et al (2008) pH-triggered reversible 'stealth' polycationic micelles. Biomacromolecules 9(1):255–262. https://doi.org/10.1021/bm701084w

Gu X et al (2013) Temperature-responsive drug delivery systems based on polyaspartamides with isopropylamine pendant groups. Soft Matter 9(30):7267–7273. https://doi.org/10.1039/c3sm50904d

Gu Z et al (2013) Injectable nano-network for glucose-mediated insulin delivery. ACS Nano 7(5):4194–4201. https://doi.org/10.1021/nn400630x

Guo X, Cheng Y, Zhao X, Luo Y, Chen J, Yuan WE (2018) Advances in redox-responsive drug delivery systems of tumor microenvironment. J Nanobiotechnol 16(1):1–10. BioMed Central Ltd. https://doi.org/10.1186/s12951-018-0398-2

Gupta S, Archana, Niranjan AK (2022) A comprehensive review on in-situ gel drug delivery system. J Drug Deliv Ther 12(4-S):245–248. https://doi.org/10.22270/jddt.v12i4-s.5539

Ha CS, Gardella JA (2005) Surface chemistry of biodegradable polymers for drug delivery systems. Chem Rev 105(11):4205–4232. https://doi.org/10.1021/CR040419Y/ASSET/CR040419Y.FP.PNG_V03

Han X et al (2022) Biomaterial-assisted biotherapy: a brief review of biomaterials used in drug delivery, vaccine development, gene therapy, and stem cell therapy. Bioactive Mater 17:29–48. KeAi Communications Co. https://doi.org/10.1016/j.bioactmat.2022.01.011

Hezaveh H, Muhamad II (2012) The effect of nanoparticles on gastrointestinal release from modified κ-carrageenan nanocomposite hydrogels. Carbohydr Polym 89(1):138–145. https://doi.org/10.1016/j.carbpol.2012.02.062

How KN, Yap WH, Lim CLH, Goh BH, Lai ZW (2020) Hyaluronic acid-mediated drug delivery system targeting for inflammatory skin diseases: a mini review. Front Pharmacol 11:1105. Frontiers Media S.A. https://doi.org/10.3389/fphar.2020.01105

Hu Q, Katti PS, Gu Z (2014) Enzyme-responsive nanomaterials for controlled drug delivery. Nanoscale 6(21):12273–12286. https://doi.org/10.1039/c4nr04249b

Huang G, Huang H (2018a) Application of hyaluronic acid as carriers in drug delivery. Drug Deliv 25(1):766–772. https://doi.org/10.1080/10717544.2018.1450910

Huang G, Huang H (2018b) Hyaluronic acid-based biopharmaceutical delivery and tumor-targeted drug delivery system. J Control Release 278:122–126. Elsevier. https://doi.org/10.1016/j.jconrel.2018.04.015

Hurler J, Berg OA, Skar M, Conradi AH, Johnsen PJ, Škalko-Basnet N (2012) Improved burns therapy: liposomes-in-hydrogel delivery system for mupirocin. J Pharm Sci 101(10):3906–3915. https://doi.org/10.1002/jps.23260

Jain NK, Tare MS, Mishra V, Tripathi PK (2015) The development, characterization and in vivo anti-ovarian cancer activity of poly(propylene imine) (PPI)-antibody conjugates containing encapsulated paclitaxel. Nanomed Nanotechnol Biol Med 11(1):207–218. https://doi.org/10.1016/j.nano.2014.09.006

Jiann Chong ET, Ng JW, Lee P-C (2022) Classification and medical applications of biomaterials—a mini review. Bio Integr. https://doi.org/10.15212/bioi-2022-0009

Jumelle C, Gholizadeh S, Annabi N, Dana R (2020) Advances and limitations of drug delivery systems formulated as eye drops. J Control Release 321:1–22. Elsevier. https://doi.org/10.1016/j.jconrel.2020.01.057

Karandish F et al (2016) Prostate-specific membrane antigen targeted polymersomes for delivering mocetinostat and docetaxel to prostate cancer cell spheroids. ACS Omega 1(5):952–962. https://doi.org/10.1021/acsomega.6b00126

Karimi M et al (2016) Temperature-responsive smart nanocarriers for delivery of therapeutic agents: applications and recent advances. ACS Appl Mater Interfaces 8(33):21107–21133. NIH Public Access. https://doi.org/10.1021/acsami.6b00371

Kedir WM, Deresa EM, Diriba TF (2022) Pharmaceutical and drug delivery applications of pectin and its modified nanocomposites. Heliyon 8(9):e10654. https://doi.org/10.1016/j.heliyon.2022.e10654

Kim JK, Kim HJ, Chung JY, Lee JH, Young SB, Kim YH (2014) Natural and synthetic biomaterials for controlled drug delivery. Arch Pharm Res 37(1):60–68. https://doi.org/10.1007/s12272-013-0280-6

Konop M, Rybka M, Drapała A (2021) Keratin biomaterials in skin wound healing, an old player in modern medicine: a mini review. Pharmaceutics 13(12). Multidisciplinary Digital Publishing Institute (MDPI). https://doi.org/10.3390/pharmaceutics13122029

Lara-Velazquez M et al (2020) Chitosan-based non-viral gene and drug delivery systems for brain cancer. Front Neurol 11. https://doi.org/10.3389/fneur.2020.00740

Leach JB, Schmidt CE (2005) Characterization of protein release from photocrosslinkable hyaluronic acid-polyethylene glycol hydrogel tissue engineering scaffolds. Biomaterials 26(2):125–135. https://doi.org/10.1016/j.biomaterials.2004.02.018

Lee KY, Mooney DJ (2012) Alginate: properties and biomedical applications. Prog Polym Sci 37(1):106–126. https://doi.org/10.1016/j.progpolymsci.2011.06.003

Lee JH, Oh H, Baxa U, Raghavan SR, Blumenthal R (2012) Biopolymer-connected liposome networks as injectable biomaterials capable of sustained local drug delivery. Biomacromolecules 13(10):3388–3394. https://doi.org/10.1021/bm301143d

Lee H et al (2015) Erratum to: human hair keratin-based BioFilm for potent application to periodontal tissue regeneration [23(3):300–308 (2015)], Macromol Res 23(9):885. Polymer Society of Korea. https://doi.org/10.1007/s13233-015-3201-3

Lemos PVF, Marcelino HR, Cardoso LG, de Souza CO, Druzian JI (2021) Starch chemical modifications applied to drug delivery systems: from fundamentals to FDA-approved raw materials. Int J Biol Macromol 184:218–234. https://doi.org/10.1016/j.ijbiomac.2021.06.077

Leong KH, Chung LY, Noordin MI, Onuki Y, Morishita M, Takayama K (2011) Lectin-functionalized carboxymethylated kappa-carrageenan microparticles for oral insulin delivery. Carbohydr Polym 86(2):555–565. https://doi.org/10.1016/j.carbpol.2011.04.070

Li J et al (2018) Chitosan-based nanomaterials for drug delivery. Molecules 23(10):1–26. https://doi.org/10.3390/molecules23102661

Li R, Peng F, Cai J, Yang D, Zhang P (2020a) Redox dual-stimuli responsive drug delivery systems for improving tumor-targeting ability and reducing adverse side effects. Asian J Pharmaceut Sci 15(3):311–325. Elsevier. https://doi.org/10.1016/j.ajps.2019.06.003

Li C, Sun W, Lu Z, Ao X, Li S (2020b) Ceramic nanocomposite membranes and membrane fouling: a review. Water Res 175. Elsevier Ltd. https://doi.org/10.1016/j.watres.2020.115674

Liechty WB, Kryscio DR, Slaughter BV, Peppas NA (2010) Polymers for drug delivery systems. Annu Rev Chem Biomol Eng 1:149–173. https://doi.org/10.1146/annurev-chembioeng-073009-100847

Lohani A, Singh G, Bhattacharya SS, Rama Hegde R, Verma A (2016) Tailored-interpenetrating polymer network beads of κ-carrageenan and sodium carboxymethyl cellulose for controlled drug delivery. J Drug Deliv Sci Technol 31:53–64. https://doi.org/10.1016/j.jddst.2015.11.005

Luo Z et al (2019) Light-induced redox-responsive smart drug delivery system by using selenium-containing polymer@MOF shell/core nanocomposite. Adv Healthc Mater 8(15):1900406. https://doi.org/10.1002/adhm.201900406

Ma X et al (2012) A functionalized graphene oxide-iron oxide nanocomposite for magnetically targeted drug delivery, photothermal therapy, and magnetic resonance imaging. Nano Res 5(3):199–212. https://doi.org/10.1007/s12274-012-0200-y

Magadala P, Amiji M (2008) Epidermal growth factor receptor-targeted gelatin-based engineered nanocarriers for DNA delivery and transfection in human pancreatic cancer cells. AAPS J 10(4):565–576. https://doi.org/10.1208/s12248-008-9065-0

Milivojevic M, Pajic-Lijakovic I, Bugarski B, Nayak AK, Hasnain MS (2019) Gellan gum in drug delivery applications. Elsevier Inc.

Misbah M et al (2021) Clinical pharmacology & biopharmaceutics rectal drug delivery system: an overview. Clin Pharmacol Biopharm 10(5)

Miyazaki S, Suisha F, Kawasaki N, Shirakawa M, Yamatoya K, Attwood D (1998) Thermally reversible xyloglucan gels as vehicles for rectal drug delivery. J Control Release 56(1–3):75–83. https://doi.org/10.1016/S0168-3659(98)00079-0

Moinuddin S et al (2019) Nasal drug delivery system: a innovative approach. Pharma Innov J 8(3):169–177. www.thepharmajournal.com

Nappini S, Fogli S, Castroflorio B, Bonini M, Baldelli Bombelli F, Baglioni P (2016) Magnetic field responsive drug release from magnetoliposomes in biological fluids. J Mater Chem B 4(4):716–725. https://doi.org/10.1039/c5tb02191j

Niu Y, Yuan X, Zhao Y, Zhang W, Ren L (2017) Temperature and pH dual-responsive supramolecular polymer hydrogels hybridized with functional inorganic nanoparticles. Macromol Chem Phys 218(9):1600540. https://doi.org/10.1002/macp.201600540

Noori A, Ashrafi SJ, Vaez-Ghaemi R, Hatamian-Zaremi A, Webster TJ (2017) A review of fibrin and fibrin composites for bone tissue engineering. Int J Nanomed 12:4937–4961. Dove Press. https://doi.org/10.2147/IJN.S124671

Numata K, Kaplan DL (2010) Silk-based delivery systems of bioactive molecules. Adv Drug Deliv Rev 62(15):1497–1508. https://doi.org/10.1016/j.addr.2010.03.009

Oechsle AM, Häupler M, Weigel F, Gibis M, Kohlus R, Weiss J (2016) Modulation of extruded collagen films by the addition of co-gelling proteins. J Food Eng 171:164–173. https://doi.org/10.1016/j.jfoodeng.2015.10.004

Ong W, Yang Y, Cruciano AC, McCarley RL (2008) Redox-triggered contents release from liposomes. J Am Chem Soc 130(44):14739–14744. https://doi.org/10.1021/ja8050469

Onofre-Cordeiro NA et al (2018) Agarose-silver particles films: effect of calcium ascorbate in nanoparticles synthesis and film properties. Int J Biol Macromol 119:701–707. https://doi.org/10.1016/j.ijbiomac.2018.07.115

Padhi JR, Nayak D, Nanda A, Rauta PR, Ashe S, Nayak B (2016) Development of highly biocompatible Gelatin & i-Carrageenan based composite hydrogels: in depth physiochemical analysis for biomedical applications. Carbohydr Polym 153:292–301. https://doi.org/10.1016/j.carbpol.2016.07.098

Pankaew P, Klumdoung P, Naemchanthara K (2015) A study of the preparation of silk sericin/chitosan composite film for future wound dressing applications. Appl Mech Mater 804:179–182. https://doi.org/10.4028/www.scientific.net/amm.804.179

Peng YY, Glattauer V, Ramshaw JAM (2017) Stabilisation of collagen sponges by glutaraldehyde vapour crosslinking. Int J Biomater 2017. https://doi.org/10.1155/2017/8947823

Phromsopha T, Baimark Y (2014) Preparation of starch/gelatin blend microparticles by a water-in-oil emulsion method for controlled release drug delivery. Int J Biomater 2014. https://doi.org/10.1155/2014/829490

Qian B, Zhao Q, Ye X (2020) Ultrasound and magnetic responsive drug delivery systems for cardiovascular application. J Cardiovasc Pharmacol 76(4):414–426. https://doi.org/10.1097/FJC.0000000000000885

Qureshi D, Nayak SK, Maji S, Kim D, Banerjee I, Pal K (2019) Carrageenan: a wonder polymer from marine algae for potential drug delivery applications. Curr Pharm Des 25(11):1172–1186. https://doi.org/10.2174/1381612825666190425190754

Rahman HS et al (2020) Novel drug delivery systems for loading of natural plant extracts and their biomedical applications. Int J Nanomed 15:2439–2483. https://doi.org/10.2147/IJN.S227805

Saeed RM, Dmour I, Taha MO (2020) Stable chitosan-based nanoparticles using polyphosphoric acid or hexametaphosphate for tandem ionotropic/covalent crosslinking and subsequent investigation as novel vehicles for drug delivery. Front Bioeng Biotechnol 8:1–21. https://doi.org/10.3389/fbioe.2020.00004

Sahithi B, Ansari SK, Hameeda SK, Sahithya G, Prasad DM, Lakshmi Y (2013) A review on collagen based drug delivery systems. Indian J Res Pharm Biotechnol 1:461–468

Sahoo N, Sahoo RK, Biswas N, Guha A, Kuotsu K (2015) Recent advancement of gelatin nanoparticles in drug and vaccine delivery. Int J Biol Macromol 81:317–331. Elsevier B.V. https://doi.org/10.1016/j.ijbiomac.2015.08.006

Saini M (2015) Implant biomaterials: a comprehensive review. World J Clin Cases 3(1):52. https://doi.org/10.12998/wjcc.v3.i1.52

Salati MA et al (2020) Agarose-based biomaterials: opportunities and challenges in cartilage tissue engineering. Polymers 12(5). Multidisciplinary Digital Publishing Institute (MDPI). https://doi.org/10.3390/POLYM12051150

Salehi S, Koeck K, Scheibel T (2020) Spider silk for tissue engineering applications. Molecules 25(3). Multidisciplinary Digital Publishing Institute (MDPI). https://doi.org/10.3390/molecules25030737

Salimi M, Mosca S, Gardner B, Palombo F, Matousek P, Stone N (2022) Nanoparticle-mediated photothermal therapy limitation in clinical applications regarding pain management. Nanomaterials 12(6):1–29. https://doi.org/10.3390/nano12060922

Sannino A et al (2004) Cellulose derivative-hyaluronic acid-based microporous hydrogels cross-linked through divinyl sulfone (DVS) to modulate equilibrium sorption capacity and network stability. Biomacromol 5(1):92–96. https://doi.org/10.1021/bm0341881

Santoro M, Tatara AM, Mikos AG (2014) Gelatin carriers for drug and cell delivery in tissue engineering. J Control Release 190:210–218. https://doi.org/10.1016/j.jconrel.2014.04.014

Sathuvan M et al (2017) κ-Carrageenan: an effective drug carrier to deliver curcumin in cancer cells and to induce apoptosis. Carbohydr Polym 160:184–193. https://doi.org/10.1016/j.carbpol.2016.12.049

Setyawati MI, Tay CY, Bay BH, Leong DT (2017) Gold nanoparticles induced endothelial leakiness depends on particle size and endothelial cell origin. ACS Nano 11(5):5020–5030. https://doi.org/10.1021/acsnano.7b01744

Sharifianjazi F et al (2021) Polymer incorporated magnetic nanoparticles: applications for magnetoresponsive targeted drug delivery. Mater Sci Eng B Solid State Mater Adv Technol 272:115358. Elsevier. https://doi.org/10.1016/j.mseb.2021.115358

Shevtsov M et al (2018) Targeting experimental orthotopic glioblastoma with chitosan-based superparamagnetic iron oxide nanoparticles (CS-DX-SPIONs). Int J Nanomed 13:1471–1482. https://doi.org/10.2147/IJN.S152461

Shukla SK, Mishra AK, Arotiba OA, Mamba BB (2013) Chitosan-based nanomaterials: a state-of-the-art review. Int J Biol Macromol 59:46–58. Elsevier. https://doi.org/10.1016/j.ijbiomac.2013.04.043

Singhvi G, Hans N, Shiva N, Kumar Dubey S (2019) Xanthan gum in drug delivery applications. Elsevier Inc.

Song J, Leeuwenburgh SCG (2014) Sustained delivery of biomolecules from gelatin carriers for applications in bone regeneration. Therapeut Deliv 5(8):943–958. Future Science Ltd London, UK. https://doi.org/10.4155/tde.14.42

Sood A, Gupta A, Agrawal G (2021) Recent advances in polysaccharides based biomaterials for drug delivery and tissue engineering applications. Carbohydr Polym Technol Appl 2:1–24. Elsevier Ltd. https://doi.org/10.1016/j.carpta.2021.100067

Sreekanth SP et al (2021) Multi-walled carbon nanotube-based nanobiosensor for the detection of cadmium in water. Environ Res 197:111148. https://doi.org/10.1016/j.envres.2021.111148

Srinivasarao DA, Lohiya G, Katti DS (2019) Fundamentals, challenges, and nanomedicine-based solutions for ocular diseases. Wiley Interdiscip Rev Nanomed Nanobiotechnol 11(4). Wiley-Blackwell. https://doi.org/10.1002/wnan.1548

Su K, Wang C (2015) Recent advances in the use of gelatin in biomedical research. Biotechnol Lett 37(11):2139–2145. https://doi.org/10.1007/s10529-015-1907-0

Su Y et al (2015) Redox-responsive polymer-drug conjugates based on doxorubicin and chitosan oligosaccharide-g-stearic acid for cancer therapy. Mol Pharm 12(4):1193–1202. https://doi.org/10.1021/mp500710x

Sun W, Saldaña MDA, Zhao Y, Dong T, Jin Y, Zhang J (2016) New application of kappa-carrageenan: producing pH-sensitive lappaconitine-loaded kappa-carrageenan microparticle using two-step self-assembly. J Appl Phycol 28(3):2041–2050. https://doi.org/10.1007/s10811-015-0712-4

Sun B, Zhang M, Shen J, He Z, Fatehi P, Ni Y (2018) Applications of cellulose-based materials in sustained drug delivery systems. Curr Med Chem 26(14):2485–2501. https://doi.org/10.2174/0929867324666170705143308

Surwit EA et al (1982) Evaluation of topically applied trans-retinoic acid in the treatment of cervical intraepithelial lesions. Am J Obstet Gynecol 143(7):821–823. https://doi.org/10.1016/0002-9378(82)90016-3

Takemoto S et al (2008) Preparation of collagen/gelatin sponge scaffold for sustained release of bFGF. Tissue Eng Part A 14(10):1629–1638. https://doi.org/10.1089/ten.tea.2007.0215

Thouvenin A, Toussaint B, Marinovic J, Gilles AL, Dufaÿ Wojcicki A, Boudy V (2022) Development of thermosensitive and mucoadhesive hydrogel for buccal delivery of (S)-ketamine. Pharmaceutics 14(10). https://doi.org/10.3390/pharmaceutics14102039

Tønnesen HH, Karlsen J (2002) Alginate in drug delivery systems. Drug Dev Ind Pharm 28(6):621–630. https://doi.org/10.1081/DDC-120003853

Troy E, Tilbury MA, Power AM, Wall JG (2021) Nature-based biomaterials and their application in biomedicine. Polymers 13(19):1–37. MDPI. https://doi.org/10.3390/polym13193321

Vallet-Regí M (2022) Evolution of biomaterials. Front Mater 9:1–5. https://doi.org/10.3389/fmats.2022.864016

Vasile C, Pamfil D, Stoleru E, Baican M (2020) New developments in medical applications of hybrid hydrogels containing natural polymers. Molecules 25(7). https://doi.org/10.3390/molecules25071539

Vidya M, Rajagopal S (2021) Silk fibroin: a promising tool for wound healing and skin regeneration. Int J Polym Sci 2021. https://doi.org/10.1155/2021/9069924

Wang H, Mukherjee S, Yi J, Banerjee P, Chen Q, Zhou S (2017) Biocompatible chitosan-carbon dot hybrid nanogels for NIR-imaging-guided synergistic photothermal-chemo therapy. ACS Appl Mater Interfaces 9(22):18639–18649. https://doi.org/10.1021/acsami.7b06062

Wang YC, Wang F, Sun TM, Wang J (2011) Redox-responsive nanoparticles from the single disulfide bond-bridged block copolymer as drug carriers for overcoming multidrug resistance in cancer cells. Bioconjug Chem 22(10):1939–1945. https://doi.org/10.1021/bc200139n

Wang C, Willner B, Willner I (2021) Redox-responsive and light-responsive DNA-based hydrogels and their applications. React Funct Polym 166:104983. https://doi.org/10.1016/j.reactfunctpolym.2021.104983

Wang Y, Wang Z, Dong Y (2022) Collagen-based biomaterials for tissue engineering. ACS Biomater Sci Eng 9(3):1132–1150. https://doi.org/10.1021/acsbiomaterials.2c00730

Way AE, Hsu L, Shanmuganathan K, Weder C, Rowan SJ (2012) PH-responsive cellulose nanocrystal gels and nanocomposites. ACS Macro Lett 1(8):1001–1006. https://doi.org/10.1021/mz3003006

Wu J, Wei W, Wang LY, Su ZG, Ma GH (2007) A thermosensitive hydrogel based on quaternized chitosan and poly(ethylene glycol) for nasal drug delivery system. Biomaterials 28(13):2220–2232. https://doi.org/10.1016/j.biomaterials.2006.12.024

Xiao F et al (2009) In vitro cyto-biocompatibility and cell detachment of temperature-sensitive dextran hydrogel. Colloids Surf B Biointerfaces 71(1):13–18. https://doi.org/10.1016/j.colsurfb.2008.12.040

Xu Q, Tanaka Y, Czernuszka JT (2007) Encapsulation and release of a hydrophobic drug from hydroxyapatite coated liposomes. Biomaterials 28(16):2687–2694. https://doi.org/10.1016/j.biomaterials.2007.02.007

Young S, Wong M, Tabata Y, Mikos AG (2005) Gelatin as a delivery vehicle for the controlled release of bioactive molecules. J Control Release 109(1–3):256–274. https://doi.org/10.1016/j.jconrel.2005.09.023

Zeplin PH, Berninger AK, Maksimovikj NC, Van Gelder P, Scheibel T, Walles H (2014) Verbesserung der Biokompatibilität von Silikonimplantaten durch Spinnenseidenbeschichtung: Immunhistochemische Untersuchungen zum Einfluss auf die Kapselbildung. Handchirurgie Mikrochirurgie Plast Chir 46(6):336–341. https://doi.org/10.1055/s-0034-1395558

Zeplin PH et al (2014) Spider silk coatings as a bioshield to reduce periprosthetic fibrous capsule formation. Adv Funct Mater 24(18):2658–2666. https://doi.org/10.1002/adfm.201302813

Zhang Y, Chan HF, Leong KW (2013) Advanced materials and processing for drug delivery: the past and the future. Adv Drug Deliv Rev 65(1):104–120. https://doi.org/10.1016/j.addr.2012.10.003

Zhang J et al (2020) Novel pH-sensitive drug-loaded electrospun nanofibers based on regenerated keratin for local tumor chemotherapy. Text Res J 90(19–20):2336–2349. https://doi.org/10.1177/0040517520919920

Zhang MX et al (2021a) Effects of chitosan-collagen dressing on wound healing in vitro and in vivo assays. J Appl Biomater Funct Mater 19:1–10. https://doi.org/10.1177/2280800021989698

Zhang W, Qin W, Li H, Wu AM (2021b) Biosynthesis and transport of nucleotide sugars for plant hemicellulose. Front Plant Sci 12. https://doi.org/10.3389/fpls.2021.723128

Zhao X, Zhou L, Li Q, Zou Q, Du C (2018) Biomimetic mineralization of carboxymethyl chitosan nanofibers with improved osteogenic activity in vitro and in vivo. Carbohydr Polym 195:225–234. https://doi.org/10.1016/j.carbpol.2018.04.090

Zhu L, Zhao L, Qu X, Yang Z (2012) PH-sensitive polymeric vesicles from coassembly of amphiphilic cholate grafted poly(l-lysine) and acid-cleavable polymer-drug conjugate. Langmuir 28(33):11988–11996. https://doi.org/10.1021/la3015767

Zhuo S, Zhang F, Yu J, Zhang X, Yang G, Liu X (2020) pH-sensitive biomaterials for drug delivery. Molecules 25(23):1–20. https://doi.org/10.3390/molecules25235649

Transdermal Drug Delivery Systems

Vipin Kumar, Nazia Praveen, Prerna Kewlani, Arvind, Amita Singh, Anurag Kumar Gautam, and Vijayakumar Mahalingam Rajamanickam

Abstract The transdermal drug delivery systems, also known as TDDS, are the most effective alternative to the oral and parenteral routes to deliver therapeutic molecules across the skin in order to produce systemic effects. TDDS possess several benefits such as self-administrable nature, escape of first-pass metabolism, minimize dose, less side effects, better dosage regimen, termination of dosing at any time, absence of gastric irritation, and improved patient compliance. In the earlier scenario, TDDS was indicated only for the transdermal patches. These patches include matrix devices, reservoir type, micro-reservoir forms, and adhesive controlled systems. In the present situation, various state of the art technologies were introduced in advancing TDDS utilization. For passive drug delivery, nanovesicles, polymeric nanocarriers adhesive-controlled and nanoemulsions were implemented. Active delivery strategies include iontophoresis, sonophoresis, electroporation, photomechanical waves, thermal ablation, and microneedle technology. The earlier transdermal patches are utilized only to deliver the small molecule drugs whereas the latest versions of TDDS provide a

V. Kumar · N. Praveen · P. Kewlani · Arvind · A. Singh · A. K. Gautam ·
V. Mahalingam Rajamanickam (✉)
Department of Pharmaceutical Sciences, Babasaheb Bhimrao Ambedkar University (A Central University), Vidya Vihar, Raibareli Road, Lucknow 226025, India
e-mail: mrvceutics@gmail.com

V. Kumar
e-mail: vipinchoudhary1103@gmail.com

N. Praveen
e-mail: ziyaalam671@gmail.com

P. Kewlani
e-mail: prernakewlani18@gmail.com

Arvind
e-mail: arvindbarana786@gmail.com

A. Singh
e-mail: a.s.rathore2197@gmail.com

A. K. Gautam
e-mail: kumargautamanurag@gmail.com

T. S. Santra and A. U. S. Shinde (eds.), *Advanced Drug Delivery*,
Studies in Mechanobiology, Tissue Engineering and Biomaterials 26,
https://doi.org/10.1007/978-981-99-6564-9_13

comprehensive selection of molecules from low molecular weight drugs to macromolecule proteins. Among them, microneedles are proven to deliver several macromolecular biomolecules such as vaccines, extracted proteins, and whole viral particles. In this chapter, we are going to know about the basics of TDDS such as mechanisms of skin permeation, factors influencing transdermal permeation, basic ingredients for transdermal patches, categories of formulations, and permeation promoters. In addition to the basic knowledge, we are also about to explore the transdermal patches available in the market and the latest technologies introduced in TDDS. In near future, we can expect several new innovative tools to be marked within the context of the transdermal delivery of drugs.

1 Introduction

The transdermal drug delivery system, abbreviated as TDDS, is also commonly referred to as 'transdermal patches' or simply as 'patches' (Kumar and Philip 2007). TDDS are self-contained, discrete dose forms that, when placed to undamaged skin, release the drug(s) into the systemic circulation at a regulated frequency through the skin. This process is known as transdermal drug delivery systems (TDDS) (Ali et al. 2015). The first transdermal system received approval from the Food and Drug Administration in 1981 (Kumar and Philip 2007). TDDS are a great substitute for oral administration and hypodermic injections (Prausnitz and Langer 2008). Because the delivery route is a more common and simple technique for administering drugs to patients, it also boosts the therapeutic advantages to patients. TDDS avoids discomfort, bruising, and bleeding as compared to injection administration, which enhances patient acceptability and compliance. They also lower the risk of needle-related disease transmission and unintentional needle injury, as well as the development of hazardous medical waste sharps (Ita 2015a). The term "TDDS" refers broadly to all medication formulations designed to be applied topically and transport the active ingredient into the bloodstream. TDDS are devices that offer regulated, continuous medication administration to the bloodstream through the skin. The medicine that is released from the TDDS is taken up by the bloodstream and delivered to the target tissue using the stratum corneum (SC), the epidermis, and the dermis as passageways. For effective transdermal delivery, the entirety of the drug should be able to travel through the layers of the skin and reach the blood stream without becoming trapped in any of the layers (Allen et al. 2012; Chien and Banga 1989). According to the commonly utilized direct administration routes, which involve injections with needles, the transdermal delivery system (TDDS) has quickly become one of the techniques of noninvasive medication delivery into the body through the skin that has received the greatest attention from researchers. Currently, numerous transdermal drug delivery methods are utilized for treating a variety of ailments, such as, pain management, hormone therapy, smoking cessation treatments, and disorders of the cardiovascular and central nervous systems (Ali et al. 2015; Roohnikan et al. 2019; Peña-Juárez et al. 2022; Leppert et al. 2018; Pastore et al. 2015). There is no loss due to first-pass metabolism, and drugs can be delivered without interference from pH, enzymes, or intestinal flora since TDDS does not require transit through the gastrointestinal tract (Akhtar et al. 2020; Pires et al. 2019; Ruby et al. 2014). In addition to reducing the risk of harm, it has been established that there is a reduction in the overall costs of medical treatment brought on by TDDS (Frei et al. 2003). In addition, TDDS can provide a sustained as well as regulated release of the medication reduce the peak medication concentration as much as possible while also minimising the related systemic toxicity (Varvel et al. 1989; Kornick et al. 2003). They offer a unique administrative flexibility, easy to apply and quick to remove. This not only implies that drug delivery can be rapidly halted if it produces localised or systemic side effects, but it also means that they are well suited for self-administration. The use of TDDS in the treatment of dermatological conditions is particularly well suited. A drug's effectiveness can be increased while its negative effects are reduced if it is applied directly to the target skin spot. The most concerning drawbacks associated with TDDS are the potential for causing skin irritation and hypersensitivity. Several different transdermal medication delivery techniques are discussed and compared in this article. In this article, we will discuss the properties of active and passive transdermal delivery systems, as well as the types of systems and the methodologies used to characterise them. In addition to this, we take a look ahead at some potential developments in the field of TDDS.

2 Advantages and Disadvantages of TDDSs

Advantages (Prausnitz and Langer 2008; Kumar and Philip 2007; Honeywell-Nguyen and Bouwstra 2005; Alkilani and Nasereddin 2022)	Disadvantages (Kumar and Philip 2007; Keleb et al. 2010; Gandhi et al. 2012; Ali et al. 2015; Prausnitz and Langer 2008; Ita 2015a)
• Maintain relatively consistent drug-release rates by maintaining moderately uniform concentrations • Increases the pace and extent of skin permeation by increasing skin temperature and hydration • Skin penetration enhancers produces lower diffusional resistance and higher transport • Stratum corneum of the skin, in particular, offers a huge surface area for drug diffusion • Transdermal administration is relatively noninvasive and aids in avoiding invasive parenteral medication • In the case of vomiting and diarrhoea, they might be used as an alternative to oral administration • Short-acting drugs last longer • Self-administered • Have the capacity to deliver medications to a particular local site	• Only relatively tiny (1 KD), lipophilic medication molecules can effectively penetrate the skin • Prolonged stay could enhance the likelihood of a localized bacterial population • Not appropriate for the medications with high doses • The size of the patch, the kind of skin, and the surroundings can all affect how well the patch adheres to the skin • Cause serious skin allergies in some patients • A lot of hydrophilic-structured medications penetrate the skin too slowly to be therapeutically effective • Poor control over a transdermal patches drug release may be the result of damage • Drugs with high melting points are unsuitable for transdermal drug administration due to their poor solubility in water and fat

3 The Skin

Skin is the important parameter for the delivery over the transdermal surface and hence the nature of the skin and pH play a vital role. Skin is considered to be the largest organ having the wider surface area. Skin has three layers that are accountable for the drug delivery which are as follows.

3.1 Epidermis

It is a multi-layered, striated and vascular in nature having the size of 20–89 μm. The thickness is based on different regions of the body i.e., it is thickest at the palm then other parts (Jayaprakash et al. 2017). The blood capillaries providing nutrition enters the epidermis through dermis and hypodermis. The epidermis is composed of keratinocytes forming 95% of the cellular volume arising from basal layer in outward direction. The different layers of epidermis are classified on the basis of keratin maturation.

a. **Stratum basale**: It is a single-layered structure that forms the epidermis's deepest sublayer and serves as the principal centre for the creation of keratin. It extends beyond the edge of the dermis layer and contains 8% of the water that is lost as a natural part of the ageing process. In addition, melanocytes can be found in this layer.
b. **Stratum spinosum**: Upward to stratum basale is a 10–20 layers of stratum spinosum in which cells of basal layers somewhat flattened to form the layer. These cells have spine like structures hence known as prickle layer having the thickness of 50–150 μm.
c. **Stratum spinosum**: The keratinocytes start keratinizing to form stratum spinosum layer. It is 2–3 layered in which nucleus and mitochondria cleaves and the cells are filled with keratin fibres. The cells are more flattened and contains less moisture.
d. **Stratum lucidum (clear layer)**: They are densely concentrated in the palms and soles of the feet (Igarashi et al. 2007).
e. **Stratum corneum**: The skin barrier is the topmost layer of epidermis, also known as the non-viable layer. This layer of the epidermis is located on the surface of the skin (Bouwstra and Gooris 2010; Pathan and Setty 2009). This layer has a thickness of between 10 and 15 μm and is made up of dead corneocytes that are surrounded by a barrier of extracellular lipid matrix (Andrews et al. 2013). These cells help facilitate the exchange of moisture and oxygen from the surrounding environment with the body, and they constitute the interface between the body and the external environment (Giacomoni et al. 2011). As corneocytes are the primary pathway for permeation, their size directly affects how much they are penetrated (larger the size, the greater the penetration). The size also depends on the area to be present i.e., it is smaller at skin but wider in arms (Hadgraft and Lane

2009). The cohesiveness between the corneocytes arises through desmosomes (Morrow et al. 2007). The stratum corneum is made up of 40% protein, most of which is keratin, 40% water, and the remaining 10% of its composition is made up of lipids in the form of ceramides, cholesterol, and saturated fatty acids (Andrews et al. 2013; Morrow et al. 2007).

3.2 Dermis

This layer is 1–2 mm in thickness and is composed of fibrous tissues. They have a substantial blood supply, which facilitates the absorption of drugs from the stratum corneum. The layer comprises of connective tissue which assists to supply oxygen to skin and detoxifies through the removal of waste products. This layer contains a lot of sebaceous glands, sweat glands, and hair follicles having storage facility (Jayaprakash et al. 2017). On average, a human's skin has 200 and 250 sweat glands and 10–70 hair follicles for every square centimetre (Andrews et al. 2013). The following sublayers are formed by the dermis.

a. **Papillary layer**: This layer distinguish the epidermis layer from dermis formed by a layer of loosely arranged connective tissues, nerve fibres, capillaries, water and fibroblasts.
b. **Reticular layer**: This is the lowest layer to connect dermis and hypodermis. It is dense and thicker in nature and the collagen fibres are aligned parallel to the skin surface (Igarashi et al. 2007).

3.3 Hypodermis

The hypodermis is also known as the subcutis. It is elastic in nature and has a large lipid capacity, both of which allow it to absorb lipids from blood vessels and nerve terminals. This layer is anywhere from 4 to 9 mm thick (Igarashi et al. 2007). It is a deepest layer having bundle of fatty tissue that supports dermis and epidermis. It also helps to provide nutrients and maintains the body temperature (Jayaprakash et al. 2017) (Fig. 1).

4 Permeation Routes via the Skin for Various Drugs

It is possible for the drugs to be absorbed via the skin in different ways, and these ways are based on the physicochemical qualities of the drug and the physiological features of the skin. The following three mechanisms have the potential to facilitate the absorption of the drug.

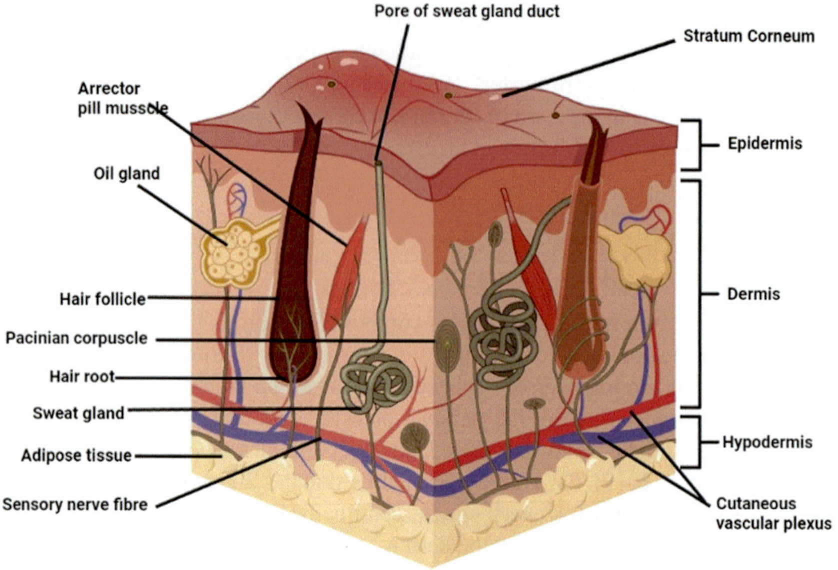

Fig. 1 A depiction in schematic form showing various layers of the skin

4.1 *Trans-follicular Pathway*

A high number of pores, oil glands, sweat glands, and hair follicles are penetrated during the diffusion process that takes place along the trans follicular pathway. This is the quickest and most extensive channel for diffusion. The absorption depends on the factors such as gland secretion including its quality and quantity.

4.2 *Trans-cellular Pathway*

When the drug is absorbed through corneocytes, majorly of hydrophilic drugs are diffused through trans cellular pathway. This pathway allows the drug to absorb through cytoplasm to diffuse in the blood.

4.3 *Inter-cellular Pathway*

The inter cellular pathway works through the matrix present between the cells. This pathway is favourable for uncharged lipophilic medications (Neupane et al. 2021) (Fig. 2).

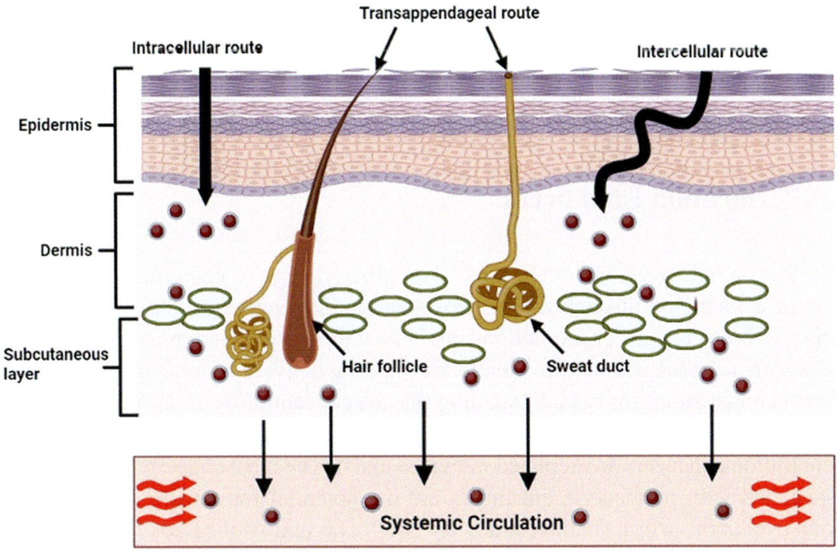

Fig. 2 Schematic representation of drug penetration pathway

5 Drug Permeation Through the Skin

The transport of the medicine is accomplished through the skin in three different ways: through the hair follicles, the sweat ducts, and the sebaceous glands. With the presence of aqueous layer hydrophilic drug can easily be penetrated while the lipophilic drug can easily diffuse through lipid membrane (Benson et al. 2019). This molecular transport of drug through channels is termed as fluxes which can be easily defined as the number of molecules to be transported at a specific period of time through a constant area. It can be given by

$$\mathbf{J} = \frac{m}{At}$$

where J is the flux, M is the mass of the drug transported, A is the cross-section area and T is the time period.

When transport goes across channels, it takes an external force to make it happen, which can be given by Fick's first law of diffusion. This force can be thought of as a driving force. This law states that the flow is directly proportional to the concentration gradient, which may be found in the equation (Jayaprakash et al. 2017)

$$\frac{dm}{dt} = \mathrm{DS}\frac{dc}{dx}$$

where dm/dt is the change in mass of material with respect to time, D is the diffusion coefficient, cm^2/s, S is the surface area, cm^2, dC is the change in concentration of material, g/cm^3.

6 Permeation Enhancers

Permeation enhancers are compounds that either change or encourage the absorption of a medicine through the skin by temporarily increasing the skin's permeability. These changes or promotions might be temporary or permanent. Permeation enhancers are utilised most frequently for the drug delivery of ionizable and impermeable medications, or to maintain drug plasma concentration, or as aids to transfer high-molecular-weight agents (Roy et al. 2017). Ideal properties and categories of penetration enhancers are depicted in Figs. 3 and 4. The most concerning drawbacks associated with permeation enhancers are the potential for causing allergic reactions over skin surface (Pathan and Setty 2009; Das and Ahmed 2017). Penetration enhancers can be classified in the following categories.

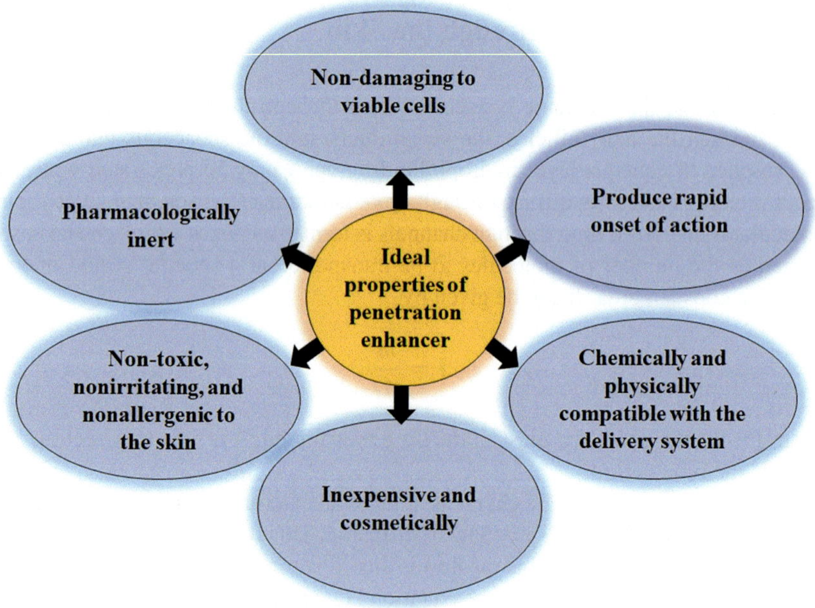

Fig. 3 Ideal characteristics of permeation enhancers

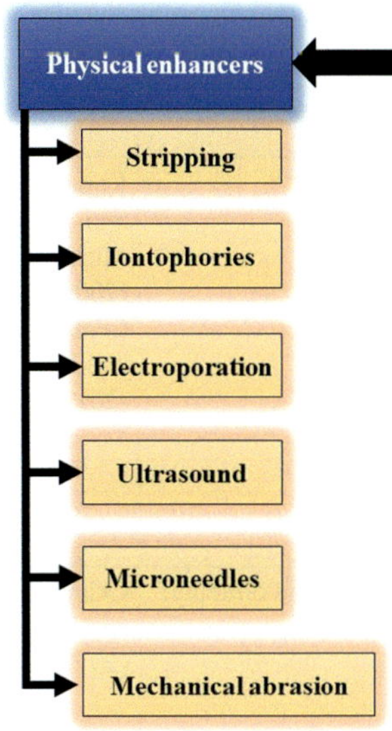

Fig. 4 Permeation enhancers used in TDDS

6.1 *Chemical Modifiers*

Chemical enhancers work by disrupting the stratum corneum, interacting with proteins present at intercellular space and improving the drug partitioning characteristics through the utilisation of co-enhancers in the stratum corneum. Improving the drug partitioning characteristics can be achieved either by modifying the confirmational change of the skin protein or enhancing the swelling properties of solvent. Various chemical enhancers are listed in Table 1 (Barry 1988).

6.2 *Physical Modifiers*

(a) **Iontophoresis**: It includes diffusion, migration and electro-osmosis of drug across the skin. The principle states that the fluid flow is independent of concentration gradient. When the skin is in normal condition it is slightly negative charged. So, the flow take place from cathode to anode, permitting the cationic drugs to penetrate easily (Green 1996).

(b) **Sonophoresis**: When the skin permeability is enhanced by the ultrasound then it is said to be sonophoresis. It is done by the effects of cavitation, connective conveyance, effects of heat and the effects that are produced mechanically (Mitragotri et al. 1995).

Table 1 List of chemicals that improve transdermal penetration of drugs along with examples and their mechanism of action

S. no.	Modifying agents	Mechanism	Examples	References
1	Alcohols	Enhances the permeation rate through lipid and protein extraction, swelling of stratum corneum or improving drug solubility	Propylene glycols, glyceryl mono-caprylate	Karande and Mitragotri (2009), Chantasart and Li (2012)
2	Amines and amides	Improving skin hydration or shaping diffusion channels for hydrophilic drugs	Urea, dimethyl acetamide and dimethyl formamide	Chantasart and Li (2012)
3	Cyclodextrins	The permeability is increased with the complexation to the lipophilic drugs	O-carboxymethyl-oethyl-β-cyclodextrin (CME-β-CD)	Pathan and Setty (2009)
4	Fatty acids	Unsaturated FA are better permeation enhancer in comparison to saturated FA	Palmitoeic acid	Sinha and Kaur (2000)

(c) **Thermal energy**: When the ultrasound energy is used to modify skin permeability, there is an increase in temperature. This approach is used to design a heating unit known as CHADD (Controlled Heat-aided Drug Delivery) which transmits thermal energy at certain interval of specific intensity (Kumar and Philip 2007).

(d) **Stripping of stratum corneum**: The use of microderm abrasion to promote skin resurfacing is one technique to cause skin disruption. This enhances the permeability rate up to 100 times and treats acne, scars, and hyperpigmentation (Kumar and Philip 2007).

(e) **Hydration of stratum corneum**: When the water content in the stratum corneum is increased the permeability is raised due to the swelling of the skin (Benson et al. 2019).

6.3 Biochemical Modifiers

Biochemical enhancers actin the following two ways to act as permeation enhancer.

(a) **Synthesis of bio-convertible pro-drugs**: Many enzymes are used for the conversion of prodrugs to its active form which can be achieved using many permeation enhancers. For this approach many steroids are designed such as S6-acyloxymethyl and 9-dialkylaminomethyl.
(b) **Co-administration of skin metabolism inhibitors**: This approach is predominantly used to enhance the permeation of drug through skin by temporarily blocking the synthesis of various ceramide, fatty acids and cholesterol that may interfere with barrier hemostasis (Roy et al. 2017).

7 Factors That Influence the Transdermal Delivery

7.1 Physicochemical Factor

(a) **Dimensions and shapes of the molecules**: Because of their great efficacy in this range, drugs with molecular weights ranging from 100 to 500 Dalton are frequently considered for delivery by the transdermal route.
(b) **Partition coefficient**: It is easier for the body to absorb drugs through the skin if they are both lipid- and water-soluble. Intracellular dosing is an option for certain pharmaceuticals if they have a high lipophilicity and a moderate partition coefficient (log K between 1 and 3). It is probable that the transcellular pathway will be more effective for hydrophilic atoms (log K 1) than the other pathways (Ita 2015b; Wiechers 1992).
(c) **Ionization**: According to the pH-partition theory, the lipid barrier is the only pathway that allows significant amounts of the unionised form of the medication to pass through (Govil 1988).
(d) **Solubility/melting point**: More hydrophilic compounds tend to penetrate the skin more slowly than lipophilic molecules. At normal and high temperatures, drugs with high melting points have low water solubility (Choudhary and Singh 2021).

7.2 Biological Factor

(a) **Skin condition**: Although healthy skin acts as a barrier on its own, it is possible for many substances, including acids and alkalis, to penetrate the skin by penetrating the barrier cells and enter the body. The complex and dense structure can be dissolved using a variety of solvents. Methanol and chloroform are two examples of solvents that can be used to remove the lipid component of the horny layer. This results in the creation of artificial shunts that make it simpler for medication molecules to pass through (Singh and Singh 1993).
(b) **Skin age**: The skin of adults and younger people is more permeable in comparison to the skin of elderly people. The concept of toxicity is based on the ratio of body surface area to body weight (Choudhary and Singh 2021).
(c) **Hydration of the skin**: Hydrated skin increases permeation and enhances drug delivery.
(d) **Body temperature and the pH level of the skin**: Increase in skin temperature increase the pace at which the skin penetrates, which is expected to increase the energy available for diffusivity. The pH regulate permeability of drug (Rastogi and Yadav 2012).

7.3 Formulation Factor

(a) **Penetration enhancer**: The addition of penetration enhancers improve the skin permeability of the drugs (Jayaswal and Sood 1987).
(b) **pH of vehicle**: The alkaline or acidic pH cause irritation to skin which affect the permeation of drug (Kavadya et al. 2018).
(c) **Release characteristic**: Solubility of drugs affects the release.

7.4 Components of TDDS

(a) **Drug**: Many medications that are subject to extensive first-pass digestion but have a limited therapeutic window or a short half-life can be delivered using transdermal patches. In addition, TDDS have recently been approved for a number of drugs, including rivastigmine for Alzheimer's disease and Parkinson's dementia, rotigotine for Parkinson's disease, methylphenidate for attention deficit hyperactivity disorder, and selegiline for depression (Jeong et al. 2021) (Fig. 5).
(b) **Drug reservoir/Polymer matrix**: The TDDS is built on a foundation of polymers, which are responsible for controlling how much medication is required to be released from the device. The formation of a polymer matrix is possible by the use of drug dispersion in a synthetic polymer base, either in liquid or solid form. The polymers that are used in TDDS need to be chemically

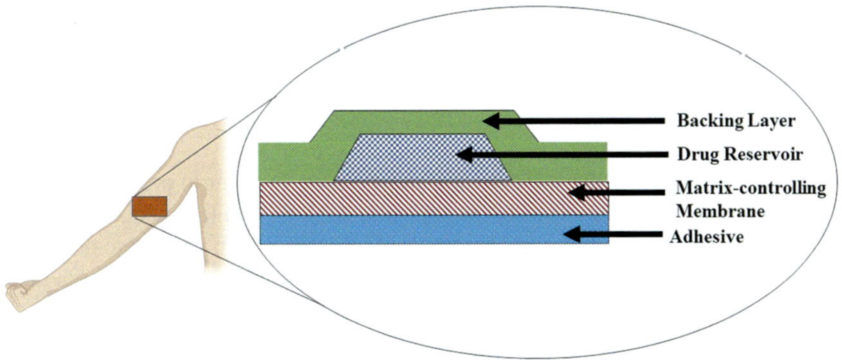

Fig. 5 A representation in schematic form of the TDDS's components

and biologically compatible with the drugs, in addition to being compatible with the other elements of the system, including penetrating enhancers and pressure-sensitive adhesive (Rastogi and Yadav 2012). In TDDS, different kinds of polymer, ranging from natural to synthetic, can be employed,

(i) **Natural polymers**: Wax, gelatin, starch, chitosan, cellulose derivatives, zein, natural rubber, other types of proteins and their derivatives are the examples of natural polymers.
(ii) **Synthetic elastomers**: Nitrile, acrylonitrile, butyl rubber, styrene-butadiene rubber, polybutadiene, hydrin rubber, polysiloxane silicone rubber, and other materials are used in TDDS.
(iii) **Synthetic polymers**: Polyvinyl alcohol, polyvinyl chloride, polyethylene, polypropylene, polyacrylate, polyurea, polyvinyl pyrrolidone, polymethyl methacrylate are used.

(c) **Membrane**: A membrane is used in the fabrication of the patch as a single layer, or it is bonded to the backing to create a pocket that will house the drug-containing matrix. Using the membrane's capacity for diffusion, one can control how much of the active ingredient or excipient is absorbed by the skin. As a rate-controlling membrane, ethylene vinyl acetate, silicone rubber, and polyurethrane are reported.
(d) **Permeation enhancer**: In order to achieve higher therapeutic levels of the drug, a penetration enhancers, such as lipids and proteins, are used to increase the permeability of the stratum corneum. They are of either chemical or physical permeation enhancers.
(e) **Backing laminates**: TDDS generally made up of polyester film or polyethylene film as backing laminates. They should be chemical resistance, flexible, non-irritant and possess good tensile strength.
(f) **Release liners**: It is the crucial packing material that can keep the patch secure while it is being applied on to the skin. It begins with a base layer, which may

be occlusive (like polyethylene or polyvinyl chloride) or it may not be occlusive (like a layer of aluminium foil) (like paper fabric).

(g) **Pressure sensitive adhesives (PSA)**: The PSA may be applied to the TDDS surface as a continuous adhesive layer or it may be applied to the edge of the TDDS.

(h) **Other excipients like solvents and plasticisers**: In order to produce the drug reservoir, a number of different solvents, such as dichloromethane, chloroform, methanol, acetone, and isopropanol, are utilised. To render the flexibility to TDDS, plasticizers such as dibutylphthalate, triethyl citrate, polyethylene glycol, and propylene glycol are added to it (Berlin 1997).

8 Types of TDDS

See Fig. 6.

8.1 Drug in Adhesive Type

The stratum corneum serves as a rate-regulating barrier in this form of drug loading. This is the earliest form of transdermal patch. The creation and distribution of Pharma Schwartz/deponit Lohmann's nitroglycerin-releasing system in Europe serves as the best example of this sort of transdermal medication delivery technology. Backing

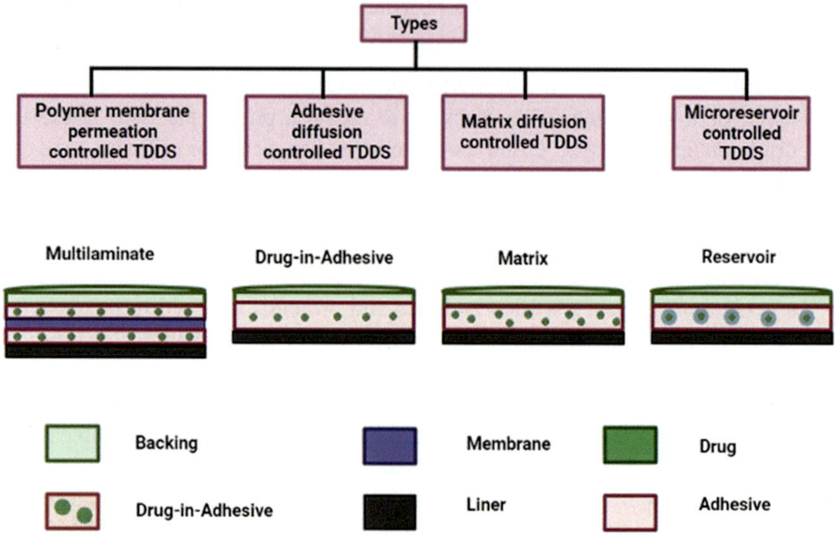

Fig. 6 Schematic illustration of various types of TDDS

membrane, adhesive containing medication, and release liner are components of the basic construction (Rastogi and Yadav 2012).

8.2 *Multi Laminate Type*

Instead of having single layer of drug in adhesive, multiple drug layers are present in this type of TDDS. Usually, the multi-drug layers are separated by a membrane.

8.3 *Matrix Type*

In this type of TDDS, the pharmaceutical agent is incorporated into the polymer matrix, which results in the drug release in a random fashion. Only the border and a small portion of the patch's interior include the sticky layer.

8.4 *Reservoir Type*

The drug is dispersed into a reservoir that has a membrane lining in this type of TDDS. The lining membrane contains adhesive in it (Wokovich et al. 2006).

8.5 *Vapour Patch*

Vapour patches are brand-new products on the market used for the release of essential oils (Eucalyptus oil) for decongestion and cold symptoms in children. There are several other types of patches intended to enhance the quality of sleep and reduce the amount of smoking available on the market today (Weiner et al. 1976).

8.6 *Microreservoir System*

Combining a matrix-dispersion device and a reservoir is what makes this type of TDDS unique. To create the drug reservoir, first the drug is dispersed throughout an aqueous solution of a water-soluble polymer. The solution is uniformly distributed throughout a lipophilic polymer, which results in the formation of millions of tiny drug reservoir spheres. This thermodynamically unstable dispersion is instantly stabilised through the use of cross-linking agents to cross-link the polymer while it is still in place to form this type of TDDS (Rizwan et al. 2009).

9 Evaluation Parameters of TDDS

See Fig. 7.

9.1 Physicochemical Assessment

(a) **Thickness**: It is possible to measure the thickness of the transdermal patch using a travelling microscope, dial gauge, screw gauge, or micrometre. These instruments can be used to measure the thickness of the patch at three separate spots on the patch where the thickness of the patch is being determined. The thickness of the patch is determined by taking the average of the three different measurements. It is possible to perform calculations in order to ascertain the variation in thickness that exists both inside and between patches (Sankar et al. 2003; Verma and Iyer 2000).

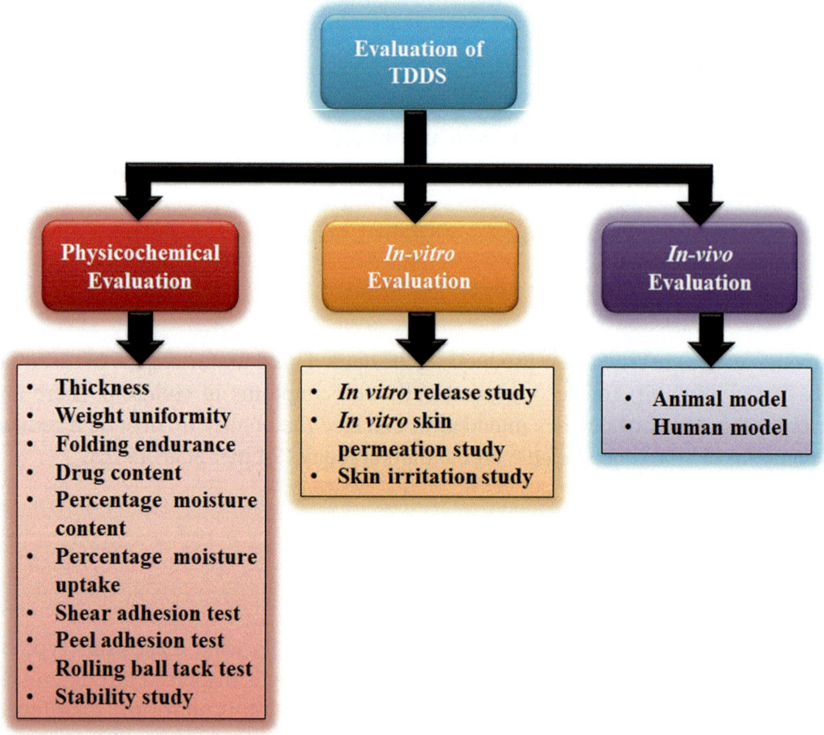

Fig. 7 Schematic representation of various evaluations of TDDS

(b) **Weight uniformity**: The patches are allowed to dry at a temperature of 60 °C before being weighed. An evaluation of the weight homogeneity of a transdermal patch is performed by cutting and weighing a 1 cm^2 part of three patches. The weight variance is then determined using this weight data. By taking the average of the three different readings, one may calculate how much weight the patch has. It is not possible for an individual's weight to differ from the average weight by a significant amount (Arijit et al. 2010; Ubaidulla et al. 2007).

(c) **Folding endurance**: In order to determine how well a patch or film can withstand when folded more than 300 times, the folding endurance of the patch or film is tested. The number of folds that a patch can maintain without becoming damaged is referred to as its folding endurance. The folding endurance of the transdermal patch is one factor that can be used to assess the patch's flexibility (Shinde et al. 2008).

(d) **Drug content**: A film with the necessary surface area and weight is first dissolved in an appropriate solvent, such as methanol or phosphate buffer with a pH of 7.4, and then filtered. With a standard curve and the appropriate dilutions, one can determine the amount of drug present in a sample by either the UV or HPLC method (Sonjoy et al. 2011; Kumar et al. 2012).

(e) **Percentage moisture content**: In order to determine the percentage of moisture present, individually weighted patches are placed in desiccators containing fused calcium chloride and left there for 24 h at room temperature (Verma and Ojha 2018; Kusum Devi et al. 2003). After waiting for 24 h, the patches are reweighed, and the percentage of moisture content is determined using the following method:

$$\textbf{Percentage moisture content} = (\textbf{Initial weight} - \textbf{Final weight}/\textbf{Initial weight}) \times \textbf{100}$$

(f) **Percentage moisture uptake**: Films that have been weighed after spending twenty-four hours in a desiccator are then subjected to a relative humidity of eighty-four percent using potassium chloride. When the films' weights have reached a stable state, the films are reweighed (Verma and Ojha 2018; Kusum Devi et al. 2003; Shaila et al. 2006). Calculating the percentage of moisture uptake is as simple as

$$\textbf{Percentage moisture uptake} = (\textbf{Final weight} - \textbf{Initial weight}/\textbf{Initial weight}) \times \textbf{100}$$

(g) **Shear adhesion test**: With the use of this test, the cohesive strength of the sticky polymer that was being tested can be determined. After placing the patch with the adhesive covering on a level surface, the required weight is then hanged directly from the patch itself. The length of time it takes to peel the patch off of the surface provides information about how effectively it is stuck to the surface.

(h) **Peel adhesion test**: A patch is adhered to a surface and then subjected to this test to measure the amount of force necessary to remove it from the surface. After applying the patch to the surface of the steel plate, the plate is turned so that it is facing in the opposite direction. The level of force that must be used in order to remove the patch is analyzed.

(i) **Rolling ball tack test**: For this test, a patch is positioned horizontally on the adhesive surface with the raised side facing upward. A steel ball with a diameter of 7/16 inch is then rolled down and inclined. The stickiness of the patch is proportional to the distance travelled by the ball (Pandey et al. 2012; Vishvakarma et al. 2012).

(j) **Stability study**: In order to evaluate how long the patch will remain functional and useable, a stability analysis is conducted. According to international conference of harmonization (ICH) criteria, stability is assessed for 6 months at 40 °C/ 75% relative humidity since the medication rapidly degrades in unstable patch formulations. At days 0 through 180, samples are collected and analysed for stability (Wade and Weller 1994).

9.2 *In-vitro Evaluations*

(a) **In vitro drug release study**: The in vitro drug release is measured with the help of USP dissolution apparatus at a speed of 50 revolutions per minute at a temperature of 37 °C. Before dipping the transdermal film into a dissolving medium that contains 900 mL of pH 7.4 phosphate buffer, it is first bonded to a glass slide using an adhesive. After this step, the slide is dipped into the dissolving medium. After twenty-four hours, a sample of five millilitres is taken, and the dissolving media is then replenished with an amount of buffer that is comparable in volume. After the sample has been analysed, the total amount of medication that has been released is calculated (Sood and Panchagnula 1999).

(b) **Skin permeation**: The skin of a male Wistar rat is used to conduct in vitro skin permeation investigations in a vertical diffusion cell with two chambers that are separated by the skin of the rat. After attaching the transdermal film to the rat's skin, a connection is made to the diffusion cell, which is located in the space between the donor and receptor compartments. In addition to the typical samples that are collected, an alternate medium of the same volume is utilised. For the purpose of computing flux, analysis of the samples is performed (Panchaguinla 1997).

(c) **Skin irritation study**: In an in vitro skin permeation test, albino rats, mice, or rabbits are used for skin irritation testing. Each of the five categories of animals comprises a total of six unique species. Group I serves as the control group, while Group II receives an adhesive tape that is commercially available (the official adhesive tape used by the USP), Group III receives a transdermal patch that does not contain any medication, Group IV receives a transdermal patch that contains medication, and Group V receives a standard irritant that is a % v/

v solution of formalin. Animal skin hairs are removed, and depending on their group, the animals receive treatment for 7 days. Every day, animals are evaluated and given grades based on their propensity to scratch, the number of scars they leave behind, and how uncomfortable they appear to be.

9.3 *In-vivo Evaluations*

(a) **Animal model**: Research conducted on animals at a smaller scale is favoured since it is more efficient in terms of both money and time compared to research conducted on humans. The mouse, the hairless rat, the hairless dog, the hairless rhesus monkey, the rabbit, and the guinea pig are the most common animal species that are utilised in the process of evaluating transdermal drug delivery techniques. Several studies have found that hairless animals perform better than hairy animals in a variety of different types of tests, including in vitro and in vivo tests. When it comes to researching transdermal drug delivery in people, the Rhesus monkey is one of the most useful models (Sandhyarani and Madhuri 2018).

(b) **Human model**: At the completion of the transdermal device development phase, the patch is administered to human volunteers, which resulted in the collection of data about pharmacokinetics and pharmacodynamics. Clinical studies have been conducted in order to assess the effectiveness, hazards, and side effects, as well as patient compliance and other criteria. Phase I clinical trials are conducted to explore the drug's safety mostly in volunteers, whereas phase II clinical studies are intended to test the drug's efficacy and side effects in patients over a shorter period of time. In phase III trials, a range of patient populations are evaluated to determine whether or not marketed patches are safe and effective. Phase IV trials, on the other hand, are conducted during post-marketing surveillance to see whether or not adverse drug responses have occurred (Singh et al. 2010; Parivesh et al. 2010; Wiechers 1992; Yamamoto et al. 1990; Al-Khamis et al. 1986).

10 Marketed Formulations and Therapeutics Application of TDDS

Several TDDS formulations are available in the market for hypertension, angina, pain relier, harmone replacement, smoking cessation, etc. Table 2. Covers the list of TDDS based marketed formulations.

Table 2 List of marketed TDDS based products

Brand name	Active agent	Indications
Androderm	Testosterone	Hypogonadism in males
Catapres TTSR	Clonidine	Hypertension
Deponit	Nitroglycerin	Angina pectoris
NictonineIIR	Nicotine	Pharmacological smoking cessation
NuPatch 100	Diclofenac diethylamine	Anti-Inflammatory
Nuvelle TS	Estrogen	Hormone replacement therapy
Matrifen R	Fentanyl	Pain relief patch
Nitodisc	Nitroglycerin	Angina pectoris
Oxytrol R	Oxybutynin	Overactive bladder
Transderm-Scop R	Scopolamine	Motion sickness
Alora	Estradiol	Postmenstrual syndrome
Duragesic R	Fentanyl	Moderate/severe pain

11 Recent Technological Advancements

When compared to the administration of medication through dermal application onto the skin, it is known that stimulation from the outside stimuli, such as electronic, mechanical, or physical stimulation, improves the skin's ability to absorb medications and macromolecules (Benson et al. 2019). Active transdermal delivery, which is TDDS enhanced by the proper tools, is known for reliably and quickly delivering medications to the skin. Additionally, this form of improved TDDS can enhance the therapeutic effectiveness of medications that are administered (Lee et al. 2018; Al Hanbali et al. 2019).

11.1 Iontophoresis

Iontophoresis has been shown to increase the release rate of several medications that have poor penetration features. This is accomplished by promoting the movement of ions across the membrane by applying an external potential difference in a smaller range (less than 0.5 mA/cm^2). Iontophoresis is able to do this because it promotes the movement of ions across the membrane. The electrochemical potential gradient approach is utilised for the purpose of facilitating the in vivo administration of drugs, regardless of whether or not the pharmaceuticals are ionic (Wang et al. 2022). Iontophoresis is a phenomenon that works differently depending on a variety of factors, including the drug formulation, the type of electrical cycle that is employed, the polarity and valency of the therapeutic molecule, and the mobility of the molecule. Iontophoresis is one of the drug delivery strategies that is less reliant on biological factors than the majority of the other drug delivery methods because it is based on an electrical current (Dhal et al. 2020).

11.2 Sonophoresis

Transdermal drug delivery can be made more effective using an ultrasonic device that produces the required range of ultrasound frequencies (Park et al. 2019). The chemical that will be utilised is coupled with a particular coupler, such as a gel or cream, in order to provide an aqueous channel via which the medicine can be supplied. This route couples ultrasonic waves to the skin, which in turn disrupts the layers of the epidermis. Ordinarily, drugs move through channels made by applying ultrasonic waves with energies ranging from 20 kHz to 16 MHz. Additionally, ultrasound raises the localized skin area's temperature and produces a thermal effect that helps in the uptake of drugs. Regardless of the solubility, dissociation and ionisation constants, and electrical properties of the medications being distributed, this method has been utilised to successfully do so for a wide range of pharmaceuticals, including insulin and mannitol, amongst others (including hydrophilicity). The actual mechanism of drug permeation by this approach is still unaware, and there are still issues with device availability, optimising exposure time and treatment cycles, and unfavourable side effects that may effect including burns (Jeong et al. 2021).

11.3 Electroporation

This method includes sending high-voltage electric pulses to the skin over brief periods of time (micro seconds), which results in the formation of microscopic pores in the subcutaneous tissue. These pores increase permeability and make it easier for drugs to be transported (Chen et al. 2020). Using closely spaced electrodes, electric pulses are introduced for painless and safe drug administration. It has been utilised to demonstrate the successful transport of both low and high molecular weight drugs, such as heparin, which is negatively charged and acts as an anticoagulant, as well as antiangiogenic peptides and oligonucleotides. This technique, which involves making small holes in the patient's skin, does not cause any discomfort and is completely risk-free. Minimal delivery loads, significant cellular disturbance, and even death of cells in some cases, heating-induced drug degradation, and denaturation of protein and other bio-macromolecular therapies are the disadvantages associated with this method (Jeong et al. 2021).

11.4 Waves of the Photomechanical Process

Photodynamic waves that are applied to the skin have the ability to permeate the SC, which enables the drug to enter the body through the temporarily established channel (Dermol-Černe et al. 2020; Lin et al. 2014). For successful transmission, a low radiation dose of approximately 5–7 J/cm^2 is used to expand the depth, between 50 and 400 μm. Ablation of a confined area is caused by the incident wave. This limited ablation demonstrated a longer rise and duration when compared to earlier techniques that involved direct ablation. As a result, the parameters of the photodynamic wave needed to be regulated in order to ensure that the product was delivered to the required depth in the skin. Within minutes, the wave from a single laser pulse caused the skin to become more porous, which made it easier for macromolecules to pass through the skin and into the body. It is possible for a single photodynamic laser pulse with a length of 23 ns to form dextran macromolecules with a weight of 40 kDa and latex particles with a size of 20 nm (Jeong et al. 2021).

11.5 Microneedle

This is among the most widely used techniques for transdermal medication delivery, and there is currently significant research in this area. This method involves utilising needles on the micron scale to puncture the epidermal layer of the skin. The goal of this approach is to facilitate the diffusion of medication. These microneedles are so small and thin that they are able to transport pharmaceuticals straight to the area of the blood capillaries where they can be actively absorbed, hence reducing the amount of discomfort felt by the patient (Zhao et al. 2020). Up to this point, the production of the microneedle has been accomplished by the use of photolithography and laser-mediated processes. Microneedles can be made of either metal or polymer by the use of laser-mediated production techniques. Using a laser, a flat metal or polymer surface is etched or ablated to create the 3D structure of a microneedle (Shakya et al. 2019; Dardano et al. 2015). The process of accurately constructing microneedles known as photolithography has the benefit of facilitating the production of needles in a variety of shapes from a wide range of materials. This technique is mostly used to fabricate silicon or dissolving/hydrogel microneedles by creating an inverse mold based on the microneedle shape using photoresist etching (Dardano et al. 2015). The prepared microneedles can take many different forms, including solid microneedles that provide a physical pathway for drug absorption, drug-coated needles that deliver drugs coated on the needles as they enter the skin, dissolving needles made of drug formulations that dissolve in the body, and naturally delivered melting needles that store drugs in hollow needles and administer them with a syringe) (Kim et al. 2012; Du et al. 2019).

11.6 TDDS with the Utilisation of Chemical Enhancers (Passive Delivery)

In order to encourage enhanced transdermal distribution and therapeutic efficacy, pharmaceuticals should have properties such as a low MW (less than 1 kDa), an affinity for both lipophilic and hydrophilic phases, a short half-life, and the inability to irritate the skin (Jeong et al. 2021). Penetration of drugs through the skin is affected by a wide variety of factors, such as differences between species, the age and location of the skin, the temperature of the skin, the condition of the skin, the area of application, the length of exposure, the amount of moisture contained in the skin, pre-treatment methods, and the physical characteristics of the penetrant. Recent research has focused on a variety of transdermal drug delivery technologies, ranging from the creation of chemical enhancers that spread drugs across skin more widely or make them more soluble in skin to the creation of novel, creative methods that expand this idea to the creation of extremely potent formulations and microemulsions (Zhang et al. 2019).

11.7 Polymeric Nanoparticles

There are different varieties of nanoparticles (NPs), ranging size from 1 to 1000 nm, is reported to possess transdermal efficacy. When a medicine is given to a patient in the form of NPs, the drug demonstrates the behaviour of having a targeted and controlled release, modifies the drug's in vivo dynamics, and prolongs its stay in the circulation, all of which improves the drug bioavailability and lessen toxicity and adverse effects. In order to fabricate NPs, polymerization and crosslinking are typically used, and biodegradable polymeric materials like chitosan and poly vinyl alcohol (PVA) are frequently employed (Kim et al. 2012).

11.8 Nano Emulsion

Lower viscosity and isotropic are two characteristics of nano emulsions (Jeong et al. 2021). A particle size range of 100–1000 nm is typical for nano emulsions, but due to the nanoscale dimensions of these systems, it has been suggested that there is an upper limit to the particle size that can be achieved. Although nano emulsions and microemulsions almost share the same droplet size range, content, and appearance, there is a substantial difference between them in terms of their structural features and their long-term thermodynamic stability (Szunerits and Boukherroub 2018). The nano emulsions' exceptionally high wettability, which ensures that they will remain in close proximity to the skin, is made possible by the nano emulsions' minute particle size, huge specific surface area, and relatively low surface tension. Nano emulsions have a high solubilization capacity in addition to other advantages such as physical stability, increased bioavailability, simplicity of preparation, low energy input during preparation, and a long shelf life. Nano emulsions offer a shorter transdermal duration and better transdermal absorption compared to topical skin remedies that are often employed. Water-in-oil (W/O) nano emulsions consist of an aqueous phase dispersed in a continuous oil phase, whilst oil-in-water (O/W) nano emulsions consist of an oil phase distributed in a continuous aqueous phase. Both types of nano emulsions are referred to as nano emulsions. There have been a number of studies that point to an increase in the utilisation of nano emulsions (O/W) as a method of delivering lipophilic medicinal components. These studies demonstrate the enormous potential of nano emulsions to support novel TDDS-based advancements in pharmaceutical applications (Jeong et al. 2019).

12 Clinical Trials Using Transdermal Delivery System

As a consequence of considerable research conducted on transdermal delivery, a great number of novel transdermal delivery formulations that are effective in treating a wide variety of disorders have been developed. Table 3 represents various transdermal delivery systems on clinical trials and ongoing researches.

Table 3 Recent clinical trials investigating TDDS

Conditions	Active agents	Trial code
Neuropathic pain, spinal stenosis	Transdermal fentanyl matrix, gabapentin	NCT01127100
Types 1 diabetes	PassPort(R) transdermal insulin delivery system	NCT00519623
Parkinson's disease	Nictoine transdermal patch	NCT01560754
Fibromyalgia syndrome	Transdermal magnesium chloride	NCT01968772
Chemotherapy-induced nausea and vomiting	Granisetron transdermal delivery system	NCT04472143
Respiratory insufficiency, analgesics, opioid	Fentanyl transdermal patch system	NCT04204967
Opioid, moderate cancer pain, transdermal fentanyl	Transdermal fentanyl	NCT04533243
Blood pressure	Device: transdermal optical imaging	NCT04539860

13 Future Prospective

The areas of regulated dosages, decreased frequency of dosing, and increased patient compliance are going to be the primary focuses of any future adjustments made to improve the efficacy of transdermal drug administration. Nicotine patches designed for transdermal administration were first introduced for the treatment of smoking dependency twenty years ago. Because of this, several transdermal patches containing nitroglycerin for the treatment of angina, estradiol for the treatment of oestrogen deficiency, oestrogen for the treatment of hormone replacement therapy, and fentanyl for the treatment of pain were developed. As a result of patent durations running their course, researchers have been motivated to come up with fresh and appealing dosage formulations for existing treatments. Because of the many breakthroughs that have been made in this field, transdermal delivery methods are quickly becoming more widespread in the phase of drug delivery. The most successful companies in the pharmaceutical industry are hard at work developing TDDS, and a wide variety of transdermal administration methods are now being researched and developed for practical applications. Innovative delivery methods such as liposomes, niosomes, nanoparticles, microspheres, and microemulsions are used in the manufacturing of transdermal drug delivery systems. These methods are used to improve the absorption of medications. The application of mechanical energy to alter the physiology of the skin or to speed up drug molecules in order to boost drug flux over the skin are two examples of the types of permeation enhancement methods that are now under development in addition to these methods. Several methods, such as electrophoresis, iontophoresis, sonophoresis, and magnetophoretic, have been suggested as ways to improve drug distribution over the skin. These methods have been suggested for use with large molecular weight pharmaceuticals as well as insoluble substances. In contrast to the oral route of administration, the technique of delivery to systemic circulation via the skin is currently the one considered the safest and most accepted by medical professionals. Transdermal administration has many major advantages, including the elimination of first-pass metabolism, steady distribution, higher patient compliance, fewer systemic medication interactions, sustained drug release, and generally greater therapeutic efficacy. Even if there are already certain goods available on the market that make use of physical techniques, there are still a lot of problems that need to be solved before large-scale manufacturing and product development can proceed. Due to the fact that there has only been one report of a phase 1 clinical trial of human proinsulin peptides coupled to gold NPs and delivered intradermally using MNs in type 1 diabetes, the use of MNs in combination with nanomedicine for transdermal drug delivery is currently restricted to proof-of-concept pre-clinical studies. It is projected that a combination of nanomedicine and MNs will expand rapidly in the future due to their focused drug delivery and their promising clinical outcomes; nevertheless, additional study is required to fully utilise the therapeutic and diagnostic potentials of this smart combination. It is probable that we are still only scratching the surface of the commercial potential offered by the market for transdermal medication delivery, taking into account the progress that has been made in terms of research into creative techniques. There are still regulatory issues that need to be addressed, as evidenced by the limited approvals granted by the FDA over the course of the past several years and the final outcomes of the products, as well as those product ideas that were never brought to market for a variety of reasons, including financial constraints, concerns regarding their level of safety, and/or concerns regarding the level of technological complexity. These issues are needed to be addressed for wider application of TDDS.

14 Conclusions

When it comes to delivering drugs through the skin and into the systemic circulation, transdermal drug delivery is the method that is most reliable and effective. In order to avoid the effects of the gastrointestinal tract, the first pass metabolism, and other sensitivities associated with various alternative drug administration routes, numerous medications, such as hormonal therapy, a wide variety of analgesics, and medications for heart disease have been developed as transdermal drug delivery systems. Because it is noninvasive, non-allergenic, has a fixed duration, ánd has a predetermined technique for the delivery of doses, TDDS makes it possible to distribute medications in a uniform manner at controlled and defined rates. The bioavailability of medications with limited absorption is being improved by a large number of new and old formulations by the use of simple methods of administration that make it possible to give high dosages over a protracted period of time. Because TDDS are becoming more common and drawing the interest of researchers, there will be an increase in the number of new medications that are available in the form of TDDS. While developing a system for the transdermal distribution of drugs, it is important to remember that the formulation of the medicine should not interfere with the normal functioning of the skin. In the future, the development of transdermal patches will benefit from increased knowledge of the anatomy and physiology of the skin. Thus, the transdermal drug delivery technology is becoming increasingly popular in the pharmaceutical industry. Nanoparticles, liposomes, nanocrystals, niosomes, and nanoemulsions are all used in the formulation for transdermal drug delivery to achieve targeted drug delivery. Despite this tactic is only applied in a fraction of the products that are now available on the market, there is huge potential for transdermal drug delivery. There are still a number of obstacles that need to be conquered, one of which is the complexity of applications that arises from the combination of transdermal devices and other drug-loaded formulations. Additionally, scaled-up, Good Manufacturing Practice (GMP) and regulatory monitoring over the manufacturing of TDDS are required. In recent years, there has been an increase in the number of research studies and products that are available in the form of TDDS in both domestic and international markets. TDDSs may play crucial role in delivering medications for cardiovascular and central nervous system diseases, diabetes, neuromuscular diseases, genetic diseases, and infectious and localised infectious diseases. TDDS may also be effectively applied for vaccination and supporting patients' via self-administration of long-term medications.

Acknowledgements The principal investigator laboratory is supported by Science and Engineering Research Board (SERB) via Start-up Research Grant (SRG) [grant no. SRG/2021/002312] and the Empowerment and Equity Opportunities for Excellence in Science (EMEQ) grant [grant no. EEQ/2022/000218] during the scripting of the book chapter. The book chapter is related to the above research grants. The authors would like to acknowledge biorender (online imaging) service providers for the assistance in drawing images.

Conflict of Interest The authors declare that there is no conflict of interest in publishing the book chapters.

References

Akhtar N et al (2020) Non-invasive drug delivery technology: development and current status of transdermal drug delivery devices, techniques and biomedical applications. Biomed Tech (Berl) 65(3):243–272

Al Hanbali OA et al (2019) Transdermal patches: design and current approaches to painless drug delivery. Acta Pharm 69(2):197–215

Ali S, Shabbir M, Nabeel Shahid M (2015) The structure of skin and transdermal drug delivery system—a review. Res J Pharm Technol 8:103

Al-Khamis KI, Davis SS, Hadgraft J (1986) Microviscosity and drug release from topical gel formulations. Pharm Res 3(4):214–217. https://doi.org/10.1023/A:1016386613239

Alkilani AZ, Nasereddin J (2022) Beneath the skin: a review of current trends and future prospects of transdermal drug delivery systems. 14(6)

Allen LV, Popovich NG, Ansel HC (2012) Ansel's pharmaceutical dosage forms and drug delivery systems

Andrews SN, Jeong E, Prausnitz MR (2013) Transdermal delivery of molecules is limited by full epidermis, not just stratum corneum. Pharm Res 30(4):1099–1109. https://doi.org/10.1007/s11095-012-0946-7

Arijit D, Sibaji G, Biplab KD, Sudip D (2010) A novel technique for treating the type II diabetes by transdermal patches prepared by using multiple polymer complexes. Int J Pharma Res Dev 9:195–204

Barry BW (1988) Action of skin penetration enhancers? The lipid protein partitioning theory. Int J Cosmet Sci 10(6):281–293. https://doi.org/10.1111/j.1467-2494.1988.tb00028.x

Benson HAE et al (2019) Topical and transdermal drug delivery: from simple potions to smart technologies. Curr Drug Deliv 16(5):444–460

Berlin CM (1997) Clinical report—alternative routes of drug administration—advantages & disadvantages (subject review). Pediatrics

Bouwstra JA, Gooris GS (2010) The lipid organisation in human stratum corneum and model systems. Open Dermatol J 4(1):10–13. https://doi.org/10.2174/1874372201004010010

Chantasart D, Li SK (2012) Structure enhancement relationship of chemical penetration enhancers in drug transport across the stratum corneum. Pharmaceutics 4(1):71–92. https://doi.org/10.3390/pharmaceutics4010071

Chen X et al (2020) Electroporation-enhanced transdermal drug delivery: effects of logP, pK(a), solubility and penetration time. Eur J Pharm Sci 151:105410

Chien YW, Banga AK (1989) Iontophoretic (transdermal) delivery of drugs: overview of historical development. J Pharm Sci 78(5):353–354

Choudhary N, Singh AP (2021) Transdermal drug delivery system: a review. Indian J Pharm Pharmacol 8(1):5–9

Dardano P et al (2015) A photolithographic approach to polymeric microneedles array fabrication. Materials (Basel) 8(12):8661–8673

Das A, Ahmed A (2017) Natural permeation enhancer for transdermal drug delivery system and permeation evaluation: a review. Asian J Pharm Clin Res 10:5

Dermol-Černe J, Pirc E, Miklavčič D (2020) Mechanistic view of skin electroporation—models and dosimetry for successful applications: an expert review. Expert Opin Drug Deliv 17(5):689–704

Dhal S, Pal K, Giri S (2020) Transdermal delivery of gold nanoparticles by a soybean oil-based oleogel under iontophoresis. ACS Appl Bio Mater 3(10):7029–7039

Du H et al (2019) Hyaluronic acid-based dissolving microneedle patch loaded with methotrexate for improved treatment of psoriasis. ACS Appl Mater Interfaces 11(46):43588–43598

Frei A et al (2003) A one year health economic model comparing transdermal fentanyl with sustained-release morphine in the treatment of chronic noncancer pain. J Pain Palliat Care Pharmacother 17(2):5–26

Gandhi K et al (2012) Transdermal drug delivery: a review

Giacomoni PU, Dong K, Collins D, Pernodet N, Yarosh D (2011) Use of reconstructed skin specimens to analyze stratum corneum remodeling and epidermal modifications. In: Skin biopsy—perspectives. InTech https://doi.org/10.5772/23023

Govil SK (1988) Drug delivery devices. In: Tyle P (ed). Marcel Dekker, New York

Green PG (1996) Iontophoretic delivery of peptide drugs. J Control Release 41(1–2):33–48. https://doi.org/10.1016/0168-3659(96)01354-5

Hadgraft J, Lane ME (2009) Transepidermal water loss and skin site: a hypothesis. Int J Pharm 373(1–2):1–3. https://doi.org/10.1016/j.ijpharm.2009.02.007

Honeywell-Nguyen PL, Bouwstra JA (2005) Vesicles as a tool for transdermal and dermal delivery. Drug Discov Today Technol 2(1):67–74

Igarashi T, Nishino K, Nayar SK (2007) The appearance of human skin: a survey. Found Trends® Comput Graph Vis 3(1):1–95. https://doi.org/10.1561/0600000013

Ita K (2015a) Transdermal delivery of drugs with microneedles-potential and challenges. Pharmaceutics 7(3):90–105

Ita KB (2015b) Chemical penetration enhancers for transdermal drug delivery—success and challenges. Curr Drug Deliv 12(6):645–651

Jayaprakash R, Hameed J, Anupriya A (2017) An overview of transdermal delivery system. Asian J Pharm Clin Res 10(10):36. https://doi.org/10.22159/ajpcr.2017.v10i10.19909

Jayaswal SB, Sood R (1987) Transdermal drug delivery system—a review. Eastern Pharm 30(357):47–50

Jeong WY et al (2019) Transdermal delivery of Minoxidil using HA-PLGA nanoparticles for the treatment in alopecia. Biomater Res 23:16

Jeong WY et al (2021) Recent advances in transdermal drug delivery systems: a review. Biomater Res 25(1):24

Karande P, Mitragotri S (2009) Enhancement of transdermal drug delivery via synergistic action of chemicals. Biochimica et Biophysica Acta (BBA) Biomembranes 1788(11):2362–2373. https://doi.org/10.1016/j.bbamem.2009.08.015

Kavadya Y, Behera B, Mohanty P, Routray D, Ghosh S, Das L (2018) Study of physiological and pathological skin changes in neonates: an east Indian perspective. Indian J Paediatr Dermatol 19(1):40. https://doi.org/10.4103/ijpd.ijpd_21_17

Keleb E, Sharma RK, Mosa EB, Aljahvi AZ (2010) Transdermal drug delivery system-design and evaluation. Int J Adv Pharm Sci 1:201–211

Kim HM et al (2012) Transdermal drug delivery using disk microneedle rollers in a hairless rat model. Int J Dermatol 51(7):859–863

Kornick CA et al (2003) Benefit-risk assessment of transdermal fentanyl for the treatment of chronic pain. Drug Saf 26(13):951–973

Kumar R, Philip A (2007) Modified transdermal technologies: breaking the barriers of drug permeation via the skin. Trop J Pharm Res 6(1):633–644. https://doi.org/10.4314/tjpr.v6i1.14641. ISSN: 1596-5996

Kumar SR, Jain A, Nayak S (2012) Development and evaluation of transdermal patches of colchicine. Pharm Lett 4(1):330–343

Kusum Devi V, Saisivam S, Maria GR, Deepti PU (2003) Design and evaluation of matrix diffusion controlled transdermal patches of verapamil hydrochloride. Drug Dev Ind Pharm 29(5):495–503. https://doi.org/10.1081/DDC-120018638

Lee H et al (2018) Device-assisted transdermal drug delivery. Adv Drug Deliv Rev 127:35–45

Leppert W et al (2018) Transdermal and topical drug administration in the treatment of pain. Molecules 23(3)

Lin CH, Aljuffali IA, Fang JY (2014) Lasers as an approach for promoting drug delivery via skin. Expert Opin Drug Deliv 11(4):599–614

Mitragotri S, Edwards DA, Blankschtein D, Langer R (1995) A mechanistic study of ultrasonically-enhanced transdermal drug delivery. J Pharm Sci 84(6):697–706. https://doi.org/10.1002/jps.2600840607

Morrow DIJ, McCarron PA, Woolfson AD, Donnelly RF (2007) Innovative strategies for enhancing topical and transdermal drug delivery. Open Drug Deliv J 1(1):36–59. https://doi.org/10.2174/1874126600701010036

Neupane R et al (2021) Transdermal delivery of chemotherapeutics: strategies, requirements, and opportunities. Pharmaceutics 13(7):960. https://doi.org/10.3390/pharmaceutics13070960

Panchaguinla R (1997) Transdermal delivery of drug. Indian J Pharmacol 29:140

Pandey D, Akhilesh D, Prabhakara P, Kamath J (2012) Transdermal drug delivery system: a novel drug delivery system. Int Res J Pharm 3(5):89–94

Parivesh S, Sumeet D, Abhishek D (2010) Design, evaluation, parameters and marketed products of transdermal patches: a review. J Pharm Res 3(2):235–240

Park J et al (2019) Enhanced transdermal drug delivery by sonophoresis and simultaneous application of sonophoresis and iontophoresis. AAPS PharmSciTech 20(3):96

Pastore MN et al (2015) Transdermal patches: history, development and pharmacology. Br J Pharmacol 172(9):2179–2209

Pathan I, Setty C (2009) Chemical penetration enhancers for transdermal drug delivery systems. Trop J Pharm Res 8(2). https://doi.org/10.4314/tjpr.v8i2.44527

Peña-Juárez MC, Guadarrama-Escobar OR, Escobar-Chávez JJ (2022) Transdermal delivery systems for biomolecules. J Pharm Innov 17(2):319–332

Pires LR et al (2019) A perspective on microneedle-based drug delivery and diagnostics in paediatrics. J Pers Med 9(4)

Prausnitz MR, Langer R (2008) Transdermal drug delivery. Nat Biotechnol 26(11):1261–1268

Rastogi V, Yadav P (2012) Transdermal drug delivery system: an overview. Asian J Pharm 6(3)

Rizwan M, Aqil M, Talegoankar S, Azeem A, Sultana Y, Ali A (2009) Enhanced transdermal drug delivery techniques: an extensive review on patents. Recent Pat Drug Deliv Formul 3(2):105–124

Roohnikan M et al (2019) A snapshot of transdermal and topical drug delivery research in Canada. Pharmaceutics 11(6)

Roy N et al (2017) On permeation enhancers: a major breakthrough in drug delivery technology

Ruby PK, Pathak SM, Aggarwal D (2014) Critical attributes of transdermal drug delivery system (TDDS)—a generic product development review. Drug Dev Ind Pharm 40(11):1421–1428

Sandhyarani G, Madhuri M (2018) Development and evaluation of fast disintegrating films and tablets of Valsartan. Int J Med Sci Pharma Res 4(1):17. https://doi.org/10.22270/ijmspr.v4i1.23

Sankar V, Johnson DB, Sivan V, Ravich V, Raghuraman S, Velrajan G, Palaniappan R, Rajasekar S (2003) Design and evaluation of nifedipine transdermal patches. Indian J Pharm Sci 65(5):510

Shaila L, Pandey S, Udupa N (2006) Design and evaluation of matrix type membrane controlled transdermal drug delivery system of nicotine suitable for use in smoking cessation. Indian J Pharm Sci 68(2):829–831. https://doi.org/10.4103/0250-474X.25711

Shakya AK et al (2019) Microneedles coated with peanut allergen enable desensitization of peanut sensitized mice. J Control Release 314:38–47

Shinde AJ, Garala KC, More HN (2008) Development and characterization transdermal therapeutics system of tramadol hydrochloride. Asian J Pharm (AJP) 2(4):265–269. https://doi.org/10.4103/0973-8398.45044

Singh S, Singh J (1993) Transdermal drug delivery by passive diffusion and iontophoresis: a review. Med Res Rev 13

Singh MC, Naik AS, Sawant SD (2010) Transdermal drug delivery systems with major emphasis on transdermal patches: a review. J Pharm Res 3(10):2537–2543

Sinha VR, Kaur MP (2000) Permeation enhancers for transdermal drug delivery. Drug Dev Ind Pharm 26(11):1131–1140. https://doi.org/10.1081/DDC-100100984

Sonjoy M, Thimmasetty J, Ratan GN, Kilarimath BH (2011) Formulation and evaluation of carvedilol transdermal patches. Int Res J Pharm 2(1):237–248

Sood A, Panchagnula R (1999) Role of dissolution studies in controlled release drug delivery systems. STP Pharma Sci 9(2):157–168

Szunerits S, Boukherroub R (2018) Heat: a highly efficient skin enhancer for transdermal drug delivery. Front Bioeng Biotechnol 6:15

Ubaidulla U, Reddy MV, Ruckmani K, Ahmad FJ, Khar RK (2007) The transdermal therapeutic system of carvedilol: effect of hydrophilic and hydrophobic matrix on in vitro and in vivo characteristics. AAPS PharmSciTech 8(1):13–20. https://doi.org/10.1208/pt0801002

Varvel JR et al (1989) Absorption characteristics of transdermally administered fentanyl. 70(6):928–934

Verma PR, Iyer SS (2000) Transdermal delivery of propranolol using mixed grades of Eudragit: design and in vitro and in vivo evaluation. Drug Dev Ind Pharm 26(4):471–476. https://doi.org/10.1081/DDC-100101257

Verma S, Ojha A (2018) Transdermal drug delivery system: a review. Int J Med Sci Pharma Res 4(1):1–9. https://doi.org/10.22270/ijmspr.v4i1.21

Vishvakarma P, Aggrawal S, Sharma R, Sharma S (2012) Transdermal drug delivery system: review. Int Res J Pharm 23(5):50–53

Wade A, Weller PJ (1994) Handbook of pharmaceutical excipients. American Pharmaceutical Publishing Association, Washington DC, pp 362–366

Wang Y et al (2022) Influencing factors and drug application of iontophoresis in transdermal drug delivery: an overview of recent progress. Drug Deliv Transl Res 12(1):15–26

Weiner E, Victor A, Johansson ED (1976) Plasma levels of d-Norgestel after oral administration. Contraception 14:563–570. Keith AD (1983) Polymer matrix consideration for transdermal devices. Drug Dev Ind Pharm 9:605–625

Wiechers J (1992) Use of chemical penetration enhancers in transdermal drug delivery–possibilities and difficulties. Acta Pharm Nord 4(2):123

Wokovich A et al (2006) Transdermal drug delivery system (TDDS) adhesion as a critical safety, efficacy and quality attribute. Eur J Pharm Biopharm: Off J Arbeitsgemeinschaft für Pharmazeutische Verfahrenstechnik e.V 64:1–8

Yamamoto T, Katakabe K, Akiyoshi K, Kan K, Asano T (1990) Topical application of glibenclamide lowers blood glucose levels in rats. Diabetes Res Clin Pract 8:19–22. https://doi.org/10.1016/0168-8227(90)90091-7

Zhang Y et al (2019) Advances in transdermal insulin delivery. Adv Drug Deliv Rev 139:51–70

Zhao Z, Chen Y, Shi Y (2020) Microneedles: a potential strategy in transdermal delivery and application in the management of psoriasis. RSC Adv 10(24):14040–14049

GPSR Compliance

The European Union's (EU) General Product Safety Regulation (GPSR) is a set of rules that requires consumer products to be safe and our obligations to ensure this.

If you have any concerns about our products, you can contact us on ProductSafety@springernature.com

In case Publisher is established outside the EU, the EU authorized representative is:

Springer Nature Customer Service Center GmbH
Europaplatz 3
69115 Heidelberg, Germany

Batch number: 10731141

Printed by Printforce, the Netherlands